T0073151

Advances in Computer Vision and Pattern Recognition

More information about this series at https://link.springer.com/bookseries/4205

Husrev Taha Sencar · Luisa Verdoliva ·
Nasir Memon
Editors

Multimedia Forensics

 Springer

Editors
Husrev Taha Sencar
Research Complex, HBKU, Education City
Qatar Computing Research Institute
Doha, Qatar

Luisa Verdoliva
University of Naples Federico II
Napoli, Italy

Nasir Memon
Center for Cyber Security
New York University
Brooklyn, NY, USA

ISSN 2191-6586   ISSN 2191-6594 (electronic)
Advances in Computer Vision and Pattern Recognition
ISBN 978-981-16-7620-8   ISBN 978-981-16-7621-5 (eBook)
https://doi.org/10.1007/978-981-16-7621-5

This Springer imprint is published by the registered company Springer Nature Singapore Pte Ltd.
The registered company address is: 152 Beach Road, #21-01/04 Gateway East, Singapore 189721, Singapore

Preface

In 2013, we edited the first edition of this book to provide an overview of challenges and developments in digital image forensics. This was the first book on the topic providing a comprehensive review of prominent research themes as well as the perspectives of legal professionals and forensic practitioners. Since then, the research in this field has expanded dramatically. On one hand, the volume of multimedia content generated and exchanged over the Internet increased hugely. Images and videos represent an ever-increasing share of the data traveling on the net and the preferred communications means for most users, especially of the younger generations. Therefore, they are also the preferred target of malicious attacks. On the other hand, with the advent of deep learning, attacking multimedia is becoming easier and easier. Ten years ago, manipulating an image in a credible way required uncommon technical skills and significant efforts. Manipulating a video, beyond trivial frame-level operations, was possible only for movie productions. This scenario has radically changed now, and multimedia manipulation is at anyone's reach. Sophisticated tools based on autoencoders or generative adversarial networks are freely available, as well as huge datasets to train them at will. Deepfakes have become commonplace and represent a real and growing menace for the integrity of information, with repercussions on various sectors of society, from the economy to journalism to politics. And these problems can only worsen as new and increasingly sophisticated forms of manipulation are taking shape.

Unsurprisingly, multimedia manipulation and forensics are becoming hot topics, not just for the research community but for society at large. Several no-profit organizations have been created to investigate these phenomena. Major IT companies, like Google and Facebook, published online large datasets of manipulated videos to foster research on these topics. Funding agencies are promoting large research projects. Especially remarkable is the Media Forensic initiative (MediFor) launched by DARPA in 2016 to support leading research groups on media integrity all over the world, which generated important outcomes in terms of methods and reference datasets. This was followed in 2020 by the Semantics Forensics initiative (SemaFor) aiming at semantic-level detectors of fake media, working jointly on text, audio, images, and video to counter next-generation attacks.

The greatly increased focus on multimedia integrity has caused an intense research effort in latest years, which involves not only the multimedia forensics community but the much larger field of computer vision. Indeed, deep learning takes the lion's share not only on the attacker's side but also on the defender's side. It provides the analyst with new powerful forensic tools for detection. Given a suitably large training set, deep learning architectures ensure a significant performance gain with respect to conventional methods and much higher robustness to post-processing and evasions techniques. In addition, deep learning can be used to address other fundamental problems, such as source identification. In fact, being able to track the provenance of a multimedia asset helps establishing its integrity and, as sensors become more sophisticated and software-oriented, source identification tools must follow suit. Likewise, neural networks, especially generative adversarial networks, are natural candidates to perform precious counter-forensics tasks. In synthesis, deep learning is part of the problem but also part of the solution, and the research activity going on in this cat-and-mouse game seems to be continuously growing.

All this said, we felt this was the right time to create an updated version of that book, with largely renovated content, to cover the rapid advancements in multimedia forensics and provide an updated review of the most promising methods and state-of-the-art techniques.

We organized the book into three main parts following a brief introduction of development in imaging technologies. The first part will provide an overview of key findings related to the attribution problem. The work pertaining to the integrity of images and videos is collated in the second part. The last part features work on combatting adversarial techniques that rise as a significant challenge to forensics methods.

Doha, Qatar   Husrev Taha Sencar
Napoli, Italy   Luisa Verdoliva
Brooklyn, USA   Nasir Memon

Contents

Symbols

iid	Independent and identically distributed
PCE	Peak to Correlation Energy
PRNU	Photo-Response Non-Uniformity
r.v.	Random variable
$Cov(.,.)$	Covariance of two r.v.s
$E(.)$	Expected value of a r.v.
ρ	Linear correlation value
$f_X(x)$	Probability density function of a r.v. X
$F_X(x)$	Cumulative distribution function of a r.v. X
μ_X, σ_X^2	Mean and variance of a r.v. X
DCT	Discrete cosine transform
DFT	Discrete Fourier transform
I	A multidimensional array that may represent an image or a video
$I(m,n)$	The value of the (m,n)-th entry of image I
$I(m,n,t)$	The value of the (m,n)-th entry of t-th frame of video I
H_0	Null hypothesis
H_1	Alternative hypothesis
P_D	Probability of Detection
P_{FA}	Probability of False Alarm
P_M	Probability of Miss
P_{CR}	Probability of Correct Rejection
P_E	Probability of Error
TPR	True Positive Rate
FPR	False Positive Rate
FNR	False Negative Rate
T_{ro}	Read-out time of a rolling shutter camera
DNN	Deep Neural Networks
$R@k$	Recall at top-k objects retrieved in a rank
VO	Vertex Overlap
EO	Edge Overlap
VEO	Graph overlap, a.k.a. vertex-edge overlap

PSNR	Peak Signal to Noise Ratio
MSE	Mean Squared Error
ASR	Attack Success Rate
CNN	Convolutional Neural Network
DNN	Deep Neural network
DL	Deep Learning
ML	Machine Learning

Notation

We use the following notational conventions.

- Concerning random variables:

  - Random variables by uppercase roman letters, *e.g.*, X, Y.
  - Particular realizations of a random variable by corresponding lowercase letters, *e.g.*, x, y.
  - A sequence of n r.v.s and its particular realizations will be, respectively, denoted by X^n and x^n.
  - Predicted random variables by the hat sign, *e.g.*, $\hat{X}$.
  - The alphabet of a random variable will be indicated by the capital calligraphic letter corresponding to the r.v. $\mathcal{X}$.

- Concerning matrices and vectors:

  - Matrices by boldface capital letters, *e.g.*, $\mathbf{A}$.
  - The entry in the i-th row and j-th column of a matrix $\mathbf{A}$ by $A(i, j)$.
  - Column vectors by boldface lowercase letters, *e.g.*, $\mathbf{x}$.
  - $< v, w >$: scalar product between vectors v and w.
  - Operators are set in normal font, for example, the transpose $\mathbf{A}^{\mathrm{T}}$.

- Concerning sets:

  - Set names by uppercase calligraphic letters, *i.e.*, $\mathcal{A}$-$\mathcal{Z}$.
  - Sets will be specified by listing their objects within curly brackets, *i.e.*, $\{\dots\}$.
  - Standard number sets by Blackboard bold typeface, *e.g.*, $\mathbb{N}$, $\mathbb{Z}$, $\mathbb{R}$, etc.

- Concerning functions:

  - Functions are typeset as a single lowercase letter, in math font, *e.g.*, $f(\mathbf{A})$.
  - Multi-letter functions, and in particularly commonly known functions, are set in lowercase normal font, *e.g.*, $\cos(x)$, $\mathrm{corr}(\mathbf{x}, \mathbf{y})$, or $j^* = \underset{j}{\mathrm{argmin}} \sum_{i=1}^{N} \mathbf{A}(i, j)$.

- To deal with specific functions from a field of functions (e.g., spherical harmonics basis functions that are indexed by two parameters), we could use comma-separated subscripts, e.g., $h_{i,j}(x)$.
- Non-math annotations are set in normal font, e.g., $f_{\text{special}}(\mathbf{A})$. The purpose is to disambiguate names from variables, like in the example $f_i^{\text{special}}(\mathbf{A})$.

- Concerning images and videos:

    - I will be exclusively used to refer to an image or a sequence of images, and I_n with $n \in \mathbb{N}$ for multiple images/videos.
    - $I(i, j)$ will be used as a row-major pixel-wise image reference denoting the pixel value at row i, column j of I.
    - Similarly, $I(i, j, t)$ denotes the value of the (i, j)-th pixel of t-th frame of video I.
    - Image pair-wise adjacency matrices will be denoted by M for single matrix and M_n with $n \in \mathbb{N}$ for multiple matrices.
    - Row-major adjacency matrix element reference $M(i, j) \in M$ for the element expressing the relation between the i-th and the j-th images of interest.

- Concerning graphs:

    - A generic graph by G for single and G_n with $n \in \mathbb{N}$ for multiple graphs.
    - $G = (V, E)$ to express a graph's set of vertices V and set of edges E.
    - Vertex names by $v_n \in V$, with $n \in \mathbb{N}$.
    - Edge names by $e_n \in E$, with $n \in \mathbb{N}$; $E = (v_i, v_j)$ to detail an edge's pair of vertices; if the edge is directed, v_i precedes v_j in such relation.

- In the context of machine learning and deep learning:

    - A generic dataset by $\mathcal{D}$.
    - Training, validation, and test datasets, respectively, by $\mathcal{D}_{tr}, \mathcal{D}_v, \mathcal{D}_{ts}$.
    - i-th sample at the input of a DNN and its corresponding label (if indicated) by (x_i, y_i).
    - Loss function used for DNN training by $\mathcal{L}$.
    - Sigmoid activation function $\sigma()$.
    - Output of a DNN by $\phi(x)$ when the input is equal to x.
    - Gradient of function ϕ by ∇_ϕ.
    - The Jacobian matrix of a multivalued function ϕ by J_ϕ.
    - The logit value (output of a CNN prior to the final softmax layer) at node i by z_i.
    - Mathematical calligraphic capital letters in function form $\mathcal{M}(\cdot)$ represent a model function (i.e., a Neural Network).

- Concerning norms:

    - L1, L2, squared L2, and L-infinity norms, respectively, by $L_1, L_2, L_2^2, L_\infty$.
    - p norm by $P.P_p$.

Part I
Present and Challenges

Chapter 1
What's in This Book and Why?

Husrev Taha Sencar, Luisa Verdoliva, and Nasir Memon

1.1 Introduction

Multimedia forensics is a societally important and technically challenging research area that will need significant effort for the foreseeable future. While the research community is growing and work like that in this book demonstrates significant progress, many challenges remain. We expect that forensically tackling ever-improving media acquisition and generation methods and countering the pace of change in media manipulation will continue to require significant technical breakthroughs in defensive technologies.

Whether performed at the individual level or class level attribution is at the heart of multimedia forensics. This form of passive forensic analysis exploits traces introduced by the acquisition or generation pipeline and subsequent editing. These traces are often subtle and invisible to humans but can be detected currently by analysis algorithms and provide mechanisms by which generation or manipulation can be automatically detected. Despite significant achievements in this area of research, advances in imaging technologies and newly introduced approaches to media synthesis have a strong potential to render existing forensic capabilities ineffective.

More critically, media manipulation has now become a pressing problem with broad implications. In the past, it required significant skill to create compelling manipulations because editing tools, such as Adobe Photoshop, required experienced users to alter images convincingly. Over the last several years, the rise of machine learning-based technologies has dramatically lowered the skill necessary to create

H. T. Sencar (✉)
Qatar Computing Research Institute, HBKU, Ar-Rayyan, Qatar
e-mail: hsencar@hbku.edu.qa

L. Verdoliva
University Federico II of Naples, Naples, Italy

N. Memon
New York University Tandon School of Engineering, New York, NY, USA

compelling manipulations. For example, Generative Adversarial Networks (GANs) can create photo-realistic faces with no skill by an end-user other than the ability to refresh a web page. Deepfake algorithms, such as autoencoders to swap faces in a video, can create manipulations much more easily than previous generation video tools. While these machine learning techniques have many positive uses, they have also been misused for darker purposes such as to perpetrate fraud, to create false personas, and to attack personal reputations.

The asymmetry in the operational setting of digital forensics where attackers have access to details and inner workings of the involved methods and tools and thereby have the freedom to tailor their actions makes the task even more difficult. It won't be surprising to see these advanced capabilities being deployed to counter forensic methods. Clearly, these are all pressing and important problems for which technical solutions must play a role.

There are a number of significant challenges that must be addressed by forensics community. Imaging technologies are constantly being improved by manufacturers at both hardware and software levels with little public information about their specifics. This implies a black-box access to these technologies and involves a significant reverse engineering effort on the part of researchers. With rapid advancements in computational imaging, this task will only get more strenuous.

Further, new manipulation techniques are becoming available at a rapid pace and each generation improves on the previous. As a result, defenders must develop ways to limit the a priori knowledge and training samples required from an attack algorithm and should assume an open world scenario where they might not have knowledge of all the attack algorithms. Meta-learning approaches for developing detection algorithms more quickly would also be highly beneficial. Manipulations in the wild are propagated via noisy channels, like social media, and so there is a need for robust detection technologies that are not fooled by launderings such as simple compression or more sophisticated replay attacks. Finally, part of the task of forensics is unwinding how the media was manipulated and who manipulated it. Consequently, approaches to understanding the provenance of manipulated media are necessary.

All of these challenges motivated us to prepare this book on multimedia forensics. Our main objective was to discuss new research directions and also present some possible solutions to existing problems. In this regard, this book provides a timely vehicle for showcasing important research that has emerged in the last years addressing media attribution, authenticity verification, and counter-forensics. We hope our content also sparks new research efforts in the fight against forgeries and misinformation.

1.2 Overviews

Our book begins with an overview of the challenges posed by media manipulation as seen from today. In their chapter, Hendrix and Morozoff provide a comprehensive view of all aspects of the problem to better emphasize the increasing need for advanced media forensics capabilities. This is followed by an examination of forensics challenges that arise from the increased adoption of computational imaging. The chapter by McCloskey assesses the ability to detect automated focus manipulations performed by computational cameras. For this, it looks at cues that can be used to distinguish optical blur from synthetic blur. The chapter also covers computational imaging research with potential future implications on forensics.

This chapter is followed by a series of chapters focusing on different dimensions of the attribution problem. Undoubtedly, one of the breakthrough discoveries in multimedia forensics is that *Photo-Response Non-Uniformity* (PRNU) of an imaging sensor, which manifests as a unique and permanent pattern introduced to all media captured by the sensor, can be used for identification and verification of the source of digital media. Our book has three chapters covering different aspects PRNU-based source camera attribution. The first chapter by Kirchner reviews the rich body of literature on camera identification from sensor noise fingerprints with an emphasis on still images from digital cameras and the evolving challenges in this domain. The second chapter by Sencar examines extension of these capabilities to video domain and outlines recently developed methods for estimating the sensor's PRNU from videos. The third chapter on this topic by Taspinar et al. provides an overview of techniques proposed to perform source matching efficiently in the presence of a large collection of media.

Another vertical in attribution domain is the source camera model identification problem. Assuming that a picture or video has been digitally acquired with a camera, the goal of this research direction is to identify the brand and model of the device used at acquisition time without relying on metadata. The chapter by Mandelli et al. investigates source camera model identification through pixel analysis and discusses wide series of methodologies proposed to solve this problem.

The remarkable progress of deep learning, in particular Generative Adversarial Networks (GANs), has led to the generation of extremely realistic fake facial content. The potential for misuse of such capabilities opens up new problem in forensics that focuses on identification of GAN fingerprints used in face image synthesis. Neves et al. provide an in-depth literature analysis of state-of-the-art detection approaches for face synthesis and manipulation as well as spoofing those detectors.

The next part of our book involves several chapters focusing on integrity and authenticity of multimedia. This part starts with a review of physics-based methods that mainly analyzes the interaction of light and objects and the geometric mapping of light and objects onto the image sensor. In his chapter, Reese reviews the major lines of research on physics-based methods and discusses their strengths and limitations.

The following chapter investigates forensic applications of the *Electric Network Frequency* (ENF) signal. Since ENF serves as an environmental signature captured by audio and video recordings made in locations where there is electrical activity, it provides an additional dimension for time–location authentication and for inferring the grid in which a recording was made. The chapter by Hajj-Ahmad et al. provides an overview of the increasing amount of research work that has been done in this field.

The long-lasting problem of image and video manipulation is revisited in the subsequent four chapters. However, distinct from conventional manipulation detection methods, Cozzolino et al. focus on the data-driven deep learning-based methods proposed in recent years. This chapter discusses in detail forensics traces these methods rely on, and describes architectural solutions used for detection and localization, together with the associated training strategies. The next chapter features the new class of forgery known as DeepFakes that involve impersonating audios and videos generated by deep neural networks. In this chapter, Lyu surveys the state-of-the-art DeepFake detection methods and evaluates solutions proposed to tackle this problem along with their pros and cons. In the third chapter of the topic, Long et al. provide an overview of the work related to video frame deletion and duplication and introduce deep learning-based approaches to tackle these two types of manipulations.

A complementary integrity verification approach is presented next. Several work demonstrated that media manipulation not only leaves forensic traces in the audio-visual content but also in the file structure. The chapter authored by Piva covers proposed forensic methods devoted to the analysis of image and video file formats to validate multimedia integrity.

The last chapter on authenticity verification introduces the problem of provenance analysis. By jointly examining multiple multimedia files, instead of making individual assessments, and evaluating pairwise relationships between them, the provenance of multimedia files is more reliably determined. The chapter by Moreira et al. addresses this problem by covering state-of-the-art analysis techniques.

The last part of our book includes two chapters on counter-forensics complementing the advances brought by deep learning to forensics. The advances brought by deep learning have also significantly expanded the capabilities of anti-forensic attackers. The first chapter by Barni et al. focuses on adversarial examples in image forensics and how they can be used in creation of attacks. It also describes possible countermeasures against them, and discusses their effectiveness. The other chapter by Stamm et al. focuses on emerging threat posed by GAN-based anti-forensic attacks in creating realistic, but completely synthetic forensic traces.

Chapter 2
Media Forensics in the Age of Disinformation

Justin Hendrix and Dan Morozoff

2.1 Media and the Human Experience

Empiricism is the notion that knowledge originates from sensory experience. Implicit in this statement is the idea that we can trust our senses. But in today's world, much of the human experience is mediated through digital technologies. Our sensory experiences can no longer be trusted a priori. The evidence before us—what we see and hear and read—is, more often than not, manipulated.

This presents a profound challenge to a species that for most of its existence has relied almost entirely on sensory experience to make decisions. A group of researchers from multiple universities and multiple disciplines that study collective behavior recently assessed that the study of the impact of digital communications technologies on humanity should be understood as a "crisis discipline," alongside the study of matters such as climate change and medicine (Bak-Coleman et al. 2021). Noting that since "our societies are increasingly instantiated in digital form," they argue the potential of digital media to perturb the ways in which we organize ourselves, deal with conflict, and solve problems is profound:

> Our social adaptations evolved in the context of small hunter-gatherer groups solving local problems through vocalizations and gestures. Now we face complex global challenges from pandemics to climate change—and we communicate on dispersed networks connected by digital technologies such as smartphones and social media.

The evidence, they say, is clear: in this new information environment, misinformation, disinformation, and media manipulation pose grave threats—from the spread of conspiracy theories (Narayanan et al. 2018), rejection of recommended public health interventions (Koltai 2020), subversion of democratic processes (Atlantic Council 2020), and even genocide (Łubiński 2020).

Media manipulation is only one of a series of problems in the digital information ecosystem—but it is a profound one. For most of modern history, trust and verifiability of media—whether printed material, audio and more recently images, video, and

J. Hendrix (✉) · D. Morozoff
New York University, Vidrovr Inc., New York, NY, USA
e-mail: jah311@nyu.edu

H. T. Sencar et al. (eds.), *Multimedia Forensics*, Advances in Computer Vision and Pattern Recognition, https://doi.org/10.1007/978-981-16-7621-5_2

other modalities—have been linked to one's ability to access and coordinate information to previously gathered facts and beliefs. Consequently, in order to deceive someone, the deceiver had to rely on the obfuscation of a media object's salient features and hope that the consumer's time and ability to perceive these differences was not sufficient. This generalization extends from boasts and lies told in the schoolyard, to false narratives circulated by media outlets, to large-scale disinformation campaigns pursued by nations during military conflicts.

A consistent property and limitation of deceptive practices is that their success is linked to two conditions. The first condition is that the ability to detect false information is linked to the amount of time we have to process and verify it before taking action on it. For instance, showing someone a brief picture of a photoshopped face is more likely to be taken as true if the viewer is given less time to examine it. The second condition is linked to the amount of time and effort required by the deceiver to create the manipulation. Hence, the impact and reach of a manipulation is limited by the resources and effort involved. A kid in school can deceive their friends with a photoshopped image, but it required Napoleon and the French Empire to flood the United Kingdom with fake pound notes during the Napoleonic wars and cause a currency crisis.

A set of technological advancements over the last 30 years has led to a tectonic shift in this dynamic. With the adoption of Internet and social media in the 1990s and 2000s, respectively, the marginal costs required to deceive a victim have plummeted. Any person on the planet with a modicum of financial capital is able to reach billions of people with their message, effectively breaking the resource dependence of a deceiver. This condition is an unprecedented one in the history of the world.

Furthermore, the maturation and spread of machine learning-powered media technologies has provided the ability to generate diverse media at scale, lowering the barriers to entry for all content creators. Today, the amount of work required to deliver mass content at Internet scale is exponentially decreasing. This is inevitably compounding the problem of trust online.

As a result, our society is approaching a precipice of far-reaching consequences. Media manipulation is easy for anyone, even if they are without significant resources or technical skills. While conversely the amount of media available to people continues to explode, the amount of time dedicated to verifying a single piece of content is also falling. As such, if we are to have any hope of adapting our human sensemaking to this digital age, we will need media forensics incorporated into our digital information environment in multiple layers. We will require automated tools to augment our human abilities to make sense of what we read, hear, and see.

2.2 The Threat to Democracy

One of the most pressing consequences of media manipulation is the threat to democracy. Concern over issues of trust and the role of information in democracies is severe. Globally, democracy has run into a difficult patch, with 2020 perhaps among its worst years in decades. The Economist Intelligence Unit's Democracy Index found that as of 2020 only "8.4% of the world's population live in a full democracy while more than a third live under authoritarian rule" (The Economist 2021). In the United States, the twin challenges of election disinformation and the fraught public discourse on the COVID-19 pandemic exposed weaknesses in the information ecosystem.

Technology—once regarded as a boon to democracy—is now regarded by some as a threat to it. For instance, a 2019 survey of 1000 experts 2 by the Pew Research Center and Elon University's Imagining the Internet Center found that, when pressed on the impact of the use of technology by citizens, civil society groups, and governments on core aspects of democracy and democratic representation, 49% of the "technology innovators, developers, business and policy leaders, researchers, and activists" surveyed said technology would "mostly weaken core aspects of democracy and democratic representation in the next decade," while 33% said technology would mostly strengthen democracies.

Perhaps the reason for such pessimism is that current interventions to manage trust in the information ecosystem are not taking place against a static backdrop. Social media has accompanied a decline of trust in institutions and the media generally, exacerbating a seemingly intractable set of social problems that create ripe conditions for the proliferation of disinformation and media manipulation.

Yet, the decade ahead will see even more profound disruptions to the ways in which information is produced, distributed, sought, and consumed. A number of technological developments are driving the disruption, including the installation of 5G wireless networks, which will enable richer media experiences, such as AR/VR and having more devices trading information more frequently. The increasing advance of various computational techniques to generate, manipulate, and target content is also problematic. The pace of technological developments suggest that, by the end of the decade, synthetic media may radically change the information ecosystem and demand new architectures and systems to aid sensemaking.

Consider a handful of recent developments:

- Natural text to speech enables technologies like Google Assistant, but is also being used for voice phishing scams.
- Fully generated realistic images from StyleGAN2—a neural network model—are being leveraged to make fake profiles in order to deceive on social media.
- Deep Fake face swapping videos are now capable of fooling many people who do not pay close attention to the video.

It is natural, in this environment, that the study of the threat itself is a thriving area of inquiry. Experts are busy compiling threats, theorizing threat models, studying the susceptibility of people to different types of manipulations, exploring what interventions build resilience to deception, helping to inform the debate about what laws and rules need to be in place, and offering input on the policies technology platforms should have in place to protect users from media manipulation.

In this chapter, we will focus on the emerging threat landscape, in order to give the reader a sense of the importance of media forensics as a discipline. While the technical work of discerning text, images, video, and audio and their myriad combinations may not seem to qualify as a "crisis discipline," we believe that it is indeed.

2.3 New Technologies, New Threats

In this section, we describe selected research trends with high potential impact on the threat landscape of synthetic media generation. We hope to ground the reader in the art of the possible at the cutting edge of research and provide some intuition to where the field is headed. We briefly explain the key contributions and why we believe they are significant for students and practitioners of media forensics. This list is not exhaustive.

These computational techniques will underpin synthetic media applications in the future.

2.3.1 End-to-End Trainable Speech Synthesis

Most speech generation systems include several processing steps that are customarily trained in isolation (e.g., text-to-spectrogram generators, vocoders). Moreover, while generative adversarial networks (GANs) have made tremendous progress on visual data, their application to human speech still remains limited. A recent paper addresses both of these issues with an end-to-end adversarial training protocol (Donahue et al. 2021). The paper describes a fully differentiable feedforward architecture for text-to-speech synthesis, which further reduces the amount of the necessary training-time supervision by eliminating the need for linguistic feature alignment. This further lowers the barrier for engineering novel speech synthesis systems, simplifying the need for data pre-processing to generate speech from textual inputs.

Another attack vector that has been undergoing rapid progress is voice cloning (Hao 2021). Voice cloning can synthetically convert a short audio recording of one person's voice to produce a very similar voice applied in various inauthentic capacities. A recent paper proposes a novel speech synthesis system, NAUTILUS, which brings text-to-speech and voice cloning under the same framework (Luong and Yamagishi 2020). This demonstrates the possibility that a single end-to-end model could be used for multiple speech synthesis tasks and could lead to unification recently seen in natural language models (e.g., BERT Devlin et al. 2019 where a single architecture beats state of the art in multiple benchmark problems).

2.3.2 GAN-Based Codecs for Still and Moving Pictures

Generative adversarial networks (GANs) have revolutionized image synthesis as they rapidly evolved from small-scale toy datasets to high-resolution photorealistic content. They exploit the capacity of deep neural networks to learn complex manifolds and enable generative sampling of novel content. While the research started from unconditional, domain-specific generators (e.g., human faces), recent work focuses on conditional generation. This regime is of particular importance for students and practitioners of media forensics as it enables control of the generation process.

One area where this technology is underappreciated with regard to its misuse potential is lossy codecs for still and moving pictures. Visual content abounds with inconsequential background content or abstract textures (e.g., grass, water, clouds) which can be synthesized in high quality instead of explicit coding. This is particularly important in limited bandwidth regimes that typically occur in mobile applications, social media, or video conferencing. Examples of recent techniques in this area include GAN-based videoconferencing compression from Nvidia (Nvidia Inc 2020) and the first high-fidelity GAN-based image codec from Google (Mentzer et al. 2020).

This may have far-reaching consequences. Firstly, effortless manipulation of video conference streams may lead to novel man-in-the-middle attacks and further blur the boundary between acceptable and malicious alterations (e.g., while replaced backgrounds are often acceptable, the faces or expressions are probably not). Secondly, the ability to synthesize an infinite number of content variations (e.g., decode the same image with different looking trees) poses a threat to reverse image search—one of the very few reliable techniques for fact checkers today.

2.3.3 Improvements in Image Manipulation

Progress in machine learning, computer vision, and graphics has been constantly lowering the barrier (i.e., skill, time, compute) for convincing image manipulation. This threat is well recognized, hence we simply report a novel manipulation technique recently developed by University of Washington and Adobe (Zhang et al. 2020). The authors propose a method for object removal from photographs that automatically detects and removes the object's shadow. This may be another blow to image forensic techniques that look for (in)consistency of physical traces in the photographs.

2.3.4 Trillion-Param Models

The historical trend of throwing more compute at a problem to see where the performance ceiling is continuous with large-scale language models (Radford and Narasimhan 2018; Kaplan et al. 2020). Following on the heels of the much popularized GPT-3 (Brown et al. 2020), Google Brain recently published their new Switch Transformer model (Fedus et al. 2021)—a $\times$ 10 parameter increase over GPT-3. Interestingly, research trends related to increasing efficiency and reducing costs are paralleling the development of these large models.

With large sparse models, it is becoming apparent that lower precision (e.g., 16-bit) can be effectively used, combined with new chip architecture improvements targeting these lower bit precisions (Wang et al. 2018; Park and Yoo 2020). Furthermore, because of the scales of data needed to train these enormous models, sparsification techniques are further being explored (Jacobs et al. 1991; Jordan and Jacobs 1993). New frameworks are also emerging that facilitate distribution of ML models across many machines (e.g., Tensorflow Mesh that was used in Google's Switch Transformers) (Shazeer et al. 2018). Finally, an important trend that is also stemming from this development is improvements in compression and distillation of models, a concept that was explored in our first symposium as a potential attack vector.

2.3.5 *Lottery Tickets and Compression in Generative Models*

Following in the steps of a landmark 2018 paper proposing the neural lottery hypothesis (Frankle and Carbin 2019), work continues in ways of discovering intelligent compression and distillation techniques to prune large models without losing accuracy. A recent paper being presented at ICLR 21 has applied this hypothesis to GANs (Chen et al. 2021). There have also been interesting approaches investigating improvements in the training dynamics of subnetworks by leveraging differential solvers to stabilize GANs (Qin et al. 2020). The Holy Grail around predicting subnetworks or initialization weights still remains elusive, and so the trend of training ever larger parameter models, and then attempting to compress/distill them, will continue until the next ceiling is reached.

2.4 New Developments in the Private Sector

To understand how the generative threat landscape is evolving, it is important to track developments from industry that could be used in media manipulation. Here, we include research papers, libraries for developers, APIs for the general public, as well as polished products. The key players include research and development divisions of large corporations, such as Google, Nvidia, and Adobe, but we also see relevant activities from the startup sector.

2.4.1 *Image and Video*

Nvidia Maxine (Nvidia Inc 2020)—Nvidia released a fully accelerated platform software development kit for developers of video conferencing services to build and deploy AI-powered features that use state-of-the-art models in their cloud. A notable feature includes GAN-based video compression, where human faces are synthesized based on facial landmarks and a single reference image. While this significantly improves user experience, particularly for low-bandwidth connections, it opens up potential for abuse.

Smart Portraits and Neural Filters in Photoshop (Nvidia Inc 2021)—Adobe introduced neural filters into their flagship photo editor, Photoshop. The addition includes smart portraits, a machine-learning-based face editing feature which brings the latest face synthesis capabilities to the masses. This development continues the trend of bringing advanced ML-power image editing to mainstream products. Other notable examples of commercially available advanced AI editing tools include Anthropic's PortraitPro (Anthropics Inc 2021) and Skylum's Luminar (Skylum Inc 2021).

ImageGPT (Chen et al. 2020) and DALL-E (Ramesh et al. 2021)—Open AI is experimenting with extending their GPT-3 language model to other data modalities. ImageGPT is an autoregressive image synthesis model capable of convincing com-

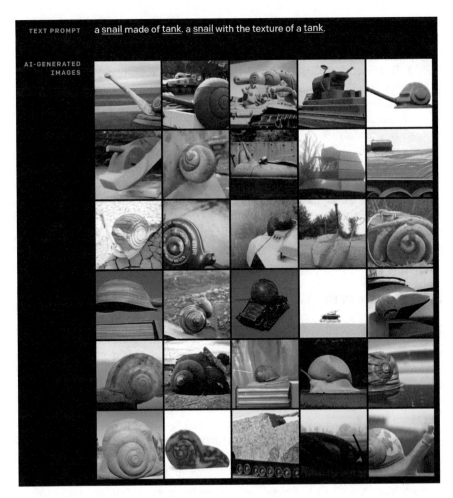

Fig. 2.1 Synthesized images generated by DALL-E based on a prompt requesting "snail-like tanks"

pletion of image prompts. DALL-E is an extension of this idea that combines visual and language tokens. It allows for image synthesis based on textual prompts. The model shows a remarkable capability in generating plausible novel content based on abstract concepts, e.g., "avocado chairs" or "snail tanks," see Fig. 2.1.

Runway ML (RunwayML Inc 2021)—Runway ML is a Brooklyn-based startup that provides an AI-powered tool for artists/creatives. Their software brings most recent open-source ML models to the masses and allows for their adoption in image/video editing workflows.

Imaginaire (Nvidia Inc 2021)—Nvidia released a PyTorch library that contains optimized implementation of several image and video synthesis methods. This tool is intended for developers and provides various models for supervised and unsupervised image-to-image and video-to-video translation.

2.4.2 Language Models

GPT-3 (Brown et al. 2020)—OpenAI developed a language model with 175-billion parameters. It performs well in few-shot learning scenarios, producing nuanced text and functioning code. Shortly after, the model has been released via a Web API for general-purpose text-to-text mapping.

Switch Transformers (Fedus et al. 2021)—Six months after the GPT-3 release, Google announced their record-breaking language model featuring over a trillion parameters. The model has been described in a detailed technical manuscript, and has been published along with a Tensorflow Mesh implementation.

CLIP (Radford et al. 2021)—Released in early January 2021 by OpenAI, CLIP is an impressive zero-shot image classifier. It pre-trained on 400 million text-to-image pairs. Given a list of descriptions, CLIP will make a prediction for which descriptor, or caption, a given image should be paired.

2.5 Threats in the Wild

In this section, we wish to provide a summary of observed disinformation campaigns and scenarios that are relevant to students and practitioners of media forensics. We would like to underscore new tactics and evolving resources and capabilities of adversaries.

2.5.1 User-Generated Manipulations

Thus far, user-generated media, aimed at manipulating public opinion, by and large, have not made use of sophisticated media manipulation capabilities beyond image, video, and sound editing software available to the general public. Multiple politically charged manipulated videos emerged and gained millions of views during the second half of 2020. Documented cases related to the US elections include manipulation of the visual background of a speaker (Reuters Staff 2021c), adding manipulated audio (Weigel 2021), cropping of a footage (Reuters Staff 2021a), manipulated editing of a video, and an attempt to spread false claim that an original video is a "deep fake" production (Reuters Staff 2021b).

In all of the cases mentioned, the videos gained millions of views within hours on Twitter, YouTube, Facebook, Instagram, TikTok, and other platforms, before they were taken down or flagged to emphasize they were manipulated. Often the videos emerged from an anonymous account on social media, but were shared widely, in part by prominent politicians and campaign accounts on Twitter, Facebook, and other platforms. Similar incidents emerged in India (e.g., an edited version of a video documenting police brutality) (Times of India Anam Ajmal 2021).

Today, any user can employ AI-generated images for creating a fake person. As recently reported in the New York Times (Hill and White 2021), Generated Photos (Generated Photos Inc 2021) can create "unique, worry-free" fake persons for "$2.99 or 1,000 people for $1000." Similarly, there's open advertising on Twitter by Rosebud AI for creating human animation (Rosebud AI Inc 2021).

Takedowns of far-right extremists and conspiracy-related accounts by several platforms, including Twitter, Facebook, Instagram, and WhatsApp announcement of changes to its privacy policies resulted in reports of migration to alternative platforms and chat services. Though it is still too early to assess the long-term impact of this change, it has the potential of impacting consumption and distribution of media, particularly among audiences that are already keen to proliferate conspiracy theories and disinformation.

2.5.2 Corporate Manipulation Services

In recent years, investigations by journalists, researchers, and social media platforms have documented several information operations that were conducted by state actors via marketing and communications firms around the world. States involved included, among others, Saudi Arabia, UAE, and Russia (CNN Clarissa Ward et al. 2021). Some researchers claim outsourcing information operations has become "the new normal," and so far the operations that were exposed did not spread original synthetic media (Grossman and Ramali 2021).

For those countries seeking the benefits of employing inauthentic content but unable to internally develop such a capability, there are plenty of manipulation companies available for hire. Regimes incapable of building a manipulation capability, deprived of the technological resources, and no longer need to worry as an entirely new industry of Advanced Persistent Manipulators (APM) (GMFUS Clint Watts 2021) have emerged in the private sector to meet the demand. Many advertising agencies, able to make manipulated content with ease, are hired by governments for public relations purposes and blur the lines between authentic and inauthentic contents.

The use of a third party by state actors tends to be limited to a targeted campaign. In 2019, reports by Twitter-linked campaigns run by the Saudi firm Smaat to the Saudi government (Twitter Inc 2021). The firm used automated tools to amplify pro-Saudi messages, using networks of thousands of accounts. The FBI has accused the firm's CEO for being the Saudi government agent in an espionage case involving two Saudi former Twitter employees (Paul 2021).

In 2020, a CNN investigation exposed an outsourced effort by Russia, through Ghana and Nigeria to impact U.S. politics. The operation targeted African American demographics in the United States using accounts on Facebook, Twitter, and Instagram. It was mainly manually executed by a dozen troll farm employees in West Africa (CNN Clarissa Ward et al. 2021).

2.5.3 Nation State Manipulation Examples

For governments seeking to utilize, acquire, or develop synthetic media tools, the resources and technology capability needed to create and harness these tools are significant, hence the rise of third-party actors is described in Sect. 2.5.2. Below is a sampling of recent confirmed or presumed nation state manipulations.

People's Republic of China

Over the last 2 years, Western social media platforms have encountered a sizeable amount of inauthentic images and video attributed or believed to be stemming from China (The Associated Press ERIKA KINETZ 2021; Hatmaker 2021). To what degree these attributions are directly connected to the Chinese government is unknown, but the incentives align to suggest the Chinese Communist Party (CCP) controlling the state, and the country's information environment creates or condones the deployment of these malign media manipulations.

Australia has made the most public stand against Chinese manipulations. In November 2020, Chinese foreign minister and leading CCP Twitter provocateur posted a fake image showing an Australian soldier with a bloody knife next to a child (BBC Staff 2021). The image sought to inflame tensions in Afghanistan where Australian soldiers have recently been found to have killed Afghan prisoners during the 2009 to 2013 time period.

Aside from this highly public incident and Australian condemnation, signs of Chinese employment of inauthentic media continue to mount. Throughout the fall of 2019 and 2020, Chinese language YouTube news channels proliferated. These new programs, created in high volume and distributed on a vast scale, periodically mix real information and real people with selectively sprinkled inauthentic content, much of which is difficult to assess as an intended manipulation or a form of narrated recreation of actual or perceived events. Graphika has reported on these "political spam networks" that distribute pro-Chinese propaganda, first targeting Hong Kong protests in 2019 (Nimmo et al. 2021), and the more recent addition of English language YouTube videos focusing on US policy in 2020 (Nimmo et al. 2021). Reports

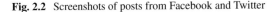

Three posts from Facebook and Twitter are direct (though unattributed) quotes from the Chinese government's official announcement of sanctions against 28 former Trump administration officials. The tweet on the right carries an image that reads "Anti-China american politicians⁵ will finally pay the price for their crazy actions".

Fig. 2.2 Screenshots of posts from Facebook and Twitter

showed the use of inauthentic or misattributed content in both visual and audio forms. Accounts used AI-generated account profile images and dubbed voice overs.

A similar network of inauthentic accounts were spotted immediately following the 2021 US presidential inauguration. Users with Deep Fake profile photos, which bear a notable similarity to AI-generated photos on *thispersondoesnotexist.com*, tweeted verbatim excerpts lifted from the Chinese Ministry of Foreign Affairs, see Fig. 2.2. These alleged Chinese accounts were also enmeshed in a publicly discussed campaign reported by The Guardian as amplifying a debunked ballot video surfacing among voter fraud allegations surrounding the U.S. presidential election.

The accounts' employment of inauthentic faces, Fig. 2.3, while somewhat sloppy, does offer a way to avoid detection as has been seen repeatedly when stolen real photos dropped into image or facial recognition software are easily detected.

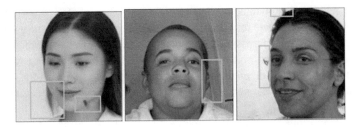

Fig. 2.3 Three examples of Deep Fake profile photos from a network of 13 pro-China inauthentic accounts on Twitter. Screengrabs of photos directly from Twitter

France

Individuals associated with the French military have been linked to a 2020 information operation, aimed at discrediting a similar Russian IO employing trolls to spread disinformation in the Central African Republic and other African countries (Graphika and The Stanford Internet Observatory 2021). The French operation used fake media to create Facebook accounts. This is an interesting example of tactical strategy that employs a fake campaign to fight disinformation.

Russia/IRA

The Russian-sponsored Internet Research Agency was found to have created a small number of fake Facebook, Twitter, and LinkedIn accounts that were amplifying a website posing as an independent news outlet, peacedata.net (Nimmo et al. 2021). While IRA troll accounts have been prolific for years, this is the first observed instance of their use of AI-generated images. Fake profiles were created to give credibility to peacedata.net, who hired unwitting freelance writers to provide left-leaning content for audiences, mostly in the US and UK, as part of a calculated, albeit ineffective, disruption campaign.

2.5.4   Use of AI Techniques for Deception 2019–2020

During the year 2019–2020, at the height of the COVID-19 pandemic, nefarious generative AI techniques gained popularity. The following are examples of various modalities employed for the purposes of deception and information operations (IO).

Ultimately, we also observe that commercial developments are rapidly advancing the capability for video and image manipulation. While resources required to train GANs and other models will remain a prohibitive factor in the near future, improvements in image codecs, larger training parameters, lowering precision, and compression/distillation create additional threat opportunities. The threat landscape will change rapidly in the next 3 to 5 years, requiring media forensics capabilities to advance quickly to keep up with the capabilities of adversaries.

GAN-Generated Images

The most prevalent documented use so far of AI capabilities in IO has been in utilizing available free tools to create fake personas online. Over the past year, the use of GAN technology has become increasingly common, in operations linked to state actors, or organizations driven by political motivation, see Fig. 2.3. The technology was mostly applied in generating profile pictures for inauthentic users on social media, relying on open-source available tools such as thispersondoesntexist.com and Generated Photos (Generated Photos Inc 2021).

Several operations used GANs to create hundreds to thousands of accounts that spammed platforms with content and comments, in an attempt to amplify messages. Some of these operations were attributed to Chinese actors and to PR firms.

Other operations limited the use of GAN-generated images to create a handful of more "believable" "fake personas" that appeared as creators of content. At least one of those efforts was attributed to Russia's IRA. In at least one case, a LinkedIn account with a GAN-generated image was used as part of a suspected espionage effort to contact US national security experts (Associated Press 2021).

Synthetic Audio
In recent years, there were at least three attempts documented, one of which was successful, to defraud companies and individuals using synthetic audio. Two cases included use of synthetic audio to impersonate the company CEO in a phone call. In 2019, a British energy company transferred over $200,000 following a fraudulent phone call, generated with synthetic audio to impersonate the parent company' CEO, requesting the money transfer (Wall Street Journal Catherine Stupp 2021). A similar attempt to defraud a US-based company failed in 2020 (Nisos 2021). Symantec has reported at least three cases in 2019 of synthetic audio being deployed to defraud private companies (The Washington Post Drew Harwell 2021).

Deception Using GPT-3
While user-generated manipulations aiming at impacting public opinion have avoided advanced technology so far, there was at least one attempt to use GPT-3 in generating automated text, by an anonymous user/bot on Reddit (The Next Web Thomas Macaulay 2021). The effort seemed to abuse a third-party app access to OpenAI API and to give hundreds of comments. In some cases, the bot published conspiracy theories and made false statements.

Additionally, recent reports noted GPT-3 has an "anti-Muslim" bias (The Next Web Tristan Greene 2021). This conclusion points to the potential damage that could be caused by apps that are using the technology, even with no malicious intentions.

IO and Fraud Using AI Techniques
Over a dozen cases of use of synthetic media for information operations and other criminal activities were recorded during the past couple of years. In December 2019, social media platforms noted the first use of AI techniques in a case of "coordinated inauthentic behavior" which was attributed to The BL, a company with ties to the publisher of The Epoch Times, according to Facebook. The company created hundreds of fake Facebook accounts using GAN techniques to create synthetic profile pictures. The use of GAN to create false portrait pictures of non-existent personas repeated itself in more than a handful of IO that were discovered during 2020, including in operations that were attributed by social media platforms to Russian IRA, to China, and to a US-based PR firm.

2.6 Threat Models

Just as the state of the art in technologies for media manipulation has advanced, so has theorizing around threat models. Conceptual models and frameworks used to understand and analyze threats and describe adversary behavior are evolving alongside what is known of the threat landscape. It is useful for students of media forensics to understand these frameworks generally, as they are used in the field by analysts,

	Manipulating the narrative		Manipulating the social network	
Positive	Engage	Messages that bring up a related but relevant topic	Back	Actions that increase the importance of the opinion leader or create a new opinion leader
	Explain	Messages that provides details on or elaborate the topic	Build	Actions that create a group or the appearance of a group
	Excite	messages that elicit a positive emotion such as joy or excitement	Bridge	Actions that build a connection between two or more groups
	Enhance	Messages that encourage the topic-group to continue with the topic	Boost	Actions that grow the size of the group or make it appear that it has grown
Negative	Dismiss	Messages about why the topic is not important	Neutralize	Actions decrease the importance of the opinion leader
	Distort	Messages that alter the main message of the topic	Nuke	Actions that lead to a group being dismantled or breaking up, or appearing to be broken up
	Dismay	Messages that elicit a negative emotion such as sadness or anger	Narrow	Actions that lead to a group becoming sequestered from other groups or marginalized
	Distract	Discussion about a totally	Neglect	Actions that reduce the size of

Fig. 2.4 BEND framework

technologists, and others who work to understand and mitigate against manipulations and deception.

It is useful to apply different types of frameworks and threat models for different kinds of threats. There is a continuum between the sub-disciplines and sub-concerns that form media and information manipulation. At times it is possible to seek to put too much under the same umbrella. For instance, when analyzing COVID-19 misinformation analysis, the processes and models that are used to tackle that are extremely different than the processes and models that we use to tackle covert influence operations that tend to be state-sponsored, or to understand attacks that are primarily financially motivated and conducted by criminal groups. It's important to think about what models apply in what circumstances, and specifically what disciplines can contribute to which models.

The below examples do not represent a comprehensive selection of threat models and frameworks, but serve as examples.

2.6.1 Carnegie Mellon BEND Framework

Developed by Dr. Kathleen Carley, the BEND framework is a conceptual system that seeks to make sense of influence campaigns, and their strategies and tactics. Carley notes that the "BEND framework is the product of years of research on disinformation and other forms of communication based influence campaigns, and communication objectives of Russia and other adversarial communities, including terror groups such as ISIS that began in late December of 2013."[1] The letters that are represented in the acronym BEND refer to elements of the framework, such as "B" which references "Back, Build, Bridge, and Boost" influence maneuvers.

2.6.2 The ABC Framework

Initially developed by Camille Francois, then Chief Innovation Officer at Graphika, the ABC framework offers a mechanism to understand "three key vectors character-istic of viral deception" in order to guide potential remedies (François 2020). In this framework:

- A is for Manipulative Actors. In Francois's framework, manipulative actors engage knowingly and with clear intent in their deceptions.
- B is for Deceptive Behavior. Behavior encompasses all of the techniques and tools a manipulative actor may use, from engaging a troll farm or a bot army to producing manipulated media.
- C is for Harmful Content. Ultimately, it is the message itself that defines a campaign and often its efficacy.

TABLE 1
The ABCDE Framework

Actor	*What kinds of actors are involved?* This question can help establish, for example, whether the case involves a foreign state actor.
Behavior	*What activities are exhibited?* This inquiry can help establish, for instance, evidence of coordination and inauthenticity.
Content	*What kinds of content are being created and distributed?* This line of questioning can help establish, for example, whether the information being deployed is deceptive.
Degree	*What is the overall impact of the case and whom does it affect?* This question can help establish the actual harms and severity of the case.
Effect	*What is the overall impact of the case and whom does it affect?* This question can help establish the actual harms and severity of the case.

Fig. 2.5 ABCDE framework

---

[1] https://link.springer.com/article/10.1007/s10588-020-09322-9.

Later, Alexandre Alaphilippe, Executive Director of the EU DisinfoLab, proposed adding a D for Distribution, to describe the importance of the means of distribution to the overall manipulation strategy (Brookings Inst Alexandre Alaphilippe 2021). And yet another letter joined the acronym courtesy of James Pamment in the fall of 2020-Pamment proposed E to evaluate the Effect of the manipulation campaign (Fig. 2.5).

2.6.3   The AMITT Framework

The Misinfosec and Credibility Coalition communities, led by Sara-Jayne Terp, Christopher Walker, John Gray, and Dr. Pablo Breuer, developed the AMITT (Adversarial Misinformation and Influence Tactics and Techniques) frameworks to codify the behaviours (tactics and techniques) of disinformation threats and responders, to enable rapid alerts, risk assessment, and analysis. AMITT proposes diagramming and treating equally both threat actor behaviors and response behaviors by the target (Gray and Terp 2019). The AMITT TTPs application of tactics, techniques, procedures, and frameworks is the disinformation version of the ATT&K model, which is an information security standard threat model. The AMITT STIX model is the disinformation version of the STIX model, which is an information security data sharing standard.

2.6.4   The SCOTCH Framework

In order to enable analysts "to comprehensively and rapidly characterize an adversarial operation or campaign," Dr. Sam Blazek developed the SCOTCH framework (Blazek 2021) (Fig. 2.6).

- S—Source—the actor who is responsible for the campaign.
- C—Channel—the platform and its affordances and features that deliver the campaign.
- O—Objective—the objective of the actor.
- T—Target—the intended audience for the campaign.
- C—Composition—the language, content, or message of the campaign.
- H—Hook—the psychological phenomena being exploited by the campaign.

Blazek offers this example of how SCOTCH might describe a campaign:

At the campaign level, a hypothetical example of a SCOTCH characterization is: Non-State Actors used Facebook, Twitter, and Low-Quality Media Outlets in an attempt to Undermine the Integrity of the 2020 Presidential Election among Conservative American Audiences using Fake Images and Videos and capturing attention by Hashtag Hijacking and Posting in Shared-Interest Facebook Groups via Cutouts (Fig. 2.7).

Planning
Preparation
Execution
Evaluation
Strategic Planning
Objective Planning
Develop People
Develop Networks
Microtargeting
Develop Content
Channel Selection
Pump Priming
Exposure
Go Physical
Persistence
Measure Effectiveness

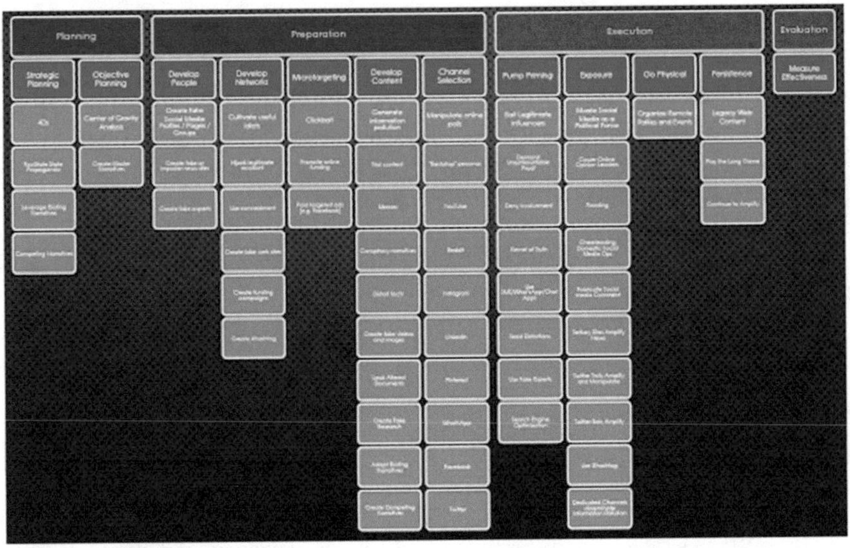

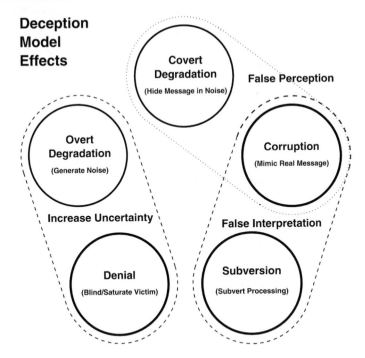

Fig. 2.6 AMITT framework

Deception Model Effects
Covert Degradation
(Hide Message in Noise)
False Perception
Overt Degradation
(Generate Noise)
Corruption
(Mimic Real Message)
Increase Uncertainty
False Interpretation
Denial
(Blind/Saturate Victim)
Subversion
(Subvert Processing)

Fig. 2.7 Deception model effects

2.6.5 Deception Model Effects

Still other frameworks are used to explain how disinformation diffuses in a population. Mathematicians have employed game theory to look at information as quantities that diffuse according to certain dynamics over networks. For instance, Carlo Kopp, Kevin Korb, and Bruce Mills model cooperation and diffusion in populations (Kopp et al. 2018) exposed to "fake news":

This model "depicts the relationships between the deception models and the components of system they are employed to compromise."

2.6.6 4Ds

Some frameworks focus on the goals manipulators. Ben Nimmo, then at the Central European Policy Institute and now at Facebook, where he leads global threat intelligence and strategy against influence operations, developed the 4Ds to describe the tactics of Russian disinformation campaigns (StopFake.org 2021). "They can be summed up in four words: dismiss, distort, distract, dismay," wrote Nimmo (Fig. 2.8).

- Dismiss may involve denying allegations made against Russia or a Russian actor.
- Distortion includes outright lies, fabrications, or obfuscations of reality.
- Distract includes methods of turning attention away from one phenomenon to another.
- Dismay is a method to get others to regard Russia as too significant a threat to confront.

2.6.7 Advanced Persistent Manipulators

Clint Watts, a former FBI Counterterrorism official, similarly thinks about the goals of threat actors and describes, in particular, nation states and other well-resourced manipulators as Advanced Persistent Manipulators.

They use combinations of influence techniques and operate across the full spectrum of social media platforms, using the unique attributes of each to achieve their objectives. They have sufficient resources and talent to sustain their campaigns, and the most sophisticated and troublesome ones can create or acquire the most sophisticated technology. APMs can harness, aggregate, and nimbly parse user data and can recon new adaptive techniques to skirt account and content controls. Finally, they know and operate within terms of service, and they take advantage of free-speech provisions. Adaptive APM behavior will therefore continue to challenge the ability of social media platforms to thwart corrosive manipulation without harming their own business model (Fig. 2.8).

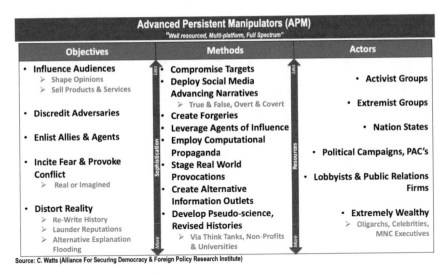

Fig. 2.8 Advanced persistent manipulators framework

2.6.8 Scenarios for Financial Harm

Jon Bateman, a fellow in the Cyber Policy Initiative of the Technology and International Affairs Program at the Carnegie Endowment for International Peace, shows how assessments of threats can be applied to specific industry areas or targets in his work on the threats of media manipulation to the financial system (Carnegie Endowment John Batemant 2020).

> In the absence of hard data, a close analysis of potential scenarios can help to better gauge the problem. In this paper, ten scenarios illustrate how criminals and other bad actors could abuse synthetic media technology to inflict financial harm on a broad swath of targets. Based on today's synthetic media technology and the realities of financial crime, the scenarios explore whether and how synthetic media could alter the threat landscape.

2.7 Investments in Countering False Media

Just as there is substantial effort at theorizing and implementing threat models to understand media manipulation, there is a growing effort in civil society, government, and the private sector to contend with disinformation. There is a large volume of work in multiple disciplines to categorize both the threats and many key indicators of population susceptibility to disinformation, including new work on resilience, dynamics of exploitative activities, and case studies on incidents and actors, including those who are not primarily motivated by malicious intent. This literature combines

TABLE 1

Ten Synthetic Media Scenarios for Financial Harm

Target	Scenario	Role of Synthetic Media	Key Malicious Technique
Individuals	1. Identity theft	Voice cloning or face-swap video is used to impersonate a wealthy individual and initiate fraudulent transactions. Alternatively, it is used to impersonate a corporate officer and gain access to databases of personal information, which can enable larger-scale identity theft.	
	2. Imposter scam	Voice cloning or face-swap video is used to impersonate a trusted government official or family member of the victim and coerce a fraudulent payment.	
	3. Cyber extortion	Synthetic pornography of the victim is used for blackmail.	
Companies	4. Payment fraud	Voice cloning or face-swap video is used to impersonate a corporate officer and initiate fraudulent transactions.	
	5. Stock manipulation via fabricated events	Voice cloning or face-swap video is used to defame a corporate leader or falsify a product endorsement, which can alter investor sentiment.	
	6. Stock manipulation via bots	Synthetic photos and text are used to construct human-like social media bots that attack or promote a brand, which can alter investor perception of consumer sentiment.	
	7. Malicious bank run	Synthetic photos and text are used to construct human-like social media bots that spread false rumors of bank weakness, which can fuel runs on cash.	
Markets	8. Malicious flash crash	Voice cloning or face-swap video is used to fabricate a market-moving event.	
Regulatory Structures	9. Fabricated government action	Voice cloning or face-swap video is used to fabricate an imminent interest rate change, policy shift, or enforcement action.	
	10. Regulatory astroturfing	Synthetic text is used to fabricate comments from the public on proposed financial regulations, which can manipulate the rulemaking process.	

Deepfake voice phishing	Fabricated private remarks	Synthetic social botnet	Narrowcast	Broadcast

Fig. 2.9 Synthetic media scenarios and financial harm (Carnegie Endowment for International Peace JON BATEMAN 2020)

with a growing list of government, civil society and media and industry interventions aimed at mitigating the harms of disinformation.

There are now hundreds of groups and organizations in the United States and around the world working on various approaches and problem sets to monitor and counter disinformation in democracies, see Fig. 2.10. This emerging infrastructure presents an opportunity to create connective tissue between these efforts, and to provide common tools and platforms that enable their work. Consider three examples:

1. The Disinfo Defense League: A project of the Media Democracy Fund, the DDL is a consortium of more than 200 grassroots organizations across the U.S. that helps to mobilize minority communities to counter disinformation campaigns (ABC News Fergal Gallagher 2021).
2. The Beacon Project: Established by the nonprofit IRI combats state-sponsored influence operations, in particular, ones from Russia. The project identifies and exposes disinformation and works with local partners to facilitate a response (Beacon Project 2021).
3. CogSec Collab: The Cognitive Security Collaborative that maps and brings together information security researchers, data scientists, and other subject-matter experts to create and improve resources for the protection and defense of the cognitive domain (Cogsec Collab 2021).

At the same time, there are significant efforts underway to contend with the implications of these developments in synthetic media and the threat vectors they create.

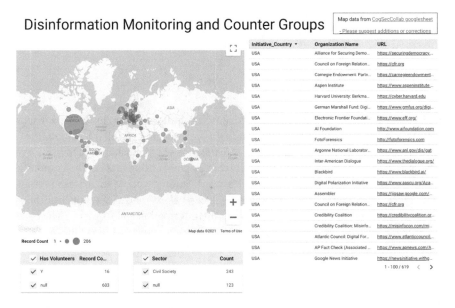

Fig. 2.10 There are now hundreds of civil society, government, and private sector groups addressing disinformation across the globe. (Cogsec Collab 2021)

2.7.1 DARPA SEMAFOR

The Semantic Forensics (SemaFor) program under the Defense Advanced Research Projects Agency (DARPA) represents a substantial research and development effort to create forensic tools and techniques to develop semantic understanding of content and identify synthetic media across multiple modalities (DARPA 2021).

2.7.2 The Partnership on AI Steering Committee on Media Integrity Working Group

The Partnership on AI has hosted an industry led effort to create similar detectors (Humprecht et al. 2020). Technology platforms and universities continue to develop new tools and techniques to determine provenance and identify synthetic media and other forms of digital media of questionable veracity. The Working Group brings together organizations spanning civil society, technology companies, media organizations, and academic institutions will be focused on a specific set of activities and projects directed at strengthening the research landscape related to new technical capabilities in media production and detection, and increasing coordination across organizations implicated by these developments.

2.7.3 JPEG Committee

In October 2020, the JPEG committee released a draft document initiating a discussion to explore "if a JPEG standard can facilitate a secure and reliable annotation of media modifications, both in good faith and malicious usage scenarios" (Joint Bi level Image Experts Group and Joint Photographic Experts Group 2021). The committee noted the need to consider revision of standardization practices to address pressing challenges, including use of AI techniques ("Deep Fakes") and use of authentic media out of context for the purposes of spreading misinformation and forgeries. The paper notes two areas of focus: (1) having a record of modifications of a file within its metadata and (2) proving the record is secured and accessible so provenance is traceable.

2.7.4 Content Authenticity Initiative (CAI)

The CAI was founded in 2019 and led by Adobe, Twitter, and The New York Times, with the goal of establishing standards that will allow better evaluation of content. In August 2020, the forum published its white paper "Setting the Standard for Digital Content Attribution" highlighting the importance of transparency around provenance of content (Adobe, New York Times, Twitter, The Content Authenticity Initiative 2021). CAI will "provide a layer of robust, tamper-evident attribution and history data built upon XMP, Schema.org, and other metadata standards that goes far beyond common uses today."

2.7.5 Media Review

Media Review is an initiative led by the Duke Reporters Lab to provide the fact checker community a schema that would allow evaluation of videos and images, through tagging of different types of manipulation (Schema.org 2021). Media Review is based on the ClaimReview project, and is currently a pending schema.

2.8 Excerpts on Susceptibility and Resilience to Media Manipulation

Like other topics in forensics, media manipulation is directly tied to psychology. Though a thorough treatment of this topic is outside the scope of this chapter, we would like to expose the reader to some foundational work in these areas of research.

The following section is a collection of critical ideas from different areas of psychological mis/disinformation research. The purpose of these is to convey powerful ideas in the authors' own words. We hope that readers will select those of interest and investigate those works further.

2.8.1 Susceptibility and Resilience

Below is a sample of recent studies on various aspects of exploitation via disinformation which are summarized and quoted. We categorize these papers by overarching topics, but note that many bear relevant findings across multiple topics.

Resilience to online disinformation: A framework for cross-national comparative research (Humprecht et al. 2020)
This paper seeks to develop a cross-country framework for analyzing disinformation effects and population susceptibility and resilience. Three country clusters are identified: media-supportive; politically consensus-driven (e.g., Canada, many Western European democracies); polarized (e.g., Italy, Spain, Greece); and low-trust, politicized, and fragmented (e.g., United States). Media-supportive countries in the first cluster are identified as most resilient to disinformation, and the United States is identified as most vulnerable to disinformation owing to its "high levels of populist communication, polarization, and low levels of trust in the news media."

Who falls for fake news? The roles of bullshit receptivity, overclaiming, familiarity, and analytic thinking (Pennycook and Rand 2020)
The tendency to ascribe profundity to randomly generated sentences – pseudo-profound bullshit receptivity – correlates positively with perceptions of fake news

accuracy, and negatively with the ability to differentiate between fake and real news (media truth discernment). Relatedly, individuals who overclaim their level of knowledge also judge fake news to be more accurate. We also extend previous research indicating that analytic thinking correlates negatively with perceived accuracy by showing that this relationship is not moderated by the presence/absence of the headline's source (which has no effect on accuracy), or by familiarity with the headlines (which correlates positively with perceived accuracy of fake and real news).

Fake news, fast and slow: Deliberation reduces belief in false (but not true) news headlines (Bago et al. 2020)
The Motivated System 2 Reasoning (MS2R) account posits that deliberation causes people to fall for fake news, because reasoning facilitates identity-protective cognition and is therefore used to rationalize content that is consistent with one's political ideology. The classical account of reasoning instead posits that people ineffectively discern between true and false news headlines when they fail to deliberate (and instead rely on intuition)... Our data suggest that, in the context of fake news, deliberation facilitates accurate belief formation and not partisan bias.

Political psychology in the digital (mis) information age: A model of news belief and sharing (Van Bavel et al. 2020)
This paper reviews complex psychological risk factors associated with belief in misinformation such as partisan bias, polarization, political ideology, cognitive style, memory, morality, and emotion. The research analyzes the implications and risks of various solutions, such as fact-checking, "pre-bunking," reflective thinking, de-platforming of bad actors, and media literacy.

Misinformation and morality: encountering fake news headlines makes them seem less unethical to publish and share (Effron and Raj 2020)
Experimental results and a pilot study indicate that "repeatedly encountering misinformation makes it seem less unethical to spread—-regardless of whether one believes it. Seeing a fake-news headline one or four times reduced how unethical participants thought it was to publish and share that headline when they saw it again – even when it was clearly labelled false and participants disbelieved it, and even after statistically accounting for judgments of how likeable and popular it was. In turn, perceiving it as less unethical predicted stronger inclinations to express approval of it online. People were also more likely to actually share repeated (vs. new) headlines in an experimental setting. We speculate that repeating blatant misinformation may reduce the moral condemnation it receives by making it feel intuitively true, and we discuss other potential mechanisms."

When is Disinformation (In) Credible? Experimental Findings on Message Characteristics and Individual Differences. (Schaewitz et al. 2020)
"...we conducted an experiment (N = 294) to investigate effects of message factors and individual differences on individuals' credibility and accuracy perceptions of disinformation as well as on their likelihood of sharing them. Results suggest that message factors, such as the source, inconsistencies, subjectivity, sensationalism, or

manipulated images, seem less important for users' evaluations of disinformation articles than individual differences. While need for cognition seems most relevant for accuracy perceptions, the opinion toward the news topic seems most crucial for whether people believe in the news and share it online."

Long-term effectiveness of inoculation against misinformation: Three longitudinal experiments. (Maertens et al. 2020)
"In 3 experiments (N = 151, N = 194, N = 170), participants played either Bad News (inoculation group) or Tetris (gamified control group) and rated the reliability of news headlines that either used a misinformation technique or not. We found that participants rate fake news as significantly less reliable after the intervention. In Experiment 1, we assessed participants at regular intervals to explore the longevity of this effect and found that the inoculation effect remains stable for at least 3 months. In Experiment 2, we sought to replicate these findings without regular testing and found significant decay over a 2-month time period so that the long-term inoculation effect was no longer significant. In Experiment 3, we replicated the inoculation effect and investigated whether long-term effects could be due to item-response memorization or the fake-to-real ratio of items presented, but found that this is not the case."

2.8.2 Case Studies: Threats and Actors

Below one may find a collection of exposed current threats and actors utilizing manipulations of varying sophistication.
The QAnon Conspiracy Theory: A Security Threat in the Making? (Combating Terrorism Center Amarnath Amarasingam, Marc-André Argentino 2021)
Analysis of QAnon shows that, while disorganized, it is a significant public security threat due to historical violent events associated with the group, ease of access for recruitment, and a "crowdsourcing" effect wherein followers "take interpretation and action into their own hands." The origins and development of the group are analyzed, including review of adjacent organizations such as Omega Kingdom Ministries. Followers' behavior is analyzed and five criminal cases are reviewed.

Political Deepfake Videos Misinform the Public, But No More than Other Fake Media (Barari et al. 2021)
"We demonstrate that political misinformation in the form of videos synthesized by deep learning ("deepfakes") can convince the American public of scandals that never occurred at alarming rates – nearly 50% of a representative sample – but no more so than equivalent misinformation conveyed through existing news formats like textual headlines or audio recordings. Similarly, we confirm that motivated reasoning about the deepfake target's identity (e.g., partisanship or gender) plays a key role in facilitating persuasion, but, again, no more so than via existing news formats. ...Finally, a series of randomized interventions reveal that brief but specific informational treatments about deepfakes only sometimes attenuate deepfakes' effects and in relatively

small scale. Above all else, broad literacy in politics and digital technology most strongly increases discernment between deepfakes and authentic videos of political elites."

Deepfakes and disinformation: exploring the impact of synthetic political video on deception, uncertainty, and trust in news (Vaccari and Chadwick 2020)
An experiment using deepfakes to measure its value as a deception tool and its impact on public trust: "While we do not find evidence that deceptive political deepfakes misled our participants, they left many of them uncertain about the truthfulness of their content. And, in turn, we show that uncertainty of this kind results in lower levels of trust in news on social media."

Coordinating a Multi-Platform Disinformation Campaign: Internet Research Agency Activity on Three US Social Media Platforms, 2015–2017 (Lukito 2020)
An analysis of the use of multiple modes as a technique to develop and deploy information attacks: [Abstract] "The following study explores IRA activity on three social media platforms, Facebook, Twitter, and Reddit, to understand how activities on these sites were temporally coordinated. Using a VAR analysis with Granger Causality tests, results show that IRA Reddit activity granger caused IRA Twitter activity within a one-week lag. One explanation may be that the Internet Research Agency is trial ballooning on one platform (i.e., Reddit) to figure out which messages are optimal to distribute on other social media (i.e., Twitter)."

The disconcerting potential of online disinformation: Persuasive effects of astro-turfing comments and three strategies for inoculation against them (Zerback et al. 2021)
This paper examines online astroturfing as part of Russian electoral disinformation efforts, finding that online astroturfing was able to change opinions and increase uncertainty even when the audience was "inoculated" against disinformation prior to exposure.

2.8.3 Dynamics of Exploitative Activities

Below are some excerpts from literature covering the dynamics and kinetics of IO activities and their relationships to disinformation.

Cultural Convergence: Insights into the behavior of misinformation networks on Twitter (McQuillan et al. 2020)
"We use network mapping to detect accounts creating content surrounding COVID-19, then Latent Dirichlet Allocation to extract topics, and bridging centrality to iden-tify topical and non-topical bridges, before examining the distribution of each topic and bridge over time and applying Jensen-Shannon divergence of topic distributions to show communities that are converging in their topical narratives." The primary topic under discussion within the data was COVID-19, and this report includes a

detailed analysis of "cultural bridging" activities among/between communities with conspiratorial views of the pandemic.

An automated pipeline for the discovery of conspiracy and conspiracy theory narrative frameworks: Bridgegate, Pizzagate and storytelling on the web (Tangherlini et al. 2020)
"Predicating our work on narrative theory, we present an automated pipeline for the discovery and description of the generative narrative frameworks of conspiracy theories that circulate on social media, and actual conspiracies reported in the news media. We base this work on two separate comprehensive repositories of blog posts and news articles describing the well-known conspiracy theory Pizzagate from 2016, and the New Jersey political conspiracy Bridgegate from 2013. ...We show how the Pizzagate framework relies on the conspiracy theorists' interpretation of "hidden knowledge" to link otherwise unlinked domains of human interaction, and hypothesize that this multi-domain focus is an important feature of conspiracy theories. We contrast this to the single domain focus of an actual conspiracy. While Pizzagate relies on the alignment of multiple domains, Bridgegate remains firmly rooted in the single domain of New Jersey politics. We hypothesize that the narrative framework of a conspiracy theory might stabilize quickly in contrast to the narrative framework of an actual conspiracy, which might develop more slowly as revelations come to light."

Fake news early detection: A theory-driven model (Zhou et al. 2020)
"The method investigates news content at various levels: lexicon-level, syntax-level, semantic-level, and discourse-level. We represent news at each level, relying on well-established theories in social and forensic psychology. Fake news detection is then conducted within a supervised machine learning framework. As an interdisciplinary research, our work explores potential fake news patterns, enhances the interpretability in fake news feature engineering, and studies the relationships among fake news, deception/disinformation, and clickbaits. Experiments conducted on two real-world datasets indicate the proposed method can outperform the state-of-the-art and enable fake news early detection when there is limited content information"

"...Among content-based models, we observe that (3) the proposed model performs comparatively well in predicting fake news with limited prior knowledge. We also observe that (4) similar to deception, fake news differs in content style, quality, and sentiment from the truth, while it carries similar levels of cognitive and perceptual information compared to the truth. (5) Similar to clickbaits, fake news headlines present higher sensationalism and lower newsworthiness while their readability characteristics are complex and difficult to be directly concluded. In addition, fake news (6) is often matched with shorter words and longer sentences."

A survey of fake news: Fundamental theories, detection methods, and opportunities (Zhou and Zafarani 2020)
"By involving dissemination information (i.e., social context) in fake news detection, propagation-based methods are more robust against writing style manipulation by malicious entities. However, propagation-based fake news detection is inefficient for fake news early detection ... as it is difficult for propagation-based models to detect

fake news before it has been disseminated, or to perform well when limited news dissemination information is available. Furthermore, mining news propagation and news writing style allow one to assess news intention. As discussed, the intuition is that (1) news created with a malicious intent, that is, to mislead and deceive the public, aims to be "more persuasive" compared to those not having such aims, and (2) malicious users often play a part in the propagation of fake news to enhance its social influence." This paper also emphasizes the need for explainable detection as a matter of public interest.

A Signal Detection Approach to Understanding the Identification of Fake News (Batailler et al. 2020)
"The current article discusses the value of Signal Detection Theory (SDT) in disentangling two distinct aspects in the identification of fake news: (1) ability to accurately distinguish between real news and fake news and (2) response biases to judge news as real versus fake regardless of news veracity. The value of SDT for understanding the determinants of fake news beliefs is illustrated with reanalyses of existing data sets, providing more nuanced insights into how partisan bias, cognitive reflection, and prior exposure influence the identification of fake news."

Sharing of fake news on social media: Application of the honeycomb framework and the third-person effect hypothesis (Talwar et al. 2020)
This publication uses the "honeycomb framework" of Kietzmann, Hermkens, McCarthy, and Silvestre (Kietzmann et al. 2011) to explain sharing behaviors related to fake news. Qualitative evaluation supports the explanatory value of the honeycomb framework, and the authors are able to identify a number of quantitative associations between demographics and sharing characteristics.

Modeling echo chambers and polarization dynamics in social networks (Baumann et al. 2020)
"We propose a model that introduces the dynamics of radicalization, as a reinforcing mechanism driving the evolution to extreme opinions from moderate initial conditions. Inspired by empirical findings on social interaction dynamics, we consider agents characterized by heterogeneous activities and homophily. We show that the transition between a global consensus and emerging radicalized states is mostly governed by social influence and by the controversialness of the topic discussed. Compared with empirical data of polarized debates on Twitter, the model qualitatively reproduces the observed relation between users' engagement and opinions, as well as opinion segregation in the interaction network. Our findings shed light on the mechanisms that may lie at the core of the emergence of echo chambers and polarization in social media."

(Mis)representing Ideology on Twitter: How Social Influence Shapes Online Political Expression (Guess 2021)
In comparing users' tweets as well as those of the accounts that they follow, alongside self-reported political affiliations from surveys, a distinction between self-reported political slant and those views expressed in tweets. This raises the notion that public

political expression may be distinct from individual affiliations, and that an individual's tweets may not be an accurate proxy for their beliefs. The authors call for a re-examination of potential social causes of/influences upon public political expression: [from Abstract] "we find evidence consistent with the hypothesis that users' public expression is powerfully shaped by their followers, independent of the political ideology they report identifying with in attitudinal surveys. Finally, we find that users' ideological expression is more extreme when they perceive their Twitter networks to be relatively like-minded."

2.8.4 Meta-Review

A modern review on the literature surrounding disinformation may be found below.

A systematic literature review on disinformation: Toward a unified taxonomical framework (Kapantai et al. 2021)
"Our online information landscape is characterized by a variety of different types of false information. There is no commonly agreed typology framework, specific categorization criteria, and explicit definitions as a basis to assist the further investigation of the area. Our work is focused on filling this need. Our contribution is twofold. First, we collect the various implicit and explicit disinformation typologies proposed by scholars. We consolidate the findings following certain design principles to articulate an all-inclusive disinformation typology. Second, we propose three independent dimensions with controlled values per dimension as categorization criteria for all types of disinformation. The taxonomy can promote and support further multidisciplinary research to analyze the special characteristics of the identified disinformation types."

2.9 Conclusion

While media forensics as a field is generally occupied by engineers and computer scientists, it intersects with a variety of other disciplines, and its importance is crucial to the ability for people to form consensus and collaboratively solve problems, from climate change to global health.

John Locke, an English philosopher and influential thinker on empiricism and the Enlightenment, once wrote that "the improvement of understanding is for two ends: first, our own increase of knowledge; secondly, to enable us to deliver that knowledge to others." Indeed, your study of media forensics takes Locke's formulation a step further. The effort you put in to understanding this field will not only increase your own knowledge, it will allow you to build an information ecosystem that delivers that knowledge to others—the knowledge necessary to fuel human discourse.

The work is urgent.

Acknowledgements This chapter relies on contributions from Sam Blazek, Adi Cohen, Stef Daley, Pawel Korus, and Clint Watts.

References

ABC News Fergal Gallagher. Minority communities fighting back against disinformation ahead of election. https://www.goodmorningamerica.com/news/story/minority-communities-fighting-back-disinformation-ahead-election-73794172

Adobe, New York Times, Twitter, The Content Authenticity Initiative. The content authenticity initiative. https://documentcloud.adobe.com/link/track?uri=urn:aaid:scds:US:2c6361d5-b8da-4aca-89bd-1ed66cd22d19

Anthropics Inc. PortraitPro. https://www.anthropics.com/portraitpro/

Associated Press, Experts: spy used AI-generated face to connect with targets. https://apnews.com/article/bc2f19097a4c4fffaa00de6770b8a60d

Atlantic Council (2020) The long fuse: Misinformation and the 2020 election. https://www.atlanticcouncil.org/event/the-long-fuse-eip-report/

Bago B, Rand DG, Pennycook G (2020) Fake news, fast and slow: deliberation reduces belief in false (but not true) news headlines. J Exp Psychol Gen 149(8):1608–1613

Bak-Coleman JB, Alfano M, Barfuss W, Bergstrom CT, Centeno MA, Couzin ID, Donges JF, Galesic M, Gersick AS, Jacquet J, Kao AB, Moran RE, Romanczuk P, Rubenstein DI, Tombak KJ, Van Bavel JJ, Weber EU (2021) Stewardship of global collective behavior. Proc Natl Acad Sci 118(27)

Barari S, Lucas C, Munger K (2021) Political deepfakes are as credible as other fake media and (sometimes) real media

Batailler C, Brannon S, Teas P, Gawronski B (2020) A signal detection approach to understanding the identification of fake news. Perspect Psychol Sci 10

Baumann F, Lorenz-Spreen P, Sokolov IM, Starnini M (2020) Modeling echo chambers and polarization dynamics in social networks. Phys Rev Lett 124:048301

BBC Staff. Australia demands China apologise for posting 'repugnant' fake image. https://www.bbc.com/news/world-australia-55126569

Beacon Project. The Beacon Project. https://www.iribeaconproject.org/who-we-are/mission

Blazek S, SCOTCH: a framework for rapidly assessing influence operations. https://www.atlanticcouncil.org/blogs/geotech-cues/scotch-a-framework-for-rapidly-assessing-influence-operations/

Brookings Inst Alexandre Alaphilippe. Adding a 'D' to the ABC disinformation framework . https://www.brookings.edu/techstream/adding-a-d-to-the-abc-disinformation-framework/

Brown TB, Mann B, Ryder N, Subbiah M, Kaplan J, Dhariwal P, Neelakantan A, Shyam P, Sastry G, Askell A, Agarwal S, Herbert-Voss A, Krueger G, Henighan T, Child R, Ramesh A, Ziegler DM, Wu J, Winter C, Hesse C, Chen M, Sigler E, Litwin M, Gray S, Chess B, Clark J, Berner C, McCandlish S, Radford A, Sutskever I, Amodei D (2020) Language models are few-shot learners

Carnegie Endowment for International Peace JON BATEMAN. Deepfakes and synthetic media in the financial system: assessing threat scenarios. https://carnegieendowment.org/2020/07/08/deepfakes-and-synthetic-media-in-financial-system-assessing-threat-scenarios-pub-82237

Carnegie Endowment John Batemant (2020) Deepfakes and synthetic media in the financial system: assessing threat scenarios. https://carnegieendowment.org/2020/07/08/deepfakes-and-synthetic-media-in-financial-system-assessing-threat-scenarios-pub-82237

Chen M, Radford A, Child R, Wu J, Jun H, Luan D, Sutskever I (2020) Generative pretraining from pixels. In: Daumé III H, Singh A (eds) Proceedings of the 37th international conference on machine learning, vol 119. Proceedings of machine learning research, pp 1691–1703. PMLR, 13–18

Chen X, Zhang Z, Sui Y, Chen T (2021) Gans can play lottery tickets too

CNN Clarissa Ward et al (2020) Russian election meddling is back – via Ghana and Nigeria – and in your feeds. https://www.cnn.com/2020/03/12/world/russia-ghana-troll-farms-2020-ward/index.html

Cogsec Collab, Global disinformation groups. https://datastudio.google.com/u/0/reporting/a8491164-6aa8-45d0-b609-c70339689127/page/ierzB

Cogsec Collab. Cogsec Collab. https://cogsec-collab.org/

Combating Terrorism Center Amarnath Amarasingam, Marc-André Argentino. The QAnon conspiracy theory: a security threat in the making? https://ctc.usma.edu/the-qanon-conspiracy-theory-a-security-threat-in-the-making/

DARPA. DARPA SEMAFOR. https://www.darpa.mil/news-events/2021-03-02

Devlin J, Chang M-W, Lee K, Toutanova K (2019) Bert: pre-training of deep bidirectional transformers for language understanding

Donahue J, Dieleman S, Bińkowski M, Elsen E, Simonyan K (2021) End-to-end adversarial text-to-speech

Effron DA, Raj M (2020) Misinformation and morality: encountering fake-news headlines makes them seem less unethical to publish and share. Psychol Sci 31(1):75–87 PMID: 31751517

Fedus W, Zoph B, Shazeer N (2021) Switch transformers: scaling to trillion parameter models with simple and efficient sparsity

François C (2020) Actors, behaviors, content: a disinformation abc. Algorithms

Frankle J, Carbin M (2019) The lottery ticket hypothesis: finding sparse, trainable neural networks

Generated Photos Inc. Generated photos. https://generated.photos/

GMFUS Clint Watts. Advanced persistent manipulators, part one: the threat to the social media industry. https://securingdemocracy.gmfus.org/advanced-persistent-manipulators-part-one-the-threat-to-the-social-media-industry/

Graphika and The Stanford Internet Observatory. More-troll kombat. https://public-assets.graphika.com/reports/graphika_stanford_report_more_troll_kombat.pdf

Gray JF, Terp S-J (2019) Misinformation: We're Four Steps Behind Its Creators. https://cyber.harvard.edu/sites/default/files/2019-11/Comparative%20Approaches%20to%20Disinformation%20-%20John%20Gray%20Abstract.pdf

Grossman S, Ramali LK, Outsourcing disinformation. https://www.lawfareblog.com/outsourcing-disinformation

Guess AM, (Mis)representing ideology on Twitter: how social influence shapesonline political expression. https://www.uzh.ch/cmsssl/ikmz/dam/jcr:995dbede-c863-4931-9ba8-bc0722b6cb59/20201116_guess.pdf

Hao K (2021) AI voice actors sound more human than ever—and they're ready to hire. https://www.technologyreview.com/2021/07/09/1028140/ai-voice-actors-sound-human/

Hill K, White NYTMJ (2020) Designed to deceive: do these people look real to you? https://www.nytimes.com/interactive/2020/11/21/science/artificial-intelligence-fake-people-faces.html

Humprecht E, Esser F, Van Aelst P (2020) Resilience to online disinformation: a framework for cross-national comparative research. Int J Press/Polit 25(3):493–516

Humprecht E, Esser F, Van Aelst P (2020) Resilience to online disinformation: a framework for cross-national comparative research. Int J Press/Polit 25(3):493–516

Jacobs RA, Jordan MI, Nowlan SJ, Hinton GE (1991) Adaptive mixtures of local experts. Neural Comput 3(1):79–87

Joint Bi level Image Experts Group and Joint Photographic Experts Group. JPEG fake media: context use cases and requirements. http://ds.jpeg.org/documents/wg1n89043-REQ-JPEG_Fake_Media_Context_Use_Cases_and_Requirements_v0_1.pdf

Jordan MI, Jacobs RA (1993) Hierarchical mixtures of experts and the em algorithm. In: Proceedings of 1993 international conference on neural networks (IJCNN-93-Nagoya, Japan), vol 2, pp 1339–1344

Kapantai E, Christopoulou A, Berberidis C, Peristeras V (2021) A systematic literature review on disinformation: toward a unified taxonomical framework. New Media Soc 23(5):1301–1326

Kaplan J, McCandlish S, Henighan T, Brown TB, Chess B, Child R, Gray S, Radford A, Wu J, Amodei D (2020) Scaling laws for neural language models

Kietzmann JH, Hermkens K, McCarthy IP, Silvestre BS (2011) Social media? get serious! understanding the functional building blocks of social media. Bus Horiz 54(3):241–251. SPECIAL ISSUE: SOCIAL MEDIA

Koltai K (2020) Vaccine information seeking and sharing: how private facebook groups contributed to the anti-vaccine movement online. IN: AoIR selected papers of internet research, 2020, Oct

Kopp C, Korb KB, Mills BI (2018) Information-theoretic models of deception: modelling cooperation and diffusion in populations exposed to "fake news". PLOS ONE 13(11):1–35

Łubiński P (2020) Social media incitement to genocide (in:) the concept of genocide in international criminal law developments after Lemkin, p 306

Lukito J (2020) Coordinating a multi-platform disinformation campaign: internet research agency activity on three u.s. social media platforms, 2015 to 2017. Polit Commun 37(2):238–255

Luong H-T, Yamagishi J (2020) Nautilus: a versatile voice cloning system

Maertens R, Roozenbeek J, Basol M, van der Linden S (2020) Long-term effectiveness of inoculation against misinformation: three longitudinal experiments. J Exp Psychol Appl 27:10

McQuillan L, McAweeney E, Bargar A, Ruch A (2020) Insights into the behavior of misinformation networks on twitter, Cultural convergence

Mentzer F, Toderici G, Tschannen M, Agustsson E (2020) High-fidelity generative image compression

Narayanan V, Barash V, Kelly J, Kollanyi B, Neudert L-M, Howard PN (2018) Polarization, partisanship and junk news consumption over social media in the us

Nimmo B, François C, Shawn Eib C, Ronzaud L, Graphika, IRA again: unlucky thirteen. https://public-assets.graphika.com/reports/graphika_report_ira_again_unlucky_thirteen.pdf

Nimmo B, Shawn Eib C, Tamora L, Graphika, Cross-Platform Spam Network Targeted Hong Kong Protests. https://public-assets.graphika.com/reports/graphika_report_spamouflage.pdf

Nimmo B, Shawn Eib C, Tamora L, Graphika, Spamouflage goes to America. https://public-assets.graphika.com/reports/graphika_report_spamouflage_goes_to_america.pdf

Nisos, The rise of synthetic audio deepfakes. https://www.nisos.com/technical-blogs/rise_synthetic_audio_deepfakes

Nvidia Inc (2020) Taking it to the MAX: adobe photoshop gets new NVIDIA AI-powered neural filters. https://blogs.nvidia.com/blog/2020/10/20/adobe-max-ai/

Nvidia Inc. Imaginaire. https://github.com/NVlabs/imaginaire

Nvidia Inc. Nvidia Maxine. https://developer.nvidia.com/MAXINE

Park E, Yoo S (2020) Profit: a novel training method for sub-4-bit mobilenet models

Paul RK, Twitter suspends accounts linked to Saudi spying case. https://www.reuters.com/article/us-twitter-saudi-idUSKBN1YO1JT

Pennycook G, Rand DG (2020) Who falls for fake news? the roles of bullshit receptivity, overclaiming, familiarity, and analytic thinking. J Personal 88(2):185–200

Qin C, Wu Y, Springenberg JT, Brock A, Donahue J, Lillicrap TP, Kohli P (2020) Training generative adversarial networks by solving ordinary differential equations

Radford A, Kim JW, Hallacy C, Ramesh A, Goh G, Agarwal S, Sastry G, Askell A, Mishkin P, Clark J, Krueger G, Sutskever I (2021) Learning transferable visual models from natural language supervision

Radford A, Narasimhan K (2018) Improving language understanding by generative pre-training

Ramesh A, Pavlov M, Goh G, Gray S, Voss C, Radford A, Chen M, Sutskever I (2021) Zero-shot text-to-image generation

Reuters Staff. Fact check: Clip of Biden taken out of context to portray him as plotting a voter fraud scheme. https://www.reuters.com/article/uk-fact-check-biden-voter-protection-not/fact-check-clip-of-biden-taken-out-of-context-to-portray-him-as-plotting-a-voter-fraud-scheme-idUSKBN27E2VH

Reuters Staff. Fact check: Donald Trump concession video not a 'confirmed deep-fake'. https://www.reuters.com/article/uk-factcheck-trump-consession-video-deep/fact-check-donald-trump-concession-video-not-a-confirmed-deepfake-idUSKBN29G2NL

Reuters Staff. Fact check: Video does not show Biden saying 'Hello Minnesota' in Florida rally. https://www.reuters.com/article/uk-factcheck-altered-sign-biden-mn/fact-check-video-does-not-show-biden-saying-hello-minnesota-in-florida-rally-idUSKBN27H1RZ

Rosebud AI Inc. Rosebud AI. https://www.rosebud.ai/

RunwayML Inc. RunwayML. https://runwayml.com/

Schaewitz L, Kluck JP, Klösters L, Krämer NC (2020) When is disinformation (in)credible? experimental findings on message characteristics and individual differences. Mass Commun Soc 23(4):484–509

Schema.org. Media review. https://schema.org/MediaReview

Shazeer N, Cheng Y, Parmar N, Tran D, Vaswani A, Koanantakool P, Hawkins P, Lee H, Hong M, Young C, Sepassi R, Hechtman B (2018) Mesh-tensorflow: deep learning for supercomputers

Skylum Inc. Luminar. https://skylum.com/luminar

StopFake.org. Anatomy of an info-war: how Russia's propaganda machine works, and how to counter it. https://www.stopfake.org/en/anatomy-of-an-info-war-how-russia-s-propaganda-machine-works-and-how-to-counter-it/

Talwar S, Dhir A, Singh D, Virk GS, Salo J (2020) Sharing of fake news on social media: Application of the honeycomb framework and the third-person effect hypothesis. J Retail Consum Serv 57:102197

Tangherlini TR, Shahsavari S, Shahbazi B, Ebrahimzadeh E, Roychowdhury V (2020) An automated pipeline for the discovery of conspiracy and conspiracy theory narrative frameworks: bridgegate, pizzagate and storytelling on the web. PLOS ONE 15(6):1–39

Techcrunch Taylor Hatmaker. Chinese propaganda network on Facebook used AI-generated faces. https://techcrunch.com/2020/09/22/facebook-gans-takes-down-networks-of-fake-accounts-originating-in-china-and-the-philippines/

The Associated Press ERIKA KINETZ. Army of fake fans boosts China's messaging on Twitter. https://apnews.com/article/asia-pacific-china-europe-middle-east-government-and-politics-62b13895aa6665ae4d887dcc8d196dfc

The Economist. Global democracy has a very bad year. https://www.economist.com/graphic-detail/2021/02/02/global-democracy-has-a-very-bad-year

The Next Web Thomas Macaulay. Someone let a GPT-3 bot loose on Reddit – it didn't end well. https://thenextweb.com/neural/2020/10/07/someone-let-a-gpt-3-bot-loose-on-reddit-it-didnt-end-well/

The Next Web Tristan Greene. GPT-3 is the world's most powerful bigotry generator. What should we do about it? https://thenextweb.com/neural/2021/01/19/gpt-3-is-the-worlds-most-powerful-bigotry-generator-what-should-we-do-about-it/

The Washington Post Drew Harwell, An artificial-intelligence first: voice-mimicking software reportedly used in a major theft. https://www.washingtonpost.com/technology/2019/09/04/an-artificial-intelligence-first-voice-mimicking-software-reportedly-used-major-theft/

Times of India Anam Ajmal. 1st in India: Twitter tags BJP IT cell chief's tweet as 'manipulated media'. https://timesofindia.indiatimes.com/india/1st-in-india-twitter-tags-bjp-it-cell-chiefs-tweet-as-manipulated-media/articleshow/79538441.cms

Twitter Inc. New disclosures to our archive of state-backed information operations. https://blog.twitter.com/en_us/topics/company/2019/new-disclosures-to-our-archive-of-state-backed-information-operations.html

Vaccari C, Chadwick A (2020) Deepfakes and disinformation: exploring the impact of synthetic political video on deception, uncertainty, and trust in news. Soc Media + Soc 6(1):2056305120903408

Van Bavel JJ, Harris EA, Pärnamets P, Rathje S, Doell K, Tucker JA (2020) Political psychology in the digital (mis)information age: a model of news belief and sharing

Wall Street Journal Catherine Stupp. Fraudsters used AI to mimic CEO's voice in unusual cyber-crime case. https://www.wsj.com/articles/fraudsters-use-ai-to-mimic-ceos-voice-in-unusual-cybercrime-case-11567157402

Wang N, Choi J, Brand D, Chen C-Y, Gopalakrishnan K (2018) Training deep neural networks with 8-bit floating point numbers

Washington Post David Weigel. Twitter flags GOP video after activist's computerized voice was manipulated. https://www.washingtonpost.com/politics/2020/08/30/ady-barkan-scalise-twitter-video/

Zerback T, Töpfl F, Knöpfle M (2021) The disconcerting potential of online disinformation: persuasive effects of astroturfing comments and three strategies for inoculation against them. New Media Soc 23(5):1080–1098

Zhang E, Martin-Brualla R, Kontkanen J, Curless B (2020) No shadow left behind: removing objects and their shadows using approximate lighting and geometry

Zhou X, Jain A, Phoha VV, Zafarani R (2020) Fake news early detection: a theory-driven model. Digit Threat Res Pract 1(2)

Zhou X, Zafarani R (2020) A survey of fake news: fundamental theories, detection methods, and opportunities. ACM Comput Surv 53(5)

Chapter 3
Computational Imaging

Scott McCloskey

Since the advent of smartphones, photography is increasingly being done with small, portable, multi-function devices. Relative to the purpose-built cameras that dominated previous eras, smartphone cameras must overcome challenges related to their small form factor. Smartphone cameras have small apertures that produce a wide depth of field, small sensors with rolling shutters that lead to motion artifacts, and small form factors which lead to more camera shake during exposure. Along with these challenges, smartphone cameras have the advantage of tight integration with additional sensors and the availability of significant computational resources. For these reasons, the field of computational imaging has advanced significantly in recent years, with academic groups and researchers from smartphone manufacturers helping these devices become more capable replacements for purpose-built cameras.

3.1 Introduction to Computational Imaging

Computational imaging (or computational photographic) approaches are characterized by the co-optimization of what and how the sensor captures light and how that signal is processed in software. Computational imaging approaches now commonly available on most smartphones include panoramic stitching, multi-frame high dynamic range (HDR) imaging, 'portrait mode' for low depth of field, and multi-frame low-light imaging ('night sight' or 'night mode').

As image quality is viewed as a differentiating feature between competing smartphone brands, there has been tremendous progress improving subjective image quality, accompanied by a lack of transparency due to the proprietary nature of the work.

S. McCloskey (✉)
Kitware, Inc., New York, NY, USA
e-mail: scott.p.mccloskey@kitware.com

© The Author(s) 2022
H. T. Sencar et al. (eds.), *Multimedia Forensics*, Advances in Computer Vision and Pattern Recognition, https://doi.org/10.1007/978-981-16-7621-5_3

Like the smartphone market more generally, computational imaging approaches used therein change very quickly from one generation of the phone to the next. This combination of different modes, and the proprietary and rapid nature of changes, all pose challenges for forensics practitioners.

This chapter investigates some of the forensics challenges that arise from the increased adoption of computational imaging and assesses our ability to detect automated focus manipulations performed by computational cameras. We then look more generally at a cue to help distinguish optical blur from synthetic blur. The chapter concludes with a look at some early computational imaging research that may impact forensics in the near future.

This chapter will not focus on the definitional and policy issues related to computational imagery, believing that these issues are best addressed within the context of a specific forensic application. The use of an automated, aesthetically driven computational imaging mode does not necessarily imply nefarious intent on the part of the photographer in all cases, but may in some. This suggests that broad-scale screening for manipulated imagery (e.g., on social media platforms) might not target portrait mode images for detection and further scrutiny, but in the context of an insurance claim or a court proceeding, higher scrutiny may be necessary.

As with the increase in the number, complexity, and ease of use of software packages for image manipulation, a key forensic challenge presented by computational imaging is the degree to which they democratize the creation of manipulated media. Prior to these developments, the creation of a convincing manipulation required a knowledgeable user dedicating a significant amount of time and effort with only partially automated tools. Much like Kodak cameras greatly simplified consumer photography with the motto "You Press the Button, We Do the Rest," computational cameras allow users to create—with the push of a button—imagery that's inconsistent with the classically understood physical limitations of the camera. Realizing that most people won't carry around a heavy camera with a large lens, a common goal of computational photography is to replicate the aesthetics of Digital Single Lens Reflex (DSLR) imagery using the much smaller and lighter sensors and optics used in mobile phones. From this, one of the significant achievements of computational imaging is the ability to replicate the shallow depth of field of a large aperture DSLR lens on a smartphone with a tiny aperture.

3.2 Automation of Geometrically Correct Synthetic Blur

Optical blur is a perceptual cue to depth (Pentland 1987), a limiting factor in the performance of computer vision systems (Bourlai 2016), and an aesthetic tool that photographers use to separate foreground and background parts of a scene. Because smartphone sensors and apertures are so small, images taken by mobile phones often appear to be in focus everywhere, including background elements that distract a viewer's attention. To draw viewers' perceptual attention to a particular object and avoid distractions from background objects, smartphones now include 'portrait

Fig. 3.1 (Left) For a close-in scene captured by a smartphone with a small aperture, all parts of the image will appear in focus. (Right) In order to mimic the aesthetically pleasing properties of a DSLR with a wide aperture, computational cameras now include a 'portrait mode' that blurs the background so the foreground stands out better

mode' which automatically manipulates the sensor image via the application of spatially varying blur. Figure 3.1 shows an example of the all-in-focus natively captured by the sensor and the resulting portrait mode image. Because there is a geometric relationship between optical blur and depth in the scene, the creation of geometrically correct synthetic blur requires knowledge of the 3D scene structure. Consistent with the spirit of computational cameras, portrait modes enable this with a combination of hardware and software. Hardware acquires the scene in 3D using either stereo cameras or sensors with the angular sensitivity of a light field camera (Ng 2006). Image processing software uses this 3D information to infer and apply the correct amount of blur for each part of the scene. The result is automated, geometrically correct optical blur that looks very convincing. A natural question, then, is whether or not we can accurately differentiate 'portrait mode'-based imagery from genuine optical blur.

3.2.1   Primary Cue: Image Noise

Regardless of how the local blurring is implemented, the key difference between optical blur and portrait mode-type processing can be found in image noise. When blur happens optically, before photons reach the sensor, only small signal-dependent noise impacts are observed. When blur is applied algorithmically to an already digitized image, however, the smoothing or filtering operation also implicitly de-noises the image. Since the amount of de-noising is proportional to the amount of local smoothing or blurring, differences in the amount of algorithmic local blur can be detected via inconsistencies between the local intensity and noise level. Two regions of the image having approximately the same intensity should also have approximately the same level of noise. If one region is blurred more than the other, or one is blurred while the other is not, an inconsistency is introduced between the intensities and local noise levels.

For our noise analysis, we extend the combined noise models of Tsin et al. (2001) and Liu et al. (2006). Ideally, a pixel produces a number of electrons E_{num} proportional to the average irradiance from the object being imaged. However, shot noise N_S is a result of the quantum nature of light and captures the uncertainty in the number of electrons stored at a collection site; N_S can be modeled as the Poisson noise. Additionally, site-to-site non-uniformities called *fixed pattern noise K* are a multiplicative factor impacting the number of electrons; K can be characterized as having mean 1 and a small spatial variance σ_K^2 over all of the collection sites. Thermal energy in silicon generates free electrons which contribute *dark current* to the image; this is modeled as an additive factor N_{DC}, modeled as the Gaussian noise. The on-chip output amplifier sequentially transforms the charge collected at each site into a measurable voltage with a scale A, and the amplifier generates zero mean read-out noise N_R with variance σ_R^2. Demosaicing is applied in color cameras to interpolate two of the three colors at each pixel, and it introduces an error that is sometimes modeled as noise. After this, the camera response function (CRF) $f(\cdot)$ maps this voltage via a non-linear transform to improve perceptual image quality. Lastly, the analog-to-digital converter (ADC) approximates the analog voltage as an integer multiple of a quantization step q. The quantization noise can be modeled as the addition of a noise source N_Q.

With these noise sources in mind, we can describe a digitized 2D image as follows:

$$D(x, y) = f\Big((K(x, y)E_{num}(x, y) + N_{DC}(x, y) + N_S(x, y) + N_R(x, y))A\Big) + N_Q(x, y) \tag{3.1}$$

The variance of the noise is given by

$$\sigma_N^2(x, y) = f'^2\Big(A^2\big(K(x, y)E_{num}(x, y) + \mathrm{E}[N_{DC}(x, y)] + \sigma_R^2\big)\Big) + \frac{q^2}{12} \tag{3.2}$$

where $\mathrm{E}[\cdot]$ is the expectation function. This equation tells us two things which are typically overlooked in the more simplistic model of noise as an additive Gaussian source:

1. The noise variance's relationship with intensity reveals the shape of the CRF's derivative f'.
2. Noise has a signal-dependent aspect to it, as evidenced by the E_{num} term.

An important corollary to this is that different levels of noise in regions of an image having different intensities is not *per se* an indicator of manipulation, though it has been taken as one in past work (Mahdian and Saic 2009). We show in our experiments that, while the noise inconsistency cue from Mahdian and Saic (2009) has some predictive power in detecting manipulations, a proper accounting for signal-dependent noise via its relationship with image intensity significantly improves accuracy.

Measuring noise in an image is, of course, ill-posed and is equivalent to the long-standing image de-noising problem. For this reason, we leverage three different approximations of local noise, measured over approximately uniform image regions: intensity variance, intensity gradient magnitude, and the noise feature of Mahdian and Saic (2009) (abbreviated NOI). Each of these is related to the image intensity of the corresponding region via a 2D histogram. This step translates subtle statistical relationships in the image to shape features in the 2D histograms which can be classified by a neural network. As we show in the experiments, our detection performance on histogram features significantly improves on that of popular approaches applied directly to the pixels of the image.

3.2.2 Additional Photo Forensic Cues

One of the key challenges in forensic analysis of images 'in the wild' is that compression and other post-processing may overwhelm subtle forgery cues. Indeed, noise features are inherently sensitive to compression which, like blur, smooths the image. In order to improve detection performance in such challenging cases, we incorporate additional forensic cues which improve our method's robustness. Some portrait mode implementations appear to operate on a JPEG image as input, meaning that the outputs exhibit cues related to double JPEG compression. As such, there is a range of different cues that can reveal manipulations in a subset of the data.

3.2.2.1 Demosaicing Artifacts

Forensic researchers have shown that the differences between the demosaicing algorithm and the differences between the physical color filter array bonded to the sensor can be detected from the image. Since focus manipulations are applied on the demosaiced images, the local smoothing operations will alter these subtle Color Filter Array (CFA) demosaicing artifacts. In particular, the lack of CFA artifacts or the detection of weak, spatially varying CFA artifacts indicates the presence of global or local tampering, respectively.

Following the method of Dirik and Memon (2009), we consider the demosaicing scheme f_d being a bilinear interpolation. We divide the image into $W \times W$ subblocks and only compute the demosaicing feature at the non-smooth blocks of pixels. Denote each non-smooth block as B_i, where $i = 1, \ldots, m_B$, and m_B is the number of non-smooth blocks in the image. The re-interpolation error of i-th sub-block for the k-th CFA pattern θ_k is defined as $\hat{B}_{i,k} = f_d(B_i, \theta_k)$ and $k = 1, \ldots, 4$. The MSE error matrix $E_i^{(2)}(k, c), c \in R, G, B$ between the blocks B and $\hat{B}$ is computed in non-smooth regions all over the image. Therefore, we define the metric to estimate the uniformity of normalized green channel column vector as

$$F = median\left(\sum_{l=1}^{4} \left|100 \times \frac{E_i^{(2)}(k, 2)}{\sum_{l=1}^{3} E_i^{(2)}(l, 2)} - 25\right|\right)$$

$$E_i^{(2)}(k, c) = 100 \times \frac{E_i(k, 2)}{\sum_{l=1}^{3} E_i(l, 2)}$$

$$E_i(k, c) = \frac{1}{W \times W} \sum_{x=1}^{W} \sum_{y=1}^{W} \left(B_i(x, y, c) - \hat{B}_{i,k}(x, y, c)\right)^2 \quad (3.3)$$

3.2.2.2 JPEG Artifact

In some portrait mode implementations, such as the iPhone, the option to save both an original and a portrait mode image of the same scene suggests that post-processing is applied *after* JPEG compression. Importantly, both the original JPEG image and the processed version are saved in the JPEG format *without resizing*. Hence, Discrete Cosine Transform (DCT) coefficients representing un-modified areas will undergo two consecutive JPEG compressions and exhibit double quantization (DQ) artifacts, used extensively in the forensics literature. DCT coefficients of locally blurred areas, on the other hand, will result from non-consecutive compressions and will present weaker artifacts.

We follow the work of Bianchi et al. (2011) and use the Bayesian inference to assign to each DCT coefficient a probability of being doubly quantized. Accumulated over each 8×8 block of pixels, the DQ probability map allows us to distinguish original areas (having high DQ probability) from tampered areas (having low DQ probability). The probability of a block being tampered can be estimated as

$$p = 1/\left(\prod_{i|m_i \neq 0} \left(R(m_i) - L(m_i)\right) * k_g(m_i) + 1\right)$$

$$R(m) = Q_1\left(\left\lceil \frac{Q_2}{Q_1}\left(m - \frac{b}{Q_2} - \frac{1}{2}\right)\right\rceil - \frac{1}{2}\right)$$

$$L(m) = Q_1\left(\left\lfloor \frac{Q_2}{Q_1}\left(m - \frac{b}{Q_2} + \frac{1}{2}\right)\right\rfloor + \frac{1}{2}\right) \quad (3.4)$$

where m is the value of the DCT coefficient; $k_g(\cdot)$ is a Gaussian kernel with standard deviation σ_e/Q_2; Q_1 and Q_2 are the quantization steps used in the first and second compression, respectively; b is the bias; and u is the unquantized DC coefficient.

3.2.3 Focus Manipulation Detection

To summarize the analysis above, we adopt five types of features: color variance (VAR), image gradient (GRAD), double quantization (ADQ) (Bianchi et al. 2011), color filter artifacts (CFA) (Dirik and Memon 2009), and noise inconsistencies (NOI) (Mahdian and Saic 2009) for refocusing detection. Each of these features is computed densely at each location in the image, and Fig. 3.2 illustrates the magnitude of these features in a feature map for an authentic image (top row) and a portrait mode image (middle row). Though there are notable differences between the feature maps in these two rows, there is no clear indication of a manipulation except, perhaps, the ADQ feature. And, as mentioned above, the ADQ cue is fragile because it depends on whether blurring is applied after an initial compression.

As mentioned in Sect. 3.2.1, the noise cues are signal-dependent in the sense that blurring introduces an inconsistency between intensity and noise levels. To illustrate this, Fig. 3.2's third row shows scatter plots of the relationship between intensity (on the horizontal axis) and the various features (on the vertical axis). In these plots, particularly the columns related to noise (Variance, Gradient, and NOI), the distinction between the statistics of the authentic image (blue symbols) and the manipulated image (red symbols) becomes quite clear. Noise is reduced in most of the images, though the un-modified foreground region (the red bowl) maintains relatively higher noise because it is not blurred. Note also that the noise levels across the manipulated image are actually *more* consistent than in the authentic image, showing that previous noise-based forensics (Mahdian and Saic 2009) are ineffective.

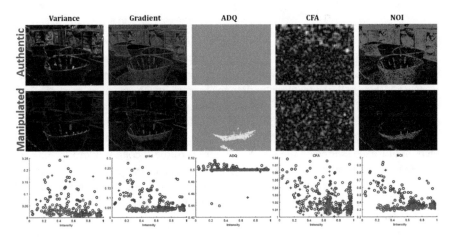

Fig. 3.2 Feature maps and histogram for authentic and manipulated images. On the first row are the authentic image feature maps; the second row shows the corresponding maps for the manipulated image. We show scatter plots relating the features to intensity in the third row, where blue sample points correspond to the authentic image, and red corresponds to a manipulated DoF image (which was taken with an iPhone)

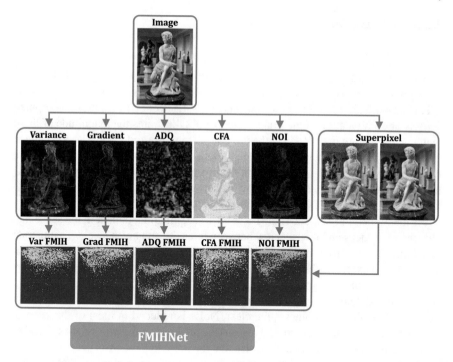

Fig. 3.3 Manipulated refocusing image detection pipeline. The example shown is an iPhone7plus portrait mode image

Figure 3.3 shows our portrait mode detection pipeline, which incorporates these five features. In order to capture the relationship between individual features and the underlying image intensity, we employ an intensity versus feature bivariate histogram—which we call the focus manipulation inconsistency histogram (FMIH). We use FMIH for all five features for defocus forgery image detection, each of which is analyzed by a neural network called FMIHNet. These five classification results are combined by a majority voting scheme to determine a final classification label.

After densely computing the VAR, GRAD, ADQ, CFA, and NOI features for each input image (shown in the first five columns of the second row of Fig. 3.3), we partition the input image into superpixels and, for each superpixel i_{sp}, we compute the mean $F(i_{sp})$ of each feature measure and its mean intensity. Finally, we generate the FMIH for each of the five figures, shown in the five columns of the third row of Fig. 3.3. Note that the FMIH are flipped vertically with respect to the scatter plots shown in Fig. 3.2. A comparison of the FMIH extracted features from the same scene captured with different cameras is shown in Fig. 3.4.

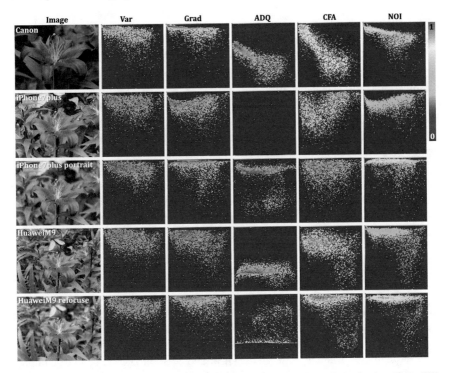

Fig. 3.4 Extracted FMIHs for the five feature measures with images captured using Canon60D, iPhone7Plus, and HuaweiMate9 cameras

3.2.3.1 Network Architectures

We have designed a FMIHNet, illustrated in Fig. 3.5, for the five histogram features. Our network is a VGG (Simonyan and Zisserman 2014) style network consisting of convolutional (CONV) layers with small receptive fields (3×3). During training, the input to our FMIHNet is a fixed-size 101×202 FMIH. The FMIHNet is a fusion of two relatively deep sub-networks: FMIHNet1 with 20 CONV layers for VAR and CFA features, and FMIHNet2 with 30 CONV layers for GRAD, ADQ, and NOI features. The CONV stride is fixed to 1 pixel; the spatial padding of the input

Fig. 3.5 Network architecture: FMIHNet1 for Var and CFA features; FMIHNet2 for Grad, ADQ, and NOI features

features is set to 24 pixels to preserve the spatial resolution. Spatial pooling is carried out by five max-pooling layers, performed over a 2×2 pixel window, with stride 2. A stack of CONV layers followed by one Fully Connected (FC) layer performs two-way classification. The final layer is the soft-max layer. All hidden layers have rectification (ReLU) non-linearity.

There are two reasons for the small 3×3 receptive fields: first, incorporating multiple non-linear rectification layers instead of a single one makes the decision function more discriminative; secondly, this reduces the number of parameters. This can be seen as imposing a regularization on a later CONV layer by forcing it to have a decomposition through the 3×3 filters.

Because most of the values in our FMIH are zeros (i.e., most cells in the 2D histogram are empty) and because we only have two output classes (authentic and portrait mode), more FC layers seem to degrade the training performance.

3.2.4 Portrait Mode Detection Experiments

Having introduced a new method to detect focus manipulations, we first demonstrate that our method can accurately identify manipulated images *even if they are geometrically correct*. Here, we also show that our method is more accurate than both past forensic methods (Bianchi et al. 2011; Dirik and Memon 2009; Mahdian and Saic 2009) and the modern vision baseline of CNN classification applied directly to the image pixels. Second, having claimed that the photometric relationship of noise cues with the image intensity is important, we will show that our FMIH histograms are a more useful representation of these cues.

To demonstrate our performance on the hard cases of geometrically correct focus manipulations, we have built a focus manipulation dataset (FMD) of images captured with a Canon 60D DSLR and two smartphones having dual lens camera-enabled portrait modes: the iPhone7Plus and the Huawei Mate9. Images from the DSLR represent real shallow DoF images, having been taken with focal lengths in the range 17–70 and f numbers in the range F/2.8-F/5.6. The iPhone was used to capture aligned pairs of authentic and manipulated images using portrait mode. The Mate9 was also used to capture authentic/manipulated image pairs, but these are only approximately aligned due to its inability to save the image both before and after portrait mode editing.

We use 1320 such images for training and 840 images for testing. The training set consists of 660 authentic images (220 from each of the three cameras) and 660 manipulated images (330 from each of iPhone7Plus and Huawei Mate9). The test set consists of 420 authentic images (140 from each of the three cameras) and 420 manipulated images (140 from each of iPhone7Plus and HuaweiMate9).

Figure 3.4 shows five sample images from FMD and illustrates the perceptual realism of the manipulated images. The first row of Table 3.1 quantifies this performance and shows that a range of CNN models (AlexNet, CaffeNet, VGG16, and VGG19) have classification accuracies in the range of 76–78%. Since our method uses five

Table 3.1 Image classification accuracy on FMD

Data	AlexNet	CaffeNet	VGG16	VGG19
Image	0.760	0.782	0.771	0.784
VAR map	0.688	0.725	0.714	0.726
GRAD map	0.733	0.767	0.740	0.769
ADQ map	0.735	0.759	0.736	0.740
CFA map	0.761	0.788	0.777	0.785
NOI map	0.707	0.765	0.745	0.760

Table 3.2 FMIH classification accuracy on FMD

Data	SVM	AlexNet	CaffeNet	VGG16	VGG19	LeNet	FMIHNet
VAR	0.829	0.480	0.480	0.475	0.512	0.635	**0.850**
GRAD	0.909	0.503	0.500	0.481	0.486	0.846	**0.954**
ADQ	0.909	0.496	0.520	0.503	0.511	0.844	**0.946**
CFA	0.882	0.510	0.520	0.481	0.510	0.871	**0.919**
NOI	0.858	0.497	0.506	0.520	0.530	0.779	**0.967**
Vote	0.942	--	--	--	--	0.888	**0.982**

different feature maps which can easily be interpreted as images, the remaining rows of Table 3.1 show the classification accuracies of the same CNN models applied to these feature maps. The accuracies are slightly lower than for the image-based classification.

In Sect. 3.2.1, we claimed that a proper accounting for signal-dependent noise via our FMIH histograms improves upon the performance of the underlying features. This is demonstrated by comparing the image- and feature map-based classification performance of Table 3.1 with the FMIH-based classification performance shown in Table 3.2. Using FMIH, even the relatively simple SVMs and LeNet CNNs deliver classification accuracies in the 80–90% range. Our FMIHNet architecture produces significantly better results than these, with our method's voting output having a classification accuracy of 98%.

3.2.5 Conclusions on Detecting Geometrically Correct Synthetic Blur

We have presented a novel framework to detect focus manipulations, which represent an increasingly difficult and important forensics challenge in light of the availability of new camera hardware. Our approach exploits photometric histogram features, with a particular emphasis on noise, whose shapes are altered by the manipulation process. We have adopted a deep learning approach that classifies these 2D histograms separately and then votes for a final classification. To evaluate this, we have produced a new focus manipulation dataset with images from a Canon60D DSLR, iPhone7Plus, and HuaweiMate9. This dataset includes manipulations, particularly from the iPhone portrait mode, that are geometrically correct due to the use of dual lens capture devices. Despite the challenge of detecting manipulations that are geometrically correct, our method's accuracy is 98%, significantly better than image-based detection with a range of CNNs, and better than prior forensics methods.

While these experiments make it clear that the detection of imagery generated by first-generation portrait modes is reliable, it's less clear that this detection capability will hold up as algorithmic and hardware improvements are made to mobile phones. This raises the question of how detectable digital blurring (i.e., blurring performed in software) is compared to optical blur.

3.3 Differences Between Optical and Digital Blur

As mentioned in the noise analysis earlier in this chapter, the shape of the Camera Response Function (CRF) shows up in image statistics. The non-linearity of the CRF impacts more than just noise and, in particular, its effect is now well-understood on motion deblurring (Chen et al. 2012; Tai et al. 2013). While that work showed how an unknown CRF is a noise source in blur estimation, we will demonstrate that it is a key signal helping to distinguish authentic edges from software-blurred gradients, particularly at splicing boundaries. This allows us to identify forgeries that involve artificially blurred edges, in addition to ones with artificially sharp transitions.

We first consider blurred edges: for images captured by a real camera, the blur is applied to scene irradiance as the sensor is exposed, and it is then mapped to image intensity via the CRF. There are two operations involved, CRF mapping and optical/manual blur convolution. By contrast, in a forged image, the irradiance of a sharp edge is mapped to intensity first and then blurred by the manual blur point spread function (PSF). The key to distinguishing these is that the CRF is a non-linear operator, so it is not commutative, i.e., applying the PSF before the CRF leads to a different result than applying the CRF first, even if the underlying signal is the same.

The key difference is that the profile of an authentic blurred edge (CRF before PSF) is asymmetric w.r.t the center location of an edge, whereas an artificially blurred edge *is* symmetric, as illustrated in Fig. 3.6. This is because, due to its non-linearity, the slope of the CRF is different across the range of intensities. CRFs typically have a larger gradient in low intensities, small gradient in high intensities, and approximately constant slope in the mid-tones. In order to capture this, we use the statistical bivariate histogram of pixel intensity versus gradient magnitude, which we call the intensity gradient histogram (IGH), and use it as the feature for synthetic blur detection around edges in an image. The IGH captures key information about the CRF without having to explicitly estimate it and, as we show later, its shape differs between natural and artificial edge profiles in a way that can be detected with existing CNN architectures.

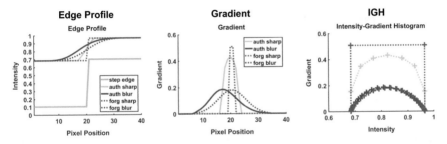

Fig. 3.6 Authentic blur (blue), authentic sharp (cyan), forgery blur (red) and forgery sharp (magenta) edge profiles (left), gradients (center), and IGHs (right)

Before starting the theoretical analysis, we clarify our notation. For simplicity, we study the role of the operations ordering on a 1D edge. We denote the CRF, which is assumed to be a non-linear monotonically increasing function, as $f(r)$, normalized to satisfy $f(0) = 0$, $f(1) = 1$. And the inverse CRF is denoted as $g(R) = f^{-1}(R)$, satisfying $g(0) = 0$, $f(g(1)) = 1$. r represents irradiance, and R intensity. We assume that the optical blur PSF is a Gaussian function, having confirmed that the blur tools in popular image manipulation software packages use a Gaussian kernel,

$$K_g(x) = \frac{1}{\sigma\sqrt{2\pi}} e^{-\frac{x^2}{2\sigma^2}} \tag{3.5}$$

When the edge in irradiance space r is a step edge, like the one shown in green in Fig. 3.6,

$$H_{step}(x) = \begin{cases} a & x < c \\ b & x \geq c \end{cases} \tag{3.6}$$

where x is the pixel location. If we use a unit step function,

$$u(x) = \begin{cases} 0 & x < 0 \\ 1 & x \geq 0 \end{cases} \tag{3.7}$$

The step edge can be represented by

$$H_{step}(x) = (b - a)u(x - c) + a \tag{3.8}$$

3.3.1 Authentically Blurred Edges

An authentically blurred (ab) edge is

$$I_{ab} = f(K_g(x) * H_{step}(x)) \tag{3.9}$$

where $*$ represents convolution. The gradient of this

$$\nabla I_{ab} = f'(K_g(x) * H_{step}(x)) \cdot \frac{d[K_g(x) * H_{step}(x)]}{dx}$$
$$= f'(K_g(x) * H_{step}(x)) \cdot K_g(x) * \frac{d H_{step}(x)}{dx}$$

Because the differential of the step edge is a delta function,

$$\frac{d H_{step}(x)}{dx} = (b - a)\delta(x - c) \tag{3.10}$$

We have

$$\nabla I_{ab} = f'(K_g(x) * H_{step}(x)) \cdot K_g(x) * (b - a)\delta(x - c)$$
$$= (b - a)f'(K_g(x) * H_{step}(x)) \cdot K_g(x - c)$$

Substituting (3.9) into above equation, we have the relationship between I_{ab} and gradient ∇I_{ab}

$$\nabla I_{ab} = (b - a)f'(f^{-1}(I_{ab})) \cdot K_g(x - c) \tag{3.11}$$

And $K_g(x - c)$ is just shifting the blur kernel to the location of the step. Because f is non-linear and its gradient is large at lower irradiance and small at higher irradiance, f' is asymmetric. Therefore, IGH of authentically blurred edge is asymmetric, as shown in blue in Fig. 3.6.

3.3.2 Authentic Sharp Edge

In real images, the assumption that a sharp edge is a step function does not hold. Some (Ng et al. 2007) assume a sigmoid function, for simplicity, whereas we choose a small Gaussian kernel to approximate the authentic sharp (as) edge.

$$I_{as} = f(K_s(x) * H_{step}(x)) \tag{3.12}$$

The IGH is the same as that of an authentic blur edge (3.11), with the size and σ of the kernel being smaller. Because the blur extent is very small, there is a small transition edge region, and the effect of CRF will not be very obvious. The IGH remains symmetric shown as cyan in Fig. 3.6.

3.3.3 Forged Blurred Edge

The model for a forged blurred (fb) edge is

$$I_{fb} = K_g(x) * f(H_{step}(x))$$

The gradient of this is

$$\nabla I_{fb} = \frac{d[K_g(x) * f(H_{step}(x))]}{dx}$$
$$= K_g(x) * [f'(H_{step}(x)) \cdot H'_{step}(x)]$$

Because

$$f'(H_{step}(x)) = (f'(b) - f'(a))u(x - c) + f'(a) \tag{3.13}$$

and

$$f(x) \cdot \delta(x) = f(0) \cdot \delta(x), \tag{3.14}$$

we have

$$\nabla I_{fb} = K_g(x) * [f'(b)\delta(x - c)]$$
$$= (b - a)f'(b)K_g(x) * \delta(x - c)$$
$$= (b - a)f'(b)K_g(x - c).$$

Clearly, ∇I_{fb} has the shape the same as the PSF kernel, which is symmetric w.r.t. the location of the step c, shown as red in Fig. 3.6.

3.3.4 Forged Sharp Edge

A forged sharp (fs) edge appears as an abrupt jump in intensity as

$$I_{fs} = f(H_{step}(x)) \tag{3.15}$$

The gradient of the spliced sharp boundary image is

$$\nabla I_{fs} = \frac{dI_{fs}}{dx} = \frac{d[f(H_{step}(x))]}{dx}$$
$$= f'(H_{step}(x)) \cdot (b - a)\delta(x - c)$$
$$= f'(b) \cdot \delta(x - c)$$

There are only two intensities a and b, both having the same (large) gradient. The IGH of a forged sharp edge will only have all pixels fall in only two bins shown as magenta in Fig. 3.6.

3.3.5 Distinguishing IGHs of the Edge Types

To validate the theoretical analysis, we show IGHs of the four types of edges in Fig. 3.7. An authentically blurred edge is blurred first, inducing intensity values between the low and high values in irradiance. This symmetric blur profile then is mapped through the CRF, becoming asymmetric.

In the forged blur edge case, by contrast, the irradiance step edge is mapped by CRF first to a step edge in intensity space. Then the artificial blur (via PhotoShop, etc.) induces values between the two intensity extrema, and the profile reflects the symmetric shape of PSF.

The forged sharp edge (from, for instance, a splicing operation) is an ideal step edge, whose nominal IGH has only two bins with non-zero counts. However, due to the existence of noise in images captured by cameras, shading, etc., the IGH of forgery sharp edge appears a rectangular shape as shown in Fig. 3.7. The horizontal line shows that the pixels fall into bins of different intensities with the same gradient, which is caused by the pixel intensity varying along the sharp edge. The two vertical lines show that the pixels fall into bins of different gradients with similar intensity values, which is caused by the pixels intensity varying on the constant color regions.

As for the authentic sharp edge, the IGH is easily confused with a forged blurred edge, in that both are symmetric. If we only consider the shape of IGH, this would lead to a large number of false alarms. To disambiguate the two, we add an additional feature: the absolute value of the center intensity and gradient for each bin. This value helps due to the fact that at the same intensity value, the gradient of the blurred edge

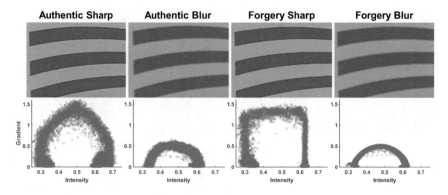

Fig. 3.7 The figure shows the four different edge classes and their IGH

is always smaller than the sharp edge. With our IGH feature, we are able to detect splicing boundaries that are hard for prior methods, such as spliced regions having constant color or those captured by the same camera.

3.3.6 Classifying IGHs

Having described the IGH and how these features differ between the four categories of edges, we now consider mechanisms to solve the inverse problem of inferring the edge category from an image patch containing an edge. Since our IGH is similar to a bag of words-type features used extensively in computer vision, SVMs are a natural classification mechanism to consider. In light of the recent success of powerful Convolutional Neural Networks (CNNs) applied directly to pixel arrays, we consider this approach, but find a limited performance that may be due to relatively sparse training data over all combinations of edge step height, blur extent, etc. To address this, we map the classification problem from the data-starved patch domain to a character shape recognition in the IGH domain, for which we leverage existing CNN architectures and training data.

3.3.6.1 SVM Classification

As in other vision applications, SVMs with the histogram intersection kernel (Maji et al. 2008) perform best for IGH classification. We unwrap our 2D IGH into a vector and append the center intensity and gradient value of each bin. Following the example of SVM-based classification on the bag of words features, we use a multi-classification scheme by training four one-vs-all models: authentic versus others, authentic sharp versus others, forged sharp versus others, and forged blur versus others. Then we combine the scores of all four models to classify the IGH as either authentic or a forgery.

3.3.6.2 CNN on Edge Patches

We evaluate whether a CNN applied directly to edge patches can out-perform methods involving our IGH. We use the very popular Caffe (Jia et al. 2014) implementation for the classification task.

We first train a CNN model on edge patches. In order to examine the difference between authentic and forged patches, we synthesized the authentic and forged blur processes on white, gray, and black edges as shown in Fig. 3.8. Given the same irradiance map, we apply the same Gaussian blur and CRF mapping in both orderings. The white region in the authentically blurred image always appears to be larger than in the forgery blur image, because the CRF has a higher slope in the low-intensity region. That means all intensities of an authentically blurred edge are brought to the

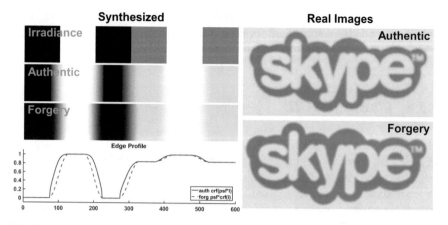

Fig. 3.8 Authentic versus forgery images and edge profiles. The synthesized images are using the same CRF and Gaussian kernel. The real images are captured by Canon 70D. The forgery image is generated by blurring a sharp image the same amount to match the authentic image

white point faster than a forgery. This effect can also be observed in real images, such as the Skype logo in Fig. 3.8, where the white letters in the real image are bolder than in the forgery. Another reason for this effect is that cameras have limited dynamic range. A forged sharp edge will appear like the irradiance map with step edges, while an authentically sharp edge would be easily confused with forgery blur since this CRF effect is only distinguishable with a relatively large amount of blur. Thus, the transition region around the edge *potentially* contains a cue for splicing detection in a CNN framework.

3.3.6.3 CNN on IGHs

Our final approach marries our IGH feature with the power of CNNs. Usually, people wouldn't use a histogram with CNN classification because spatial arrangements are important for other tasks, e.g., object recognition. But, as our analysis has shown, the cues relevant to our problem are found at the pixel level, and our IGH can be used to eliminate various nuisance factors: the orientation of the edge, the height of the step, etc. This has an advantage of reducing the training data dimensionality and, thus, the large number of training data needed to produce accurate models. Lacking an ImageNet-scale training set for forgeriesis a key advantage of our method.

In the sense that a quantized IGH looks like various U-shaped characters, our approach reduces the problem of edge classification into a handwritten character recognition problem, which has been well studied (LeCun et al. 1998). Since we are only interested in the shape of the IGH, the LeNet model (LeCun et al. 1998) is very useful for the IGH recognition task.

3.3.7   Splicing Logo Dataset

To validate our IGH-based classification approach, we built our own Splicing Logo Dataset (SpLogo) containing 1533 authentic images and 1277 forged blur images of logos with different colors and different amounts of blur. All the images are taken by a Canon70D. Optical blur is controlled via two different settings: one is by changing the focal plane (lens moving) and the other is by changing the aperture with the focal plane slightly off the logo plane. The logos are printed on a sheet of paper, and the focal plane is set to be parallel to the logo plane to eliminate the effect of depth-related defocus. Next, the digitally blurred images are generated to match the amount of optical blur through a optimization routine.

3.3.8   Experiments Differentiating Optical and Digital Blur

We use a histogram intersection kernel (Maji et al. 2008) for our SVM and LeNet (LeCun et al. 1998) for CNN with a $1e^{-6}$ base learning rate and $1e^6$ maximum number of iterations. The training set contains 1000 authentic and 1000 forged patches.

Table 3.3 compares the accuracy of the different approaches described in Sect. 3.3.6. Somewhat surprisingly, the CNN applied directly to patches does the worst job of classification. Among the classification methods employing our hand-crafted feature, CNN classifiers obtain better results than SVM, and adding the absolute value of intensity and gradient to IGH increases classification accuracy.

3.3.9   Conclusions: Differentiating Optical and Digital Blur

These experiments show that, at a patch level, the IGH is an effective tool to detect digital blur around edge regions. How these patch-level detections relate to a higher-level forensics application will differ based on the objectives. For instance, to detect image splicing (with or without subsequent edge blurring), it would suffice to find strong evidence of one object with a digitally blurred edge. To detect globally blurred images, on the other hand, it would require evidence that all of the edges in the image are digitally blurred. In either case, the key to our detection method is that a non-linear CRF leads to differences in pixel-level image statistics depending on whether

Table 3.3 Patch classification accuracy for SpLogo data and different classification approaches

Classifier	Image	IGHnoV	IGH
SVM	–	0.972	0.972
CNN	0.896	0.972	**0.99**

it is applied before or after blurring. Our IGH feature captures these statistics and provides a way to eliminate nuisance variables such as edge orientation and step height so that CNN methods can be applied despite a lack of a large training set.

3.4 Additional Forensic Challenges from Computational Cameras

Our experiments with blurred edges show that it's possible to differentiate blur created optically from blur added post-capture via signal processing. The key to doing so is understanding that the non-linearities of the CRF are not commutative with blur, which is modeled as a linear convolution. But what about the CRF itself? Our experiments involved imagery from a small number of devices, and past work (Hsu and Chang 2010) has shown that different devices have different CRF shapes. Does that mean that we need a training set of imagery from all different cameras, in order to be invariant to different CRFs? In subsequent work (Chen et al. 2019), we showed that CRFs from nearly 200 modern digital cameras had very similar shapes. While the CRFs between camera models differed enough to recognize the source camera when a calibration target (a color checker chart) was present, we later found that the small differences did not support reliable source camera recognition on 'in the wild' imagery. For forensic practitioners, the convergence of CRFs represents a mixed bag: their similarity makes IGH-like approaches more robust to different camera types, but reduces the value of CRFs as a cue to source camera model identification.

Looking forward, computational cameras present additional challenges to forensics practitioners. As described elsewhere in this book, the development of deep neural networks that generate synthetic imagery is an important tool in the fight against misinformation. While Generative Adversarial Network (GAN)-based image detectors are quite effective at this point, they may become less so as computational cameras increasingly incorporate U-Net enhancements (Ronneberger et al. 2015) that are architecturally similar to GANs. It remains to be seen whether U-Net enhanced imagery will lead to increased false alarms from GAN-based image detectors, since the current generation of evaluation datasets don't explicitly include such imagery.

To the extent that PRNU and CFA-based forensics are important tools, new computational imaging approaches challenge their effectiveness. Recent work from Google (Wronski et al. 2019), for instance, uses a tight coupling between multi-frame capture and image processing to remove the need for demosaicing and breaks the sensor-to-image pixel mapping that's needed for PRNU to work. As these improvements make their way into mass market smartphone cameras, assumptions about the efficacy of classical forensic techniques will need to be re-assessed.

References

Bianchi T, De Rosa A, Piva A (2011) Improved dct coefficient analysis for forgery localization in jpeg images. In: 2011 IEEE international conference on acoustics, speech and signal processing (ICASSP). IEEE, pp 2444–2447

Bourlai T (ed) (2016) Face recognition across the imaging spectrum. Springer

Chen X, Li F, Yang J, Yu J (2012) A theoretical analysis of camera response functions in image deblurring. In: European conference on computer vision. Springer, pp 333–346

Chen C, McCloskey S, Yu J (2019) Analyzing modern camera response functions. In: IEEE winter conference on applications of computer vision, WACV 2019, Waikoloa Village, HI, USA, January 7-11, 2019. IEEE, pp 1961–1969

Dirik AE, Memon N (2009) Image tamper detection based on demosaicing artifacts. In: 2009 16th IEEE international conference on image processing (ICIP). IEEE, pp 1497–1500

Hsu Y-F, Chang S-F (2010) Camera response functions for image forensics: an automatic algorithm for splicing detection. IEEE Trans Inf Forensics Secur 5(4):816–825

Jia Y, Shelhamer E, Donahue J, Karayev S, Long J, Girshick R, Guadarrama S, Darrell T (2014) Caffe: convolutional architecture for fast feature embedding. arXiv:1408.5093

LeCun Y, Bottou L, Bengio Y, Haffner P (1998) Gradient-based learning applied to document recognition. Proc IEEE 86(11):2278–2324

Liu C, Freeman WT, Szeliski R, Bing Kang S (2006) Noise estimation from a single image. In: 2006 IEEE computer society conference on computer vision and pattern recognition, vol 1. IEEE, pp 901–908

Mahdian B, Saic S (2009) Using noise inconsistencies for blind image forensics. Image Vis Comput 27(10):1497–1503

Maji S, Berg AC, Malik J (2008) Classification using intersection kernel support vector machines is efficient. In: IEEE conference on computer vision and pattern recognition, 2008. CVPR 2008. IEEE, pp 1–8

Ng R (2006) Digital light field photography. PhD thesis, Stanford, CA, USA. AAI3219345

Ng T-T, Chang S-F, Tsui M-P (2007) Using geometry invariants for camera response function estimation. In: 2007 IEEE conference on computer vision and pattern recognition. IEEE, pp 1–8

Pentland AP (1987) A new sense for depth of field. IEEE Trans Pattern Anal Mach Intell PAMI–9(4):523–531

Ronneberger O, Fischer P, Brox T (2015) U-net: convolutional networks for biomedical image segmentation. In: Medical image computing and computer-assisted intervention (MICCAI), vol 9351. LNCS. Springer, pp 234–241. arXiv:1505.04597 [cs.CV]

Simonyan K, Zisserman A (2014) Very deep convolutional networks for large-scale image recognition. arXiv:1409.1556

Tai Y-W, Chen X, Kim S, Joo Kim S, Li F, Yang J, Yu J, Matsushita Y, Brown MS (2013) Nonlinear camera response functions and image deblurring: theoretical analysis and practice. IEEE Trans Pattern Anal Mach Intell 35(10):2498–2512

Tsin Y, Ramesh V, Kanade T (2001) Statistical calibration of ccd imaging process. In: Eighth IEEE international conference on computer vision, 2001. ICCV 2001. Proceedings, vol 1. IEEE, pp 480–487

Wronski B, Garcia-Dorado I, Ernst M, Kelly D, Krainin M, Liang C-K, Levoy M, Milanfar P (2019) Handheld multi-frame super-resolution. ACM Trans Graph 38(4)

Part II
Attribution

Chapter 4
Sensor Fingerprints: Camera Identification and Beyond

Matthias Kirchner

Every imaging sensor introduces a certain amount of noise to the images it captures—slight fluctuations in the intensity of individual pixels even when the sensor plane was lit absolutely homogeneously. One of the breakthrough discoveries in multimedia forensics is that photo-response non-uniformity (PRNU), a multiplicative noise component caused by inevitable variations in the manufacturing process of sensor elements, is essentially a sensor fingerprint that can be estimated from and detected in arbitrary images. This chapter reviews the rich body of literature on camera identification from sensor noise fingerprints with an emphasis on still images from digital cameras and the evolving challenges in this domain.

4.1 Introduction

Sensor noise fingerprints have been a cornerstone of media forensics ever since Lukáš et al. (2005) observed that digital images can be traced back to their sensor based on unique noise characteristics. Minute manufacturing imperfections are believed to make every sensor physically unique, leading to the presence of a weak yet deterministic sensor pattern noise in images captured by the camera (Fridrich 2013). This fingerprint, commonly referred to as photo-response non-uniformity (PRNU), can be estimated from images captured by a specific camera for the purpose of source camera identification. As illustrated in Fig. 4.1, a noise signal is extracted in this process at test time from a probe image of unknown provenance, which can then be compared against pre-computed fingerprint estimates from a set of candidate cameras.

M. Kirchner (✉)
Kitware, Inc., Clifton Park, NY, USA
e-mail: matthias.kirchner@kitware.com

© The Author(s) 2022
H. T. Sencar et al. (eds.), *Multimedia Forensics*, Advances in Computer Vision and Pattern
Recognition, https://doi.org/10.1007/978-981-16-7621-5_4

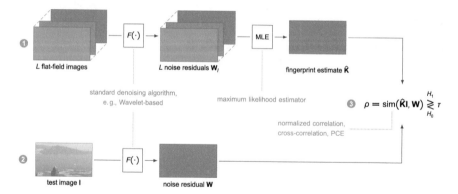

Fig. 4.1 Basic camera identification from PRNU-based sensor noise fingerprints (Fridrich 2013). (1) A camera fingerprint $\hat{\mathbf{K}}$ is estimated from a number flatfield images taken by the camera of interest. (2) At test time, a noise residual $\mathbf{W}$ is extracted from a probe image of unknown provenance. (3) A detector decides whether or not the probe image originates from the camera of interest by evaluating a suitable similarity score, ρ

Because PRNU emerges from physical noise-like properties of individual sensor elements, it has a number of attractive characteristics for source device attribution (Fridrich 2013). The fingerprint signal appears random and is of high dimensionality, which makes the probability of two sensors having the same fingerprint extremely low (Goljan et al. 2009). At the same time, it can be assumed that all common imaging sensor types exhibit PRNU, that each sensor output contains a PRNU component, except for completely dark or saturated images, and that PRNU fingerprints are stable over time (Lukás et al. 2006). Finally, various independent studies have found that the fingerprint is highly robust to common forms of post-processing, including lossy compression and filtering.

The goal of this chapter is not to regurgitate the theoretical foundations of the subject at length, as these have been discussed coherently before, including in a dedicated chapter in the first edition of this book (Fridrich 2013). Instead, we hope to give readers an overview of the ongoing research and the evolving challenges in the domain while keeping the focus on conveying the important concepts.

So why is it that there is still an abundance of active research when extensive empirical evidence (Goljan et al. 2009) has already established the feasibility of highly reliable PRNU-based consumer camera identification at scale? The simple answer is technological progress. Modern cameras, particularly those installed in smartphones, go to great lengths to produce visually appealing imagery. Imaging pipelines are ever evolving, and computational photography challenges our understanding of what a "camera-original" image looks like. On the flip side, many of these new processing steps interfere with the underlying assumptions at the core of PRNU-based camera identification, which requires strictly that the probe image and the camera fingerprint are spatially aligned with respect to the camera sensor elements. Techniques such as lens distortion correction (Goljan and Fridrich 2012), electronic image stabilization (Taspinar et al. 2016), or high dynamic range imaging (Shaya et al. 2018) have all been found to impede camera identification if not accounted for through spatial resynchronization. Robustness to low resolution and strong compression is another concern (van Houten and Geradts 2009; Chuang et al. 2011; Goljan et al. 2016; Altinisik et al. 2020, i. a.) that has been gaining more and more practical relevance due the widespread sharing of visual media through online social networks (Amerini et al. 2017; Meij and Geradts 2018). At the same time, the remarkable success of PRNU-based camera identification has also surfaced concerns for the anonymity of photographers, who may become identifiable through the analysis and combination of information derived from one or multiple images. As a result, the desire to protect anonymous image communication, e.g., in the case of journalism, activism, or legitimate whistle-blowing, has brought counter-forensic techniques (Böhme and Kirchner 2013) to suppress traces of origin in digital images to the forefront.

We discuss these and related topics with a specific focus on still camera images in more detail below. Our treatment of the subject here is complemented by dedicated chapters on video source attribution (Chap. 5) and large scale camera identification (Chap. 6). An overview of computational photography is given in Chap. 3. Sections 4.2 and 4.3 start with the basics of sensor noise fingerprint estimation and camera identification, before Sect. 4.4 delves into the challenges related to spatial sensor misalignment. Section 4.5 focuses on recent advances in image manipulation localization based on PRNU fingerprints. Counter-forensic techniques for fingerprint removal and copying are discussed in Sect. 4.6, while Sect. 4.7 highlights early attempts to apply tools from the field of deep learning to domain-specific problems. A brief overview of relevant public datasets in Sect. 4.8 follows, before Sect. 4.9 concludes the chapter.

4.2 Sensor Noise Fingerprints

State-of-the-art sensor noise forensics assumes a simplified imaging model for single-channel images $\mathbf{I}(m, n), 0 \leq m < M, 0 \leq n < N$,

$$\mathbf{I} = \mathbf{I}^{(o)}(1 + \mathbf{K}) + \boldsymbol{\theta}, \qquad (4.1)$$

in which the multiplicative PRNU factor $\mathbf{K}$ modulates the noise-free image $\mathbf{I}^{(o)}$, while $\boldsymbol{\theta}$ comprises a variety of additive noise components (Fridrich 2013). Ample empirical evidence suggests that signal $\mathbf{K}$ is a unique and robust camera fingerprint (Goljan et al. 2009). It can be estimated from a set of L images taken with the specific sensor of interest. The standard procedure relies on a denoising filter $F(\cdot)$ to obtain a noise residual,

$$\mathbf{W}_l = \mathbf{I}_l - F(\mathbf{I}_l), \qquad (4.2)$$

from the l-th image $\mathbf{I}_l$, $0 \leq l < L$. The filter acts mainly as a means to increase the signal-to-noise ratio between the signal of interest (the fingerprint) and the observed image. To date, most works still resort to the wavelet-based filter as adopted by Lukáš et al. (2006) for its efficiency and generally favorable performance, although a number of studies have found that alternative denoising algorithms can lead to moderate improvements (Amerini et al. 2009; Cortiana et al. 2011; Al-Ani and Khelifi 2017; Chierchia et al. 2014, i. a.). In general, it is accepted that noise residuals obtained from off-the-shelf filters are imperfect by nature and that they are contaminated non-trivially by remnants of image content. Salient textures or quantization noise may exacerbate the issue. For practical applications, a simplified modeling assumption

$$\mathbf{W}_l = \mathbf{K}\mathbf{I}_l + \boldsymbol{\eta}_l \qquad (4.3)$$

with i. i. d. Gaussian noise $\boldsymbol{\eta}_l$ leads to a maximum likelihood estimate (MLE) $\hat{\mathbf{K}}$ of the PRNU fingerprint of the form (Fridrich 2013)

$$\hat{\mathbf{K}} = \frac{\sum_l \mathbf{W}_l \mathbf{I}_l}{\sum_l \mathbf{I}_l^2} . \qquad (4.4)$$

In this procedure, it is assumed that all images $\mathbf{I}_l$ are spatially aligned so that the pixel-wise operations are effectively carried out over the same physical sensor elements.

If available, it is beneficial to use flatfield images for fingerprint estimation to minimize contamination from image content. The quality of the estimate in Eq. (4.4) generally improves with L, but a handful of homogeneously lit images typically suffices in practice. Uncompressed or raw sensor output is preferable over compressed images, as it will naturally reduce the strength of unwanted nuisance signals in Eq. (4.1). When working with raw output from a sensor with a color filter array (CFA), it is advisable to subsample the images based on the CFA layout (Simon et al. 2009).

Practical applications warrant a post-processing step to clean the fingerprint estimate from non-unique artifacts, which may otherwise increase the likelihood of false fingerprint matches. Such artifacts originate from common signal characteristics that occur consistently across various devices, for instance, due to the distinctively structured layout of the CFA or block-based JPEG compression. It is thus strongly recommended to subject $\hat{\mathbf{K}}$ to zero-meaning and frequency-domain Wiener filtering, as detailed in Fridrich (2013). Additional post-processing operations have been discussed in the literature (Li 2010; Kang et al. 2012; Lin and Li 2016, i. a.), although their overall merit is often marginal (Al-Ani and Khelifi 2017). Other non-trivial non-unique artifacts have been documented as well. Cases of false source attribution linked to lens distortion correction have been reported when images from a different camera were captured at a focal length that was prominently featured during fingerprint estimation (Goljan and Fridrich 2012; Gloe et al. 2012). Non-unique artifacts introduced by advanced image enhancement algorithms in modern devices are a subject of ongoing research (Baracchi et al. 2020; Iuliani et al. 2021).

For multi-channel images, the above procedures can be applied to each color channel individually before averaging the obtained signals into a single-channel fingerprint estimate (Fridrich 2013).

4.3 Camera Identification

For a given probe image $\mathbf{I}$ of unknown provenance, camera identification can be formulated as a hypothesis testing problem:

$H_0 : \mathbf{W} = \mathbf{I} - F(\mathbf{I})$ does not contain the fingerprint of interest, $\mathbf{K}$

$H_1 : \mathbf{W}$ does contain the fingerprint $\mathbf{K}$;

i.e., the probe is attributed to the tested camera if H_1 holds. In practice, the test can be decided by evaluating a similarity measure ρ,

$$\rho = \text{sim}(\mathbf{W}, \hat{\mathbf{K}}\mathbf{I}) \gtrless_{H_0}^{H_1} \tau . \tag{4.5}$$

for a suitable threshold τ. Under the modeling assumptions adopted in the literature (Fridrich 2013), the basic building block for this is the normalized cross-correlation (NCC), which is computed over a grid of shifts $s = (s_1, s_2), 0 \le s_1 < M, 0 \le s_2 < N$, for two matrices $\mathbf{A}, \mathbf{B}$ as

$$\text{NCC}_{\mathbf{A},\mathbf{B}}(s_1, s_2) = \frac{\sum_{m,n} \left(\mathbf{A}(m, n) - \bar{\mathbf{A}}\right) \left(\mathbf{B}(m + s_1, n + s_2) - \bar{\mathbf{B}}\right)}{\|\mathbf{A} - \bar{\mathbf{A}}\| \|\mathbf{B} - \bar{\mathbf{B}}\|} . \tag{4.6}$$

We assume implicitly that matrices $\mathbf{A}$ and $\mathbf{B}$ are of equal dimension and that zero-padding has been applied to assert this where necessary. In practice, the above expression is evaluated efficiently in the frequency domain. It is common in the field to approximate Eq. (4.6) by working with the circular cross-correlation, i.e., by operating on the FFTs of matrices $\mathbf{A}, \mathbf{B}$ without additional zero-padding. Taking the maximum NCC over a set S of admissible shifts as similarity,

$$\rho_{\text{ncc}}^{(S)} = \max_{s \in S} \text{NCC}_{\mathbf{W}, \hat{\mathbf{K}}\mathbf{Y}}(s) , \tag{4.7}$$

can conveniently account for potential sensor misalignment between the tested fingerprint and the probe as they would result from different sensor resolutions and/or cropping. Peak-to-correlation energy (PCE) has been proposed as an alternative that mitigates the need for sensor-specific thresholds

$$\rho_{\text{pce}}^{(S)} = \frac{(MN - |\mathcal{N}|) \cdot \left(\rho_{\text{ncc}}^{(S)}\right)^2}{\sum_{s \notin \mathcal{N}} \text{NCC}_{\mathbf{W}, \hat{\mathbf{K}}\mathbf{Y}}^2(s)} . \tag{4.8}$$

In the equation above, $\mathcal{N}$ denotes a small neighborhood around the peak NCC. It is commonly set to a size of 11×11 (Goljan et al. 2009). In practice, camera identification must account for the possibility of mismatching sensor orientations in the probe image and in the fingerprint estimate, so the test for the presence of the fingerprint should be repeated with one of the signals rotated by 180° if the initial test stays below the threshold.

The choice of set $\mathcal{S}$ crucially impacts the characteristics of the detector, as a larger search grid will naturally increase the variance of detector responses on true negatives. If it can be ruled out a priori that the probe image underwent cropping, the "search" can be confined to $\mathcal{S} = \{(0, 0)\}$ to reduce the probability of false alarms. Interested readers will find a detailed error analysis for this scenario in Goljan et al. (2009), where the authors determined a false alarm rate of 2.4×10^{-5} while attributing 97.62% of probes to their correct source in experiments with more than one million images from several thousand devices. The PCE threshold for this operating point is 60, which is now used widely in the field as a result.

A number of variants of the core camera identification formulation can be addressed with only a little modification (Fridrich 2013). For instance, it can be of interest to determine whether two arbitrary images originate from the same device, without knowledge or assumptions about associated camera fingerprints (Goljan et al. 2007). In a related scenario, the goal is to compare and match a number of fingerprint estimates, which becomes particularly relevant in setups with the objective of clustering a large set of images by their source device (Bloy 2008; Li 2010; Amerini et al. 2014; Marra et al. 2017, i.a.).

Specific adaptations also exist for testing against large databases of camera fingerprints, where computational performance becomes a relevant dimension to monitor in practice. This includes efforts to reduce the dimensionality of camera fingerprints (Goljan et al. 2010; Bayram et al. 2012; Valsesia et al. 2015, i.a.) and protocols for efficient fingerprint search and matching (Bayram et al. 2015; Valsesia et al. 2015; Taspinar et al. 2020, i.a.). We refer the reader to Chap. 6 in this book which discusses these techniques in detail.

4.4 Sensor Misalignment

Camera identification from PRNU-based sensor fingerprints can only succeed if the probe image and the tested fingerprint are spatially aligned. While cross-correlation can readily account for translation and cropping, additional steps become inevitable if more general geometric transformations have to be considered. This includes, for instance, combinations of scaling and cropping when dealing with the variety of image resolutions and aspect ratios supported by modern devices (Goljan and Fridrich 2008; Tandogan et al. 2019). Specifically, assuming that a probe image $\mathbf{I}$ underwent a geometric transform $T_{\mathbf{u}^*}(\cdot)$ with parameters $\mathbf{u}^*$, we want to carry out the above basic procedures on $\mathbf{I} \leftarrow T_{\mathbf{u}^*}^{-1}(\mathbf{I})$. If the parameters of the transform are unknown, the problem essentially translates into a search over a set $\mathcal{U}$ of admissible candidate transforms $\mathbf{u}$. In practice, the maximum similarity over all candidate transforms will determine whether or not an image contains the tested fingerprint,

$$\rho = \max_{\mathbf{u} \in \mathcal{U}} \text{sim}\left(T_{\mathbf{u}}^{-1}(\mathbf{I}) - F\left(T_{\mathbf{u}}^{-1}(\mathbf{I})\right), \hat{\mathbf{K}} T_{\mathbf{u}}^{-1}(\mathbf{I})\right) \gtrless_{H_0}^{H_1} \tau, \tag{4.9}$$

and the detector threshold should be adjusted accordingly compared to the simpler case above to maintain a prescribed false alarm rate. Unfortunately, this search can quickly become computationally expensive. A number of approximations have been proposed to speed up the search (Goljan and Fridrich 2008), including applying the inverse transform to a precomputed noise residual $\mathbf{W}$ instead of recomputing the noise residual for each candidate transform, and evaluating $\text{sim}(T_{\mathbf{u}}^{-1}(\mathbf{WI}), \hat{\mathbf{K}})$ instead of $\text{sim}(T_{\mathbf{u}}^{-1}(\mathbf{W}), \hat{\mathbf{K}} T_{\mathbf{u}}^{-1}(\mathbf{I}))$.[1] As for the search itself, a coarse-to-fine grid search is recommended by Goljan and Fridrich (2008) for potentially scaled and cropped images, while Mandelli et al. (2020) adopt a particle swarm optimization technique. Gradient-based search methods generally do typically not apply to the problem due to the noise-like characteristics of detector responses outside of a very sharp peak around the correct candidate transform (Fridrich 2013).

Sensor misalignment may not only be caused by "conventional" image processing. With camera manufacturers constantly striving to improve the visual image quality that their devices deliver to the customer, advanced in-camera processing and the rise of computational photography in modern imaging pipelines can pose significant challenges in that regard. One of the earliest realizations along those lines was that computational lens distortion correction can introduce non-linear spatial misalignment that needs to be accounted for (Goljan and Fridrich 2012; Gloe et al. 2012). Of particular interest is lens radial distortion, which lets straight lines in a scene appear curved in the captured image. This type of distortion is especially prominent for zoom lenses as they are commonly available for a wide variety of consumer cameras. For a good trade-off between lens size, cost and visual quality, cameras correct for lens radial distortion through warping the captured image according to a suitable compensation model that effectively inverts the nuisance warping introduced by the lens. This kind of post-processing will displace image content relative to the sensor elements. It impairs camera identification because the strength of lens radial distortion depends on the focal length, i.e., images taken by the same camera at different focal lengths will undergo different warping in the process of lens distortion correction. As a result, a probe image captured at a certain focal length that was not (well) represented in the estimation of the camera fingerprint $\hat{\mathbf{K}}$ in Eq. (4.4) may not be associated with its source device inadvertently.

A simple first-order parametric model to describe and invert radially symmetric barrel/pincushion distortion facilitates camera identification via a coarse-to-fine search over a single parameter (Goljan and Fridrich 2012), similar to the handling of general geometric transformations in Eq. (4.9) above. The model makes a number

---

[1] The two expressions are equivalent when used in combination with the PCE whenever it can be assumed that $T_{\mathbf{u}}^{-1}(\mathbf{W}) \cdot T_{\mathbf{u}}^{-1}(\mathbf{I}) \approx T_{\mathbf{u}}^{-1}(\mathbf{WI})$.

of simplifying assumptions that may not always hold in practice to avoid a higher dimensional search space, including the assumed concurrence of the optical center and the image center. Crucially, the practical applicability of such an approach first and foremost depends on the validity of the assumed distortion correction model in real cameras. In most circumstances, it is ultimately not fully known which design choices camera manufacturers make, and we are not aware of published large-scale evaluations that span a significant number of different devices/lenses. There are incidental observations of missed detections that suggest deviations from a purely radial distortion correction model (Gloe et al. 2012; Goljan and Fridrich 2014), however.

One of the reasons why interest in the effects of lens distortion correction and their remedies seems to have waned over the past years may be the enormous gain in the popularity of smartphones and similar camera-equipped mobile devices. These devices operate under very different optical and computational constraints than "conventional" consumer cameras and have introduced their own set of challenges to the field of PRNU-based camera identification. The source attribution of video data, and in particular the effects of electronic image stabilization (EIS), have arguably been the heavyweight in this research domain, and we direct readers to Chap. 5 of this book for a dedicated exposition. In the context of this chapter, it suffices to say that EIS effectively introduces gradually varying spatial misalignment to sequences of video frames, which calls for especially efficient computational correction approaches (Taspinar et al. 2016; Iuliani et al. 2019; Mandelli et al. 2020; Altinisik and Sencar 2021, i.a.).

Other types of in-camera processing are more and more moving into the focus as well. For instance, high dynamic range (HDR) imaging (Artusi et al. 2017) is routinely supported on many devices today and promises enhanced visual quality, especially also under challenging light conditions. In smartphones, it is usually realized by fusing a sequence of images of a scene, each taken at a different exposure setting in rapid succession. This requires registration of the images to mitigate global misalignment due to camera motion and local misalignment due to moving objects in the scene. In practice, the HDR image will be a content-dependent weighted mixture of the individual exposures, and different regions in the image may have undergone different geometric transformations. Without additional precautions, camera identification may fail under these conditions (Shaya et al. 2018). While it seems infeasible to express such complex locally varying geometric transformations in a parametric model, it is possible to conduct a search over candidate transformation on local regions of the image. Empirical evidence from a handful of smartphones suggests that the contributions of the individual exposures can be locally synchronized via cross-correlation, after correcting for a global, possibly anisotropic rescaling operation (Hosseini and Goljan 2019). Future research will have to determine whether such approach generalizes.

Sensor misalignment can be expected to remain a practical challenge, and it is likely to take on new forms and shapes as imaging pipelines continue to evolve. We surmise that the nature of misalignments will broaden beyond purely geometrical characteristics with the continued rise of computational photography and camera manufacturers allowing app developers access to raw sensor measurements. Empirical reports of impaired camera identification across mismatching imaging pipelines may be taken as first cautionary writing on the wall (Joshi et al. 2020).

4.5 Image Manipulation Localization

When a region of an image is replaced with content from elsewhere, the new content will lack the characteristic camera PRNU fingerprint one would expect to find otherwise. This is true irrespective of whether the inserted content has been copied from within the same image, or from a different image. Recasting camera identification as a local test for the presence of an expected fingerprint thus allows for the detection and localization of image manipulations (Chen et al. 2008; Fridrich 2013).

A straightforward approach is to examine the probe image $\mathbf{I}$ by sliding an analysis window of size $B \times B$ over the probe image, and to assign a binary label $\mathbf{Y}(m, n) \in \{-1, 1\}$,

$$\mathbf{Y}(m, n) = \mathrm{sgn}\left(\rho(m, n) - \tau\right), \qquad (4.10)$$

to the window centered around location (m, n), with $\rho(m, n)$ obtained by evaluating Eq. (4.5) for the corresponding analysis window. The resulting binary map Y will then be indicative of local manipulations, with $\mathbf{Y}(m, n) = -1$ corresponding to the absence of the fingerprint in the respective neighborhood. The literature mostly resorts to the normalized correlation for this purpose, $\rho = \rho_{\mathrm{ncc}}^{(\{0\})}$. It can be computed efficiently in one sweep for all sliding windows by implementing the necessary summations as linear filtering operations on the whole image.

The localization of small manipulated regions warrants sufficiently small analysis windows, which impacts the ability to reliably establish whether or not the expected fingerprint is present negatively. The literature often finds a window size of $B = 64$ as a reasonable trade-off between resolution and accuracy (Chierchia et al. 2014; Chakraborty and Kirchner 2017; Korus and Huang 2017). A core problem for analysis windows that small is that the measured local correlation under H_1 depends greatly on local image characteristics. One possible remedy is to formulate a camera-specific correlation predictor $\hat{\rho}(m, n)$ that uses local image characteristics to predict how strongly the noise residual in a particular analysis window is expected to correlate with the purported camera fingerprint under H_1 (Chen et al. 2008). The decision whether to declare the absence of the tested fingerprint can then be conditioned on the expected correlation.

Adopting the rationale that more conservative decisions should be in place when the local correlation cannot be expected to take on large values per se, an adjusted binary labeling rule decides $\mathbf{Y}(m, n) = -1$ iff $\rho(m, n) \leq \tau$ and $\hat{\rho}(m, n) > \lambda$, where threshold λ effectively bounds the probability of missed fingerprint detection under H_1 (Chen et al. 2008). To ensure that the binary localization map mostly contains connected regions of a minimum achievable size, a pruning and post-processing step with morphological filters is recommended.

(a) pristine image   (b) manipulated image

(c) local correlation ρ (d) binary localization map

Fig. 4.2 Image manipulation localization. The local correlation between the camera fingerprint and the noise residual extracted from the manipulated image is lowest (darker) in the manipulated region. It was computed from sliding windows of size 64×64. The binary localization map (overlayed on top of the manipulated image) was obtained with a conditional random field approach that evaluates the difference between measured and predicted correlation, $\rho - \hat{\rho}$ (Chakraborty and Kirchner 2017). Images taken from the Realistic Tampering Dataset (Korus and Huang 2017)

Significant improvements have been reported when explicitly accounting for the observation that local decisions from neighboring sliding windows are interdependent. A natural formulation follows from approaching the problem in a global optimization framework with the objective of finding the optimal mapping

$$\mathbf{Y}^* = \arg \max_{\mathbf{Y}} p\left(\mathbf{Y}|\rho, \hat{\boldsymbol{\rho}}\right) . \tag{4.11}$$

This sets the stage for rewarding piecewise constant label maps via a variety of probabilistic graphical modeling techniques such as Markov random fields (Chierchia et al. 2014) and conditional random fields (Chakraborty and Kirchner 2017; Korus and Huang 2017). Figure 4.2 gives an example result.

A general downside of working with fixed-sized sliding windows is that $\rho(m, n)$ will naturally change only very gradually, which makes the detection of very small manipulated regions challenging (Chierchia et al. 2011). Multi-scale reasoning over various analysis window sizes can mitigate this to some extent (Korus and Huang 2017). Favorable results have also been reported for approaches that incorporate image segmentation (Korus and Huang 2017; Zhang et al. 2019; Lin and Li 2020) or guided filtering (Chierchia et al. 2014) to adaptively adjust analysis windows based on image characteristics.

Surprisingly, very few notable updates to the seminal linear correlation predictor $\hat{\rho}$ from simple intensity, flatness, and texture features by Chen et al. (2008) have surfaced in the literature, despite its paramount role across virtually all PRNU-based localization approaches and a generally rather mixed performance (Quan and Li 2021). We highlight here the replacement of the original linear regression model with a feed-forward neural network by Korus and Huang (2017), and the observation that a more accurate prediction can be achieved when the camera's ISO speed is taken into account (Quan and Li 2021). Attempts to leverage a deep learning approach to obtain potentially more expressive features Chakraborty (2020) are commendable but require a more thorough evaluation for authoritative conclusions.

Overall, image manipulation localization based on camera sensor noise has its place in the broader universe of digital media forensics when there is a strong prior belief that the probe image indeed originates from a specific camera. If the manipulated region is sufficiently small, this can be established through conventional full-frame camera identification. Non-trivial sensor misalignment as discussed in Sect. 4.4 can be expected to complicate matters significantly for localization, but we are not aware of a principled examination of this aspect to date.

4.6 Counter-Forensics

The reliability and the robustness of camera identification based on sensor noise have been under scrutiny ever since seminal works on sensor noise forensics surfaced over 15 years ago. As a result, it is widely accepted that PRNU fingerprints survive a variety of common post-processing operations, including JPEG compression and resizing. Counter-forensics focuses on more deliberate attempts to impair successful camera identification by acknowledging that there are scenarios where intelligent actors make targeted efforts to induce a certain outcome of forensic analyses (Böhme and Kirchner 2013). In this context, it is instructive to distinguish between two major goals of countermeasures, fingerprint removal and fingerprint copying (Lukáš et al. 2006; Gloe et al. 2007). The objective of fingerprint removal is the suppression of a camera's fingerprint to render source identification impossible. This can be desirable in efforts of protecting the anonymity of photographers, journalists, or legitimate whistleblowers in threatening environments (Nagaraja et al. 2011). Fingerprint copying attempts to make an image plausibly appear as if it was captured by a different camera, which is typically associated with nefarious motives. It strictly implies the suppression of the original fingerprint and it is thus generally a harder problem. The success of such counter-forensic techniques is to a large degree bound by the admissible visual quality of the resulting image. If an image purports to be camera-original but has suffered from noticeable degradation, it will likely raise suspicion. If anonymity is of utmost priority, strong measures that go along with a severe loss of image resolution are more likely acceptable.

Existing fingerprint removal methods can be categorized under two general approaches (Lukáš et al. 2006). Methods of the first category are side-informed

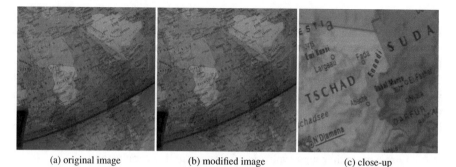

(a) original image   (b) modified image   (c) close-up

Fig. 4.3 Fingerprint removal with PatchMatch replaces local image content with visually similar content from elsewhere in the image (Entrieri and Kirchner 2016). The PCE, computed considering all possible cross-correlation shifts, decreases from 5,617 for the original image to 32 after the modification. The "anonymized" image has a PSNR of 38.3 dB. Original image size: 2,000×2,000 pixels

in the sense that they use an estimate of the sensor noise fingerprint to ensure a detector output below the identification threshold. Flatfielding—a denoising technique that targets the general imaging model in Eq. (4.1)—is known to remove the multiplicative noise term $\mathbf{K}$ effectively, but it ideally requires access to the raw sensor measurements (Lukás et al. 2006; Gloe et al. 2007). Adaptive fingerprint removal techniques explicitly attempt to minimize Eq. (4.5) by finding a noise sequence that cancels out the multiplicative fingerprint term in Eq. (4.3) (Karaküçük and Dirik 2015; Zeng et al. 2015). This works best when exact knowledge of the detector (and thus the images used to estimate $\hat{\mathbf{K}}$) is available.

Uninformed techniques make less assumptions and directly address the robustness of the sensor noise fingerprint. Methods of this category apply post-processing to the image until the noise pattern is too corrupted to correlate with the fingerprint. No specific knowledge of the camera, the camera's fingerprint, or the detector is assumed in this process. However, the high robustness of the sensor fingerprint makes this a non-trivial problem (Rosenfeld and Sencar 2009), and solutions may often come with a more immediate loss of image quality compared to side-informed methods. One promising direction is to induce irreversible sensor misalignment. Seam-carving—a form of content-adaptive resizing that shrinks images by removing low-energy "seams" (Avidan and Shamir 2007)—is an effective candidate operation in this regard (Bayram et al. 2013), although a considerable amount of seams must be removed to successfully desynchronize the sensor fingerprint (Dirik et al. 2014). This bears a high potential for the removal of "important" seams, degrading image quality and resolution. In many ways, recomposing the image entirely instead of removing a large number of seams is thus a more content-preserving alternative. The idea here is to replace local content with content from elsewhere in the image with the objective of finding replacements that are as similar to the original content as possible while lacking the telltale portion of the fingerprint. A modified version of the PatchMatch algorithm (Barnes et al. 2009) has been demonstrated to produce viable results (Entrieri and Kirchner 2016), while a later variant employed inpainting for this purpose (Mandelli et al. 2017). Figure 4.3 showcases an example of successful fingerprint removal with the PatchMatch algorithm. All three strategies, seam-carving, PatchMatch, and inpainting, can reliably prevent PRNU-based camera identification from a single image. An aggregation of fingerprint traces from multiple "anonymized" images from the same camera can reestablish the link to the common source device to some extent, however (Taspinar et al. 2017; Karaküçük and Dirik 2019).

In a fingerprint copy attack, a nefarious actor Eve operates with the goal of making an arbitrary image $\mathbf{J}$ look as if it was captured by an innocent user Alice's camera. Eve may obtain an estimate $\hat{\mathbf{K}}_E$ of Alice's camera fingerprint from a set of publicly available images and leverage the multiplicative nature of the PRNU to obtain (Lukás et al. 2006)

$$\mathbf{J}' = \mathbf{J}(1 + \alpha \hat{\mathbf{K}}_E), \qquad (4.12)$$

where the scalar factor $\alpha > 0$ determines the fingerprint strength. Attacks of this type have been demonstrated to be effective, in the sense that they can successfully mislead a camera identification algorithm in the form of Eq. (4.5). The attack's success generally depends on a good choice of α: too low values mean that the bogus image $\mathbf{J}'$ may not be assigned to Alice's camera; a too strong embedding will make the image appear suspicious (Goljan et al. 2011; Marra et al. 2014). In practical scenarios, Eve may have to apply further processing to make her forgery more compelling, e.g., removing the genuine camera fingerprint, synthesizing color filter interpolation artifacts (Kirchner and Böhme 2009), and removing or adding traces of JPEG compression (Stamm and Liu 2011).

Under realistic assumptions, it is virtually impossible to prevent Eve from forcing a high similarity score in Eq. (4.5). All is not lost, however. Alice can utilize that noise residuals computed with practical denoising filters are prone to contain remnants of image content. The key observation here is that the similarity between a noise residual $\mathbf{W}_I$ from an image $\mathbf{I}$ taken with Alice's camera and the noise residual $\mathbf{W}_{J'}$ due to a common attack-induced PRNU term will be further increased by some shared residual image content, if $\mathbf{I}$ contributed to Eve's fingerprint estimate $\hat{\mathbf{K}}_E$. The so-called triangle test (Goljan et al. 2011) picks up on this observation by also considering the correlation between Alice's own fingerprint and both $\mathbf{W}_I$ and $\mathbf{W}_{J'}$ to determine whether the similarity between $\mathbf{W}_I$ with $\mathbf{W}_{J'}$ is suspiciously large. A pooled version of the test establishes whether any images in a given set of Alice's images have contributed to $\hat{\mathbf{K}}_E$ (Barni et al. 2018), without determining which. Observe that in either case Alice may have to examine the entirety of images ever made public by her. On Eve's side, efforts to create a fingerprint estimate $\hat{\mathbf{K}}_E$ in a procedure that deliberately suppresses telltale remnants of image content can thwart the triangle test's success (Caldelli et al. 2011; Rao et al. 2013; Barni et al. 2018), thereby only setting the stage for the next iteration in the cat-and-mouse game between attacks and defenses.

The potentially high computational (and logistical) burden and the security concerns around the triangle test can be evaded in a more constrained scenario. Specifically, assume that Eve targets the forgery of an uncompressed image but only has access to images shared in a lossy compression format when estimating $\hat{\mathbf{K}}_E$. Here, it can be sufficient to test for the portion of the camera fingerprint that is fragile to lossy compression to establish that Eve's image $\mathbf{J}'$ does not contain a complete fingerprint (Quiring and Kirchner 2015). This works because the PRNU in uncompressed images is relatively uniform across the full frequency spectrum, whereas lossy compression mainly removes high-frequency information from images. As long as Alice's public images underwent moderately strong compression, such as JPEG at a quality factor of about 85, no practicable remedies for Eve to recover the critically missing portion of her fingerprint estimate are known at the time of this writing (Quiring et al. 2019).

4.7 Camera Fingerprints and Deep Learning

The previous sections have hopefully given the reader the impression that research around PRNU-based camera fingerprints is very much alive and thriving, and that new (and old) challenges continue to spark the imagination of academics and practitioners alike. Different from the broader domain of media forensics, which is now routinely drawing on deep learning solutions, only a handful of works have made attempts to apply ideas from this rapidly evolving field to the set of problems typically discussed in the context of device-specific (PRNU-based) camera fingerprints. We can only surmise that this in part due to the robust theoretical foundations that have defined the field and that have ultimately led to the wide acceptance of PRNU-based camera fingerprints in practical forensic casework, law enforcement, and beyond. Data-driven "black-box" solutions may thus appear superfluous to many.

However, one of the strengths that deep learning techniques can bring to the rigid framework of PRNU-based camera identification is in fact their very nature: they are data-driven. The detector in Sect. 4.3 was originally derived under a specific set of assumptions with respect to the imaging model in Eq. (4.1) and the noise residuals in Eq. (4.3). There is good reason to assume that real images will deviate from these simplified models to some degree. First and foremost, noise residuals from a single image will always suffer from significant and non-trivial distortion, if we accept that content suppression is an ill-posed problem in the absence of viable image models (and possibly even of the noise characteristics itself (Masciopinto and Pérez-González 2018)). This opens the door to potential improvements from data-driven approaches, which ultimately do not care about modeling assumptions but rather learn (hopefully) relevant insights from the training data directly.

One such approach specifically focuses on the extraction of a camera signature from a single image $\mathbf{I}$ at test time (Kirchner and Johnson 2019). Instead of relying on a "blind" denoising procedure as it is conventionally the case, a convolutional neural network (CNN) can serve as a flexible non-linear optimization tool that learns how to obtain a better approximation of $\mathbf{K}$. Specifically, the network is trained to extract a noise pattern $\tilde{\mathbf{K}}$ to minimize $\|\hat{\mathbf{K}} - \tilde{\mathbf{K}}\|_2^2$, as the pre-computed estimate $\hat{\mathbf{K}}$ is the best available approximation of the actual PRNU signal under the given imaging

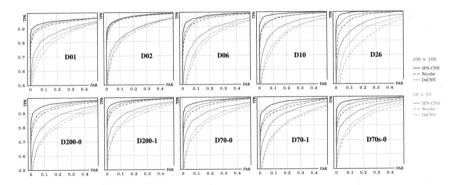

Fig. 4.4 Camera identification from noise signals computed with a deep-learning based fingerprint extractor (coined "SPN-CNN") (Kirchner and Johnson 2019). The ROC curves were obtained by thresholding the normalized correlation, $\rho_{ncc}^{(0)}$, between the extracted SPN-CNN noise patterns, $\tilde{\mathbf{K}}$, and "conventional" camera fingerprints, $\hat{\mathbf{K}}$, for patches of size 100×100 (blue) and 50×50 (orange). Device labels correspond to devices in the VISION (Shullani et al. 2017) (top) and Dresden Image Databases (Gloe and Böhme 2010) (bottom). Curves for the standard Wavelet denoiser and an off-the-shelf DnCNN denoiser (Zhang et al. 2017) are included for comparison

model. The trained network replaces the denoiser $F(\cdot)$ at test time, and the fingerprint similarity is evaluated directly for $\hat{\mathbf{K}}$ instead of $\hat{\mathbf{K}}\mathbf{I}$ in Eq. (4.5). Notably, this breaks with the tradition of employing the very same denoiser for both fingerprint estimation and detection. Empirical results suggest that the resulting noise signals have clear benefits over conventional noise residuals when used for camera identification, as showcased in Fig. 4.4. A drawback of this approach is that it calls for extraction CNNs trained separately for each relevant camera (Kirchner and Johnson 2019).

The similarity function employed by the detector in Eq. (4.5) is another target for potential improvement. If model assumptions do not hold, the normalized cross-correlation may no longer be a good approximation of the optimal detector (Fridrich 2013), and a data-driven approach may be able to reach more conclusive decisions. Along those lines, a Siamese network structure trained to compare spatially aligned patches from a fingerprint estimate $\hat{\mathbf{K}}$ and the noise residual $\mathbf{W}$ from a probe image $\mathbf{I}$ was reported to greatly outperform a conventional PCE-based detector across a range of image sizes (Mandelli et al. 2020). Different from the noise extraction approach by Kirchner and Johnson (2019), experimental results suggest that no device-specific training is necessary. Both approaches have not yet been subjected to more realistic settings that include sensor misalignment.

While the two works discussed above focus on specific building blocks in the established camera identification pipeline, we are not currently aware of an end-to-end deep learning solution that achieves source attribution at the level of individual devices at scale. There is early evidence, however, to suggest that camera model traces derived from a deep model can be a beneficial addition to conventional camera fingerprints, especially when the size of the analyzed images is small (Cozzolino et al. 2020).

CNNs have also been utilized in the context of counter-forensics, where the two-fold objective of minimizing fingerprint similarity while maximizing image quality almost naturally invite data-driven optimization solutions to the problem of side-informed fingerprint removal. An early proposal trains an auto-encoder inspired anonymization operator by encoding the stated objectives in a two-part cost function (Bonettini et al. 2018). The solution, which has to be retrained for each new image, relies on a learnable denoising filter as part of the network, which is used to extract noise residuals from the "anonymized" images during training. The trained network is highly effective at test time as long as the detector uses the same denoising function, but performance dwindles when a different denoising filter, such as the standard Wavelet-based approach, is used instead. This limits the practical applicability of the fingerprint removal technique for the time being.

Overall, it seems almost inevitable that deep learning will take on a more prominent role across a wide range of problems in the field of (PRNU-based) device identification. We have included these first early works here mainly to highlight the trend, and we invite the reader to view them as important stepping stones for future developments to come.

4.8 Public Datasets

Various publicly available datasets have been compiled over time to advance research on sensor-based device identification, see Table 4.1. Not surprisingly, the evolution of datasets since the release of the trailblazing Dresden Image Database (Gloe and Böhme 2010) mirrors the general consumer shift from dedicated digital cameras to mobile devices. There are now also several diverse video datasets with a good coverage across different devices available. In general, a defining quality of a good dataset for source device identification is not only the number of available images per unique device, but also whether multiple instances of the same device model are represented. This is crucial for studying the effects of model-specific artifacts on false alarms, which may remain unidentified when only one device per camera model is present. Other factors worth considering may be whether the image/video capturing protocol controlled for certain exposure parameters, or whether all devices were used to capture the same set (or type) of scenes.

Although it has become ever more easy to gather suitably large amounts of data straight from some of the popular online media sharing platforms, dedicated custom-made research datasets offer the benefit of a well-documented provenance while fostering the reproducibility of research and mitigating copyright concerns. In addition, many of these datasets include, by design, a set of flatfield images to facilitate the computation of high-quality sensor fingerprints. However, as we have seen throughout this chapter, a common challenge in this domain is to keep pace with the latest technological developments on the side of camera manufacturers. This translates into a continued need for updated datasets to maintain practical relevance and timeliness. Results obtained on an older dataset may not hold up well on data from newer

Table 4.1 Public datasets with a special focus on source device identification. The table lists the number of available images and videos per dataset, the number of unique devices and camera models covered, as well as the types of cameras included (C: consumer, D: DSLR, M: mobile device)

Dataset	Year	Images	Videos	Devices	Models	Cameras
Dresden Gloe and Böhme (2010)[*]	2010	14k+		73	25	C D
RAISE Dang-Nguyen et al. (2015)	2015	8k+		3	3	D
VISION Shullani et al. (2017)[**]	2017	11k+	0.6k	35	30	M
HDR Shaya et al. (2018)	2018	5k+		23	22	M
SOCRatES Galdi et al. (2019)	2019	9k+	1k	103	60	M
video-ACID Hosler et al. (2019)	2019		12k+	46	36	C D M
NYUAD-MMD Taspinar et al. (2020)	2020	6k+	0.3k	78	62	M
Daxing Tian et al. (2019)	2020	43k+	1k+	90	22	M
Warwick Quan et al. (2020)[*]	2020	58k+		14	11	C D
Forchheim Hadwiger and Riess (2020)[**]	2020	3k+		27	25	M

[*]Includes multiple images of the same scene with varying exposure settings
[**]Provides auxiliary data from sharing the base data on various social network sites

devices due to novel sensor features or acquisition pipelines. A good example is the recent release of datasets with a specific focus on high dynamic range (HDR) imaging (Shaya et al. 2018; Quan et al. 2020), or the provision of annotations for videos that underwent electronic image stabilization (Shullani et al. 2017). With the vast majority of media now shared in significantly reduced resolution through online social network platforms or messaging apps, some of the more recent datasets also consider such common modern-day post-processing operations explicitly (Shullani et al. 2017; Hadwiger and Riess 2020).

4.9   Concluding Remarks

Camera-specific sensor noise fingerprints are a pillar of media forensics, and they are unrivaled when it comes to establishing source device attribution. While our focus in this chapter has been on still camera images (video data is covered in Chap. 5 of this book), virtually all imaging sensors introduce the same kind of noise fingerprints as we discussed here. For example, line sensors in flatbed scanners received early attention (Gloe et al. 2007; Khanna et al. 2007), and recent years have also seen biometric sensors move into the focus (Bartlow et al. 2009; Kauba et al. 2017; Ivanov and Baras 2017, 2019, i. a.).

Although keeping track of advances by device manufacturers and novel imaging pipelines is crucial for maintaining this status, an active research community has so far always been able to adapt to new challenges. The field has come a long way over the past 15 years, and new developments such as the cautious cross-over into the world of deep learning promise a continued potential for fruitful exploration. New perspectives and insights may also arise from applications outside the realm of media forensics. For example, with smartphones as ubiquitous companions in our everyday life, proposals to utilize camera fingerprints as building blocks to multi-factor authentication let users actively provide their device's sensor fingerprint via a captured image to be granted access to a web server (Valsesia et al. 2017; Quiring et al. 2019; Maier et al. 2020, i. a.). This poses a whole range of interesting practical challenges on its own, but it also invites a broader discussion about what camera fingerprints mean to an image-saturated world. At the minimum, concerns over image anonymity must be taken seriously in situations that call for it, and so we also see counter-forensics as part of the bigger picture unequivocally.

Acknowledgements Parts of this work were supported by AFRL and DARPA under Contract No. FA8750-16-C-0166. Any findings and conclusions or recommendations expressed in this material are solely the responsibility of the authors and do not necessarily represent the official views of AFRL, DARPA, or the U.S. Government.

References

Al-Ani M, Khelifi F (2017) On the SPN estimation in image forensics: a systematic empirical evaluation. IEEE Trans Inf Forensics Secur 12(5):1067–1081

Altinisik E, Sencar HT (2021) Source camera verification for strongly stabilized videos. IEEE Trans Inf Forensics Secur 16:643–657

Altinisik E, Tasdemir K, Sencar HT (2020) Mitigation of H.264 and H.265 video compression for reliable PRNU estimation. IEEE Trans Inf Forensics Secur 15:1557–1571

Amerini I, Caldelli R, Cappellini V, Picchioni F, Piva A (2009) Analysis of denoising filters for photo response non uniformity noise extraction in source camera identification. In: 16th international conference on digital signal processing, DSP 2009, Santorini, Greece, July 5-7, 2009. IEEE, pp 1–7

Amerini I, Caldelli R, Crescenzi P, del Mastio A, Marino A (2014) Blind image clustering based on the normalized cuts criterion for camera identification. Signal Process Image Commun 29(8):831–843

Amerini I, Caldelli R, Del Mastio A, Di Fuccia A, Molinari C, Rizzo AP (2017) Dealing with video source identification in social networks. Signal Process Image Commun 57:1–7

Artusi A, Richter T, Ebrahimi T, Mantiuk RK (2017) High dynamic range imaging technology [lecture notes]. IEEE Signal Process Mag 34(5):165–172

Avidan S, Shamir A (2007) Seam carving for content-aware image resizing. ACM Trans Graph 26(3):10

Baracchi D, Iuliani M, Nencini AG, Piva A (2020) Facing image source attribution on iphone X. In: Zhao X, Shi Y-Q, Piva A, Kim HJ (eds) Digital forensics and watermarking - 19th international workshop, IWDW 2020, Melbourne, VIC, Australia, November 25–27, 2020, Revised Selected Papers, vol 12617. Lecture notes in computer science. Springer, pp 196–207

Barni M, Nakano-Miyatake M, Santoyo-Garcia H, Tondi B (2018) Countering the pooled triangle test for prnu-based camera identification. In: 2018 IEEE international workshop on information forensics and security, WIFS 2018, Hong Kong, China, December 11-13, 2018. IEEE, pp 1–8

Barni M, Santoyo-Garcia H, Tondi B (2018) An improved statistic for the pooled triangle test against prnu-copy attack. IEEE Signal Process Lett 25(10):1435–1439

Bartlow N, Kalka ND, Cukic B, Ross A (2009) Identifying sensors from fingerprint images. In: IEEE conference on computer vision and pattern recognition, CVPR workshops 2009, Miami, FL, USA, 20–25 June 2009. IEEE Computer Society, pp 78–84

Bayram S, Sencar HT, Memon ND (2013) Seam-carving based anonymization against image & video source attribution. In: 15th IEEE international workshop on multimedia signal processing, MMSP 2013, Pula, Sardinia, Italy, September 30 - Oct. 2, 2013. IEEE, pp 272–277

Bayram S, Sencar HT, Memon ND (2012) Efficient sensor fingerprint matching through fingerprint binarization. IEEE Trans Inf Forensics Secur 7(4):1404–1413

Bayram S, Sencar HT, Memon ND (2015) Sensor fingerprint identification through composite fingerprints and group testing. IEEE Trans Inf Forensics Secur 10(3):597–612

Bloy Greg J (2008) Blind camera fingerprinting and image clustering. IEEE Trans Pattern Anal Mach Intell 30(3):532–534

Böhme R, Kirchner M (2013) Counter-forensics: attacking image forensics. In: Sencar HT, Memon N (eds) Digital image forensics: there is more to a picture than meets the eye. Springer, pp 327–366

Bonettini N, Bondi L, Mandelli S, Bestagini P, Tubaro S, Guera D (2018) Fooling PRNU-based detectors through convolutional neural networks. In: 26th European signal processing conference, EUSIPCO 2018, Roma, Italy, September 3-7, 2018. IEEE, pp 957–961

Caldelli R, Amerini I, Novi A (2011) An analysis on attacker actions in fingerprint-copy attack in source camera identification. In: 2011 ieee international workshop on information forensics and security, WIFS 2011, Iguacu Falls, Brazil, November 29 - December 2, 2011. IEEE Computer Society, pp 1–6

Chakraborty S (2020) A CNN-based correlation predictor for prnu-based image manipulation localization. In: Alattar AM, Memon ND, Sharma G (eds) Media watermarking, security, and forensics 2020, Burlingame, CA, USA, 27-29 January 2020

Chakraborty S, Kirchner M (2017) PRNU-based image manipulation localization with discriminative random fields. In: Alattar AM, Memon ND (eds) Media Watermarking, Security, and Forensics 2017, Burlingame, CA, USA, 29 January 2017 - 2 February 2017, pages 113–120. Ingenta, 2017

Chen M, Fridrich JJ, Goljan M, Lukáš J (2008) Determining image origin and integrity using sensor noise. IEEE Trans Inf Forensics Secur 3(1):74–90

Chierchia G, Cozzolino D, Poggi G, Sansone C, Verdoliva L (2014) Guided filtering for PRNU-based localization of small-size image forgeries. In: IEEE international conference on acoustics, speech and signal processing, ICASSP 2014, Florence, Italy, May 4-9, 2014. IEEE, pp 6231–6235

Chierchia G, Parrilli S, Poggi G, Verdoliva L, Sansone C (2011) PRNU-based detection of small-size image forgeries. In: 17th international conference on digital signal processing, DSP 2011, Corfu, Greece, July 6-8, 2011. IEEE, pp 1–6

Chierchia G, Poggi G, Sansone C, Verdoliva L (2014) A Bayesian-MRF approach for PRNU-based image forgery detection. IEEE Trans Inf Forensics Secur 9(4):554–567

Chuang W-H, Su H, Wu M (2011) Exploring compression effects for improved source camera identification using strongly compressed video. In: Macq B, Schelkens P (eds) 18th IEEE international conference on image processing, ICIP 2011, Brussels, Belgium, September 11-14, 2011. IEEE, pp 1953–1956

Connelly B, Eli S, Adam F, Goldman Dan B (2009) Patchmatch: a randomized correspondence algorithm for structural image editing. ACM Trans Graph 28(3):24

Cortiana A, Conotter V, Boato G, De Natale FGB (2011) Performance comparison of denoising filters for source camera identification. In: Memon ND, Dittmann J, Alattar AM, Delp EJ III (eds) Media forensics and security III, San Francisco Airport, CA, USA, January 24-26, 2011, Proceedings, SPIE Proceedings, vol 7880. SPIE, p 788007

Cozzolino D, Marra F, Gragnaniello D, Poggi G, Verdoliva L (2020) Combining PRNU and noiseprint for robust and efficient device source identification. EURASIP J Inf Secur 2020:1

Dang-Nguyen D-T, Pasquini C, Conotter V, Boato G (2015) RAISE: a raw images dataset for digital image forensics. In: Ooi WT, Feng W-c, Liu F (eds), Proceedings of the 6th ACM multimedia systems conference, MMSys 2015, Portland, OR, USA, March 18-20, 2015. ACM, pp 219–224

Dirik AE, Sencar HT, Memon ND (2014) Analysis of seam-carving-based anonymization of images against PRNU noise pattern-based source attribution. IEEE Trans Inf Forensics Secur 9(12):2277–2290

Entrieri J, Kirchner M (2016) Patch-based desynchronization of digital camera sensor fingerprints. In: Alattar AM, Memon ND (eds) Media watermarking, security, and forensics 2016, San Francisco, California, USA, February 14-18, 2016. Ingenta, pp 1–9

Fridrich J (2013) Sensor defects in digital image forensic. In: Sencar HT, Memon N (eds) Digital Image forensics: there is more to a picture than meets the eye. Springer, Berlin, pp 179–218

Galdi C, Hartung F, Dugelay J-L (2019) SOCRatES: A database of realistic data for source camera recognition on smartphones. In: De Marsico M, di Baja GS, Fred ALN (eds) Proceedings of the 8th international conference on pattern recognition applications and methods, ICPRAM 2019, Prague, Czech Republic, February 19-21, 2019. SciTePress, pp 648–655

Gloe T, Böhme R (2010) The Dresden image database for benchmarking digital image forensics. J Digit Forensic Pract 3(2–4):150–159

Gloe T, Franz E, Winkler A (2007) Forensics for flatbed scanners. In: Delp EJ III, Wong PW (eds.) Security, steganography, and watermarking of multimedia contents IX, San Jose, CA, USA, January 28, 2007, SPIE Proceedings, vol 6505. SPIE, p. 65051I

Gloe T, Kirchner M, Winkler A, Böhme R (2007) Can we trust digital image forensics? In: Lienhart R, Prasad AR, Hanjalic A, Choi S, Bailey BP, Sebe N (eds) Proceedings of the 15th international conference on multimedia 2007, Augsburg, Germany, September 24-29, 2007. ACM, pp 78–86

Gloe T, Pfennig S, Kirchner M (2012) Unexpected artefacts in PRNU-based camera identification: a 'Dresden image database' case-study. In: Li C-T, Dittmann J, Katzenbeisser S, Craver S (eds) Multimedia and security workshop, MM&Sec 2012, Coventry, United Kingdom, September 6-7, 2012. ACM, pp 109–114

Goljan M, Chen M, Comesaña P, Fridrich JJ (2016) Effect of compression on sensor-fingerprint based camera identification. In: Alattar AM, Memon ND (eds) Media watermarking, security, and forensics 2016, San Francisco, California, USA, February 14-18, 2016. Ingenta, pp 1–10

Goljan M, Chen M, Fridrich JJ (2007) Identifying common source digital camera from image pairs. In: Proceedings of the international conference on image processing, ICIP 2007, September 16-19, 2007, San Antonio, Texas, USA. IEEE, pp 125–128

Goljan M, Fridrich JJ, Filler T (2009) Large scale test of sensor fingerprint camera identification. In: Delp EJ, Dittmann J, Memon ND, Wong PW (eds) Media forensics and security I, part of the IS&T-SPIE electronic imaging symposium, San Jose, CA, USA, January 19-21, 2009, Proceedings, SPIE Proceedings, vol 7254. SPIE, p. 72540I

Goljan M, Fridrich JJ, Filler T (2010) Managing a large database of camera fingerprints. In: Memon ND, Dittmann J, Alattar AM, Delp EJ (eds) Media forensics and security II, part of the IS&T-SPIE electronic imaging symposium, San Jose, CA, USA, January 18-20, 2010, Proceedings, SPIE Proceedings, vol 7541. SPIE, pp 754108

Goljan M, Fridrich JJ (2008) Camera identification from cropped and scaled images. In: Delp EJ III, Wong PW, Dittmann J, Memon ND (eds) Security, forensics, steganography, and watermarking of multimedia contents X, San Jose, CA, USA, January 27, 2008, SPIE Proceedings, vol 6819. SPIE, pp 68190E

Goljan M, Fridrich JJ (2012) Sensor-fingerprint based identification of images corrected for lens distortion. In: Memon ND, Alattar AM, Delp EJ III (eds) Media watermarking, security, and forensics 2012, Burlingame, CA, USA, January 22, 2012, Proceedings, SPIE Proceedings, vol 8303. SPIE, p 83030H

Goljan M, Fridrich JJ (2014) Estimation of lens distortion correction from single images. In: Alattar AM, Memon ND, Heitzenrater C (eds) Media watermarking, security, and forensics 2014, San Francisco, CA, USA, February 2, 2014, Proceedings, SPIE Proceedings, vol 9028. SPIE, p 90280N

Goljan M, Fridrich JJ, Chen M (2011) Defending against fingerprint-copy attack in sensor-based camera identification. IEEE Trans Inf Forensics Secur 6(1):227–236

Hadwiger B, Riess C (2021) The Forchheim image database for camera identification in the wild. In: Del Bimbo A, Cucchiara R, Sclaroff S, Farinella GM, Mei T, Bertini M, Escalante HJ, Vezzani R (eds) Pattern recognition. ICPR international workshops and challenges - virtual event, January 10-15, 2021, Proceedings, Part VI, Lecture Notes in Computer Science, vol 12666. Springer, pp 500–515

Hosler BC, Zhao X, Mayer O, Chen C, Shackleford JA, Stamm MC (2019) The video authentication and camera identification database: a new database for video forensics. IEEE Access 7:76937–76948

Hosseini MDM, Goljan M (2019) Camera identification from HDR images. In: Cogranne R, Verdoliva L, Lyu S, Troncoso-Pastoriza JR, Zhang X (eds) Proceedings of the ACM workshop on information hiding and multimedia security, IH&MMSec 2019, Paris, France, July 3-5, 2019. ACM, pp 69–76

Iuliani M, Fontani M, Piva A (2019) Hybrid reference-based video source identification. Sensors 19(3):649

Iuliani M, Fontani M, Piva A (2021) A leak in PRNU based source identification - questioning fingerprint uniqueness. IEEE Access 9:52455–52463

Ivanov VI, Baras JS (2017) Authentication of swipe fingerprint scanners. IEEE Trans Inf Forensics Secur 12(9):2212–2226

Ivanov VI, Baras JS (2019) Authentication of area fingerprint scanners. Pattern Recognit 94:230–249

Joshi S, Korus P, Khanna N, Memon ND (2020) Empirical evaluation of PRNU fingerprint variation for mismatched imaging pipelines. In: 12th IEEE international workshop on information forensics and security, WIFS 2020, New York City, NY, USA, December 6-11, 2020. IEEE, pp 1–6

Kang X, Li Y, Qu Z, Huang J (2012) Enhancing source camera identification performance with a camera reference phase sensor pattern noise. IEEE Trans Inf Forensics Secur 7(2):393–402

Karaküçük A, Dirik AE (2015) Adaptive photo-response non-uniformity noise removal against image source attribution. Digit Investigat 12:66–76

Karaküçük A, Dirik AE (2019) PRNU based source camera attribution for image sets anonymized with patch-match algorithm. Digit Investig 30:43–51

Kauba C, Debiasi L, Uhl A (2017) Identifying the origin of iris images based on fusion of local image descriptors and PRNU based techniques. In: 2017 IEEE international joint conference on biometrics, IJCB 2017, Denver, CO, USA, October 1-4, 2017. IEEE, pp 294–301

Khanna N, Mikkilineni AK, Chiu GT-C, Allebach JP, Delp EJ (2007) Scanner identification using sensor pattern noise. In: Delp EJ III, Wong PW (eds) Security, steganography, and watermarking of multimedia contents IX, San Jose, CA, USA, January 28, 2007. SPIE Proceedings, vol 6505, p 65051K. SPIE

Kirchner M, Böhme R (2009) Synthesis of color filter array pattern in digital images. In: Delp EJ, Dittmann J, Memon ND, Wong PW (eds) Media forensics and security I, part of the IS&T-SPIE electronic imaging symposium, San Jose, CA, USA, January 19-21, 2009, Proceedings, SPIE Proceedings, vol 7254. SPIE, p 72540K

Kirchner M, Johnson C (2019) SPN-CNN: boosting sensor-based source camera attribution with deep learning. In: IEEE international workshop on information forensics and security, WIFS 2019, Delft, The Netherlands, December 9-12, 2019. IEEE, pp 1–6

Knight S, Moschou S, Sorell M (2009) Analysis of sensor photo response non-uniformity in RAW images. In: Sorell M (ed) Forensics in telecommunications, information and multimedia, second international conference, e-forensics 2009, Adelaide, Australia, January 19–21, 2009, Revised Selected Papers, vol 8. Lecture Notes of the Institute for Computer Sciences. Springer, Social Informatics and Telecommunications Engineering, pp 130–141

Korus P, Huang J (2017) Multi-scale analysis strategies in prnu-based tampering localization. IEEE Trans Inf Forensics Secur 12(4):809–824

Li C-T (2010) Source camera identification using enhanced sensor pattern noise. IEEE Trans Inf Forensics Secur 5(2):280–287

Li C-T (2010) Unsupervised classification of digital images using enhanced sensor pattern noise. International symposium on circuits and systems (ISCAS 2010), May 30 - June 2, 2010. France. IEEE, Paris, pp 3429–3432

Lin X, Li C-T (2016) Preprocessing reference sensor pattern noise via spectrum equalization. IEEE Trans Inf Forensics Secur 11(1):126–140

Lin X, Li C-T (2020) PRNU-based content forgery localization augmented with image segmentation. IEEE Access 8:222645–222659

Lukás J, Fridrich JJ, Goljan M (2005) Determining digital image origin using sensor imperfections. In: Said A, Apostolopoulos JG (eds) Electronic Imaging: Image and Video Communications and Processing 2005, San Jose, California, USA, 16–20 January 2005, SPIE Proceedings, vol 5685. SPIE

Lukáš J, Fridrich JJ, Goljan M (2006) Digital camera identification from sensor pattern noise. IEEE Trans Inf Forensics Secur 1(2):205–214

Maier D, Erb H, Mullan P, Haupert V (2020) Camera fingerprinting authentication revisited. In: 23rd international symposium on research in attacks, intrusions and defenses ({RAID} 2020), pp 31–46

Mandelli S, Bestagini P, Verdoliva L, Tubaro S (2020) Facing device attribution problem for stabilized video sequences. IEEE Trans Inf Forensics Secur 15:14–27

Mandelli S, Bondi L, Lameri S, Lipari V, Bestagini P, Tubaro S (2017) Inpainting-based camera anonymization. In: 2017 IEEE international conference on image processing, ICIP 2017, Beijing, China, September 17-20, 2017. IEEE, pp 1522–1526

Mandelli S, Cozzolino D, Bestagini P, Verdoliva L, Tubaro S (2020) CNN-based fast source device identification. IEEE Signal Process Lett 27:1285–1289

Marra F, Poggi G, Sansone C, Verdoliva L (2017) Blind PRNU-based image clustering for source identification. IEEE Trans Inf Forensics Secur 12(9):2197–2211

Marra F, Roli F, Cozzolino D, Sansone C, Verdoliva L (2014) Attacking the triangle test in sensor-based camera identification. In: 2014 IEEE international conference on image processing, ICIP 2014, Paris, France, October 27-30, 2014. IEEE, pp 5307–5311

Masciopinto M, Pérez-González F (2018) Putting the PRNU model in reverse gear: findings with synthetic signals. In: 26th European signal processing conference, EUSIPCO 2018, Roma, Italy, September 3-7, 2018. IEEE, pp 1352–1356

Meij C, Geradts Z (2018) Source camera identification using photo response non-uniformity on WhatsApp. Digit Investig 24:142–154

Nagaraja S, Schaffer P, Aouada D (2011) Who clicks there!: anonymising the photographer in a camera saturated society. In: Chen Y, Vaidya J (eds) Proceedings of the 10th annual ACM workshop on Privacy in the electronic society, WPES 2011, Chicago, IL, USA, October 17, 2011. ACM, pp 13–22

Quan Y, Li C-T, Zhou Y, Li L (2020) Warwick image forensics dataset for device fingerprinting in multimedia forensics. In: IEEE international conference on multimedia and expo, ICME 2020, London, UK, July 6-10, 2020. IEEE, pp 1–6

Quan Y, Li C-T (2021) On addressing the impact of ISO speed upon PRNU and forgery detection. IEEE Trans Inf Forensics Secur 16:190–202

Quiring E, Kirchner M (2015) Fragile sensor fingerprint camera identification. In: 2015 IEEE international workshop on information forensics and security, WIFS 2015, Roma, Italy, November 16-19, 2015. IEEE, pp 1–6

Quiring E, Kirchner M, Rieck K (2019) On the security and applicability of fragile camera fingerprints. In: Sako K, Schneider SA, Ryan PYA (eds) Computer security - ESORICS 2019–24th European symposium on research in computer security, Luxembourg, September 23–27, 2019, Proceedings, Part I, vol 11735. Lecture notes in computer science. Springer, pp 450–470

Rao Q, Li H, Luo W, Huang J (2013) Anti-forensics of the triangle test by random fingerprint-copy attack. In: Computational visual media conference

Rosenfeld K, Sencar HT (2009) A study of the robustness of PRNU-based camera identification. In: Delp EJ, Dittmann J, Memon ND, Wong PW (eds), Media forensics and security I, part of the IS&T-SPIE electronic imaging symposium, San Jose, CA, USA, January 19-21, 2009, Proceedings, SPIE Proceedings, vol 7254. SPIE, p 72540M

Shaya OA, Yang P, Ni R, Zhao Y, Piva A (2018) A new dataset for source identification of high dynamic range images. Sensors 18(11):3801

Shullani D, Fontani M, Iuliani M, Shaya OA, Piva A (2017) VISION: a video and image dataset for source identification. EURASIP J Inf Secur 2017:15

Stamm MC, Liu KJR (2011) Anti-forensics of digital image compression. IEEE Trans Inf Forensics Secur 6(3–2):1050–1065

Tandogan SE, Altinisik E, Sarimurat S, Sencar HT (2019) Tackling in-camera downsizing for reliable camera ID verification. In: Alattar AM, Memon ND, Sharma G (eds) Media watermarking, security, and forensics 2019, Burlingame, CA, USA, 13–17 January 2019. Ingenta

Taspinar S, Mohanty M, Memon ND (2016) Source camera attribution using stabilized video. In: IEEE international workshop on information forensics and security, WIFS 2016, Abu Dhabi, United Arab Emirates, December 4-7, 2016. IEEE, pp 1–6

Taspinar S, Mohanty M, Memon ND (2017) PRNU-based camera attribution from multiple seam-carved images. IEEE Trans Inf Forensics Secur 12(12):3065–3080

Taspinar S, Mohanty M, Memon ND (2020) Camera fingerprint extraction via spatial domain averaged frames. IEEE Trans Inf Forensics Secur 15:3270–3282

Taspinar S, Mohanty M, Memon ND (2020) Camera identification of multi-format devices. Pattern Recognit Lett 140:288–294

Tian H, Xiao Y, Cao G, Zhang Y, Xu Z, Zhao Y (2019) Daxing smartphone identification dataset. IEEE. Access 7:101046–101053

Valsesia D, Coluccia G, Bianchi T, Magli E (2015) Compressed fingerprint matching and camera identification via random projections. IEEE Trans Inf Forensics Secur 10(7):1472–1485

Valsesia D, Coluccia G, Bianchi T, Magli E (2015) Large-scale image retrieval based on compressed camera identification. IEEE Trans Multim 17(9):1439–1449

Valsesia D, Coluccia G, Bianchi T, Magli E (2017) User authentication via PRNU-based physical unclonable functions. IEEE Trans Inf Forensics Secur 12(8):1941–1956

van Houten W, Geradts ZJMH (2009) Source video camera identification for multiply compressed videos originating from youtube. Digit Investig 6(1–2):48–60

Zeng H, Chen J, Kang X, Zeng W (2015) Removing camera fingerprint to disguise photograph source. In: 2015 IEEE international conference on image processing, ICIP 2015, Quebec City, QC, Canada, September 27-30, 2015. IEEE, pp 1687–1691

Zhang W, Tang X, Yang Z, Niu S (2019) Multi-scale segmentation strategies in PRNU-based image tampering localization. Multim Tools Appl 78(14):20113–20132

Zhang K, Zuo W, Chen Y, Meng D, Zhang L (2017) Beyond a Gaussian denoiser: residual learning of deep CNN for image denoising. IEEE Trans Image Process 26(7):3142–3155

Chapter 5
Source Camera Attribution from Videos

Husrev Taha Sencar

Photo-response non-uniformity (PRNU) is an intrinsic characteristic of a digital imaging sensor, which manifests as a unique and permanent pattern introduced to all media captured by the sensor. The PRNU of a sensor has been proven to be a viable identifier for source attribution and has been successfully utilized for identification and verification of the source of digital media. In the past decade, various approaches have been proposed for the reliable estimation, compact representation, and faster matching of PRNU patterns. These studies, however, mainly featured photographic images, and the extension of source camera attribution capabilities to the video domain remains a problem. This is primarily because the steps involved in the generation of a video are much more disruptive to the PRNU pattern, and therefore, its estimation from videos involves several additional challenges. This chapter examines these challenges and outlines recently developed methods for estimating the sensor's PRNU from videos.

5.1 Introduction

The PRNU is caused by variations in size and material properties of the photosensitive elements that comprise a sensor. These essentially affect the response of each picture element under the same amount of illumination. Therefore, the process of identification boils down to quantifying the sensitivity of each picture element. This is realized through an estimation procedure using a set of pictures acquired by the sensor (Chen et al. 2008). To determine whether a media is captured by a given sensor,

H. T. Sencar (✉)
Qatar Computing Research Institute, HBKU, Ar-Rayyan, Qatar
e-mail: hsencar@hbku.edu.qa

© The Author(s) 2022
H. T. Sencar et al. (eds.), *Multimedia Forensics*, Advances in Computer Vision and Pattern Recognition, https://doi.org/10.1007/978-981-16-7621-5_5

the estimated sensitivity profile, i.e., the PRNU pattern, from the media in question is compared to a reference PRNU pattern obtained in advance using a correlation-based measure, most typically using the peak-to-correlation energy (PCE) (Vijaya Kumar 1990).[1] In essence, the reliability and accuracy of a source attribution decision strongly depend on the fact that the PRNU-bearing raw signal at the sensor output has to pass through several steps of in-camera processing before the media is generated. These processing steps have a disruptive effect on the inherent PRNU pattern.

The imaging sub-systems used by digital cameras largely remain proprietary to the manufacturer; therefore, it is quite difficult to access their details. At a higher level, however, the imaging pipeline in a camera includes various stages, such as acquisition, pre-processing, color-processing, post-processing, and image and video coding. Figure 5.1 shows the basic processing steps involved in capturing a video, most of which are also utilized during the acquisition of a photograph. When generating a video, an indispensable processing step is the downsizing of the full-frame sensor output to reduce the amount of data that needs processing. This may be realized at acquisition by sub-sampling pixel readout data (trough binning Zhang et al. 2018 or skipping Guo et al. 2018 pixels), as well as during color-processing by downsampling color interpolated image data. Another key processing step employed by modern day cameras at the post-processing stage is image stabilization. Since many videos are not captured while the camera is in a fixed position, all modern cameras perform image stabilization to correct perspective distortions caused by handheld shooting or other vibrations. Stabilization can be followed by other post-processing steps, such as noise reduction and dynamic range adjustment, which may also include transformations, resulting in further weakening of the PRNU pattern. Finally, the sequence of post-processed pictures is encoded into a standard video format for effective storage and transfer.

Video generation involves at least three additional steps compared to the generation of photos, including downsizing, stabilization, and video coding. When combined together, these operations have a significant adverse impact on PRNU estimation in two main respects. The first relates to geometric transformations applied during downsizing and stabilization, and the second concerns information loss largely caused by resolution reduction and compression. These detrimental effects could be further compounded by out-of-camera processing. Video editing tools today provide a wide variety of sophisticated processing options over cameras as they are less bounded by computational resources and real-time constraints. Hence, the estimation of sensor PRNU from videos requires a thorough understanding of the implications of these operations and the development of effective mitigations against them.

---

[1] Details on the estimation of the PRNU pattern from photographic images can be found in the previous two chapters of this book.

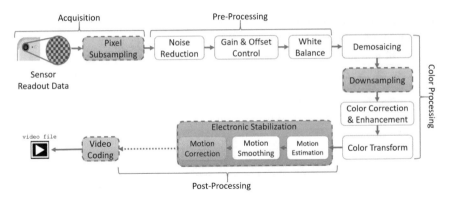

Fig. 5.1 The processing steps involved in the imaging pipeline of a camera when generating a video. The highlighted boxes are specific to video capture

5.2   Challenges in Attributing Videos

The now well-established PRNU estimation approach was devised considering in-camera processing relevant to the generation of photographic images. There are, however, several obstacles that prevent its direct application to the sequence of pictures that comprise a video. These difficulties can be summarized as follows.

Video frames have lower resolution than the full-sensor resolution typically used for acquiring photos. Therefore, a reference PRNU pattern estimated from photos provides a more comprehensive characteristic of the sensor. Hence, the use of a photo-based reference pattern for attribution requires the mismatch with the dimensions of video frames to be taken into account. Alternatively, one might consider creating a separate reference PRNU pattern at all frame sizes. This, however, may not always be possible as videos can be captured at many resolutions (Tandogan et al. 2019). Moreover, in the case of smartphones, this may vary from one camera app to another based on the preference of developers among a set of supported resolutions. Overall, the downsizing operation in a camera involves various proprietary hardware and software mechanisms that involve both cropping and resizing operations. Therefore, source attribution from videos requires the determination of those device-dependent downsizing parameters. When these parameters are not known a priori, PRNU estimation also has to incorporate the search for these parameters.

The other challenge concerns the fact that observed PCE values in matching PRNU patterns are much lower in videos compared to photos. This decrease in PCE values is primarily caused by the downsizing operation and video compression. More critically, low PCE values are encountered even when the PRNU pattern is estimated from multiple frames of the video whose source is in question, i.e., reference-to-reference matching. When performing reference-to-reference matching, downsizing can be ignored as a factor as long as the resizing factor is higher than $\frac{1}{6}$ (Tandogan et al. 2019), and the compression is the main concern. In this setting, at medium-to-low compression levels (2 Mbps to 600 Kbps coding bit rates), the average PCE value drops significantly as compression becomes more severe (where PCE values drop from around 2000 to 40) (Altinisik et al. 2020). Alternatively, when PRNU patterns from video frames are individually matched with the reference pattern (i.e., frame-to-reference matching), even downsizing by a factor of two causes a significant reduction in measured PCE values (Bondi et al. 2019). Frame-to-reference matching at HD-sized frames has been found to yield PCE values mostly around 20 (Altinisik and Sencar 2021).

Correspondingly, a further issue concerns the difficulty of setting a decision threshold for matching. Large-scale tests performed on photographic images show that setting the PCE value to 60 as a threshold yields extremely low false matches when the correct-match rate is quite high. In contrast, as demonstrated in the results of earlier works, where decision thresholds of 40–100 (Iuliani et al. 2019) and 60 (Mandelli et al. 2020) are utilized when performing frame-to-reference matching, such threshold values on video frames yield much lower attribution rates.

Another major challenge to video source attribution is posed by image stabilization. When performed electronically, stabilization requires determining and counteracting undesired camera motion. This involves the application of frame or sub-frame level geometric transformations to align the content in successive pictures of a video. From the PRNU estimation standpoint, this requires registration of PRNU patterns by inverting stabilization transformations in a blind manner. This both increases the computational complexity of PRNU estimation and reduces accuracy due to an increase in false-positive matches generated during the search for unknown transformation parameters.

Although in-camera processing steps introduce artifacts that obstruct correct attribution, they can also be utilized to perform more reliable matching. For example, biases introduced to the PRNU estimate by the demosaicing operation and blockiness caused by compression are known to introduce periodic structures onto the estimated PRNU pattern. These artifacts can essentially be treated as pilot signals to detect clues about the processing history of a media after the acquisition. In the case of photos, the linear pattern associated with the demosaicing operation has been shown to be effective in determining the amount of shift, rotation, and translation, with weaker presence in newer cameras (Goljan 2018). In the case of videos, however, the linear pattern is observed to be even weaker, most likely due to the application of in-camera downsizing and more aggressive compression of video frames compared to photos. Therefore, it cannot be reliably utilized in identifying global or local transformations. In a similar manner, since video coding uses variable block sizes determined adaptively during encoding, as opposed to a fixed block size used by still image coding, blockiness artifact is also not useful in reducing the computational complexity of determining the subsequent processing.

Obviously, the most important factor that underlies these challenges is the inability to access details of the in-camera processing steps due to strong competition between manufacturers. In the lack of such knowledge and with continued advances in camera technology, research in this area has to treat in-camera processing largely as a black-box and propose PRNU estimation methods that can cope with this ambiguity. As a result, the estimation of the PRNU from videos is likely to incur higher computational complexity and yield lower attribution accuracy.

5.3 Attribution of Downsized Media

Cameras capture images and videos at a variety of resolutions. This is typically performed to accommodate different viewing options and support different display and print qualities. However, downsizing becomes necessary when taking a burst of shots or capturing a video in order to reduce the amount of data that needs processing. Furthermore, to perform complex operations like image stabilization, typically, the sensor is cropped and only the center portion is converted into an image.

The downsizing operation can be realized through different mechanisms. In the most common case, cameras capture data at full-sensor resolution which is converted into an image and then resampled in software to the target resolution. An alternative or precursor to sub-sampling is on-sensor binning or averaging which effectively obtains larger pixels. This may be performed either at the hardware level during the acquisition by electrically connecting neighboring pixels or by averaging pixel values digitally immediately after analog-to-digital conversion. Therefore, as a result of performing resizing, pixel binning, and sensor cropping, photos and videos captured at different resolutions are likely to yield non-matching PRNUs.

The algorithm for how a camera performs downsizing of full-frame sensor output when capturing a photo or video is implemented as part of the in-camera processing pipeline and is proprietary to every camera maker. Therefore, the viable approach to understand downsizing behavior is to acquire media at different resolutions and compare the resulting PRNUs to determine the best method for matching PRNU patterns at different resolutions.

Given that smartphones and tablets have become the defacto camera today, an important concern is whether the use of the default camera interface, i.e., the native camera app, to capture photos and videos only exposes certain default settings and does not reveal all possible options that may potentially be selected by other camera apps that utilize the built-in camera hardware in a device.

To address this limitation, Tandogan et al. (2019) used a custom camera app to discover the features of 21 Android-based smartphone cameras. This approach also allowed direct control over various camera settings, such as the application of electronic zoom, deployment of video stabilization, compression, and use of high dynamic range (HDR) imaging, which may interfere with the sought after downsizing behavior. Most notably, it was determined that cameras can capture photos at 15–30 resolutions and videos at 10–20 resolutions, with newer cameras offering more options. A great majority of these resolutions were in 4:3 and 16:9 aspect ratios. To determine the active sensor pixels used during video acquisition, videos were captured at all supported frame resolutions and a photo was taken while recording continued. This setting led to the finding that each photo is taken at the highest resolution of the same aspect ratio, which possibly indicates that frames are downscaled by a fixed factor before being encoded into a video.

The maximum resolution of a video is typically much smaller than the full-sensor size. Therefore, a larger amount of cropping and scaling has to be performed on video frames. Hence, the two most important questions that relate to in-camera downsizing are about its impact on the reliability of PRNU matching and how matching should be performed when two PRNU patterns are of different sizes.

5.3.1   The Effect of In-Camera Downsizing on PRNU

Investigating the effects of downsizing on camera ID verification, (Tandogan et al. 2019) determined that an important consideration is knowledge of the specifics of the downsizing operation. When downscaling is performed using the well-known bilinear, bicubic, and Lanczos interpolation filters, the resulting PCE values for all resizing factors higher than $\frac{1}{12}$ (which reduces the number of pixels in the original image by a factor of 12^2) is found to yield PCE values higher than the commonly used decision threshold of 60. However, when there is a mismatch between the method used for downscaling the reference PRNU pattern and the downsizing method deployed by a camera, the matching performance dropped very fast. In this setting, even at the scaling factor of $\frac{1}{4}$, estimated PRNUs could not be matched. Hence, it can be deduced that when the underlying method is known, downsizing does not pose a significant obstacle to source attribution despite significant data loss. When it is not known, however, in-camera downsizing is disruptive to successful attribution even at relatively low downsizing factors.

5.3.2   Media with Mismatching Resolutions

The ability to reliably match PRNU patterns extracted from media at different resolutions depends on knowledge of how downsizing is performed. In essence, this reduces to determining the amount of cropping and scaling applied to media at lower resolution. Although implementation details of these operations may remain unknown, the area and position of cropping and an approximate scaling factor can be identified through a search. An important concern is whether the search for downsizing parameters should be performed in the spatial domain or the PRNU domain.

Since downsizing is performed in the spatial domain, the search has to be ideally performed in the spatial domain where identified parameters are validated based on the match of the estimated PRNU with the reference PRNU. In this search setting, each trial for the parameters in the spatial domain has to be followed by PRNU estimation. As downsizing parameters are determined through a brute-force search and the search space for the parameters can be large, this may be computationally expensive. This is indeed a greater concern when it comes to inverting stabilization transformations because the search has to be performed for each frame individually rather than once for the camera model. Hence, an alternative approach is to perform the search directly using the PRNU patterns, removing the need for repetitive PRNU estimation. However, since the PRNU estimation process is not linear in nature, this change in the search domain is likely to introduce a degradation in performance. To determine this, Altinisik et al. (2021) applied random transformations to a set of video frames, estimated PRNU patterns, inverted the transformation, and evaluated the match with the corresponding reference patterns. The resulting PCE values when compared to performing transformation and inversion in the spatial domain (to also take into account the effect of transformation-related interpolation) showed that the search of parameters in the PRNU domain yielded mostly acceptable results.

Source attribution under geometric transformations has been previously studied. Considering scaled and cropped photographic images, Goljan et al. (2008) demonstrated that through a search of transformation parameters, the alignment between PRNU patterns can be re-established. For this, the PRNU pattern obtained from the image in question is upsampled in discrete steps and matched with the reference PRNU at all shifts. The parameter values that yield the highest PCE are identified as the correct scaling factor and the cropping position. In a similar manner, a PRNU pattern extracted from a higher resolution video can be compared with PRNU patterns extracted from low-resolution videos. With this goal, Refs. Tandogan et al. (2019), Iuliani et al. (2019), and Taspinar et al. (2020) examined in-camera downsizing behavior to enable reliable matching of PRNU patterns at different resolutions. Essentially to determine how scaling and cropping are performed, the high-resolution PRNU pattern is downscaled and a search is performed to determine cropping boundaries by computing the normalized cross-correlation (NCC) between the two patterns. This process is repeated by downsizing the high-resolution pattern in decrements and performing the search until the alignment is established. Alternatively, Bellavia et al. (2021) proposed a content alignment approach. That is, instead of aligning PRNU patterns extracted from photos and pictures of a video captured independently, they used pictures and videos of a static scene and estimated the scaling and cropping parameters through keypoint registration.

The presence of electronic stabilization poses two complications to this process. First, since stabilization transformations vary from frame to frame, the above search may at best (i.e., only when stabilization transformation is in the form of a frame level affine transformation) identify the combined effect, and downsizing-related cropping and scaling can only be approximately determined. Second, the search has to be performed at the frame level (frame-to-reference matching) where PRNU strength will be significantly weaker as opposed to using a reference pattern obtained from multiple frames (reference-to-reference matching). Therefore, the process will be more error prone. Ultimately, since downsizing-related parameters (i.e., scaling ratio and cropping boundaries) vary at the camera model level, the above approach can be used to generate a dictionary of parameters that specify how in-camera downsizing is performed using videos acquired under controlled conditions to prevent interference from stabilization.

5.4   Mitigation of Video Coding Artifacts

As a raw video is a sequence of images, its size is impractically large to store or transfer. Video compression exploits the fact that frames in a video sequence are highly correlated in time and aims to reduce the spatial and temporal redundancy so that as few bits as possible are used to represent the video sequence. Consequently, compression-related information loss causes much more severe artifacts in videos compared to those in photos. Currently, the most prevalent and popular video compression standards are H.264/AVC and its recent descendent H.265/HEVC.

The ability to cope with video coding artifacts crucially requires addressing two problems. The first relates to the disruptive effects of a filtering operation, i.e., the loop filtering, deployed by video codecs to suppress artifacts arising due to block-based operation of video coding standards. Proposed solutions to cope with this have either tried to compensate for such effects in post-processing or to eliminate its effects by intervening in the decoding process. The second problem concerns compression-related information loss. Several approaches have been proposed to combat the adverse effects of compression by incorporating frame and block level encoding information to the PRNU estimation.

5.4.1   Video Coding from Attribution Perspective

The widely used PRNU estimation method tailored for photographic images in fact takes into account compression artifacts. To combat this, the Fourier domain representation of the PRNU pattern is Wiener filtered to remove all structural patterns such as those caused by blockiness. However, video coding has several important differences from still image coding (Richardson 2011; Sze et al. 2014), and the PRNU estimation approach for videos should take these differences into account.

Block-based operation: Both H.264 and H.265 video compression standards operate by dividing a picture into smaller blocks. Each coding standard has different naming and size limits for those blocks. In H.264, the largest blocks are called macroblocks and they are 16×16 pixels in size. A macroblock can be further divided into smaller sub-macroblock regions that are as small as 4×4 pixels. H.265 format is an improved version of H.264, and it is designed with parallel processing in mind. The sizes of container blocks in H.265 can vary from 64×64 to 16×16 pixels. These container blocks can be recursively divided into smaller blocks (down to the size of 8×8 pixels) which may be further divided into sub-blocks of 4×4 pixels during prediction. For the sake of brevity, we refer to all block types as macroblocks even though the term macroblock no longer exists in the H.265 standard.

Frame types: A frame in an H.264 or H.265 video can have three types: I, P, and B. An I frame is self-contained and temporally independent. It can be likened to a JPEG image in this sense. However, they are quite different in many aspects including variable block sizes, use of prediction, integer transformation derived from a form of discrete sine transform (DST), variable quantization, and deployment of a loop filter. The macroblocks in P frame can use previous I or P frame macroblocks to find the best prediction. Encoding of a B frame provides the same flexibility. Moreover, future I and P frames can be utilized when predicting B frame macroblocks. A video is composed of a sequence of frames that may exhibit a fixed or varying pattern of I, P, and B frames.

Block types: Similar to frames, macroblocks can be categorized into three types, namely, I, P, and B. An I macroblock is intra-coded with no dependence on future or previous frames, and I frames only contain this type of block. A P macroblock can be predicted from a previous P or I frame, and P frames can contain both types of blocks. Finally, a B macroblock can be predicted using one or two frames of I or P types. A B frame can contain all three types of blocks. If a block happens to have the same motion vectors as its preceding block and few or no residuals, this block is skipped since it can be constructed at the decoder side using the preceding block information. This type of block has the inherent properties of its preceding blocks and they are called skipped blocks. Skipped blocks might appear in P and B frames only.

Rate-Distortion optimization: During encoding, each picture block is predicted either from the neighboring blocks of the same frame (intra-prediction) or from the blocks of the past or future frames (inter-prediction). Therefore, each coded block contains the position information of the reference picture block (motion vector) and the difference between the current block and the reference block (residual). This residual matrix is first transformed to the frequency domain and then quantized. This is also the stage where information loss takes place.

There is ultimately a trade-off between the amount of data used for representing a picture (data rate) and the quality of the reconstructed picture. Finding an optimum balance has been a research topic for a long time in video coding. The approach taken by the H.264 standard is based on rate-distortion optimization (RDO). According to the RDO algorithm, the relationship between the rate and distortion is defined as

$$J = D + \lambda R, \tag{5.1}$$

where D is the distortion introduced to a picture block, R is the number of bits required for its coding, λ is a Lagrangian multiplier computed as a function of the quantization parameter QP, and J is the rate-distortion (RD) value to be minimized. The value of J obviously depends on several coding parameters such as block type (I, P or B), intra-prediction mode (samples used for extrapolation), the number of pair of motion vectors, sub-block size (4×4, 8×4, etc.), and the QP. Each parameter combination is called *coding mode* of the block, and the encoder picks the mode m that yields the minimum J as given in

$$m = \underset{i}{\operatorname{argmin}} J_i, \tag{5.2}$$

where the index i ranges over all possible coding modes. It must be noted that the selection for the optimum mode in Eq. (5.2) is performed on the encoder side. That is, the encoder knows all variables involved in the computation of Eq. (5.1). The decoder, in contrast, has only the knowledge of the block type, the rate (R), and the λ value and is oblivious to D and J values.

Overall due to the RDO, at one end of the distortion spectrum are the intra-coded blocks which undergo light distortion and yield the highest number of bits, and at the other end are the skipped blocks which have the same motion vectors and parameters (i.e., the quantization parameter and the quantized transform coefficients) as their spatially preceding blocks, thereby requiring only a few bits to encode at the expense of a higher distortion.

Quantization: In video coding, the transformation and quantization steps are designed to minimize computational complexity so that they can be performed using limited-precision integer arithmetic. Therefore, an integer transform is used, which is then merged with quantization in a single step so that it can be realized efficiently on hardware. As a result of this optimization, the actual quantization step size cannot be selected by the user but is controlled indirectly as a function of the above-stated quantization parameter, QP. In practice, the user mostly decides on the bitrate of coded video and the encoder varies the quantization parameter so that the intended rate can be achieved. Thus, the quantization parameters might change several times when coding a video. Moreover, a uniform quantizer step is applied to every coefficient in a 4×4 or 8×8 block by default. However, frequency-dependent quantization can also be performed depending on the selected compression profile.

Loop filtering: Block-wise quantization performed during encoding yields a blockiness effect on the output pictures. Both coding standards deploy an in-loop filtering operation to suppress this artifact by applying a spatially variant low pass filter to smooth the block boundaries. The strength of the filter and the number of pixels affected by filtering depend on several constraints such as being at the boundary of a macroblock, the amount of quantization applied, and the gradient of image samples across the boundary. It must be noted that the loop filter is also deployed during encoding to ensure synchronization between encoder and decoder. In essence, the decoder reconstructs each macroblock by adding a residual signal to a reference block identified by the encoder during prediction. In the decoder, however, all the previously reconstructed blocks would have been filtered. To not create such a discrepancy, the encoder also performs prediction assuming filtered block data rather than using the original macroblocks. The only exception to this is the intra-frame prediction where extrapolated neighboring pixels are taken from an unfiltered block.

Overall, the reliability of an estimated PRNU pattern from a picture of a video depends on the amount of distortion introduced during coding. This distortion is mainly induced by the RDO algorithm at the block level and by the loop filtering at block boundaries depending on the strength of compression. Therefore, the disruptive effects of these operations must be countered.

5.4.2   *Compensation of Loop Filtering*

The strength of the blocking effect at the block boundaries of a video frame is crucially determined by the level of compression. In high-quality videos, such an artifact will be less apparent, so from a PRNU estimation point of view, the presence of loop filtering will be of less concern. However, for low-quality videos, the application of loop filtering will result in the removal of a significant portion of the PRNU pattern, thereby making source attribution less reliable (Tan et al. 2016).

Earlier research explored this problem by focusing on the removal of blocking and ringing noise (Chen et al. 2007). Assuming most codecs use fixed 16×16 sized macroblocks, a frequency domain method was proposed to remove PRNU components that exhibit periodicity. To better suppress compression artifacts, Hyun et al. (2012) proposed an alternative approach utilizing a minimum average correlation energy (MACE) filter which essentially minimizes the average correlation plane energy, thereby yielding higher PCE values. By applying a MACE filter to the estimated pattern, a 10% improvement in identification accuracy is compared to the results of Chen et al. (2007). Later, Altinisik et al. (2018) took a proactive approach that involves intervention in the decoding process as opposed to performing post-estimation processing.

A simplified decoding schema for H.264 and H.265 codecs is given in Fig. 5.2. In this model, the coded bit sequence is run through entropy decoding, dequantization, and inverse transformation steps before the residual block and motion vectors are obtained. Finally, video frames are reconstructed through the recursive motion compensation loop. The loop filtering is the last processing step before the visual presentation of a frame. Despite the fact that the loop filter is primarily related to the decoder side, it is also deployed on the encoder side to be compatible with the decoder. On the encoder side, a decoded picture buffer is used for storing reference frames and those pictures are all loop filtered. Inter-prediction is carried out on loop filtered pictures.

It can be seen in Fig. 5.2 that for intra-prediction, unfiltered blocks are used. Since I type macroblocks that appear in all frames are coded using intra-frame prediction, bypassing the loop filter during decoding will not introduce any complications. However, for P and B type macroblocks since the encoder performs prediction assuming filtered blocks, simply removing the loop filter at the decoder will not yield a correct reconstruction. To compensate for this behavior that affects P and B frames, the decoding process must be modified to reconstruct both filtered and non-filtered versions of each macroblock. Filtered macroblocks must be used for reconstructing future macroblocks, and non-filtered ones need to be used for PRNU estimation.[2]

It must be noted that due to this modification, video frames will now exhibit blocking artifacts. However, since (sub-)macroblock sizes are not deterministic and partitioning of macroblocks into smaller blocks is mainly decided by the RDO algo-

---

[2] An implementation of loop filter compensation capability for H.264 and H.265 decoding modules in the FFMPEG libraries can be obtained at https://github.com/VideoPRNUExtractor/LoopFilterCompensation.

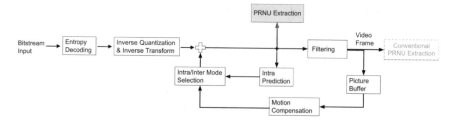

Fig. 5.2 Simplified schema of decoding for H.264 and H.265 codecs. In the block diagram, for H.264, the filtering block represents the deblocking loop filter and for H.265, it involves a similar deblocking filter cascaded with the sample adaptive offset filter

rithm, PRNU estimates containing blocking artifacts are very unlikely to exhibit a structure that is persistent across many consecutive frames. Hence, estimation of a reference PRNU pattern from several video frames will suppress these blocking artifacts through averaging.

5.4.3 Coping with Quantization-Related Weakening of PRNU

The quantization operation applied to prediction error for picture blocks is the key contributing factor to the weakening of the underlying PRNU pattern. A relative comparison between the effects of video and still image compression on PRNU estimation is provided in Altinisik et al. (2020) in terms of the change in the mean squared error (MSE) and PCE values computed between a set of original and compressed images. Table 5.1 shows the compression parameter pairs, i.e., QP for H.264 and QF for the JPEG, which yield similar MSE and PCE values. It can be seen that H.264 coding, in comparison to JPEG coding, maintains a relatively high image quality, despite rapidly weakening the inherent PRNU pattern. Essentially, these measurements show that what would be considered medium level in video compression is equivalent to heavy JPEG compression. This essentially indicates why the conventional method of PRNU estimation is not very effective on videos.

Several studies have proposed ways to tackle this problem. In van Houten and Geradts (2009), the authors studied the effectiveness of the PRNU estimation and detection process on singly and multiply compressed videos. For this purpose, videos captured by webcams were compressed using a variety of codecs at varying resolutions and matched with the reference PRNU patterns estimated from raw videos. The results of that study highlight the dependence on the knowledge of encoding settings for successful attribution and the non-linear relationship between quality and the detection statistic.

Chuang et al. (2011) explored the use of different types of frames in PRNU estimation and found that I frames yield more reliable PRNU patterns than predicted P and B frames and suggested giving a higher weight to the contribution of I frames

Table 5.1 A comparative evaluation of JPEG and H.264 encoding in terms of MSE and PCE

Equalized MSE		Equalized PCE	
QF_{JPEG}	$QP_{H.264}$	QF_{JPEG}	$QP_{H.264}$
92	5	92	5
92	10	90	10
91	15	84	15
90	18	77	18
89	21	66	21
87	24	50	24
82	27	33	27
77	30	22	30
67	35	10	35
61	40	3	40
57	45	1	45
57	50	1	50

during estimation. In Altinisik et al. (2018), macroblock level compression information is incorporated into estimation by masking out heavily compressed blocks. A similar masking approach was proposed by Kouokam et al. (2019) in which blocks that lack high-frequency content in the prediction error are eliminated from PRNU estimation. Later, Altinisik et al. (2020) examined the relationship between the quantization parameter of a block and the reliability of the PRNU pattern to introduce a block level weighting scheme that adjusts the contribution of each block accordingly. To better utilize block level parameters, the same authors in Altinisik et al. (2021) considered different PRNU weighting schemes based on encoding block type and coding rate of a block to provide a more reliable estimation of the PRNU pattern.

Since encoding is performed at the block level, its weakening effect on the PRNU pattern must also be compensated at the block level by essentially weighting the contribution of each block in accordance with the level of distortion it has undergone. It must be emphasized here that the block-based PRNU weighting approach disregards the picture content and only utilizes decoding parameters. Hence, other approaches proposed to enhance the strength of PRNU by suppressing content interference (Li 2010; Kang et al. 2012; Lin and Li 2016) and improving the denoising performance (Al-Ani et al. 2015; Zeng and Kang 2016; Kirchner and Johnson 2019) are complementary to it. The PRNU weighting approach can be incorporated into the conventional PRNU estimation method (Chen et al. 2008) by simply introducing a block-wise mask for each picture. When obtaining a reference pattern from an unstabilized (or very lightly stabilized) video, this can be formulated as

$$K = \frac{\sum_{i=1}^{N} I_i \times W_i \times M_i}{\sum_{i=1}^{N} (I_i)^2 \times M_i}, \tag{5.3}$$

where K is the sensor-specific PRNU factor computed using a set of video pictures $I_1, \ldots, I_N$; W_i represents the noise residue obtained after denoising picture I_i; and M_i is a mask to appropriately weigh the contribution of each block based on different block level parameters. It must be noted that the weights that comprise the mask can take values higher than one; therefore, to prevent a blow-up, the mask term is also included in the denominator as a normalization factor. When the video is stabilized, however, Eq. (5.3) cannot be used since each picture would have been subjected to a different stabilization transformation. Therefore, each frame needs to be first inverse transformed and Eq. (5.3) must be evaluated only after applying the identified inverse transformation to both I_i and the mask M_i.

Ultimately, the distortion induced to a block as a result of the RDO serves as the best estimator for the strength of the PRNU. This distortion term, however, is unknown at the decoder. Therefore, when devising a PRNU weighting scheme, available block level coding parameters must be used. This essentially leads to several weighting schemes based on masking out skipped blocks, the quantization strength, and the coding bitrate of a block. Developing a formulation that relates these block level parameters directly to the strength of the PRNU is in general analytically intractable. Therefore, one needs to rely on measurements obtained by changing coding parameters in a controlled manner and determine how the estimated PRNU pattern affects the accuracy of the attribution.

Eliminating Skipped Blocks: Skip blocks do not transmit any block residual information. Hence, they provide larger bitrate savings at the expense of a higher distortion when compared to other block types. As a consequence, the PRNU pattern extracted from skipped blocks will be the least reliable. Nevertheless, the distortion introduced by substituting a skip block with a reference block cannot be arbitrarily high as it is bounded by the RDO algorithm, Eq. (5.1). This weighting scheme, therefore, deploys a binary mask M where all pixel locations corresponding to a skipped block are set to zero and those corresponding to coded blocks are assigned a value of one. A major concern with the use of this scheme is that at lower bitrates, skip blocks are expected to be much more frequently encountered than other types of blocks, thereby potentially leaving very few blocks to reliably estimate a PRNU pattern. The rate of the skipped blocks in a video, i.e., the ratio of the number of skipped blocks to the total number of blocks, depending on the coding bitrate is investigated in Altinisik et al. (2021). Accordingly, it is determined that for videos with a bitrate lower than 1 Mbps, more than 70% of the blocks are skipped during coding. This finding overall indicates that unless the video is extremely short in duration or encoded at a very low bitrate, elimination of skipped blocks will not significantly impair the PRNU estimation.

Quantization-Based PRNU Weighting: Quantization of the prediction residue associated with a block is the main reason behind the weakening of the PRNU pattern during compression. To exploit this, Altinisik et al. (2020) used the strength of quantization as the basis of the weighting scheme and empirically obtained a relation between the quantization parameter QP and the PCE, which reflects the reliability of the estimated PRNU pattern. The underlying relationship is obtained by re-encoding a set of high-quality videos at all possible, i.e., 51, QP levels. Then, average PCE

Fig. 5.3 Quantization-based weighting functions for PRNU blocks depending on QP values when skipped blocks are included (red curve) and eliminated (blue curve)

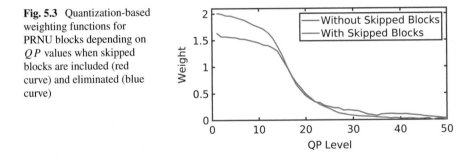

values between video pictures coded at different QP values and the camera reference pattern are computed separately for each video. To suppress the camera-dependent variation in PCE values, the obtained average PCE values are normalized with respect to the PCE value obtained for a fixed QP value. The resulting normalized average PCE values are further averaged across all videos to obtain a camera-independent relationship between PCE and QP values. Finally, the obtained QP-PCE relationship is converted into a weighting function by taking into account the formulation of the PCE metric as shown in Fig. 5.3 (red curve). Correspondingly, the frame level mask M in Eq. (5.3) is obtained by filling all pixel locations corresponding to a block with a particular QP value with the respective values on this curve.

Quantization-Based PRNU Weighting Without Skipped Blocks: Another weighting scheme that follows the previous two schemes is the quantization-based weighting of non-skipped blocks. Repeating the same evaluation process for quantization-based PRNU weighting while excluding skipped blocks, the weighting function given in Fig. 5.3 (blue curve) is obtained (Altinisik et al. 2021). As can be seen, both curves exhibit similar characteristics, although in this case, the weights vary over a smaller range. This indicates that the variation in the strength of the PRNU estimated across all coded blocks is less variable. This is mainly because eliminating skipped blocks result in higher PCE values at higher QP values where compression is more severe and skipped blocks are more frequent. Therefore, when the resulting QP-PCE values are normalized with respect to a fixed PCE value, it yields a more compact weight function. This new weight function is used in a similar way when creating a mask for each video picture with one difference that for skipped blocks the corresponding mask values are set to zero regardless of the QP value of the block.

Coding Rate (λR)-Based PRNU Weighting: Although the distortion introduced to a block can be inferred from its type or through the strength of quantization it has undergone (QP), the correct evaluation of the reliability of the PRNU ultimately requires incorporating the RDO algorithm. Since the terms D and J, in Eq. (5.1), are unknown at the decoder, Altinisik et al. (2021) considered using λR which can be determined using the number of bits spent for coding a block and the block's QP. Similar to previous schemes, the relationship between λR and PCE is empirically determined by re-encoding the raw-quality videos at 42 different bitrates varying

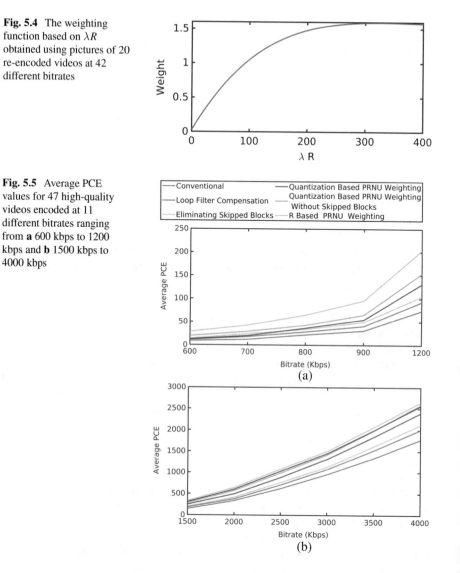

Fig. 5.4 The weighting function based on λR obtained using pictures of 20 re-encoded videos at 42 different bitrates

Fig. 5.5 Average PCE values for 47 high-quality videos encoded at 11 different bitrates ranging from **a** 600 kbps to 1200 kbps and **b** 1500 kbps to 4000 kbps

between 200 Kbps and 32 Mbps and obtaining a camera-independent weighting function as displayed in Fig. 5.4 to create a mask for weighting PRNU blocks.

Comparison: Tests performed on a set of videos coded at different bitrates showed that block-based PRNU weighting schemes yield a better estimate of the PRNU both in terms of the average improvement in PCE values and in the number of videos that yield a PCE value above 60 when compared to the conventional method, even after performing loop filter compensation (Altinisik et al. 2021). It can be observed that the improvement in PCE values due to PRNU weighting becomes more visible at bitrates higher than 900 Kbps as shown in Fig. 5.5. Overall, at all bitrates, the rate associated with a block is determined to be a more reliable estimator for the

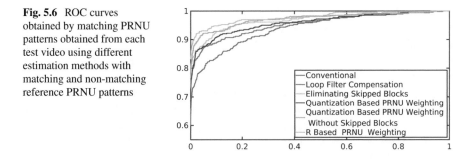

Fig. 5.6 ROC curves obtained by matching PRNU patterns obtained from each test video using different estimation methods with matching and non-matching reference PRNU patterns

strength of extracted PRNU pattern. Similarly, the performance of the QP-based weighting scheme consistently improves when skipped blocks are eliminated from the estimation process.

It was also found that at very high compression rates (at coding rates less than 0.024 bits per pixel per frame), elimination of skipped blocks performs comparably to a coding rate-based weighting scheme. This finding essentially indicates that at high compression rates, PRNU patterns extracted from coded blocks are all equally weakened, and a block-based weighting approach does not provide an additional gain. Tests also showed that the improvement in the performance due to block-based PRNU weighting does not come at the expense of an increased number of false-positive matches as demonstrated by the resulting ROC curves presented in Fig. 5.6. The gap between the two curves that correspond to the conventional method (tailored for photographic images) and the use of rate-based PRNU weighting scheme also reveals the importance of incorporating video coding parameters into PRNU estimation.

5.5 Tackling Digital Stabilization

Most cameras deploy digital stabilization while recording a video. With digital stabilization, frames captured by the sensor are moved and warped to align with one another through in-camera processing. Stabilization can also be applied externally with greater freedom in processing on a computer using one of several video editing tools or in the cloud, such as performed by YouTube. In all cases, however, the application of such frame level processing introduces asynchronicity among PRNU patterns of consecutive frames of a video.

Attribution of digitally stabilized videos, thus, requires an understanding of the specifics of how stabilization is performed. This, however, is a challenging task as the inner workings and technical details of processing steps of camera pipelines are usually not revealed. Nevertheless, at a high level, the three main steps of digital stabilization involve camera motion estimation, motion smoothing, and alignment of video frames according to the corrected camera motion. Motion estimation is performed either by describing the geometric relationship between consecutive frames through a parametric model or through tracking key feature points across frames to obtain feature trajectories (Xu et al. 2012; Grundmann et al. 2011). With sensor-rich devices such as smartphones and tablets, data from motion sensors are also utilized to improve the estimation accuracy (Thivent et al. 2018). This is followed by the application of a smoothing operation to estimate camera motion or the obtained feature trajectories to eliminate the unwanted motion. Finally, each frame is warped according to the smoothed motion parameters to generate the stabilized video.

The most critical factor in stabilization depends on whether the camera motion is represented by a two-dimensional (2D) or three-dimensional (3D) model. Early methods mainly relied on the 2D motion model that involves application of full-frame 2D transformations, such as affine or projective models, to each frame during stabilization. Although this motion model is effective for scenes far away from the camera where parallax is not a concern, it cannot be generalized to more complicated scenes captured under spatially variant camera motion. To overcome 2D modeling limitations, more sophisticated methods have considered 3D motion models. However, due to difficulties in 3D reconstruction, which requires depth information, these methods introduce simplifications to 3D structure and rely heavily on the accuracy of feature tracking (Liu et al. 2009, 2014; Kopf 2016; Wang et al. 2018). Most critically, these methods involve the application of spatially variant warping to video frames in a way that preserves the content from distortions introduced by such local transformations.

In essence, digital image stabilization tries to align content in successive pictures through geometric registration, and the details of how this is performed depend on the complexity of camera motion during capture. This may vary from the application of a simple Euclidean transformation (scale, rotation, and shift applied individually or in combination) to spatially varying warping transformation in order to compensate for any type of perspective distortion. As these transformations are applied on a per-frame basis and the variance of camera motion is high enough to easily remove pixel-to-pixel correspondences among frames, alignment or averaging of frame level PRNU patterns will not be very effective in estimating a reference PRNU pattern. Therefore, performing source attribution in a stabilized video requires blindly determining and inverting those transformations applied to each frame.

At the simplest level, an affine motion model can be assumed. However, the deficiency of affine transformation-based stabilization solutions have long been a motivation for research in the field of image stabilization. Furthermore, real-time implementation of stabilization methods has always been a concern. With the continuous development of smartphones and each year's models packing more computing power, it is hard to make an authoritative statement on the state of stabilization algorithms. In order to develop a better understanding, we performed a test using the Apple iMovie video editing tool that runs on Mac OS computers and iOS mobile devices. With this objective, we shot a video by panning the camera around a still indoors scene while stabilization and electronic zoom were turned off. The video was then stabilized by iMovie at 10% stabilization setting which determines the maximum amount of cropping that can be applied to each frame during alignment.

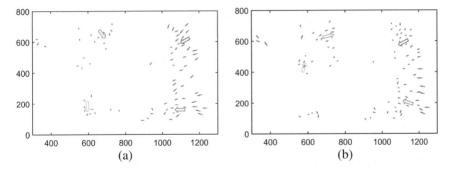

Fig. 5.7 KLT feature point displacements in two sample frames in a video stabilized using the iMovie editing tool at 0.1 stabilization setting. Displacements are measured in reference to stabilized frames

To evaluate the nature of warping applied to video frames, we extracted Kanade–Lucas–Tomasi (KLT) reference feature points, which are frequently used to estimate the motion of key points (Tomasi and Kanade 1991). Then, displacements of KLT points in pre- and post-stabilized video frames were measured. Figure 5.7 shows the optical flows estimated using KLT points for two sample frames. It can be seen that KLT points in a given locality move similarly, mostly inwards due to cropping and scaling. However, a single global transformation that will cover the movement of all points seems unlikely. In fact, our attempts to determine a single warp transformation to map key points in successive frames failed with only 4–5, out of the typical 50, points resulting in a match. This finding overall supports the assumption that digital image stabilization solutions available in today's cameras may be deploying sophisticated methods that go beyond frame level affine transformations.

To take into account more complex stabilization transformations, projective transformations can be utilized. A projective transformation can be represented by a 3 × 3 matrix with 8 free parameters that specify the amount of rotation, scaling, translation, and projection applied to a point in a two-dimensional space. In this sense, affine transformations form a subset of all such transformations without the two-parameter projection vector. Since a transformation is performed by multiplying the coordinate vector of a point with the transformation matrix, its inversion requires determination of all these parameters. This problem is further exacerbated when transformations are applied in a spatially variant manner. Since in that case different parts of a picture will have undergone different transformations, transform inversion needs to be performed at the block level where encountered PCE values will be much lower.

5.5.1 Inverting Frame Level Stabilization Transformations

Attribution for images subjected to geometric transformations has been previously studied earlier in a broader context. Considering scaled and cropped photographic images, Goljan et al. (2008) proposed a brute-force search for the geometric transform parameters to re-establish alignment between PRNU patterns. To be able to achieve this, the PRNU pattern obtained from the image in question is upsampled in discrete steps and matched with the reference PRNU at all shifts. The values that yield the highest PCE are identified as the correct scaling factor and the cropping position. More relevantly, by focusing on panoramic images, Karakucuk et al. (2015) investigated source attribution under more complex geometric transformations. Their work showed the feasibility of estimating inverse transform parameters considering projective transformations.

In the case of stabilized videos, Taspinar et al. (2016) proposed determining the presence of stabilization in a video by extracting reference PRNU patterns from the beginning and end of a video and by testing the match of the two patterns. If stabilization is detected, one of the I frames is designated as a reference and it is attempted to align other I frames to it through a search of inverse affine transformations to correct for the applied shift and rotation. The pattern obtained from the aligned I frames is then matched with a reference PRNU pattern obtained from a non-stabilized video by performing another search.

Iuliani et al. (2019) introduced another source verification method similar to Taspinar et al. (2016) by also considering mismatches in resolution, thereby effectively identifying stabilization and downsizing transformations in a combined manner. To perform source verification, 5–10 I frames are extracted and corresponding PRNU patterns are aligned with the reference PRNU pattern by searching for the correct amount of scale, shift, and cropping applied to each frame. Those frames that yield a matching statistic above some predetermined PCE value are combined together to create an aligned PRNU pattern. Tests performed on videos captured by 8 cameras in the VISION dataset using the first fiveframes of each video revealed that 86% of videos captured by cameras that support stabilization in the reduced dataset can be correctly attributed to their source with no false positives. This method was also shown to be effective on a subset of videos downloaded from YouTube with an overall accuracy of 87.3%.

To extract a reference PRNU pattern from a stabilized video, Mandelli et al. (2020) presented a method considering stabilization transformations in the form of affine transformations. In this approach, PRNU estimates obtained from each frame are matched with other frames in a pair-wise manner to identify those translated with respect to each other. Then the largest group of frames that yield a sufficient match are combined together to obtain an interim reference PRNU pattern, and the remaining frames are aligned with respect to this pattern. Alternatively, when the reference PRNU pattern at a different resolution is already known, then it is used as a reference and PRNU patterns of all other frames are matched by searching for transformation parameters. For verification, five I frames extracted from the stabilized video are matched to this reference PRNU pattern. If the resulting PCE values for at least one of the frames is observed to be higher than the threshold, a match is assumed to be achieved. The results obtained on the VISION dataset show that the method is effective in successfully attributing 71% and 77% of videos captured by cameras that support stabilization with 1% false-positive rate, respectively, when 5 and 10

I frames are used. Alternatively, if the reference PRNU pattern is extracted from photos, rather than flat videos, under the same conditions, attribution rates are found to increase to 87% for 5 frames and to 91% for 10 frames.

A note concerning the use of videos in the VISION dataset is the lack of labeling for videos. The above-mentioned studies (i.e., Iuliani et al. 2019 and Mandelli et al. 2020) considered all videos captured by cameras that support stabilization as stabilized and reported performance results accordingly. However, the number of weakly and strongly stabilized videos in the VISION dataset is smaller than the overall number of 257 videos captured by such cameras. In fact, it is demonstrated in Altinisik and Sencar (2021) that 105 of those videos are not stabilized or very lightly stabilized (as they can be attributed using the conventional method after compensation of the loop filtering), 108 are stabilized using frame level affine transformations, and the remaining 44 are more strongly stabilized. Hence, by including the 105 videos in their evaluation, these approaches inadvertently yielded higher attribution performance for stabilized videos.

5.5.2 Inverting Spatially Variant Stabilization Transformations

This form of stabilization may cause different parts of a frame to undergo different warpings. Therefore, tackling it requires inverting the stabilization transformation while being restricted to operate at sub-frame level. To address this problem, Altinisik et al. (2021) proposed a source verification method that allows an efficient search of a large range of projective transformations by operating at the block level. Considering different block sizes, they determined that blocks of size less than 500×500 pixels yield extremely low PCE values and cannot be used to reliably identify transformation parameters. Therefore, in their search, a block of a video frame is inverse transformed repetitively and the transformation that yields the highest PCE between the inverse transformed block and the reference PRNU pattern is identified.

During the search, rather than changing transformation parameters blindly, which may lead to unlikely transformations and necessitate interpolation to convert non-integer coordinates to integer values, only transformations that move corner vertices of the block within a search window are considered. It is assumed that coordinates of each corner of a selected block may move independently within a window of 15×15 pixels, i.e., spanning a range of ± 7 pixels in both coordinates in respect of the original position. Considering the possibility that a block might have also been subjected to a translation (e.g., introduced by a cropping) not contained within the searched space of transformations, each inverse transformed block is also searched within a shift range of ± 50 pixels in all directions in the reference pattern.

To accelerate the search, instead of performing a pure random search over all transformation space, the method adopts a three-level hierarchical grid search. With this approach, the search space over rotation, scale, and projection is coarsely sampled. In the first level, each corner coordinate is moved by ± 4 pixels (in all directions) over a coarse grid to identify five transformations (out of 9^4 possibilities) that yield the highest PCE values. A higher resolution search is then performed by the same process over neighboring areas of the identified transformations on a finer grid by changing the corner coordinates of transformed blocks ± 2 and, again, retaining only the five transformations producing the five highest values. Finally, in the third level, coarse transformations determined in the previous level are further refined by considering all neighboring pixel coordinates (around a ± 1 range) to identify the most likely transformations needed for inverting the warping transformation due to stabilization. This overall reduced the number of transformations from 15^8 to 11×9^4 possibilities. A pictorial depiction of the grid partitioning of the transformation space is shown in Fig. 5.8. The search complexity is further lowered by introducing a two-level grid search approach that provides an order of magnitude decrease in computation, from 11×9^4 to 11×9^3 transformations, without a significant change in the matching performance.

The major concern with increasing the degree of freedom when inverting stabilization transformation is the large number of transformations that must be considered to correctly identify each transformation. Essentially, this causes an increase in false-positive matches, as multiple inverse transformed versions of a PRNU block (or frame) are generated and matched with the reference pattern. Therefore, there is a need for measures to eliminate spurious matches arising due to incorrect transformations. With this objective, (Altinisik and Sencar 2021) also imposed block and sub-block level constraints to validate the correctness of identified inverse transformations and demonstrated their effectiveness.

The results of this approach are used to verify the source of stabilized videos in three datasets. Results showed that the introduced method is able to correctly attribute 23–30% of the stabilized videos in the three datasets that cannot be attributed using

Fig. 5.8 Three-level hierarchical grid partitioning of transformation space. Coarse grid points that yield high PCE values are more finely partitioned for subsequent search. The arrows show a sample trace of search steps to identify a likely transformation point for one of the corner vertices of a selected block

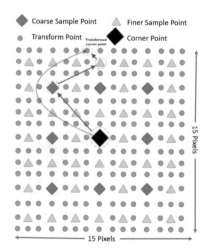

Table 5.2 Characteristics of datasets to study stabilization

Dataset	# of Cameras	# of unstabilized videos	# of weakly stabilized videos	# of strongly stabilized videos
VISION	14	105	108	44
iPhone SE-XR	8	0	19	22
APS	7	0	8	15

earlier proposed approaches (that assume an affine motion model for stabilization), without any false attributions while utilizing only 10 frames from each video.

5.6 Datasets

There is a lack of large-scale datasets to enable test and development of video source attribution methods. This is mainly because there are very few resources on the Internet that provide original videos compared to photographic images. In almost all public video sharing sites, video data are re-encoded to reduce the storage and transmission requirements. Furthermore, video metadata typically exclude source camera information, adding another layer of complication to data collection from open sources.

The largest available dataset is the VISION dataset, which includes a collection of photos and videos captured by 35 different camera models (Shullani et al. 2017). However, this dataset is, however, not comprehensive enough to cover the diversity of compression and stabilization behavior of available cameras. To tackle the compression problem, (Altinisik et al. 2021; Tandogan et al. 2019) utilized a set of raw-quality videos captured by 20 smartphone cameras through a custom camera app under controlled conditions. These videos were then externally compressed under different settings.[3] To study the effects of stabilization, (Altinisik and Sencar 2021) introduced two other datasets in addition to the VISION dataset. One of these includes media captured by two iPhone camera models (the iPhone SE-XR dataset) and the other includes a set of externally stabilized videos using the Adobe Premiere Pro video processing tool (the APS dataset).[4] Table 5.2 provides the stabilization characteristics of the videos in those datasets.

---

[3] These videos are available at https://github.com/VideoPRNUExtractor.

[4] The iPhone SE-XR and the APS datasets are also available at the above link.

5.7   Conclusions and Outlook

The generation of a video involves several processing steps that are growing in both sophistication and number. These processing steps are known to be very disruptive to PRNU estimation, and there is little public information about their specifics. Therefore, developing a PRNU estimation method for videos that is as accurate and efficient as it is for photographic images is an extremely challenging task.

Current research focusing on this problem has mainly attempted to mitigate down-sizing, stabilization, and video coding operations. Overall, the reported results of these studies have shown that it is difficult to generate a reference PRNU pattern from a video, specifically for stabilized and low-to-medium compressed videos. Therefore, almost all approaches to attribution of stabilized videos assume a source verification setting where a given video is matched against a known camera. That is, a reference PRNU pattern of the sensor is assumed to be available.

In a source verification setting, with the increase in the number of frames in a video, the likelihood of correctly attributing a video also increases. However, this does not remove the need to devise a PRNU estimation method. To emphasize this, we tested how the conventional PRNU estimation method (developed for photos, without the Wiener filtering component) performs when the source of all frames comprising strongly stabilized videos in the three public datasets mentioned above is verified using the corresponding camera reference PRNU patterns. These results showed that out of more than 80K frames tested, only two frames yielded a PCE value above the commonly accepted threshold of 60 (78 and 491) while a great majority yielded PCE values less than 10.

Ultimately, the reliable attribution of videos depends on the severity of information loss mainly caused by video coding and the ability to invert geometric transformations mostly stemming from stabilization. For highly compressed videos, with bitrates lower than 900 kbps, the PRNU is significantly weakened. Similarly, the use of smaller blocks or low-resolution frames, with sizes below 500×500 pixels, do not yield reliable PRNU patterns. Thus, PCE values corresponding to true and false matching cases largely overlap.

Nonetheless, the most challenging aspect of source attribution remains stabilization. Since most modern smartphone cameras activate stabilization when the camera is set to video mode, before video recording starts, its disruptive effects should be considered to be always present. Overall, coping with stabilization induces a significant computational overhead to PRNU estimation which is mainly determined by the assumed degree of freedom in the search for applied transformation parameters. More critically, this warping inversion step amplifies the number of false attributions, which forces the use of higher threshold values for correct matching, thereby further reducing the accuracy. This underscores the need to develop methods for an effective search of transformation space to determine the correct transformation and transformation validation mechanisms to eliminate false-positive matches as a research priority.

Overall, the PRNU-based source attribution of photos and videos is an important capability in the arsenal of media forensics tools. Despite significant work in this field, reliable attribution of a video to its source camera remains a clear problem. Therefore, further research is needed to increase the applicability of this capability into practice.

References

Al-Ani M, Khelifi F, Lawgaly A, Bouridane A (2015) A novel image filtering approach for sensor fingerprint estimation in source camera identification. In: 12th IEEE international conference on advanced video and signal based surveillance, AVSS 2015, Karlsruhe, Germany, August 25–28, 2015. IEEE Computer Society, pp 1–5

Altinisik E, Sencar HT (2021) Source camera verification for strongly stabilized videos. IEEE Trans Inf Forensics Secur 16:643–657

Altinisik E, Tasdemir K, Sencar HT (2018) Extracting PRNU noise from H.264 coded videos. In: 26th european signal processing conference, EUSIPCO 2018, Roma, Italy, September 3-7, 2018. IEEE, pp 1367–1371

Altinisik E, Tasdemir K, Sencar HT (2020) Mitigation of H.264 and H.265 video compression for reliable PRNU estimation. IEEE Trans Inf Forensics Secur 15:1557–1571

Altinisik E, Tasdemir K, Sencar HT (2021) Prnu estimation from encoded videos using block-basedweighting. In: Media watermarking, security, and forensics. Society for Imaging Science and Technology

Bellavia F, Fanfani M, Colombo C, Piva A (2021) Experiencing with electronic image stabilization and PRNU through scene content image registration. Pattern Recognit Lett 145:8–15

Bondi L, Bestagini P, Pérez-González F, Tubaro S (2019) Improving PRNU compression through preprocessing, quantization, and coding. IEEE Trans Inf Forensics Secur 14(3):608–620

Chen M, Fridrich JJ, Goljan M, Lukás J (2008) Determining image origin and integrity using sensor noise. IEEE Trans Inf Forensics Secur 3(1):74–90

Chen M, Fridrich JJ, Goljan M, Lukás J (2007) Source digital camcorder identification using sensor photo response non-uniformity. In: Delp III EJ, Wong PW (eds) Security, steganography, and watermarking of multimedia contents IX, San Jose, CA, USA, January 28, 2007, SPIE proceedings, vol 6505. SPIE, p 65051G

Chuang W-H, Su H, Wu M (2011) Exploring compression effects for improved source camera identification using strongly compressed video. In: Macq B, Schelkens P (eds) 18th IEEE international conference on image processing, ICIP 2011, Brussels, Belgium, September 11–14, 2011. IEEE, pp 1953–1956

Goljan M (2018) Blind detection of image rotation and angle estimation. In: Alattar AM, Memon ND, Sharma G (eds) Media watermarking, security, and forensics 2018, Burlingame, CA, USA, 28 January 2018–1 February 2018. Ingenta

Goljan M, Fridrich JJ (2008) Camera identification from cropped and scaled images. In: Delp III EJ, Wong PW, Dittmann J, Memon ND (eds) Security, forensics, steganography, and watermarking of multimedia contents X, San Jose, CA, USA, January 27, 2008, SPIE proceedings, vol 6819. SPIE, pp 68190E

Grundmann M, Kwatra V, Essa IA (2011) Auto-directed video stabilization with robust L1 optimal camera paths. In: The 24th IEEE conference on computer vision and pattern recognition, CVPR 2011, Colorado Springs, CO, USA, 20–25 June 2011. IEEE Computer Society, pp 225–232

Guo J, Gu H, Potkonjak M (2018) Efficient image sensor subsampling for dnn-based image classification. In: Proceedings of the international symposium on low power electronics and design, ISLPED 2018, Seattle, WA, USA, July 23-25, 2018. ACM, pp 40:1–40:6

Hyun D-K, Choi C-H, Lee H-K (2012) Camcorder identification for heavily compressed low resolution videos. In: Computer science and convergence. Springer, pp 695–701

Iuliani M, Fontani M, Shullani D, Piva A (2019) Hybrid reference-based video source identification. Sensors 19(3):649

Kang X, Li Y, Zhenhua Q, Huang J (2012) Enhancing source camera identification performance with a camera reference phase sensor pattern noise. IEEE Trans Inf Forensics Secur 7(2):393–402

Karaküçük A, Emir Dirik A, Sencar HT, Memon ND (2015) Recent advances in counter PRNU based source attribution and beyond. In: Alattar AM, Memon ND, Heitzenrater C (eds) Media watermarking, security, and forensics 2015, San Francisco, CA, USA, February 9–11, 2015, Proceedings, SPIE proceedings, vol 9409. SPIE, p 94090N

Kirchner M, Johnson C (2019) SPN-CNN: boosting sensor-based source camera attribution with deep learning. In: IEEE international workshop on information forensics and security, WIFS 2019, Delft, The Netherlands, December 9–12, 2019. IEEE, pp 1–6

Kopf J (2016) 360° video stabilization. ACM Trans Graph 35(6):195:1–195:9

Kouokam EK, Dirik AE (2019) Prnu-based source device attribution for youtube videos. Digit Investig 29:91–100

Li C-T (2010) Source camera identification using enhanced sensor pattern noise. IEEE Trans Inf Forensics Secur 5(2):280–287

Lin X, Li C-T (2016) Enhancing sensor pattern noise via filtering distortion removal. IEEE Signal Process Lett 23(3):381–385

Liu F, Gleicher M, Jin H, Agarwala A (2009) Content-preserving warps for 3d video stabilization. ACM Trans Graph 28(3):44

Liu S, Yuan L, Tan P, Sun J (2014) Steadyflow: spatially smooth optical flow for video stabilization. In: 2014 IEEE conference on computer vision and pattern recognition, CVPR 2014, Columbus, OH, USA, June 23–28, 2014. IEEE Computer Society, pp 4209–4216

Mandelli S, Bestagini P, Verdoliva L, Tubaro S (2020) Facing device attribution problem for stabilized video sequences. IEEE Trans Inf Forensics Secur 15:14–27

Richardson IE (2011) The H. 264 advanced video compression standard. Wiley

Shullani D, Fontani M, Iuliani M, Al Shaya O, Piva A (2017) VISION: a video and image dataset for source identification. EURASIP J Inf Secur 2017:15

Sze V, Budagavi M, Sullivan GJ (2014) High efficiency video coding (hevc). In: Integrated circuit and systems, algorithms and architectures, vol 39. Springer, p 40

Tan TK, Weerakkody R, Mrak M, Ramzan N, Baroncini V, Ohm J-R, Sullivan GJ (2016) Video quality evaluation methodology and verification testing of HEVC compression performance. IEEE Trans Circuits Syst Video Technol 26(1):76–90

Tandogan SE, Altinisik E, Sarimurat S, Sencar HT (2019) Tackling in-camera downsizing for reliable camera ID verification. In: Alattar AM, Memon ND, Sharma G (eds) Media watermarking, security, and forensics 2019, Burlingame, CA, USA, 13-17 January 2019. Ingenta

Taspinar S, Mohanty M, Memon ND (2020) Camera identification of multi-format devices. Pattern Recognit Lett 140:288–294

Taspinar S, Mohanty M, Memon ND (2016) Source camera attribution using stabilized video. In: IEEE international workshop on information forensics and security, WIFS 2016, Abu Dhabi, United Arab Emirates, December 4–7, 2016. IEEE, pp 1–6

Thivent DJ, Williams GE, Zhou J, Baer RL, Toft R, Beysserie SX (2018) Combined optical and electronic image stabilization, May 22. US Patent 9,979,889

Tomasi C, Kanade T (1991) Detection and tracking of point. Technical report, features. Technical Report CMU-CS-91-132, Carnegie, Mellon University

van Houten W, Geradts ZJMH (2009) Source video camera identification for multiply compressed videos originating from youtube. Digit Investig 6(1–2):48–60

Vijaya Kumar BVK, Laurence Hassebrook (1990) Performance measures for correlation filters. Appl Opt 29(20):2997–3006

Wang Z, Zhang L, Huang H (2018) High-quality real-time video stabilization using trajectory smoothing and mesh-based warping. IEEE Access 6:25157–25166

Xu J, Chang H, Yang S, Wang M (2012) Fast feature-based video stabilization without accumulative global motion estimation. IEEE Trans Consumer Electron 58(3):993–999

Zeng H, Kang X (2016) Fast source camera identification using content adaptive guided image filter. J Forensic Sci 61(2):520–526

Zhang J, Jia J, Sheng A, Hirakawa K (2018) Pixel binning for high dynamic range color image sensor using square sampling lattice. IEEE Trans Image Process 27(5):2229–2241

Chapter 6
Camera Identification at Large Scale

Samet Taspinar and Nasir Memon

Chapters 3 and 4 have shown that PRNU is a very effective solution for source camera verification, i.e., linking a visual object to its source camera (Lukas et al. 2006). However, in order to determine the source camera, there has to be a suspect camera in the possession of a forensics analyst which is often not the case. On the other hand, in the source camera identification problem, the goal is to match a PRNU fingerprint to a large database. This capability is needed when one needs to attribute one or more images from an unknown camera to a large number of images in a large image repository to find other images that may have been taken from the same camera. For example, consider that a legal entity acquires some illegal visual objects (such as child pornography) without any suspect camera available. Consider also that the owner of the camera shares images or videos on social media such as Facebook, Flickr, or YouTube. In such a scenario, it becomes crucial to find a way to link the illegal content to those social media accounts. Given that these social media contains billions of accounts, camera identification at large scale becomes a very challenging task.

This chapter will focus on how databases that support camera identification can be created and the methodologies to search query images on those databases. Along with that, the time complexities, strengths, and drawbacks of different methods will also be presented in this chapter.

S. Taspinar
Overjet, Boston, MA, USA
e-mail: samet@overjet.ai

N. Memon (✉)
New York University Tandon School of Engineering, New York, NY, USA
e-mail: memon@nyu.edu

H. T. Sencar et al. (eds.), *Multimedia Forensics*, Advances in Computer Vision and Pattern Recognition, https://doi.org/10.1007/978-981-16-7621-5_6

6.1   Introduction

As established in Chaps. 3 and 4, once a fingerprint, K, is obtained from a camera, the PRNU noise obtained from other query visual objects can be tested against the fingerprint to determine if they are captured by the same camera. Each of these comparisons is a verification task (i.e., 1-to-1).

On the other hand, in the identification task, the source of a query image or video is searched within a fingerprint collection. The collection could be compiled from social media such as Facebook, Flickr, and YouTube. Visual objects from each camera in such a collection can be clustered together, and their fingerprints extracted to create a known camera fingerprint database. This chapter will discuss identification on a large scale which can be seen as a sequence of multiple verification tasks (i.e., 1-to-many). Specifically, we will focus on structuring large databases of images or camera fingerprints and the search efficiency of different methods using such databases.

Notice that search can mainly be done in two ways: querying one or more images from a camera against a database of camera fingerprints or querying a fingerprint against a database of images. This chapter will focus on the first case where we will assume there is only a single query image whose source camera is under question. However, the workflow is the same for the latter case as well.

Over the past decade, various approaches have been proposed to speed up camera identification using large databases of camera fingerprints. These methods can be grouped under two categories: (i) techniques for decreasing the cost of pairwise comparisons and (ii) decreasing number of comparisons made. Along with these a third category aims at combining the strengths of two approaches to create a superior method.

We will assume that neither query images nor the images used for computing the fingerprint are geometrically transformed unless otherwise stated. As described in Chap. 3, some techniques can be used to reverse the geometric transformations performed before searching the database. However, not all algorithms proposed in this chapter will work, and geometric transformations may cause a failure in the methodology. We will also assume that there is no camera brand/model information available as metadata typically easy to forge and often does not exist when visual objects are obtained from social media. However, when reliable metadata is available, it can speed up search further by restricting to the relevant models in the obvious manner.

6.2   Naive Methods

This section will present the most basic methods for source camera identification. Consider that images or videos from N known cameras are obtained. Their fingerprints are extracted as a one-time operation and stored in a fingerprint dataset,

$D = K_1, K_2, \ldots K_N$ to form a fingerprint dataset. This step is common for all the methods presented in this and other camera identification methods.

6.2.1   Linear Search

The most straightforward search method for source camera identification is linear search (brute force). A query visual object with n pixels, whose fingerprint estimate is K_q, is then compared with all the fingerprints until (i) it matches with one of them (i.e., at least one H_1 case) (ii) no more fingerprints left (i.e., all comparisons are H_0 cases)

$$
\begin{aligned}
H_0 &: K_i \neq K_q \text{ (non-matching fingerprints)} \\
H_1 &: K_i = K_q \text{ (matching fingerprints)}
\end{aligned}
\tag{6.1}
$$

So the complexity of the linear search is $O(nN)$ for a single query image. Since modern cameras typically have more than $10\,M$ pixels, and the number of cameras in a comprehensive database could be many millions, the computational cost of this method becomes exceedingly large.

6.2.2   Sequential Trimming

Another simple method is to trim all images to a fixed length such that all fingerprints become the same length k where $k \leq n$ (Goljan et al. 2010). The main benefits of this method are it would speed up the search by n/k times, and the memory and disk usage drop by the same ratio.

Given a camera fingerprint, K_i, and a fingerprint of a query object, K_q, their correlation c_q is independent of the number of remaining pixels after trimming (i.e., k). However, as k drops, the probability of false alarm, P_{FA} increases, and the probability of detection, P_D decreases for a fixed decision threshold (Fridrich and Goljan 2010). Therefore, the sequential trimming method may have only limited speed up if the goal is to retain performance.

Moreover, this method still requires the query fingerprint to be correlated with N fingerprints in the database (i.e., linear with the number of cameras in the database). However, the time the computation complexity drops to $O(kN)$ as fingerprint is trimmed from n to k elements.

6.3 Efficient Pairwise Correlation

As presented in Chap. 3, the complexity of a single correlation is proportional to the number of pixels in a fingerprint pair. Since modern cameras contain millions of pixels and a large database can contain billions of images, using naive search methods is prohibitively expensive for source camera identification.

The previous section shows that the time complexity of a brute-force search is proportional to the number of fingerprints and number of pixels in each fingerprint (i.e., $O(n \times N)$). Similarly, sequential trimming can only speed up the search a few times. Hence, further improvements for identification are required.

This section presents the first approach mentioned in Sect. 6.1 that decreases the time complexity of a pairwise correlation.

6.3.1 Search over Fingerprint Digests

The first method that addresses the above problem is searching over fingerprint digests (Goljan et al. 2010). This method reduces the dimensionality of each fingerprint by selecting the k largest (in terms of magnitude) fingerprint elements along with their spatial location from a fingerprint F.

Fingerprint Digest

Given a fingerprint, F, with n fingerprint elements, the proposed method picks the top k largest of them as well as their indices. Suppose that the values of these k elements are $V_F = \{v_1, v_2, \ldots v_k\}$ and their locations are $L_F = \{l_1, l_2, \ldots l_k\}$ for $1 \leq l_i \leq n$ where $V_F = F[L_F]$.

The digest of each fingerprint can be extracted and stored along with the original fingerprint. An important aspect is determining the number of pixels, k, in a digest. Picking a large number of pixels will not yield a high speedup, whereas a low number will result in many false positives.

Search step

Suppose that the fingerprint of the query image is X and it is correlated with one of the digests in the database V_F whose indices are L_F. In the first step, the digest of X is estimated as

$$V_X = X[L_F] \tag{6.2}$$

Then, $corr(V_F, V_X)$ is obtained. If the correlation is lower than a preset threshold, τ, then X and F are said to be from different sources. Otherwise, the original fingerprint X and F are correlated to decide if they originate from the same source camera. Overall, this method speeds up approx. k/n times where $k << n$. Therefore, a significant speedup can be achieved.

6.3.2   Pixel Quantization

The previous section presented that search in large databases is proportional to the number of images in a database, N and the number of elements each fingerprint has, n. To be more precise, together with these variables, the number of bits of each fingerprint element also plays a role in the time complexity of the search. Taking the number of bits into account, the time complexity of a brute-force search becomes $O(b \times n \times N)$ where b is the number of bits of each element. Therefore, another avenue for speeding up pairwise fingerprint comparison is by quantizing the fingerprint elements.

Bayram et al. (2012) proposed a fingerprint binarization method where each fingerprint element is represented only by its sign, and the magnitude is discarded. Hence, each element is represented by either 0 or 1. Given a camera fingerprint, X, with n elements, its binary form, X^B, can be represented as

$$X_i^B = \begin{cases} 0, & X_i < 0 \\ 1, & X_i >= 0 \end{cases} \tag{6.3}$$

where X_i^B is the ith element in binary form for $i \in \{1, \ldots n\}$.

This way, the efficiency of pairwise fingerprint matching can be improved by a factor of b. The original work (Lukas et al. 2006) uses 64-bit double precision for each fingerprint element. Therefore, the binary quantization method can save storage usage and computation efficiency by a factor 64 as compared to Lukas et al. (2006). However, binary quantization comes with an expense of a decrease in the correct detection.

Alternatively, one can use the quantization approach but with more bits instead of a single bit. This way, the metrics can be preserved. When the bit depth of two fingerprints is reduced from 64-bit to 8-bit format, the pairwise correlation changes marginally (i.e., less than 10^{-4}) but nearly 8 times speed up can be achieved as the complexity of Pearson correlation is proportional to the number of bits.

6.3.3   Downsizing

Another simple yet effective method is creating multiple downsized versions of a fingerprint where along with the original fingerprint L other copies of the fingerprint are stored (Yaqub et al. 2018; Tandoğan et al. 2019). In these methods, a camera fingerprint whose resolution is $r \times c$, is downsized by a scaling factor s in the first level. Then, in the next level, the scaled fingerprint is further resized by s, and so on. Hence, the resolution of a fingerprint in Lth level the resolution becomes $\frac{r}{s^L} \times \frac{c}{s^L}$.

Since, multiple copies of a fingerprint is stored in this method, the storage usage increases by $\frac{r \times c}{s^2 + s^4 + \ldots s^{2L}}$. The authors show that the optimal value for s is 2 as it causes the least interpolation artefact, and L is 3 level. Moreover, Lanczos-3 interpolation results in best accuracy as it best approximates the Sinc kernel (Lehman et al. 1999). In these settings, the storage requirement increases by $\approx$33%.

Matching

Given a query fingerprint and a dataset of N fingerprints (for each one of them 3 other resized versions are stored), a pairwise comparison can be done as follows:

The query fingerprint is resized to L3 size (i.e., by $1/8$) and compared by L3 version of ith fingerprint. If the comparison of L3 query and reference fingerprint pair is below a preset threshold, the query image is not captured by the ith camera, and the query fingerprint is then compared by $(i + 1)$th reference fingerprint. Otherwise, L2 fingerprint are compared. The same process is continued for other levels as well. The preset PCE threshold is set to 60, as presented in Goljan et al. (2009).

The results show that when cropping is not involved, this method can achieve up to 53 times speed up with a small drop in the true positive rate.

6.3.4 Dimension Reduction Using PCA and LDA

It is known that sensor noise is high-dimensional data (i.e., more than millions of pixels). This data contains redundant and interfering components such as demosaicing, JPEG compression, and other in-camera image processing operations. Because of these artifacts are the probability of match may decrease, and a significant slow down happens. Therefore, removing those redundant or interfering signals helps improve the matching accuracy and the efficiency of pairwise comparison.

Li et al. propose to use Principal Component Analysis (PCA) and Linear Discriminant Analysis (LDA) to compress camera fingerprints (Li et al. 2018). The proposed method first uses PCA-based denoising (Zhang et al. 2009) to create a better representation of camera fingerprints. Then LDA is utilized to decrease the dimensionality further as well as further improve matching accuracy.

Fingerprint Extraction

Suppose that there are n images, $\{I_1, \ldots I_n\}$, captured by a camera. These images are cropped from their centers such that each becomes $N \times N$ resolution. First, PRNU noise, x_i, extracted from these images using one of extraction methods (Lukas et al. 2006; Dabov et al. 2009). In this step, the noise signals are flattened to create column vectors. Hence, the size of x_i becomes $N^2 \times 1$. They are then horizontally stacked to created the training matrix, X as follows:

$$X = [x_1, x_2, \ldots x_n] \tag{6.4}$$

Feature extraction via PCA

The mean of X, μ, becomes

$$\mu = \frac{1}{n} \sum_{i=1}^{n} x_i$$

The training set is normalized by subtracting the mean from each image by $\bar{x}_i = x_i - \mu$. Hence, the normalized training set $\bar{X}$ becomes

$$\bar{X} = [\bar{x}_1, \bar{x}_2, \ldots \bar{x}_n] \tag{6.5}$$

The covariance matrix, Σ of $\bar{X}$ can be calculated as $\Sigma = E[\bar{X}\bar{X}^T] \approx \frac{1}{n}\bar{X}\bar{X}^T$.

PCA is performed for finding orthonormal vectors, u_k and their eigenvalues, λ_k. The dimensionality of $\bar{X}$ is very high (i.e., $N^2 \times n$). Hence the computational complexity of PCA is prohibitively high. Instead of direct computation, singular value decomposition (SVD) is applied, which is a close approximation of PCA. SVD can be written as

$$\Sigma = \Phi \Lambda \Phi^T \tag{6.6}$$

where $\Phi = [\phi_1, \phi_2 \ldots \phi_m]$ is the $m \times m$ orthonormal eigenvector matrix and $\Lambda = diag\{\lambda_1, \lambda_2 \ldots \lambda_n\}$ are eigenvalues.

Along with creating decorrelated eigenvectors, PCA allow to reduce dimensionality. Using the first d of m eigenvectors, transformation matrix $M_{pca} = \{\phi_1, \phi_2 \ldots \phi_d\}$ for $d < m$. Then the transformed dataset, $\hat{Y}$ becomes $\hat{Y} = M_{pca}^T \hat{X}$ whose size is $d \times n$. Generally speaking, most of the discriminative information of the SPN signal will concentrate on the first few eigenvectors. In this research, the authors kept only d eigenvalues to preserve 99% of the variance, which is used to create a compact version of the PRNU noise. Moreover, thanks to this dimension reduction, the "contamination" caused by other in-camera operations can be decreased. The authors call this a "PCA feature".

Feature extraction via LDA

LDA is used for two purposes: (i) dimension reduction (ii) increasing the distance between different classes. Since the training set is already labeled, LDA can be used to reduce dimension and increase separability of different classes. This can be done by finding a transformation matrix, M_{lda} that maximizes the ratio of the determinant of between-class scatter matrix, S_b to within-classes scatter matrix S_w:

$$M_{lda} = \underset{J}{argmax} = \left| \frac{J^T S_b J}{J^T S_w J} \right| \tag{6.7}$$

where $S_w = \sum_{j=1}^{c} \sum_{i=1}^{L} (y_i - \mu_j)(y_i - \mu_j)^T$. There are c distinct classes and jth class has L_j samples, y_i is the ith sample of the class j and μ_j is the mean of the class j. On the other hand S_b is defined as $S_b = \frac{1}{c} \sum_{j=1}^{c} \sum_{i=1}^{L} (\mu_j - \mu)(\mu_j - \mu)^T$ where μ is the average of all samples.

Using the "LDA feature" extractor, M_{lda}, a $c - 1$ dimensional vector can be obtained as

$$
\begin{aligned}
z &= M_{lda}^T [(M_{pca}^d)^T X] \\
&= ((M_{pca}^d)^T (M_{lda}^T) X
\end{aligned}
\tag{6.8}
$$

So, PCA and LDA operations can be combined using only a single operation, i.e., $M_e = ((M_{pca}^d)^T (M_{lda}^T)$. Typically $(c - 1)$ is lower than d, which can further compress the training data.

Reference Fingerprint Estimation

Consider that jth camera, c_j, which has captured image $I_1, I_2, \ldots I_{L_j}$ and their PRNU noise patterns are $x_1, x_2, \ldots x_{L_j}$, respectively.

The reference fingerprint for c_j can be obtained using (6.8). So, using (6.8), LDA features are extracted (i.e., $z_j = M_e^T X$).

Source Matching

The authors present two different methodologies: Algorithm 1 and 2. These algorithms make use of y^d (PCA-SPN) and z (LDA-SPN), respectively. Both of these compressed fingerprints have much lower dimensionality compared to the original fingerprint, x.

The source camera identification process is straightforward for both algorithms. "One-time" fingerprint compression step is used to set up the databases, the PCA-SPN of query fingerprint, y_q^d, is correlated by each fingerprint in the database. Similarly, for LDA-SPN, z_q is correlated with fingerprint in the more compressed dataset. Decision thresholds τ_y and τ_z can be setup based on desired false acceptance rate (Fridrich and Goljan 2010).

6.3.5 PRNU Compression via Random Projection

Valsesia et al. (2015a, b) utilized the idea of applying random projections (RP) followed by binary quantization to reduce fingerprint dimension for camera identification on large databases.

Dimension Reduction

Given a dataset, D which contains N n-dimensional fingerprints, the goal is to reduce each fingerprint to m-dimensions with slight or no information loss where $m << n$. RP is a method to reduce original fingerprints using a random matrix $\Phi \in R^{m \times n}$ so that the original dataset, $D \in R^{n \times N}$, can be compressed to a lower dimensional subspace, $A \in R^{m \times N}$, as follows:

$$A = \Phi D \qquad (6.9)$$

Johnson et al. present that given an n dimensional data, there exists an m dimensional Euclidean space via a transformation that retains all pairwise distances within a factor of $1 + \epsilon$ (Lindenstrauss 1984).

A linear mapping, f, can be represented by a random matrix $\Phi \in R^{m \times n}$. The elements of Φ can be drawn from certain probability distributions such as Gaussian or Rademacher (Achlioptas 2003). However, one of the main drawbacks of generating $m \times n$ random numbers. Given that n is on the order of several million, generating and storing such a random matrix is not feasible as it would require too much memory. Moreover, full matrix multiplication is carried out for the dimension reduction of each fingerprint. This problem can be overcome by using circulant matrices. Instead of generating an entire $m \times n$ matrix, only the first row is generated in Gaussian distribution, and the rest of the matrix can be obtained by circular shift. The multiplication after generating Φ can be done using Fast Fourier Transform (FFT). Thanks to FFT, the dimension reduction can be efficiently done with a complexity of $O(N \times n \times logn)$ as opposed to $O(N \times m \times n)$.

Further improvement can be achieved by binarization. Instead of a floating value, each element can be represented by a single bit as

$$A = sign(\Phi D) \qquad (6.10)$$

The compressed dataset A is now a matrix consisting of N vectors, each of which is a separate fingerprint, K_i as

$$A = \{y_i : y_i = sign(\Phi K_i)\}, i = \{1, \ldots, N\} \qquad (6.11)$$

Search

Now that database of N fingerprints is set up, the authors present two methods to identify the source of a query fingerprint: linear search and hierarchical search. In these search methods, a query fingerprint, $\hat{K} \in R^n$ is first transformed in similar manner as the database (i.e., random projection followed by binary quantization) as follows:

$$\hat{y} = sign(\Phi \hat{K}) \qquad (6.12)$$

Since the query fingerprint and the N fingerprints in the dataset are compressed to their binary form, pairwise correlation can be done using Hamming distance instead of Pearson correlation coefficient.

Linear Search

The most straightforward approach for search is by brute-force approach. In this method, the compressed query fingerprint, $\hat{y}$ is compared against each fingerprint in A, y_i. Therefore, similar to the search over fingerprint digests (Sect. 6.3.1), the computational complexity of the method is proportional to the number of fingerprints, N (i.e., $O(N)$). Moreover, this method also provides the benefit achieved by the binary quantization (Sect. 6.3.2). However, it is crucial to note that the number of pixels used for a fingerprint digest is significantly lower than the random projection. While fingerprint digests typically require up to a few $10\,K$ pixels, this method typically requires $0.5\,M$ pixels.

Hierarchical Search

A first step improvement for linear search is creating a hierarchical search schema where two different sets of compressed datasets are created. In the first step a coarse version of the compressed fingerprints m_1 elements ($m_1 << m$) correlated with a query fingerprint. This step works as a pre-filter that each coarse query fingerprint is correlated with the coarse versions of the N fingerprints. This step reduces the search space to N' ($N' << N$). In the second step, the m random projections of the N' fingerprint are loaded into memory and correlated with the m random projections of the query fingerprint. This way, as opposed to loading $m \times N$ bits into memory, $m_1 \times N + m \times N'$ bits are loaded.

One way to do this is to pick m_1 random indices such as the first m_1 of them. However, this method is not robust enough to achieve any improvement over the linear search presented above. An alternative method is to pick indices with the largest magnitude similar to fingerprint digests (Goljan et al. 2010).

6.3.6 Preprocessing, Quantization, Coding

Another method that is proposed to compress PRNU fingerprints is through preprocessing, quantization, and finally entropy coding (Bondi et al. 2018). The preprocessing step consists of decimation and random projection. In this step, the number of fingerprint elements is reduced. The quantization step aims to reduce the number of bits per fingerprint element. Finally, entropy coding is another step to further decrease the number of elements, which is similar to Huffman coding.

Preprocessing Step

The first step in this method is decimation. Given a fingerprint K with r rows, and c columns, this step decimates the fingerprint over d times. The columns are grouped by d elements to create $\lceil \frac{c}{d} \rceil$. Then the same operation is done along the rows. Thus, the resolution of the output fingerprint becomes $\lceil \frac{r}{d} \rceil \times \lceil \frac{c}{d} \rceil$.

For decimation, the authors propose to use a 1×5 bicubical interpolation as

$$K_d = \begin{cases} 1.5|x|^3 - 2.5|x|^2 + 1, & \text{if } |x| \le 1 \\ -0.5|x|^3 + 2.5|x|^2 - 4|x| + 2, & \text{if } 1 < |x| \le 2 \\ 0, & \text{otherwise} \end{cases} \tag{6.13}$$

K_d is flattened to create a column vector. Then, Random Projection (RP) is applied to K_d to produce r_P^* which contains P elements.

Dead-Zone Quantization

Although binary quantization is shown to be an effective method to reduce bitrate, the authors present a slightly different approach to divide to quantize r_P^*. Instead of checking the sign of a fingerprint element, the authors divide the value range into 3 ranges, and depending on where an element lies, its value is assigned as one of $\{0, 1, -1\}$ as follows:

$$r^\delta(i) = \begin{cases} +1, & \text{if } r * (i) > \sigma\delta \\ 0, & \text{if } -\sigma\delta \leq r * (i) \leq \sigma\delta \\ -1, & \text{if } r * (i) < -\sigma\delta \end{cases} \qquad (6.14)$$

where σ is the standard deviation of r_P^* and δ is a factor for determining sparsity of the vector. Then $+1$'s are assigned to 10 and -1's to 11 to create a bit-stream, r_P^δ from r_P^*.

Entropy Coding

The last step of this pipeline is to apply an arithmetic entropy coding to the bit-stream r_P^δ. Here δ is a tunable variable that by increasing it, more sparse vector can be created.

6.4   Decreasing the Number of Comparisons

The methodologies presented in the previous section aim to speed up pairwise comparisons. Although they can significantly decrease the cost of a fingerprint search, this improvement is not sufficient given the sheer number of cameras available. Often, a large database can contain many millions of cameras, and alternative methods may be required to carry over search in large datasets. This section presents an alternative methodology that improves searching in large databases by reducing pairwise comparisons.

6.4.1   Clustering by Cameras

An obvious way to reduce the search space is by clustering camera fingerprints by their brands or models. A query fingerprint can be compared with only the camera fingerprints from the same brand/model.

As presented in Chapter 4, a fingerprint dataset, $D = \{K_1, K_2, \ldots K_N\}$ can be divided into smaller groups as $D = \{D_1, D_2, \ldots D_m\}$ where D_i is a set that consists of a single camera model.

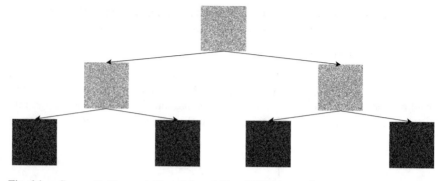

Fig. 6.1 **a** Composite Fingerprint-based Search Tree (CFST): query is matched against the tree of composite fingerprints with fingerprints at the leaves

This way, a query fingerprint from the same camera model as D_i can be searched within that group and a speedup of $\frac{|D|}{|D_i|}$ can be achieved where $|d|$ is the length of the list d.

One of the main strengths of this approach is that it can be combined with any other search methodology presented in this chapter.

6.4.2   Composite Fingerprints

Bayram et al. (2012) proposed a group testing approach to organize large databases and utilize it to decrease the number of comparisons when searching for the source of a camera fingerprint (or an image). In group testing methods, multiple objects are combined to create a composite object. When a composite object is compared with a query object, a negative decision indicates that none of the objects in the group matches with the query object. On the other hand, a positive decision indicates one or more objects in the composite may match with the query object.

Building a Composite Tree

The proposed method is proposed to generate composite fingerprints, which creates unordered binary trees. A leaf node on these trees is a single fingerprint, whereas the internal nodes are the composition of their descendant nodes.

Suppose that there are N fingerprints, $\{F_1, F_2 \dots F_N\}$ which will be used to generate a composite tree as in Fig. 6.1. The composite tree in the figure is formed of 4 fingerprints (i.e., $\{F_1, F_2, F_3, F_4\}$), and the composite fingerprints are defined as

$$C_{1:N} = \frac{1}{\sqrt{N}} \sum_{j=1}^{N} F_j \tag{6.15}$$

The reason for normalizing the composite fingerprint by $\frac{1}{\sqrt{N}}$ is to ensure the composite has unit variance.

Notice that one of the main drawbacks of this approach is that it doubles the storage usage (i.e., for N fingerprints, a composite tree contains $2 * N - 1$).

Source Identification

Suppose that a database D consists of N fingerprints of length n, i.e., $D = \{F_1, F_2 \ldots F_N\}$. Now consider a query fingerprint, F_q whose source is under questioning. To determine whether F_q is captured by any camera in the composite tree formed of $\{F_1, F_2 \ldots F_N\}$, F_q and $C_{1:N}$ are correlated.

Given a preset threshold, τ, the correlation of F_q and $C_{1:N}$ can result in the following:

- H_0^c: $corr(F_q, C_{1:N}) < \tau$: none of the fingerprints in $C_{1:N}$ match with F_q
- H_1^c: $corr(F_q, C_{1:N}) \geq \tau$, one or more fingerprints in $C_{1:N}$ may match with F_q.

The decision of the preset threshold, τ depends on the number of fingerprints in a composite (i.e., N in this case).

If the decision of the correlation with $C_{1:N}$ is positive, F_q is correlated with its children (i.e., $C_{1:N/2}$ and $C_{N/2+1:N}$). This process is recursively done until a match is found or all pairwise comparisons are found to be negative.

6.5 Hybrid Methods

In this section, other methodologies combining multiple of the previous methods. These methods improve …

6.5.1 Search over Composite-Digest Search Tree

In this work, Taspinar et al. (2017) leverages the individual strengths of the search over fingerprint digest (Sect. 6.3.1) and composite fingerprint (Sect. 6.4) to make the search on large databases more efficient.

In this method, two-level composite-digest search trees are created as in Fig. 6.2. Each root node is a composite fingerprint which is the mean of the original fingerprints in the tree, and each leaf node is a digested version of individual fingerprints.

Similar to Sect. 6.4, as the number of fingerprints increases in a composite fingerprint, the share of each fingerprint will decrease, which will result in a less reliable decision. Therefore, creating a composite from so many fingerprints will not be a viable option. Because this will lead to more incorrect decisions (i.e., lower true positive and higher false positive rates). Therefore, a more viable approach is to divide the large dataset into smaller subgroups.

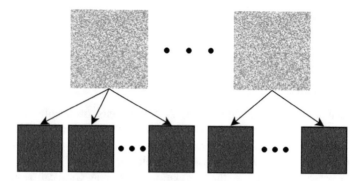

Fig. 6.2 Composite-Digest Search Tree (CDST): one-level CFST with fingerprint digests at the leaves instead of fingerprints

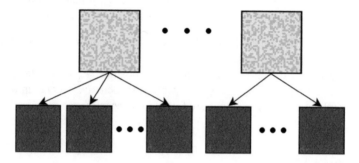

Fig. 6.3 Full Digest Search Tree (FDST): one-level CFST comprised of composite digests and fingerprint digests

The search stage for this approach is performed in two steps. In the first stage, a query fingerprint is compared with composite fingerprints. If the decision is negative, it indicates that the entire subset doesn't contain the source camera of the query image. Otherwise, a linear search is done over the fingerprint digests (leaf nodes) as opposed to the linear full-length versions of the fingerprints.

6.5.2 Search over Full Digest Search Tree

To further improve the efficiency of the previous method, Taspinar et al. present the search over a full digest search tree where the tree structure and search method are the same except that composite fingerprint are also digested as in Fig. 6.3 (Taspinar et al. 2017).

In this method, composite fingerprints contain more fingerprint elements than individual fingerprint digests. The reason for this is a composite fingerprint obtained from n fingerprints is contaminated by $n - 1$ other fingerprints when it is correlated by a fingerprint obtained from one of its child cameras. Hence, typically a higher number of fingerprint elements are used for composite fingerprint digests.

6.6 Conclusion

This chapter presented source camera identification on large databases, which is a crucial task when a crime image or video such as child pornography is found, and there is no suspect camera available. To tackle this problem, visual objects can be collected from social media, and a fingerprint estimate can be extracted from each camera in the set.

Assuming that there is no geometric transformation on visual objects, or the transformations are reverted through a registration step, this chapter focused on determining if one of the cameras in the dataset has captured the query visual objects.

The methods are grouped under three main categories: reducing the complexity of pairwise fingerprint correlation, decreasing the number of correlations, and hybrid methods that use two methods to achieve both of the ways of speeding up.

Various techniques have been presented in this chapter that can help speed up the search.

References

Achlioptas D (2003) Database-friendly random projections: johnson-lindenstrauss with binary coins. J Comput Syst Sci 66(4):671–687

Bayram S, Sencar HT, Memon N (2012) Efficient sensor fingerprint matching through fingerprint binarization. IEEE Trans Inf Forensics Secur 7(4):1404–1413

Bondi L, Bestagini P, Perez-Gonzalez F, Tubaro S (2018) Improving prnu compression through preprocessing, quantization, and coding. IEEE Trans Inf Forensics Secur 14(3):608–620

Dabov K, Foi A, Katkovnik V, Egiazarian K (2009) Bm3d image denoising with shape-adaptive principal component analysis. In: SPARS'09-signal processing with adaptive sparse structured representations

Fridrich J, Goljan M (2010) Derivation of rocs for composite fingerprints and sequential trimming. Technical Report, Department of Electrical and Computer Engineering, Binghamton University, Binghamton, NY, USA

Goljan M, Fridrich J, Filler T (2009) Large scale test of sensor fingerprint camera identification. In: Media forensics and security, vol 7254. International Society for Optics and Photonics, p 72540I

Goljan M, Fridrich J, Filler T (2010) Managing a large database of camera fingerprints. In: Media forensics and security II, vol 7541. International Society for Optics and Photonics, p 754108

Lehmann TM, Gonner C, Spitzer K (1999) Survey: interpolation methods in medical image processing. IEEE Trans Med Imaging 18(11):1049–1075

Li R, Li C-T, Guan Yu (2018) Inference of a compact representation of sensor fingerprint for source camera identification. Pattern Recogn 74:556–567

Lindenstrauss WJJ (1984) Extensions of lipschitz maps into a hilbert space. Contemp Math 26:189–206

Lukas J, Fridrich J, Goljan M (2006) Digital camera identification from sensor pattern noise. IEEE Trans Inf Forensics Secur 1(2):205–214

Tandoğan SE, Altınışık E, Sarimurat S (2019) Sencar HT (2019) Tackling in-camera downsizing for reliable camera id verification. Electron Imag 5:545–1

Taspinar S, Sencar HT, Bayram S, Memon N (2017) Fast camera fingerprint matching in very large databases. In: 2017 IEEE international conference on image processing (ICIP). IEEE, pp 4088–4092

Valsesia D, Coluccia G, Bianchi T, Magli E (2015a) Compressed fingerprint matching and camera identification via random projections. IEEE Trans Inf Forensics Secur 10(7):1472–1485

Valsesia D, Coluccia G, Bianchi T, Magli E (2015b) Large-scale image retrieval based on compressed camera identification. IEEE Trans Multimedia 17(9):1439–1449

Yaqub W, Mohanty M, Memon N (2018) Towards camera identification from cropped query images. In: 25th ICIP. IEEE, pp 3798–3802

Zhang L, Lukac R, Xiaolin W, Zhang D (2009) Pca-based spatially adaptive denoising of cfa images for single-sensor digital cameras. IEEE Trans Image Process 18(4):797–812

Chapter 7
Source Camera Model Identification

Sara Mandelli, Nicolò Bonettini, and Paolo Bestagini

Every camera model acquires images in a slightly different way. This may be due to differences in lenses and sensors. Alternatively, it may be due to the way each vendor applies characteristic image processing operations, from white balancing to compression. For this reason, images captured with the same camera model present a common series of artifacts that enable to distinguish them from other images. In this chapter, we focus on source camera model identification through pixel analysis. Solving the source camera model identification problem consists in identifying which camera model has been used to shoot the picture under analysis. More specifically, assuming that the picture has been digitally acquired with a camera, the goal is to identify the brand and model of the device used at acquisition time without the need to rely on metadata. Being able to attribute an image to the generating camera model may help forensic investigators to pinpoint the original creator of images distributed online, as well as to solve copyright infringement cases. For this reason, the forensics community has developed a wide series of methodologies to solve this problem.

7.1 Introduction

Given a digital image under analysis, we may ask ourselves a wide series of different questions about its origin. We may be interested in knowing whether the picture has been downloaded from a certain website. We may want to know which technology is used to digitalize the image (e.g., if it comes from a camera equipped with a rolling shutter or a scanner relying on a linear sensor). We may be curious about the model of camera used to shot the picture. Alternatively, we may need very specific details about the precise device instance that was used at image inception time. Despite all of these questions are related to source image attribution, they are very different in nature.

S. Mandelli · N. Bonettini · P. Bestagini (✉)
Politecnico di Milano, Milan, Italy
e-mail: paolo.bestagini@polimi.it

© The Author(s) 2022   133
H. T. Sencar et al. (eds.), *Multimedia Forensics*, Advances in Computer Vision and Pattern
Recognition, https://doi.org/10.1007/978-981-16-7621-5_7

Therefore, answering all of them is far from being an easy task. For this reason, the multimedia forensics community typically tackles one of these problems at a time.

In this chapter, we are interested in detecting the camera model that is used to acquire an image under analysis. Identifying the camera model which is used to acquire a photograph is possible thanks to the many peculiar traces left on the image at shooting time. To better understand which are the traces we are referring to, in this section we provide the reader with some background on the standard image acquisition pipeline. Finally, we provide the formal definition of the camera model identification problem considered in this chapter.

7.1.1 Image Acquisition Pipeline

We are used to shoot photographs everyday with our smartphones and cameras in a glimpse of an eye. Fractions of seconds pass from the moment we trigger the shutter to the moment we visualize the shot that we took. However, in this tiny amount of time, the camera performs a huge amount of operations.

The digital image acquisition pipeline is not unique, and may differ depending on the vendor, the device model and the available on-board technologies. However, it is reasonable to assume that a typical digital image acquisition pipeline is composed of a series of common steps (Ramanath et al. 2005), as shown in Fig. 7.1.

Light rays pass through a lens that focus them on the sensor. The sensor is typically a Charge-Coupled Device (CCD) or Complementary Metal-Oxide Semiconductor (CMOS), and can be imagined as a matrix of small elements geometrically organized on a plane. Each element represents a pixel, and returns a different voltage depending on the intensity of the light that hits it. Therefore, the higher the amount of captured light, the higher the output voltage, and the brighter the pixel value.

As these sensors react to light intensity, different strategies to capture color information may be applied. If multiple CCD or CMOS sensors are available, prisms can be used to split the light into different color components (typically red, green, and blue) that are directed to the different sensors. In this way, each sensor captures the intensity of a given color component, thus a color image can be readily obtained combining the output of each sensor. However, multiple sensors are typically available only on high-end devices, making this pipeline quite uncommon.

A more customary way of capturing color images consists in making use of a Color Filter Array (CFA) (or Bayer filter). This is a thin array of color filters placed

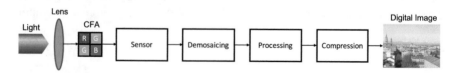

Fig. 7.1 Typical steps of a common image acquisition pipeline

on top of the sensor. Due to these filters, each sensor's element is hit only by light in a narrow wavelength band corresponding to a specific color (typically red, green or blue). This means that the sensor returns the intensity of green light for certain pixels, the intensity of blue light for other pixels, and the intensity of red light for the remaining pixels. Which pixels capture which color depends on the shape of the CFA, which is a vendor choice. At this point, the output of the sensor consists of three partially sampled color layers in which only one color value is recorded at each pixel location. Missing color information, like the blue and red components for pixels that only acquired green light, are retrieved via interpolation from neighboring cells with the available color components. This procedure is known as debayering or demosaicing, and can be implemented using proprietary interpolation techniques.

After this raw version of a color image is obtained, a list of additional operations are in order. For instance, as lenses may introduce some kinds of optical distortion, most notably barrel distortion, pincushion distortion or combinations of them, it is common to apply some digital correction that may introduce forensic traces. Additionally, white balancing and color correction are other operations that are often applied and may be vendor-specific. Finally, lossy image compression is typically applied by means of JPEG standard, which again may vary with respect to compression quality and vendor-specific implementation choices.

Since a few years ago, these processing steps were the main sources of camera model artifacts. However, with the rapid proliferation of computational photography techniques, modern devices implement additional custom functionalities. This is the case of bokeh images (also known as portrait images) synthetically obtained through processing. These are pictures in which the background is digitally blurred with respect to the foreground object to obtain an artistic effect. Moreover, many devices implement the possibility of shooting High Dynamic Range (HDR) images, which are obtained by combining multiple exposures into a single one. Additionally, several smartphones are equipped with multiple cameras and produce pictures by mixing their outputs. Finally, many vendors introduce the possibility of shooting photographs with special filter effects that may enhance the picture in different artistic ways. All of these operations are custom and add traces to the pool of artifacts that can be exploited as a powerful asset for forensic analysis.

7.1.2 Problem Formulation

The problem of source camera model identification consists in detecting the model of the device used to capture an image. Although the definition of this problem seems pretty straightforward, depending on the working hypothesis and constraints, the problem may be cast in different ways. For instance, the analyst may only have access to a finite set of possible camera models. Alternatively, the forensic investigator may want to avoid using metadata. In the following, we provide the two camera model identification problem formulations that we consider in the rest of the chapter.

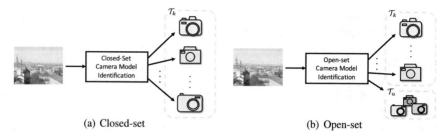

Fig. 7.2 Representation of closed-set (**a**) and open-set (**b**) camera model identification problem

In both formulations, we only focus on pixel-based analysis, i.e., we do not consider the possibility of relying on metadata information.

Closed-set Identification

Closed-set camera model identification refers to the problem of detecting which is the camera model used to shoot a picture within a set of known devices, as shown in Fig. 7.2a. In this scenario, the investigator assumes that the image under analysis has been taken with a device within a family of devices she/he is aware of. If the image does not come from any of those devices, the investigator will wrongly attribute the image to one of those known devices, no matter what.

Formally, let us define the set of labels of known camera models as $\mathcal{T}_k$. Moreover, let us define **I** as a color image acquired by the device characterized with label $t \in \mathcal{T}_k$. The goal of closed-set camera model identification is to provide an estimate $\hat{t}$ of t given the image **I** and the set $\mathcal{T}_k$. Notice that $\hat{t} \in \mathcal{T}_k$ by construction. Therefore, if the hypothesis that $t \in \mathcal{T}_k$ does not hold in practice, this approach is not applicable, as the condition $\hat{t} \in \mathcal{T}_k$ would imply that $\hat{t} \neq t$.

Open-set Identification

In many scenarios, it is not realistic to assume that the analyst has full control over the complete set of devices that may have been used to acquire the digital image under analysis. In this case, it is better to resort to the open-set problem formulation. The goal of open-set camera model identification is twofold, as shown in Fig. 7.2b. Given an image under analysis, the analyst aims

- To detect if the image comes from the set of known camera models or not.
- To detect the specific camera model, if the image comes from a known model.

This is basically a generalization of the closed-set formulation that accommodates for the analysis of images coming from unknown devices.

Formally, let us define the set of labels of known models as $\mathcal{T}_k$, and the set of labels of unknown models (i.e., all the other existing camera models) as $\mathcal{T}_u$. Moreover, let us define **I** as a color image acquired by the device characterized with label $t \in \mathcal{T}_k \bigcup \mathcal{T}_u$. The goal of open-set camera model identification is

- To estimate whether $t \in \mathcal{T}_k$ or $t \in \mathcal{T}_u$.
- To provide an estimate $\hat{t}$ of t in case $t \in \mathcal{T}_k$. In this case also $\hat{t} \in \mathcal{T}_k$.

Despite this problem formulation looks more realistic than its closed-set counterpart, the vast majority of camera model literature only copes with the closed-set problem. This is mainly due to the difficulty in well modeling the unknown set of models.

7.2 Model-Based Approaches

As mentioned in Sect. 7.1.1, multiple operations performed during image acquisition may be characteristic of a specific camera brand and model. This section provides an overview of camera model identification methods that work by specifically leveraging those traces. These methods are known as model-based methods, as they assume that each artifact can be modeled, and this model can be exploited to reverse engineer the used device. Methods that model each step of the acquisition chain are historically the first ones being developed in the literature (Swaminathan et al. 2009).

7.2.1 Color Filter Array (CFA)

As CCD/CMOS sensor elements of digital cameras are sensitive to the received light intensity and not to specific colors, the CFA is usually introduced in order to split the incoming light into three corresponding color components. Then, a three-channel color image can be obtained by interpolating the pixel information associated with the color components filtered by the CFA. There are several works in the literature that exploit specific characteristics associated with the CFA configuration, i.e., the specific arrangement of color filters in the sensor plane and the CFA interpolation algorithm to retrieve information about the source camera model.

CFA Configuration
Even without investigating the artifacts introduced by demosaicing, we can exploit information from the Bayer configuration to infer some model-specific features. For example, Takamatsu et al. (2010) developed a method to automatically identify CFA patterns from the distribution of image noise variances. In Kirchner (2010), the Bayer configuration was estimated by restoring the raw image (i.e., prior to demosaicing) from the output interpolated image. Authors of Cho et al. (2011) showed how to estimate the CFA pattern from a single image by extracting pixel statistics on 2×2 image blocks.

CFA Interpolation
In an image acquisition pipeline, the demosaicing step inevitably injects some inter-pixel correlations into the interpolated image. Existing demosaicing algorithms differ

in the size of their support region, in the way they select the pixel neighborhood, and in their assumptions about the image content and adaptability to this Gunturk et al. (2005), Menon and Calvagno (2011). Over the years, the forensics community has largely exploited these interpolation traces to discriminate among different camera models.

The first solution dates back to 2005 (Popescu and Farid 2005). The authors estimated the inter-pixel correlation weights from a small neighborhood around a given pixel, treating each color channel independently. They also observed that different interpolation algorithms present different correlation weights. Similarly, authors of Bayram et al. (2005) exploited the correlation weights estimated as shown in Popescu and Farid (2005) and combined them with frequency domain features. In Bayram et al. (2006), the authors improved upon their previous results by treating smooth and textured image regions differently. Their choice was motivated by the different treatment of distinct demosaicing implementations on high-contrast regions.

In 2007, Swaminathan et al. (2007) estimated the demosaicing weights only on interpolated pixels, i.e., without accounting for pixels which are relatively invariant to the CFA interpolation process. Authors found the CFA pattern by fitting linear filtering models and selecting the one which minimized the interpolation error. As done in Bayram et al. (2006), they considered three diverse estimations according to the local texture of the images. Interestingly, the authors' results pointed out some similarity in interpolation patterns among camera models from the same manufacturer.

For the first time, Cao and Kot (2009) did not explore each color channel independently, but instead exposed cross-channel pixel correlations caused by demosaicing. Authors reported that many state-of-the-art algorithms often employ color difference or hue domains for demosaicing, and this inevitably injects a strong correlation across image channels. In this vein, authors extended the work by Swaminathan et al. (2007) by estimating the CFA interpolation algorithm using a partial second-order derivative model to detect both intra-channel and cross-channel demosaicing correlations. They estimated these correlations by grouping pixels into 16 categories based on their CFA positions. In 2010, the same authors tackled the camera model identification problem on mobile phone devices, showing how CFA interpolation artifacts enable to achieve excellent classification results on dissimilar models, while confusing cameras of the same or very similar models (Cao and Kot 2010).

In line with Cao and Kot (2009), also Ho et al. (2010) exploited cross-channel interpolation traces. The authors measured the inter-channel differences of red and blue colors with respect to the green channel, and converted these into the frequency domain to estimate pixel correlations. They were motivated by the fact that many demosaicing algorithms interpolate color differences instead of color channels (Gunturk et al. 2005). Similarly, Gao et al. (2011) worked with variances of cross-channel differences.

Differently from previous solutions, in 2015, Chen and Stamm (2015) proposed a new framework to deal with camera model identification, inspired to the rich models proposed by Fridrich and Kodovský (2012) for steganalysis. Authors pointed out that previous solutions were limited by their essential use of linear or local linear parametric models, while modern demosaicing algorithms are both non-linear and adaptive (Chen and Stamm 2015). The authors proposed to build a rich model of demosaicing by grouping together a set of non-parametric submodels, each capturing specific partial information on the interpolation algorithm. The resulting rich model could return a much more comprehensive representation of demosaicing compared to previous strategies.

Authors of Zhao and Stamm (2016) focused on controlling the computation cost associated with the solution proposed in Swaminathan et al. (2007), which relied on least squares estimation and could become impractical when a consistent number of pixels was employed. The authors proposed an algorithm to find the pixel set that yields the best estimate of the parametric model, still keeping the computational complexity feasible. Results showed that the proposed method could achieve higher camera model classification accuracy than Swaminathan et al. (2007), at a fixed computational cost.

7.2.2 Lens Effects

Every digital camera includes a sophisticated optical system to project the acquired light intensity on small CCD or CMOS sensors. Projection from the lens on to the sensor inevitably injects some distortions in the acquired image, usually known as optical aberrations. Among them, the most common optical aberrations are radial lens distortion, chromatic aberration, and vignetting. The interesting point from a forensics perspective is that different camera models use different optical systems, thus they reasonably introduce different distortions during image acquisition. For this reason, the forensics literature has widely exploited optical aberration as a model-based trace.

Radial Lens Distortion
Radial lens distortion is due to the fact that lenses in consumer cameras usually cannot magnify all the acquired regions with a constant magnification factor. Thus, different focal lengths and magnifications appear in different areas. This lens imperfection causes radial lens distortion which is a non-linear optical aberration that renders straight lines in the real image as curved lines on the sensor. Barrel distortion and pincushion distortion are the two main distortion forms we usually find in digital images. In San Choi et al. (2006), the authors measured the level of distortion of an image and used the distortion parameters as a feature to discriminate among different source camera models. Additionally, they investigated the impact of optical zoom on the reliability of the method, noticing that a classification beyond mid-range focal lengths can be problematic, due to the vanishing of distortion artifacts.

Chromatic Aberration
Chromatic aberration stems from wavelength-dependent variations of the refractive index of a lens. This phenomenon causes a spread of the color components over the sensor plane. The consequence of chromatic aberration is that color fringes appear

in high-contrast regions. We can identify axial chromatic aberration and lateral chromatic aberration. The former accounts for the variations of the focal point along the optical axis; the latter indicates the relative displacement of different light components along the sensor plane.

In 2007, Lanh et al. (2007) applied the model proposed in Johnson and Farid (2006) to estimate the lateral chromatic aberration using small patches extracted from the image center. Then, they fed the estimated parameters to a Support Vector Machine (SVM) classifier for identifying the source camera model. Later, Gloe et al. (2010) proposed to reduce the computational cost of the previous solution by locally estimating the chromatic aberration. The authors also pointed out some issues due to nonlinear phenomena occurring in modern lenses. Other studies were carried on in Yu et al. (2011), where authors pointed out a previously overlooked interaction between chromatic aberration and focal distance of lenses. Authors were able to obtain a stable chromatic aberration pattern distinguishing different copies of the same lens.

Vignetting
Vignetting is the phenomenon of light intensity fall-off around the corners of an image with respect to the image center. Usually, wide-aperture lenses are more prone to vignetting, because fewer light rays reach the sensor's edges.

The authors of Lyu (2010) estimated the vignetting pattern from images adopting a generalization of the vignetting model proposed in Kang and Weiss (2000). They exploited statistical properties of natural images in the derivative domain to perform a maximum likelihood estimation. The proposed method was tested among synthetically generated and real vignetting, showing far better results for lens model identification on synthetic data. The lower accuracy on real scenarios can be due to difficulties in correctly estimating the vignetting pattern whenever highly textured images are considered.

7.2.3 Other Processing and Defects

The traces left by the CFA configuration, the demosaicing algorithm, and the optical aberrations due to lens defects are only a few of the footprints that forensic analysts can exploit to infer camera model-related features. We report here some other model-based processing operations and defects that carry information about the camera model.

Sensor Dust
In 2008, Dirik et al. (2008) exploited the traces left by dust particles on the sensor of digital single-lens reflex cameras to perform source camera identification. The authors estimated the dust pattern (i.e., its location and shape) from the camera or from a number of images shot by the camera. The proposed methodology was robust

to both JPEG compression and downsizing operations. However, the authors pointed out some issues in correctly estimating the dust pattern for complex and non-smooth images, especially for low focal length values.

Noise Model

Modeling the noise pattern of acquired images can represent a valuable feature to distinguish among cameras of the same or different models.

Here, Photo Response Non Uniformity (PRNU) is arguably one of the most influential contributions to multimedia forensics. PRNU is a multiplicative noise pattern that occurs in any sensor-recorded digital image, due to imperfections in the sensor manufacturing process. In Lukás et al. (2006), Chen et al. (2008) the authors provided a complete modeling of the digital image at the sensor output as a function of the incoming light intensity and noise components. They discovered the PRNU noise to represent a powerful and robust device-related fingerprint, able to uniquely identify different devices of the same model. Since this chapter deals with model-level identification granularity, we do not further analyze the potential of PRNU for device identification.

Contrarily to PRNU-based methods, which model only the multiplicative noise term due to sensor imperfections, other methods focused on modeling the entire noise corrupting the digital image and exploited this noise to tackle the camera model identification task. For instance, Thai et al. (2014) worked with raw images at the sensor output. The authors modeled the complete noise contribution of natural raw images (known as the heteroscedastic noise Foi et al. 2009) with only two parameters, and used it as a fingerprint to discriminate camera models. Contrarily to PRNU noise, heteroscedastic noise could not separate different devices of the same camera model, thus it is more appropriate for model-level identification granularity.

The heteroscedastic noise consists of a Poisson-distributed component which accounts for the photon shot noise and dark current, and a Gaussian-distributed term which addresses other stationary noise sources like read-out noise (Foi et al. 2009). The proposed approach in Thai et al. (2014) involved the estimation of the two characteristic noise parameters per camera model considering 50 images shot by the same model. However, the method presented some limitations: first, it requires raw image format without post-processing or compression operations; second, non-linear processes like gamma correction modify the heteroscedastic noise model; finally, changes in ISO sensitivity and demosaicing operations worsen the detector performances.

In 2016, the authors improved upon their previous solution, solving the camera model identification from TIFF and JPEG compressed images and taking into account the non-linear effect of gamma correction (Thai et al. 2016). They proposed a generalized noise model starting from the previously exploited heteroscedastic noise but including also the effects of subsequent processing and compression steps. This way, each camera model could be described by three parameters. The authors tested their methodology over 18 camera models.

7.3   Data-Driven Approaches

Contrarily to model-based methods presented in Sect. 7.2, in the last few years we have seen the widespread adoption of data-driven approaches to camera model identification. All solutions that extract knowledge and insights from data without explicitly modeling the data behavior using statistical models fall into this category. These data-driven approaches have greatly outperformed multiple model-based solutions proposed in the past. While model-based solutions usually focus on a specific component of the image acquisition pipeline, data-driven approaches can capture model traces left by the interplay among multiple components. For instance, image noise characteristics do not only originate from sensor properties and imperfections, but are the result of CFA interpolation and other internal processing operations (Kirchner and Gloe 2015).

We can further divide data-driven approaches into two broad categories:

1. Methods based on hand-crafted features, which derive properties of data by extracting suitable data descriptors like repeated image patterns, texture, and gradient orientations.
2. Methods based on learned features, which directly exploit raw data, i.e., without extracting any descriptor, to learn distinguishable data properties and to solve the specific task.

7.3.1   Hand-Crafted Features

Over the years, the multimedia forensics community has often imported image descriptors from other research domains to solve the camera model identification problem. For instance, descriptors developed for steganalysis and image classification have been widely used to infer forensic traces on images. The reason behind their application in the forensics field is that, in general, these descriptors enable an effective representation of images and contain relevant information to distinguish among different image sources.

Surprisingly, the very first proposed solution to solve the camera model identification problem in a blind setup exploited hand-crafted features (Kharrazi et al. 2004). The authors extracted descriptors about color channel distributions, wavelet statistics and image quality metrics, then they fed these to an SVM classifier to distinguish among few camera models.

In the following, we illustrate the existing solutions based on hand-crafted feature extraction.

Local Binary Patterns
Local binary patterns (Ojala et al. 2002) represent a good example of a general image descriptor that can capture various image characteristics. In a nutshell, local binary patterns can be computed as the difference between a central pixel and its

local neighborhood, binary quantized and coded to produce an histogram carrying information about the local inter-pixel relations.

In 2012, the authors of Xu and Shi (2012) were inspired by the idea that the entire image acquisition pipeline could generate localized artifacts in the final image, and these characteristics could be captured by the uniform grayscale invariant local binary patterns (Ojala et al. 2002). The authors extracted features for red and green color channel in the spatial and wavelet domains, resulting in a 354-dimension feature per image. As commonly done in the majority of works previous to the wide adoption of neural networks, the authors fed these features to an SVM classifier to classify image sources among 18 camera models.

DCT Domain Features

Since the vast majority of cameras automatically stores the acquired images in JPEG format, many forensics works approach camera model identification by exploiting some JPEG-related features. In particular, many researches focused on extracting model-related features in the Discrete Cosine Transform (DCT) domain.

In 2009, Xu et al. (2009) extracted the absolute value of the quantized 8×8 DCT coefficient blocks, then computed the difference between the element blocks along four directions to look for inter-coefficient correlations. They fed these to Markov transition probability matrices to identify statistical differences inside images of distinct camera models. The elements of transition probability matrices were thresholded and fed as hand-crafted features to an SVM classifier.

DCT domain features have been exploited also in Wahab et al. (2012), where the authors extracted some conditional probability features from the absolute values of three 8×8 DCT coefficient blocks. From these blocks, they only used the 4×4 upper left frequencies which demonstrated to be most significant for the model identification task. The authors fed an SVM classifier to distinguish among 4 camera models.

More recently, the authors of Bonettini et al. (2018) have shown that JPEG eigen-algorithm features can be used for camera model identification as well. These features capture the differences among DCT coefficients obtained after multiple JPEG compression steps. A standard random forest classifier is used for the task.

Color Features

Camera model identification can be faced also by looking at some model-specific image color features. Such artifacts can be found according to the specific properties of the used CFA and color correction algorithms.

As previously reported, in 2004, Kharrazi et al. (2004) assumed that certain traits and patterns should exist between the RGB bands of an image, regardless of the scene content depicted. Therefore, they extracted 21 image features which aimed at capturing cross-channel correlations and energy ratios together with intra-channel average value, pixel neighbor distributions and Wavelet statistics. In 2009, this set of features was enlarged with 6 additional color features by Gloe et al. (2009), which included the dependencies between the average values of the color channels.

Specifically, the authors extended the feature set by computing the norms of the difference between the original and the white-point corrected image.

Image Quality Metrics Features
In 2002, Avcibas et al. (2002) proposed to extract some image quality metrics to infer statistical properties from images. This feature set included measures based on pixel differences, correlations, edges, spectral content, context, and human visual system. The authors proposed to exploit these features for steganalysis applications.

In 2004, Kharrazi et al. (2004) employed a subset of these metrics to deal with the camera model identification task. The authors motivated their choice by noticing that different cameras produce images of different quality, in terms of sharpness, light intensity and color quality. They computed 13 image quality metrics, dividing them into three main categories: those based on pixel differences, those based on correlations and those based on spectral distances. While Kharrazi et al. (2004) averaged these metrics over the three color channels, Çeliktutan et al. (2008) evaluated each color band separately resulting in 40 features.

Wavelet Domain Features
Wavelet-domain features are known to be effective to analyze the information content of images and noise components, enabling to capture interesting insights on their statistical properties for what concerns different resolutions, orientations, and spatial positions (Mallat 1989).

These features have been exploited as well to solve the camera model identification task. For instance, Kharrazi et al. (2004) exploited 9 Wavelet-based color features. Later on, Çeliktutan et al. (2008) added 72 more features by evaluating the four moments of the Wavelet coefficients over multiple sub-bands, and by predicting sub-band coefficients from their neighborhood.

Binary Similarity Features
Like image quality metrics, also binary similarity features were first proposed to tackle steganalysis problems and then adopted in forensics applications. For instance, Avcibas et al. (2005) investigated statistical features extracted from different bit planes of digital images, measuring their correlations and their binary texture characteristics.

Binary similarities were exploited in 2008 by Çeliktutan et al. (2008), which considered the characteristics of the neighborhood bit patterns among the less significant bit planes. The authors investigated histograms of local binary patterns by accounting for occurrences within a single bit plane, across different bit planes and across different color channels. They resulted in a feature vector of 480 elements.

Co-occurrence Based Local Features
Together with binary similarity features and image quality metrics, also rich models, or co-occurrence-based local features (Fridrich and Kodovský 2012), were imported from steganalysis to forensics. Three main steps compose the extraction of co-occurrences of local features (Fridrich and Kodovský 2012): (i) the computation

of image residuals via high-pass filtering operations; (ii) the quantization and truncation of residuals; (iii) the computation of the histogram of co-occurrences.

In this vein, Chen and Stamm (2015) built a rich model of a camera demosaicing algorithm inspired by Fridrich and Kodovský (2012) to perform camera model identification. The authors generated a set of diverse submodels so that every submodel conveyed different aspects of the demosaicing information. Every submodel was built using a particular baseline demosaicing algorithm and computing co-occurrences over a particular geometric structure. By merging all submodels together, the authors obtained a feature vector of 1,372 elements to feed an ensemble classifier.

In 2017, Marra et al. (2017) extracted a feature vector of 625 co-occurrences from each color channel and concatenated the three color bands, resulting in 1,875 elements. The authors investigated a wide range of scenarios that might occur in real-world forensic applications, considering both the closed-set and the open-set camera model identification problems.

Methods with Several Features

In many state-of-the-art works, several different kinds of features were extracted from images instead of just one. The motivation lied in the fact that merging multiple features could better help finding model-related characteristics, thus improving the source identification performance.

For example, (as already reported) Kharrazi et al. (2004) combined color-related features with Wavelet statistics and image quality metrics, ending with 34-element features. Similarly, Gloe (2012) used the same kinds of features but extended the set up to 82 elements. On the other hand, Çeliktutan et al. (2008) exploited image quality metrics, Wavelet statistics, and binary similarity measures, resulting in a total of 592 features.

Later on, in 2018, Sameer and Naskar (2018) investigated the dependency of source classification performance with respect to illumination conditions. To infer model-specific traces, they extracted image quality metrics and Wavelet features from images.

Hand-crafted Features in Open-set Problems

The open-set identification problem started to be investigated in 2012 by Gloe (2012), which proposed two preliminary solutions based on the extraction of 82-element hand-crafted features from images. The first approach was based on a one-class SVM, trained for each available camera model. The second proposal trained a binary SVM in one-versus-one fashion, for all pairs of known camera models. However, the achieved results were unconvincing and revealed practical difficulties to handle unknown models. Nonetheless, this represented a very first attempt to handle unknown source models and highlighted the need of further investigations on the topic.

Later in 2017, Marra et al. (2017) explored the open-set scenario considering two case studies: (i) the limited knowledge situation, in which only images from one camera model were available; (ii) the zero knowledge situation, in which authors had no clue on the model used to collect images. In the former case, the authors aimed at detecting whether images came from the known model or from an unknown one; in the latter case, the goal was to retrieve a similarity among images shot by the same model.

To solve the first issue, the authors followed a similar approach to Gloe (2012) by training a one-class SVM on the co-occurrence based local features extracted from known images. They solve the second task by looking for K nearest neighbor features in the testing image dataset using a kd-tree-based search algorithm. The performed experiments reported acceptable results but definitively underlined the urgency of new strategies to handle realistic open-set scenarios.

7.3.2 Learned Features

We have recently assisted in the rapid rise of solutions based on learned forensic features which have quickly replaced hand-crafted forensic feature extraction. For instance, considering the camera model identification task, we can directly feed digital images to a deep learning paradigm in order to learn model-related features to associate images with their original source. Actually, we can differentiate between two kinds of learning frameworks:

1. Two-step sequential learning frameworks, which first learn significant features from data and successively learn how to classify data according to the extracted features.
2. End-to-end learning frameworks, which directly classify data by learning some features in the process.

Among deep learning frameworks, Convolutional Neural Networks (CNNs) are now the widespread solution to face several multimedia forensics tasks. In the following, we report state-of-the-art solutions based on learned features which tackle the camera model identification task.

Two-step Sequential Learning Frameworks

One of the first contributions to camera model identification with learned features was proposed in 2017 by Bondi et al. (2017a). The authors proposed a data-driven approach based on CNNs to learn model-specific data features. The proposed methodology was based on two steps: first, training a CNN with image color patches of 64×64 pixels to learn a feature vector of 128 elements per patch; second, training an SVM classifier to distinguish among 18 camera models from the Dresden Image Database (Gloe and Böhme 2010). In their experimental setup, the authors outperformed state-of-the-art methods based on hand-crafted features (Marra et al. 2017; Chen and Stamm 2015).

In the same year, Bayar and Stamm (2017) proposed a CNN-based approach robust to resampling and recompression artifacts. In their paper, the authors proposed an end-to-end learning approach (completely based on CNNs) as well as a two-step learning approach. In the latter scenario, they extracted a set of deep features from the penultimate CNN layer and fed an Extremely Randomized Trees (ET) classifier, purposely trained for camera model identification. The method was tested on 256×256 image patches (retaining only the green color channel) from 26 camera models of the Dresden Image Database (Gloe and Böhme 2010). To mimic realistic scenarios, JPEG compression and resampling were applied to the images. The two-step approach returned slightly better performances than the end-to-end approach.

The same authors also started to investigate open-set camera model identification using deep learning (Bayar and Stamm 2018). They compared different two-step approaches based on diverse classifiers, showing that the ET classifier outperformed the other options.

Meanwhile, Ferreira et al. (2018) proposed a two-step sequential learning approach in which the feature extraction phase was composed of two parallel CNNs, an Inception-ResNet (Szegedy et al. 2017) and an Xception (Chollet 2017) architecture. First, the authors selected 32 patches of 229×229 pixels per image according to an interest criterion. Then, they separately trained the two CNNs returning 256-element features each. They merged these features and passed them through a secondary shallow CNN for classification. The proposed method was tested over the IEEE Signal Processing Cup 2018 dataset (Stamm et al. 2018; IEEE Signal Processing Cup 2018 Database 2021). The authors considered data augmentations as well, including JPEG compression, resizing, and gamma correction.

In 2019, Rafi et al. (2019) proposed another two-step pipeline. The authors first trained a 201-layer DenseNet CNN (Huang et al. 2017) with image patches of 256×256 pixels. They proposed the DenseNet architecture since dense connections could help in detecting minute statistical features related to the source camera model. Moreover, the authors augmented the training set by applying JPEG compression, gamma correction, random rotation, and empirical mode decomposition (Huang et al. 1998). The trained CNN was used to extract features from image patches of different sizes; these features were concatenated and employed to train a second network for classification. The proposed method was trained over the IEEE Signal Processing Cup 2018 dataset (Stamm et al. 2018; IEEE Signal Processing Cup 2018 Database 2021) but tested on the Dresden Image Database as well (Gloe and Böhme 2010).

Following the preliminary analysis performed by Bayar and Stamm (2018), Júnior et al. (2019) proposed an in-depth study on open-set camera model identification. The authors pursued a two-steps learning approach, comparing several feature extraction algorithms and classifiers. Together with hand-crafted feature extraction methodologies, the CNN-based approach proposed in Bondi et al. (2017a) was investigated. Not much surprisingly, this last approach revealed to be the best effective for feature extraction.

An interesting two-step learning framework was proposed by Guera et al. (2018). The authors explored a CNN-based solution to estimate the reliability of a given image patch, i.e., its likelihood to be used for model identification. Indeed, saturated or dark regions might not contain sufficient information on the source model, thus could be discarded to improve the attribution accuracy. The authors borrowed the CNN architecture from Bondi et al. (2017a) as feature extractor and built a multi-layer network for reliability estimation. They compared an end-to-end learning approach with two different sequential learning strategies, showing that two-step approaches largely outperformed the end-to-end solution. The authors suggested the lower performance of the end-to-end approach could be due to the limited amount of training data, probably not enough to learn all the CNN parameters in end-to-end fashion. The proposed method was tested on the same Dresden's models employed in Bondi et al. (2017a). By discarding unreliable patches from the process, an improvement on 8% was measured on the identification accuracy.

End-to-end Learning Frameworks

The very first approach to camera model identification with learned features was proposed in 2016 by Tuama et al. (2016). Differently from the above presented works, the authors developed a complete end-to-end learning framework, where feature extraction and classification were performed in a unified framework exploiting a single CNN. The proposed method was quite innovative considering that the literature was based on feature extraction and classification steps. In their work, the authors proposed to extract 256×256 color patches from images, to compute a residual through high-pass filtering, and then to feed a shallow CNN architecture which returned a classification score for each camera model. The method was tested on 26 models from the Dresden Image Database (Gloe and Böhme 2010) and 6 further camera models from a personal collection.

In 2018, Yao et al. (2018) proposed a similar approach to Bondi et al. (2017a), but the authors trained a CNN using an end-to-end framework. 64×64 color patches were extracted from images, fed to a CNN and then passed through majority voting to classify the source camera model. The authors evaluated their methodology on 25 models from the Dresden Image Database (Gloe and Böhme 2010), investigating also the effects of JPEG compression, noise addition, and image re-scaling on a reduced set of 5 devices.

Another end-to-end approach was proposed in 2020 by Rafi et al. (2021), which put particular emphasis on the use of CNNs as pre-processing block prior to a classification network. The pre-processing CNN was composed of a series of "RemNant" blocks, i.e., 3-layer convolutional blocks followed by batch normalization. The output to the pre-processing step contained residual information about the input image and preserved model-related traces from a wide range of spatial frequencies. The whole network, named "RemNet", included the pre-processing CNN and a shallow CNN for classification. The authors tested their methodology on 64×64 image patches, considering 18 models from the Dresden Image Database (Gloe and Böhme 2010) and the IEEE Signal Processing Cup 2018 dataset (Stamm et al. 2018; IEEE Signal Processing Cup 2018 Database 2021).

A very straightforward end-to-end learning approach was deployed in Mandelli et al. (2020), where the authors did not aim at exploring novel strategies to boost the achieved identification accuracy, but investigated how JPEG compression artifacts can affect the results. The authors compared four state-of-the-art CNN architectures in different training and testing scenarios related to JPEG compression. Experiments were computed on 512×512 image patches from 28 camera models of the Vision

Image Dataset (Shullani et al. 2017). Results showed that being careful to the JPEG grid alignment and to the compression quality factor of training/testing images is of paramount importance, independently on the chosen network architecture.

Learned Features in Open-set Problems

The majority of the methods presented in previous sections tackled closed-set camera model identification, leaving open-set challenges to future investigations. Indeed, the open-set literature still counts very few works and lacks proper algorithm comparisons with respect to closed-set investigations. Among learned feature methodologies, only Bayar and Stamm (2018) and Júnior et al. (2019) tackled open-set classification. Here, we illustrate the proposed methodologies in greater more detail.

In 2018, Bayar and Stamm (2018) proposed two different learning frameworks for open-set model identification. The common paradigm behind the approaches was two-step sequential learning, where the feature extraction block included all the layers of a CNN that precede the classification layer. Following their prior work (Bayar and Stamm 2017), the authors trained a CNN for camera model identification and selected the resulting deep features associated with the second-to-last layer to train a classifier. The authors investigated 4 classifiers: (i) one fully connected layer followed by the softmax function, which is the standard choice in end-to-end learning frameworks; (ii) an ET classifier; (iii) an SVM-based classifier; (iv) a classifier based on the cosine similarity distance.

In the first approach, the authors trained the classifier in closed-set fashion over the set of known camera models. In the testing phase, they thresholded the maximum score returned by the classifier. If that score exceeded a predefined threshold, the image was associated with a known camera model, otherwise it was declared to be of unknown provenance.

In the second approach, known camera models were divided in two disjoint sets: "known-known" models and "known-unknown" models. A closed-set classifier was trained over the reduced set of "known-known" models, while a binary classifier was trained to distinguish "known-known" models from "known-unknown" models. In the testing phase, if the binary classifier returned the "known-unknown" class, the image was linked to an unknown source model, otherwise, the image was associated with a known model.

In any case, when the image was determined to belong to a known model, the authors estimated the actual source model as in a standard closed-set framework.

Experimental results were evaluated on 256×256 grayscale image patches selected from 10 models of the Dresden Image Database (Gloe and Böhme 2010) and other 15 personal devices. The authors showed that the softmax-based classifier performed worst among the 4 proposals. The ET classifier achieved the best performance for both approaches both in identifying the source model in closed-set fashion and in known-vs-unknown model detection.

An in-depth study on the open-set problem has been pursued by Júnior et al. (2019). The authors thoroughly investigated this issue by comparing 5 feature extraction methodologies, 3 training protocols and 12 open-set classifiers. Among feature extractors, they selected hand-crafted based frameworks (Chen and Stamm 2015;

Marra et al. 2017) together with learned features (Bondi et al. 2017a). As for training protocols, they considered the two strategies presented by Bayar and Stamm (2018) and one additional strategy which included more "known-unknown" camera models. Several open-set classifiers were explored, ranging from different SVM-based approaches to one classifier proposed in the past by the same authors and to those previously investigated by Bayar and Stamm (2018).

The training dataset included 18 camera models from the Dresden Image Database (Gloe and Böhme 2010), i.e., the same models used by Bondi et al. (2017a) in closed-set scenario, as known models, and the remaining Dresden's models as "known-unknown" to be used for the additional proposed training protocol. Furthermore, 35 models from the Image Source Attribution UniCamp Dataset (Image Source Attribution UniCamp Dataset 2021) and 250 models from the Flickr image hosting service were used to simulate unknown models.

As performed by Bondi et al. (2017a), the authors extracted 32 non-overlapping 64×64 patches from images. Results demonstrated the highest effectiveness of CNN-based learned features over hand-crafted features. Many open-set classifiers returned high performances and, as reported in Bayar and Stamm (2018), ET classifier was one of the best performing. Interestingly, the authors noticed that the third additional training protocol that required more "known-unknown" models was outperformed by a simpler solution, corresponding to the second training approach proposed by Bayar and Stamm (2018).

Learned Forensics Similarity

An interesting research work which parallels camera model identification was proposed in 2018 by Mayer and Stamm (2018). In their paper, the authors proposed a two-step learning framework that did not aim at identifying the source camera model of an image, but aimed at determining whether two images (or image patches) were shot by the same camera model.

The main novelty did not lie in the feature extraction step, which was CNN-based as many other contemporaneous works, but in the second step developing a learned forensics similarity measure between the two compared patches. This was implemented by training a multi-layer neural network that mapped the features from patches into a single similarity score. If the score overcame a predefined threshold, the two patches were said to come from the same model.

The authors did not limit the comparison over known camera models, but also demonstrated the effectiveness of the method on unknown models. In particular, they trained the two networks over disjoint camera model sets, exploiting grayscale image patches with size 256×256, selected from 20 models of the Dresden Image Database (Gloe and Böhme 2010) and 45 further models. Results showed that authors could accurately detect a model similarity even if both the two images came from unknown sources.

7.4   Datasets and Benchmarks

This section provides a template with the fundamental characteristics an image dataset should have in order to explore camera model identification. Then, it shows an overview of available datasets in the literature. It also explains how to properly deal with these datasets for a fair evaluation.

7.4.1   Template Dataset

The task of camera model identification assumes that we are not interested in retrieving information on the specific device used for capturing a photograph neither on the scene content depicted. Given these premises, we can define some "good" features a template dataset should include (Kirchner and Gloe 2015):

- Images of similar scenes taken with different camera models.
- Multiple devices per camera model.

Collecting several images with similar scenes shot by different models avoids any possible bias on the results due to the depicted scene content. Moreover, capturing images from multiple devices of the same camera model accounts for realistic scenarios. Indeed, we should expect to investigate query images shot by devices never seen at algorithm development phase, even though their camera models were known.

A common danger that needs to be avoided is to confuse model identification with device identification. Hence, as we want to solve the source identification task at model-based granularity, we must not confuse device-related traces with model-related ones. In other words, query images coming from an unknown device of a known model should not be confused as coming from an unknown source. In this vein, a dataset should consist of multiple devices from the same model. This enables to evaluate a feature set or a classifier for its ability to detect models independently from individual devices. Including more devices per model in the image dataset allows to check for this requirement and helps keeping the problem at bay.

7.4.2   State-of-the-art Datasets

The following datasets are frequently used or have been specifically designed for camera model identification:

- Dresden Image Database (Gloe and Böhme 2010).
- Vision Image Dataset (Shullani et al. 2017).
- Forchheim Image Database (Hadwiger and Riess 2020).
- Dataset for Camera Identification on HDR images (Shaya et al. 2018).
- Raise Image Dataset (Nguyen et al. 2015).

Table 7.1 Main datasets' characteristics

Dataset	No. of models	No. of images	Image formats
Dresden (Gloe and Böhme 2010)	26	18,456	JPEG, Uncompressed-Raw
Vision (Shullani et al. 2017)	28	34,427	JPEG
Forchheim (Hadwiger and Riess 2020)	25	23,106	JPEG
HDR (Shaya et al. 2018)	21	5,415	JPEG
Raise (Nguyen et al. 2015)	3	8,156	Uncompressed-Raw
Socrates (Galdi et al. 2019)	60	9,700	JPEG
IEEE Cup (Stamm et al. 2018; IEEE Signal Processing Cup 2018 Database 2021)	10	2,750	JPEG

- Socrates Dataset (Galdi et al. 2019);
- the dataset for the IEEE Signal Processing Cup 2018: Forensic Camera Model Identification Challenge (Stamm et al. 2018; IEEE Signal Processing Cup 2018 Database 2021).

To highlight the differences among the datasets in terms of camera models and images involved, Table 7.1 summarizes the main datasets' characteristics. In the following, we illustrate in detail the main features of each dataset.

Dresden Image Database

The Dresden Image Database (Gloe and Böhme 2010) has been designed with the specific purpose of investigating the camera model identification problem. This is a publicly available dataset, including approximately 17,000 full-resolution natural images stored in the JPEG format with the highest available JPEG quality and 1,500 uncompressed raw images. The image acquisition process has been carefully designed in order to provide image forensics analysts a dataset which satisfies the two properties reported in Sect. 7.4.1. Images were captured under controlled conditions from 26 different camera models considering up to 5 different instances per model. For each motif, at least 3 scenes were captured by varying the focal length.

Thanks to the careful design process and to the considerable amount of images included, the Dresden Image Database has become over the years one of the most used image datasets for tackling image forensics investigations. The use of this dataset as a benchmark has favored the spreading of research works on the topic, easing the comparison between different methodologies and their reproducibility. Focusing on the research on the camera model identification task, the Dresden Image Database has been used several times by state-of-the-art works (Bondi et al. 2017b; Marra et al. 2017; Tuama et al. 2016).

Vision Image Dataset

The Vision dataset (Shullani et al. 2017) is a recent image and video dataset, purposely designed for multimedia forensics investigations. Specifically, Vision dataset has been designed to follow the trend on image and video acquisition and social sharing. In the last few years, photo-amateurs have rapidly transitioned to hand-held devices as preferred mean to capture images and videos. Then, the acquired content is typically shared on social media platforms like WhatsApp or Facebook. In this vein, Vision dataset collects almost 12,000 native images captured by 35 modern smartphones/tablets, including also their related social media version.

The Vision dataset well satisfies the first requirement presented in Sect. 7.4.1 about capturing images of similar scenes taken with different camera models. Moreover, it represents a substantial improvement with respect to Dresden Image Database, since it collects images from modern devices and social media platforms. However, a minor limitation regards the second requirement provided in Sect. 7.4.1: there are only few camera models with two or more instances. Indeed, over the 35 available modern devices, we have 28 different camera models.

In the literature, Vision dataset has been used many times for investigations on the camera model identification problem. Among them, we can cite (Cozzolino and Verdoliva 2020; Mandelli et al. 2020).

Forchheim Image Database
The Forchheim Image Database (Hadwiger and Riess 2020) consists of more than 23,000 images of 143 scenes shot by 27 smartphone cameras of 25 models and 9 brands. It has been proposed to cleanly separate scene content and forensic traces, and to support realistic post-processing like social media recompression. Indeed, six different qualities are provided per image, collecting different copies of the same original image passed through social networks.

Dataset for Camera Identification on HDR Images
The proposed dataset collects standard dynamic range and HDR images captured in different conditions, including various capturing motions, scenes, and devices (Shaya et al. 2018). It has been proposed to investigate the source identification problem on HDR images, which usually introduce some difficulties due to their complexity and wider dynamic range. 23 mobile devices were used for capturing 5,415 images in different scenarios.

Raise Image Dataset
The Raise Image Dataset (Nguyen et al. 2015) concerns 8,156 high-resolution raw images, depicting various subjects and scenarios. 3 different camera of diverse models were employed.

Socrates Dataset
The Socrates Dataset (Galdi et al. 2019) has been built to investigate the source camera identification problem on images and videos coming from smartphones. Images and videos have been collected directly by smartphone owners, ending up with about 9,700 images and 1000 videos captured with 104 different smartphones of 15 different makes and about 60 different models.

Dataset for the IEEE Signal Processing Cup 2018: Forensic Camera Model Identification Challenge

The IEEE Signal Processing Cup (SP Cup) is a student competition in which undergraduate students form teams to work on real-life challenges. In 2018, the camera model identification goal was selected as the topic for the SP Cup (Stamm et al. 2018). Participants were provided with a dataset consisting of JPEG images from 10 different camera models (including point-and-shoot cameras, cell phone cameras, and digital single-lens reflex cameras), with 200 images captured using each camera model. In addition, also post-processed operations (e.g., JPEG recompression, cropping, contrast enhancement) were applied to the images. The complete dataset can be downloaded from IEEE DataPort (IEEE Signal Processing Cup 2018 Database 2021).

7.4.3 Benchmark Protocol

The benchmark protocol commonly followed by forensics researchers is to divide the available images into three disjoint image datasets: the training dataset $\mathcal{I}_t$, the validation dataset $\mathcal{I}_v$ and the evaluation dataset $\mathcal{I}_e$. This split ensures that images seen during the training process (i.e., images belonging to either $\mathcal{I}_t$ or $\mathcal{I}_v$) are never used in testing phase, thus they do not introduce any bias in the results.

Moreover, it is reasonable to assume that query images under analysis might not be acquired with the same devices seen in training phase. Therefore, what is usually done is to pick a selection of devices to be used for the training process, namely the training device set $\mathcal{D}_t$. Then, the proposed method can be evaluated over the total amount of available devices. A realistic and challenging scenario is to include only one device per known camera model in the training set $\mathcal{D}_t$.

7.5 Case Studies

This section is devoted to numerical analysis of some selected methods in order to showcase the capabilities of modern camera model identification algorithms. Specifically, we consider a set of baseline CNNs and co-occurrences of image residuals (Fridrich and Kodovský 2012) analyzing both closed-set and open-set scenarios. The impact of JPEG compression is also investigated.

7.5.1 Experimental Setup

To perform our experiments, we select natural JPEG compressed images from the Dresden Image Database (Gloe and Böhme 2010) and the Vision Image Dataset

(Shullani et al. 2017). Regarding the Dresden Image Database, we consider images from "Nikon_D70" and "Nikon_D70s" camera models as coming from the same model, as the differences between these two models are negligible due to a minor version update (Gloe and Böhme 2010; Kirchner and Gloe 2015). We pick the same number of images per device, and use as reference the device with the lowest image cardinality. We end up with almost 15,000 images from 54 different camera models, including 108 diverse devices.

We work in a patch-wise fashion, extracting N patches with size $P \times P$ pixels from each image. We investigate four different networks. Two networks are selected from the recently proposed EfficientNet family of models (Tan and Le 2019), which achieves very good results both in computer vision and multimedia forensics tasks. Specifically, we select EfficientNetB0 and EfficientNetB4 models. The other networks are known in literature as ResNet50 (He et al. 2016) and XceptionNet (Chollet 2017). Following a common procedure in CNN training, we initialize the network weights using those trained on ImageNet database (Deng et al. 2009). All CNNs are trained using cross-entropy loss and Adam optimizer with default parameters. The learning rate is initialized to 0.001 and is decreased by a factor 10 whenever the validation loss does not improve for 10 epochs. The minimum accepted learning rate is set to 10^{-8}. We train the networks for at most 500 epochs, and training is stopped if loss does not decrease for more than 50 epochs. The model providing the best validation loss is selected.

At test time, classification scores are always fed to the softmax function. In the closed-set scenario, we assign the query image to the camera model associated with the highest softmax score. We use the average accuracy of correct predictions as evaluation metrics. In the open-set scenario, we evaluate results as a function of the detection accuracy of "known" versus "unknown" models. Furthermore, we also provide the average accuracy of correct predictions over the set of known camera models. In other words, given that a query image was taken with one camera model belonging to the known class, we evaluate the classification accuracy as in a closed-set problem reduced to the "known" categories.

Concerning the dataset split policy, we always keep 80% of the images for training phase, further divided in 85%/15% for training set $\mathcal{I}_t$ and validation set $\mathcal{I}_v$, respectively. The remaining 20% of the images are used in the evaluation set $\mathcal{I}_e$. All tests have been run on a workstation equipped with one Intel® Xeon Gold 6246 (48 Cores @3.30 GHz), RAM 252 GB, one TITAN RTX (4608 CUDA Cores @1350 MHz), 24 GB, running Ubuntu 18.04.2. We resort to Albumentation (Buslaev et al. 2020) as data augmentation library for applying JPEG compression to images, and we use Pytorch (Paszke et al. 2019) as Deep Learning framework.

7.5.2 Comparison of Closed-Set Methods

In this section, we compare closed-set camera model identification methodologies. We do not consider model-based approaches, since data-driven methodologies have extensively outperformed them in the last few years (Bondi et al. 2017a; Marra et al. 2017). In particular, we compare 4 different CNN architectures with a well-known state-of-the-art method based on hand-crafted feature extraction. Specifically, we extracted the co-occurrences of image residuals as suggested in Fridrich and Kodovský (2012). Indeed, it has been shown that exploiting these local features provides valuable insights on the camera model identification task (Marra et al. 2017).

Since data-driven methods need to be trained on a specific set of images (or image-patches), we consider different scenarios in which the characteristics of the training dataset change.

Training with Variable Patch-sizes, Proportional Number of Patches per Image
In this setup, we work with images from both the Dresden Image Database and the Vision Image Dataset. We randomly extract N patches per image with size $P \times P$ pixels. We consider patch-sizes $P \in \{256, 512, 1024\}$. As first experiment, we would like to maintain a constant number of image pixels seen in training phase. In doing so, we can compare the methods' performance under the same number of input pixels. Hence, the smaller the patch-size, the more image patches are provided. In case of $P = 256$, we randomly extract $N = 40$ patches per image; for $P = 512$, we randomly extract $N = 10$ patches per image; when $P = 1024$, we randomly extract $N = 3$ patches per image. The number of input image pixels remains constant in the first two situations, while the last scenario includes few pixels more.

Co-occurrence Based Local Features. We extract co-occurrences features of 625 elements (Fridrich and Kodovský 2012) from each analyzed patch independent of the input patch-size P. More precisely, we extract features based on the third order filter named "s3-spam14hv", as suggested in Marra et al. (2017). We apply this filter to the luminance component of the input patches.

Then, to associate the co-occurrences features with one camera model, we train a 54-classes classifier composed of a shallow neural network. The classifier is defined as

- a fully connected layer with 625 input channels, i.e., the dimension of co-occurrences, and 256 output channels;
- a dropout layer with 0.5 as dropout ratio;
- a fully connected layer with 256 input channels and 54 output channels.

We train this elementary network using Adam optimization with initial learning rate of 0.1, following the same paradigm described in Sect. 7.5.1. Results on evaluation images are shown in Table 7.2.

Table 7.2 Accuracy of camera model identification in closed-set scenario using co-occurrences features

Patch-size P	256	512	1024
Accuracy (%)	68.77	78.81	82.77

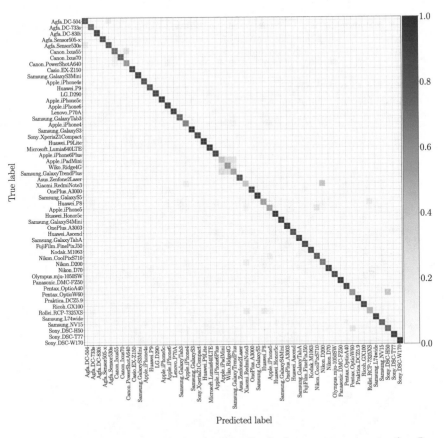

Fig. 7.3 Confusion matrix achieved in closed-set classification by co-occurrences extraction, $P = 1024$

Notice that the larger the patch-size, the higher the achieved accuracy. Even though the number of image pixels seen by the classifier in training phase is essentially constant, the features extracted from the patches become more significant as the pixel region fed to the co-occurrence extraction grows. Figure 7.3 shows the achieved confusion matrix for $P = 1024$.

CNNs. CNN results are shown in Table 7.3. Differently from co-occurrences, CNNs seem to be considerably less dependent on the specific patch-size, as long as the number of image pixels fed to the network remains the same. All the network models perform slightly worse on larger patches (i.e., $P = 1024$). This lower performance may originate from a higher difficulty of the networks in converging during the training process. Larger patches inevitably require reduced batch-size during training. Less samples in the batch means less results' average in one training epoch and reduced CNN capabilities in converging to the best parameters.

Table 7.3 Accuracy of closed-set camera model identification using 4 different CNN architectures as a function of the patch-size. The number of extracted patches per image varies such that the overall image pixels seen by CNNs is almost constant

Patch-size P	256	512	1024
EfficientNetB0 (%)	94.68	94.60	94.28
EfficientNetB4 (%)	94.29	94.57	93.20
ResNet50 (%)	93.71	92.48	91.70
XceptionNet (%)	93.74	94.21	91.23

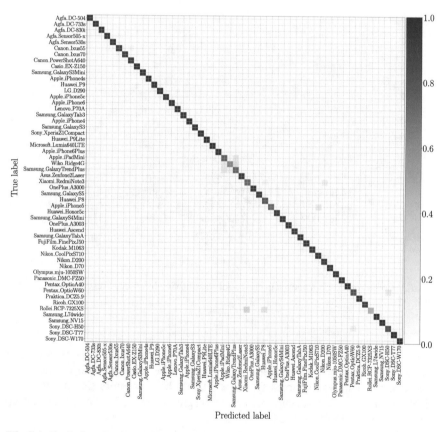

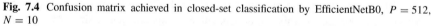

Fig. 7.4 Confusion matrix achieved in closed-set classification by EfficientNetB0, $P = 512$, $N = 10$

Overall, results on CNNs strongly outperform co-occurrences. Fixing a patch-size, all the CNNs return very similar results. The EfficientNet family of models slightly outperforms the other CNN architectures, as it has been shown several times in the literature (Tan and Le 2019).

To provide an example of the results, Fig. 7.4 shows the achieved confusion matrix by EfficientNetB0 architecture for $P = 512$, $N = 10$.

Fig. 7.5 Closed-set
classification accuracy
achieved by CNNs as a
function of the number of
extracted patches per image,
patch-size $P = 256$

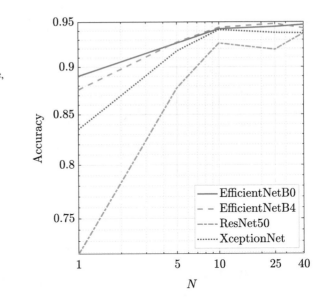

Training with Fixed Patch-sizes, Variable Number of Patches per Image

In this setup, we work again with images from both the Dresden Image Database and
the Vision Image Dataset. Differently from the previous setup, we do not evaluate
results of co-occurrences since they are significantly outperformed by the CNN-
based framework. Instead of exploring diverse patch-sizes, we now fix the patch-size
$P = 256$. We aim at investigating how CNN performance changes by varying the
number of extracted patches per image. N can vary among $\{1, 5, 10, 25, 40\}$.

Figure 7.5 depicts the closed-set classification accuracy results. For small values
of N (i.e., $N < 10$), all the 4 networks enhance their performance as the number of
extracted patches increases. When N further increases, CNNs achieve a performance
plateau and accuracy does not change too much across $N \in \{10, 25, 40\}$.

Training with Variable Patch-size, Fixed Number of Patches per Image

We now fix the number of extracted patches per image to $N = 10$. Then, we vary
the patch-size P among $\{64, 128, 256, 512\}$. The goal is to explore whether results
change as a function of the patch-size.

Figure 7.6 depicts the results. All the CNNs achieve accuracies beyond 90% when
patch-size is larger or equal to 256. When the data fed to the network are too small,
CNNs are not able to estimate well the classification parameters.

Investigations on the Influence of JPEG Grid Alignment

When editing a photograph or uploading a picture over a social media platform, it may
be cropped with respect to its original size. After the cropping, a JPEG compression
is usually applied to the image. These operations may de-synchronize the 8×8
characteristic pixel grid of the original JPEG compression. Usually, a new 8×8 grid

Fig. 7.6 Closed-set classification accuracy achieved by CNNs as a function of the patch-size P, the number of extracted patches per image is fixed to $N = 10$

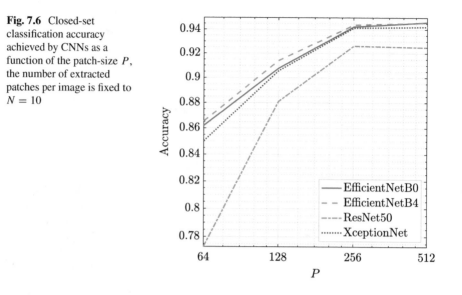

non-aligned with the original one is generated. When training a CNN to solve an image classification problem, the presence of JPEG grid misalignment on images can strongly impact the performance. In this section, we summarize the experiments performed in Mandelli et al. (2020) to investigate on the influence of JPEG grid alignment on the closed-set camera model identification task.

During training and evaluation steps, we investigate two scenarios:

1. working with JPEG compressed images whose JPEG grid is aligned to the 8×8 pixel grid starting from upper-left corner;
2. working with JPEG compressed images whose JPEG grid starts in random pixel position.

To simulate these scenarios, we first compress all the images with the maximum JPEG Quality Factor (QF), i.e., QF = 100. This process generates images with JPEG lattice and does not impair the image visual quality. The first scenario includes images cropped such that the extracted image patches are always aligned to the 8×8 pixel grid. In the latter scenario, we extract patches in random positions. Differently from previous sections, here we are considering a reduced image dataset, the same used in Mandelli et al. (2020) from which we are reporting the results. Images are selected only from the Vision dataset and $N = 10$ squared patches of 512×512 pixels are extracted from the images.

Figure 7.7 shows the closed-set classification accuracy for all CNNs as a function of the considered scenarios. In particular, Fig. 7.7a depicts results in case we train and test on randomly cropped patches; Fig. 7.7b shows what happens when training on JPEG-aligned images and testing on randomly cropped images; Fig. 7.7c explores training on randomly cropped images and testing on JPEG-aligned images; Fig. 7.7d draws results in case we train and test on JPEG-aligned images. It is worth

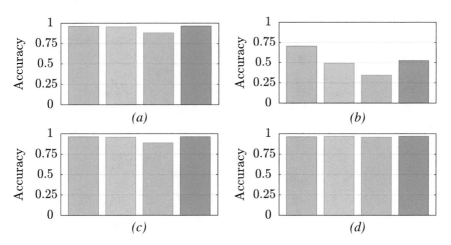

Fig. 7.7 Closed-set classification accuracy achieved by CNNs as a function of JPEG grid alignment in training and/or evaluation phases. The bars represent, respectively, from left to right: ▨ EfficientNetB0; ▨ EfficientNetB4; ▨ ResNet50; ▨ XceptionNet. In **a**, we train and test on randomly cropped patches; in **b**, we train on JPEG-aligned and test on randomly cropped patches; in **c**, we train on randomly cropped and test on JPEG-aligned patches; in **d**, we train and test on JPEG-aligned patches

noticing that being careful to the JPEG grid alignment is paramount for achieving good accuracy. When training a detector on JPEG-aligned patches and testing on JPEG-misaligned ones, results drop consistently as shown in Fig. 7.7b.

Investigations on the Influence of JPEG Quality Factor
In this section, we aim at investigating how the QF of JPEG compression affects CNN performance in closed-set camera model identification. As done in the previous section, we illustrate the experimental setup provided in Mandelli et al. (2020), where the effect of JPEG compression quality is studied for images belonging to the Vision dataset. We JPEG-compress the images with diverse QFs, namely QF $\in$ $\{50, 60, 70, 80, 90, 99\}$. Then, we extract $N = 10$ squared patches of 512×512 pixels from images. Following previous considerations, we randomly extract the patches both for the training and evaluation datasets.

To explore the influence of JPEG QF on CNN results, we train the network in two ways:

1. we use only images of the original Vision Image dataset;
2. we perform some training data augmentation. Half of the training images are taken from the original Vision Image dataset; the remaining part is compressed with a JPEG QF picked from the reported list.

Notice that the second scenario assumes some knowledge on the JPEG QF of the evaluation images, and can thus improve the achieved classification results.

Closed-set classification accuracy achieved by the CNNs is reported in Fig. 7.8. In particular, we show results as a function of the QF of the evaluation images.

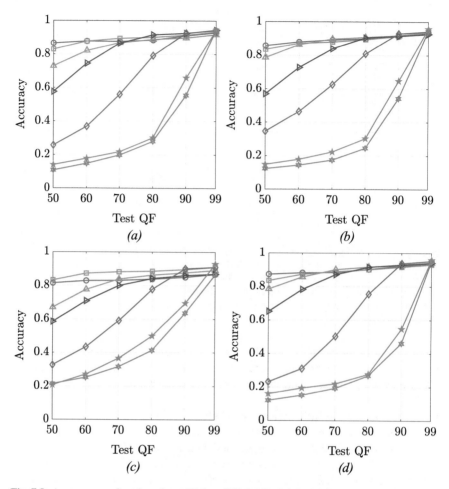

Fig. 7.8 Accuracy as a function of test QF for **a** EfficientNetB0, **b** EfficientNetB4, **c** ResNet50, **d** XceptionNet. The curves are drawn in accordance to the following legend: ● train on QF = 50; ■ train on QF = 60; ▲ train on QF = 70; ▶ train on QF = 80; ◆ train on QF = 90; ★ train on QF = 99; ✶ No augmentation

The gray curve represents the first training scenario, i.e., in absence of training data augmentation. Note that this setup always draws the worst results. Also training on data augmented with QF = 99 almost corresponds to absence of augmentation and achieves acceptable results only when the test QF matches the training one. On the contrary, training on data augmented with low JPEG QFs (i.e., QF ∈ {50, 60}) returns acceptable results for all the possible test QFs. For instance, by training on data compressed with QF = 50, evaluation accuracy can improve from 0.2 to more than 0.85.

7.5.3 Comparison of Open-Set Methods

In the open-set scenario some camera models are unknown, so we cannot train CNNs on the complete device set as information can be extracted only from the known devices. Let us represent the original camera model set with $\mathcal{T}$: known models are denoted as $\mathcal{T}_k$, unknown models are denoted as $\mathcal{T}_u$.

To perform open-set classification, two main training strategies can be pursued (Bayar and Stamm 2018; Júnior et al. 2019):

1. Consider all the available set $\mathcal{T}_k$ as "known" camera models. Train one closed-set classifier over $\mathcal{T}_k$ camera models.
2. Divide the set of $\mathcal{T}_k$ camera models into two disjoint sets: the "known-known" set $\mathcal{T}_{k_k}$, and the "known-unknown" set $\mathcal{T}_{k_u}$. Train one binary classifier to identify $\mathcal{T}_{k_k}$ camera models from $\mathcal{T}_{k_u}$ camera models.

Results are evaluated accordingly to two metrics (Bayar and Stamm 2018):

1. the accuracy of detection between known camera models and unknown camera models;
2. the accuracy of closed-set classification among the set of known camera models. This metric is validated only for images belonging to camera models in set $\mathcal{T}_k$.

Our experimental setup considers only EfficientNetB0 as network architecture, since EfficientNet family of models (both EfficientNetB0 and EfficientNetB4) always reports higher or comparable accuracies with respect to other architectures. We choose EfficientNetB0 which is lighter than EfficientNetB4 to reduce training and testing time. We exploit images from both the Dresden Image Database and the Vision Image Dataset. Among the pool of 54 camera models, we randomly extract 36 camera models for the "known" set $\mathcal{T}_k$ and leave the remaining 18 to the "unknown" set $\mathcal{T}_u$. Following considerations, we randomly extract $N = 10$ squared patches per image with patch-size $P = 512$.

Training One Closed-set Classifier over "Known" Camera Models
We train one closed-set classifier over models belonging to $\mathcal{T}_k$ set by following the same training protocol previously seen for closed-set camera model identification. In testing phase, we proceed with two consequential steps:

1. detect if the query image is taken by a "known" camera model or an "unknown" model;
2. if the query image comes from a "known" camera model, identify the source model.

Regarding the first step, we follow the approach suggested in Bayar and Stamm (2018). Given a query image, if the maximum score returned by the classifier exceeds a predefined threshold, we assign the image to the category of "known" camera models. Otherwise, the image is associated with an "unknown" model. Following this procedure, the closed-set classification accuracy across models in $\mathcal{T}_k$ is 95%. Figure 7.9 shows the confusion matrix achieved by closed-set classification.

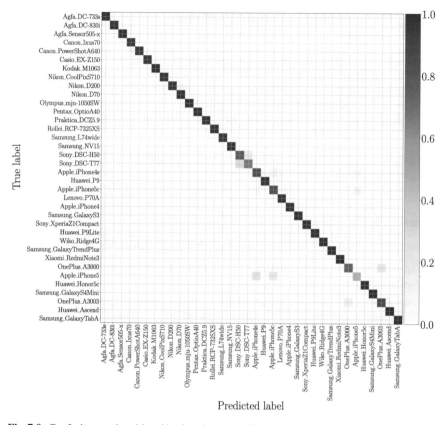

Fig. 7.9 Confusion matrix achieved in closed-set classification by training one classifier (in closed-set fashion) over the 36 camera models in $\mathcal{T}_k$. The average closed-set accuracy is 95%

The detection accuracy varies with the threshold used to classify "known" versus "unknown" camera models. Figure 7.10 depicts the ROC curve achieved by the proposed method. Specifically, we see the behavior of True Positive Rate (TPR) as a function of the False Positive Rate (FPR). The maximum accuracy value is 86.19%.

Divide the Set of Known Camera Models into "Known-known" and "Known-unknown"

In this experiment, we divide the set of known camera models into two disjoint sets: the set of "known-known" models $\mathcal{T}_{k_k}$ and the set of "known-unknown" models $\mathcal{T}_{k_u}$. Then, we train a binary classifier which discriminates between the two classes. In testing phase, all the images assigned to the "known-unknown" class will be classified as unknown. Notice that by training one binary classifier we are able to recognize as "known" category only images taken from camera models in $\mathcal{T}_{k_k}$. Therefore, even if camera models belonging to $\mathcal{T}_{k_u}$ are known, their images will be classified as unknown in testing phase.

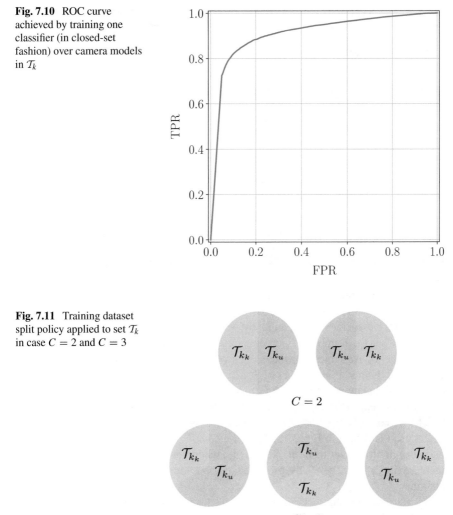

Fig. 7.10 ROC curve achieved by training one classifier (in closed-set fashion) over camera models in $\mathcal{T}_k$

Fig. 7.11 Training dataset split policy applied to set $\mathcal{T}_k$ in case $C = 2$ and $C = 3$

To overcome this limitation, we explore the possibility of training a binary classifier in different setups. In practice, we divide the known camera models' set $\mathcal{T}_k$ into C disjoint subsets containing images of $|\mathcal{T}_k|/C$ models, where $|\cdot|$ is the cardinality of the set. We define every disjoint set as $\mathcal{T}_{k_c}$, $c \in [1, C]$. Then, for each disjoint set $\mathcal{T}_{k_c}$, we train one binary classifier telling known models and unknown models apart. Set $\mathcal{T}_{k_c}$ represents the "known-known" class, while the remaining sets are joined and form the "known-unknown" class. Figure 7.11 depicts the split policy applied to $\mathcal{T}_k$ for $C = 2$ and $C = 3$. For instance, when $C = 3$, we end up with 3 different training setups: the first considers $\mathcal{T}_{k_1}$ as known class and models in $\{\mathcal{T}_{k_2}, \mathcal{T}_{k_3}\}$ as unknowns;

Fig. 7.12 Maximum
detection accuracy achieved
in the ROC curve by training
C classifiers over camera
models in $\mathcal{T}_k$

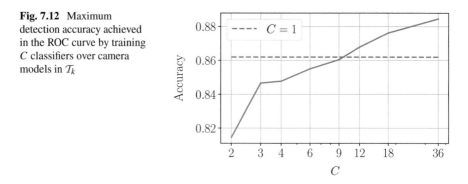

the second considers $\mathcal{T}_{k_2}$ as known class and models in $\{\mathcal{T}_{k_1}, \mathcal{T}_{k_3}\}$ as unknowns; the
third considers $\mathcal{T}_{k_3}$ as known class and models in $\{\mathcal{T}_{k_1}, \mathcal{T}_{k_2}\}$ as unknowns.

In testing phase, we assign a query image to the "known-known" class only if at
least one classifier returned a score associated with the "known-known" class which
is sufficiently confident. In practice, we threshold the maximum score associated
to the "known-known" class returned by the C classifiers. Whenever the maximum
score overcomes the threshold, we assign the query image to a known camera model;
otherwise we associate it with an unknown model.

Notice that the number of classifiers can vary from $C = 2$ to $C = 36$ (i.e., the
total number of available known camera models). In our experiments, we consider C
equal to all the possible divisors of 36, i.e., $C \in \{2, 3, 4, 6, 9, 12, 18, 36\}$. In order
to work with balanced training datasets, we always fix the number of images for each
class to be equal to the image cardinality of the known set $\mathcal{T}_{k_k}$.

The maximum detection accuracy achieved as a function of the number of classifiers used is shown in Fig. 7.12. The case $C = 1$ corresponds to training only one
classifier in closed-set fashion. Notice that when the number of classifiers starts to
increase (i.e., when $C > 9$), the proposed methodology can outperform the closed-set
classifier.

We also evaluate the accuracy achieved in closed classification for images belong-
ing to known models in $\mathcal{T}_k$. For each disjoint set $\mathcal{T}_{k_c}$, we can train a $|\mathcal{T}_k|/C$-class clas-
sifier in closed-set fashion. If one of the C binary classifiers assigns the query image
to the $\mathcal{T}_{k_c}$ set, we can identify the source camera model by exploiting the related
closed-set classifier. This procedure is not needed in case $C = |\mathcal{T}_k|$, i.e., $C = 36$.
In this case, we can exploit the maximum score returned by the C binary classifiers. If the maximum score is related to the "known-known" class, the estimated
source camera model is associated with the binary classifier providing the highest
score. If the maximum score is related to the "known-unknown" class, the estimated
source camera model is unknown. For example, Fig. 7.13 shows the confusion matrix
achieved in closed-set classification by training 36 classifiers. The average closed-set
accuracy is 99.56%.

Exploiting $C = 36$ binary classifiers outperforms the previously presented closed-
set classifier, both in known-vs-unknown detection and closed-set classification. This

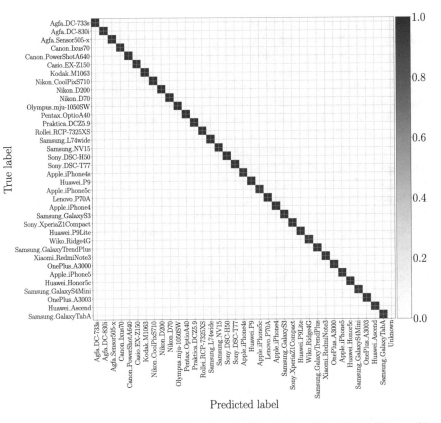

Fig. 7.13 Confusion matrix achieved in closed-set classification by training 36 classifier, considering the 36 camera models in T_k. The average closed-set accuracy is 99.56%

comes at the expense of training 36 binary classifiers instead of one C-class classifier. However, each binary classifier is trained over a reduced amount of images, i.e., twice the number of images of the "known-known" class. On the contrary, the C-class classifier is trained over the total amount of known images. Therefore, the required computation time by training 36 classifiers does not significantly increase and it is still acceptable.

7.6   Conclusions and Outlook

In this chapter, we have reported multiple examples of camera model identification algorithms developed in the literature. From this report and the showcased experimental results, it is possible to understand that the camera model identification problem can be well solved in some situations, but not always.

For instance, good solutions exist for solving the camera model identification task in a closed-set scenario, when images have not been further post-processed or edited. Given a set of known camera models and a pool of training images, standard CNNs can solve the task in a reasonable amount of time with high accuracy.

However, this setup is too optimistic. In practical situations, it is unlikely that images never undergo any processing step after acquisition. Moreover, it is not realistic to assume that an analyst has access to all possible camera models needed for an investigation. In this situation, further research is needed. As an example, compression artifacts and other processing operations like cropping or resizing can strongly impact on the achieved performance. For this reason, a deeper analysis on images shared through social media networks is worthy of investigations.

Moreover, open-set scenarios are still challenging even considering original (i.e., not post-processed) images. The proposed methods in the literature are not able to approach closed-set accuracy yet. As the open-set scenario is more realistic than its closed-set counterpart, further work in this direction should be done.

Furthermore, the newest smartphone technologies based on HDR images, multiple cameras, super-dense sensors and advanced internal processing (e.g., "beautify" filters, AI-guided enhancements, etc.) can further complicate the source identification problem, due to an increased image complexity (Shaya et al. 2018; Iuliani et al. 2021). For instance, it has been already shown that sensor-based methodologies to solve the camera attribution problem may suffer for the portrait mode employed in smartphones of some particular brands (Iuliani et al. 2021). When portrait modality is selected to capture a human subject in foreground and blur the remaining background, sensor traces risk to be hindered (Iuliani et al. 2021).

Finally, we have not considered scenarios involving attackers. On one hand, antiforensic techniques could be applied to images in order to hide camera model traces. Alternatively, working with data-driven approaches, it is important to consider possible adversarial attacks to the used classifiers. In the light of these considerations, forensic approaches for camera model identification in modern scenarios still have to be developed. Such future algorithmic innovations can be further fostered with novel datasets that include these most recent challenges.

References

Avcibas I, Sankur B, Sayood K (2002) Statistical evaluation of image quality measures. J Electron Imaging 11(2):206–223

Avcibas I, Kharrazi M, Memon ND, Sankur B (2005) Image steganalysis with binary similarity measures. EURASIP J Adv Signal Process 17:2749–2757

Bayar B, Stamm MC (2017) Augmented convolutional feature maps for robust cnn-based camera model identification. In:2017 IEEE international conference on image processing, ICIP 2017, Beijing, China, September 17–20, 2017. IEEE, pp 4098–4102

Bayar B, Stamm MC (2018) Towards open set camera model identification using a deep learning framework. In: 2018 IEEE international conference on acoustics, speech and signal processing, ICASSP 2018, Calgary, AB, Canada, April 15–20, 2018. IEEE, pp 2007–2011

Bayram S, Sencar HT, Memon ND, Avcibas I (2005) Source camera identification based on CFA interpolation. In: Proceedings of the 2005 international conference on image processing, ICIP 2005, Genoa, Italy, September 11–14, 2005. IEEE, pp 69–72

Bayram S, Sencar HT, Memon N, Avcibas I (2006) Improvements on source camera-model identification based on cfa interpolation. Proc WG 11(9)

Bondi L, Baroffio L, Guera D, Bestagini P, Delp EJ, Tubaro S (2017a) First steps toward camera model identification with convolutional neural networks. IEEE Signal Process Lett 24(3):259–263

Bondi L, Lameri S, Guera D, Bestagini P, Delp EJ, Tubaro S (2017b) Tampering detection and localization through clustering of camera-based CNN features. In: 2017 IEEE conference on computer vision and pattern recognition workshops, CVPR workshops 2017, Honolulu, HI, USA, July 21–26, 2017. IEEE Computer Society, pp 1855–1864

Bonettini N, Bondi L, Bestagini P, Tubaro S (2018) JPEG implementation forensics based on eigen-algorithms. In: 2018 IEEE international workshop on information forensics and security, WIFS 2018, Hong Kong, China, December 11–13, 2018. IEEE, pp 1–7

Buslaev A, Iglovikov VI, Khvedchenya E, Parinov A, Druzhinin M, Kalinin AA (2020) Albumentations: fast and flexible image augmentations. Information 11(2):125

Cao H, Kot AC (2009) Accurate detection of demosaicing regularity for digital image forensics. IEEE Trans Inf Forensics Secur 4(4):899–910

Cao H, Kot AC (2010) Mobile camera identification using demosaicing features. In: International symposium on circuits and systems (ISCAS 2010), May 30–June 2, 2010, Paris, France. IEEE, pp 1683–1686

Çeliktutan O, Sankur B, Avcibas I (2008) Blind identification of source cell-phone model. IEEE Trans Inf Forensics Secur 3(3):553–566

Chen M, Fridrich JJ, Goljan M, Lukás J (2008) Determining image origin and integrity using sensor noise. IEEE Trans Inf Forensics Secur 3(1):74–90

Chen C, Stamm MC (2015) Camera model identification framework using an ensemble of demosaicing features. In: 2015 IEEE international workshop on information forensics and security, WIFS 2015, Roma, Italy, November 16–19, 2015. IEEE, pp 1–6

Choi C-H, Choi J-H, Lee H-K (2011) Cfa pattern identification of digital cameras using intermediate value counting. In: Proceedings of the thirteenth ACM multimedia workshop on Multimedia and security, pp 21–26

Chollet F (2017) Xception: deep learning with depthwise separable convolutions. In: 2017 IEEE conference on computer vision and pattern recognition, CVPR 2017, Honolulu, HI, USA, July 21–26, 2017. IEEE Computer Society, pp 1800–1807

Cozzolino D, Verdoliva L (2020) Noiseprint: a cnn-based camera model fingerprint. IEEE Trans Inf Forensics Secur 15:144–159

Deng J, Dong W, Socher R, Li L-J, Li K, Li F-F (2009) Imagenet: a large-scale hierarchical image database. In: 2009 IEEE computer society conference on computer vision and pattern recognition (CVPR 2009), 20–25 June 2009, Miami, Florida, USA. IEEE Computer Society, pp 248–255

Dirik AE, Sencar HT, Memon ND (2008) Digital single lens reflex camera identification from traces of sensor dust. IEEE Trans Inf Forensics Secur 3(3):539–552

Ferreira A, Chen H, Li B, Huang J (2018) An inception-based data-driven ensemble approach to camera model identification. In: 2018 IEEE international workshop on information forensics and security, WIFS 2018, Hong Kong, China, December 11–13, 2018. IEEE, pp 1–7

Foi A (2009) Clipped noisy images: heteroskedastic modeling and practical denoising. Signal Proc 89(12):2609–2629

Fridrich JJ, Kodovský J (2012) Rich models for steganalysis of digital images. IEEE Trans Inf Forensics Secur 7(3):868–882

Galdi C, Hartung F, Dugelay J-L (2019) SOCRatES: a database of realistic data for source camera recognition on smartphones. In: De Marsico M, di Baja GS, Fred ALN (eds) Proceedings of the 8th international conference on pattern recognition applications and methods, ICPRAM 2019, Prague, Czech Republic, February 19–21, 2019. SciTePress, pp 648–655

Gao S, Xu G, Hu RM (2011) Camera model identification based on the characteristic of CFA and interpolation. In: Shi Y-Q, Kim H-J, Pérez-González F (eds) Digital forensics and watermarking - 10th international workshop, IWDW 2011, Atlantic City, NJ, USA, October 23–26, 2011, Revised Selected Papers, vol 7128 of Lecture Notes in computer science. Springer, pp 268–280

Gloe T (2012) Feature-based forensic camera model identification. Trans Data Hiding Multim Secur 8:42–62

Gloe T, Böhme R (2010) The Dresden image database for benchmarking digital image forensics. J Digit Forensic Pract 3(2–4):150–159

Gloe T, Borowka K, Winkler A (2009) Feature-based camera model identification works in practice. In: Katzenbeisser S, Sadeghi A-R (eds) Information hiding, 11th international workshop, IH 2009, Darmstadt, Germany, June 8–10, 2009, Revised Selected Papers, vol 5806 of Lecture notes in computer science. Springer, pp 262–276

Gloe T, Borowka K, Winkler A (2010) Efficient estimation and large-scale evaluation of lateral chromatic aberration for digital image forensics. In: Memon ND, Dittmann J, Alattar AM, Delp EJ (eds) Media forensics and security II, part of the IS&T-SPIE electronic imaging symposium, San Jose, CA, USA, January 18–20, 2010, Proceedings, vol 7541 of SPIE Proceedings. SPIE, p 754107

Guera D, Zhu F, Yarlagadda K, Tubaro S, Bestagini P, Delp EJ (2018) Reliability map estimation for cnn-based camera model attribution. In: 2018 IEEE winter conference on applications of computer vision, WACV 2018, Lake Tahoe, NV, USA, March 12–15, 2018. IEEE Computer Society, pp 964–973

Gunturk BK, Glotzbach JW, Altunbasak Y, Schafer RW, Mersereau RM (2005) Demosaicking: color filter array interpolation. IEEE Signal Proc Mag 22(1):44–54

Hadwiger B, Riess C (2020) The forchheim image database for camera identification in the wild. arXiv:abs/2011.02241

He K, Zhang X, Ren S, Sun J (2016) Deep residual learning for image recognition. In: 2016 IEEE conference on computer vision and pattern recognition, CVPR 2016, Las Vegas, NV, USA, June 27–30, 2016. IEEE Computer Society, pp 770–778

Ho JS, Au OC, Zhou J, Guo Y (2010) Inter-channel demosaicking traces for digital image forensics. In: Proceedings of the 2010 IEEE international conference on multimedia and Expo, ICME 2010, 19–23 July 2010, Singapore. IEEE Computer Society, pp 1475–1480

Huang NE, Shen Z, Long SR, Wu MC, Shih HH, Zheng Q, Yen N-C, Tung CC, Liu HH (1998) The empirical mode decomposition and the hilbert spectrum for nonlinear and non-stationary time series analysis. Proc R Soc Lond Ser A: Math Phys Eng Sci 454(1971):903–995

Huang G, Liu Z, van der Maaten L, Weinberger KQ (2017) Densely connected convolutional networks. In: 2017 IEEE conference on computer vision and pattern recognition, CVPR 2017, Honolulu, HI, USA, July 21–26, 2017. IEEE Computer Society, pp 2261–2269

IEEE Signal Processing Cup 2018 Database - Forensic Camera Model Identification. https://cutt.ly/acK1lg2. Accessed 06 April 2021

Image Source Attribution UniCamp Dataset. http://www.recod.ic.unicamp.br/~filipe/dataset. Accessed 09 April 2021

Iuliani M, Fontani M, Piva A (2021) A leak in PRNU based source identification. Questioning fingerprint uniqueness. IEEE Access 9:52455–52463

Johnson MK, Farid H (2006) Exposing digital forgeries through chromatic aberration. In: Voloshynovskiy S, Dittmann J, Fridrich JJ (eds) Proceedings of the 8th workshop on multimedia & security, MM&Sec 2006, Geneva, Switzerland, September 26–27, 2006. ACM, pp 48–55

Júnior PRM, Bondi L, Bestagini P, Tubaro S, Rocha A (2019) An in-depth study on open-set camera model identification. IEEE Access 7:180713–180726

Kang SB, Weiss RS (2000) Can we calibrate a camera using an image of a flat, textureless lambertian surface? In: Vernon D (ed) Computer vision - ECCV 2000, 6th European conference on computer vision, Dublin, Ireland, June 26–July 1, 2000, Proceedings, Part II, vol 1843 of Lecture notes in computer science. Springer, pp 640–653

Kharrazi M, Sencar HT, Memon ND (2004) Blind source camera identification. In: Proceedings of the 2004 international conference on image processing, ICIP 2004, Singapore, October 24–27, 2004. IEEE, pp 709–712

Kirchner M (2010) Efficient estimation of CFA pattern configuration in digital camera images. In: Memon ND, Dittmann J, Alattar AM, Delp EJ (eds) Media forensics and security II, part of the IS&T-SPIE electronic imaging symposium, San Jose, CA, USA, January 18–20, 2010, Proceedings, vol 7541 of SPIE Proceedings. SPIE, p 754111

Kirchner M, Gloe T (2015) Forensic camera model identification. In: Handbook of digital forensics of multimedia data and devices. Wiley-IEEE Press, pp 329-374

Lanh TV, Emmanuel S, Kankanhalli MS (2007) Identifying source cell phone using chromatic aberration. In: Proceedings of the 2007 IEEE international conference on multimedia and Expo, ICME 2007, July 2–5, 2007, Beijing, China. IEEE Computer Society, pp 883–886

Lukás J, Fridrich JJ, Goljan M (2006) Digital camera identification from sensor pattern noise. IEEE Trans Inf Forensics Secur 1(2):205–214

Lyu S (2010) Estimating vignetting function from a single image for image authentication. In: Campisi P, Dittmann J, Craver S (eds) Multimedia and security workshop, MM&Sec 2010, Roma, Italy, September 9–10, 2010. ACM, pp 3–12

Mallat S (1989) A theory for multiresolution signal decomposition: the wavelet representation. IEEE Trans Pattern Anal Mach Intell 11(7):674–693

Mandelli S, Bonettini N, Bestagini P, Tubaro S (2020) Training cnns in presence of JPEG compression: multimedia forensics vs computer vision. In: 12th IEEE international workshop on information forensics and security, WIFS 2020, New York City, NY, USA, December 6–11, 2020. IEEE, pp 1–6

Marra F, Poggi G, Sansone C, Verdoliva L (2017) A study of co-occurrence based local features for camera model identification. Multim Tools Appl 76(4):4765–4781

Mayer O, Stamm MC (2018) Learned forensic source similarity for unknown camera models. In: 2018 IEEE international conference on acoustics, speech and signal processing, ICASSP 2018, calgary, AB, Canada, April 15–20, 2018. IEEE, pp 2012–2016

Menon D, Calvagno G (2011) Color image demosaicking: an overview. Signal Proc Image Commun 26(8–9):518–533

Nguyen DTD, Pasquini C, Conotter V, Boato G (2015) RAISE: a raw images dataset for digital image forensics. In: Ooi WT, Feng W-C, Liu F (eds) Proceedings of the 6th ACM multimedia systems conference, MMSys 2015, Portland, OR, USA, March 18–20, 2015. ACM, pp 219–224

Ojala T, Pietikäinen M, Mäenpää T (2002) Multiresolution gray-scale and rotation invariant texture classification with local binary patterns. IEEE Trans Pattern Anal Mach Intell 24(7):971–987

Paszke A, Gross S, Massa F, Lerer A, Bradbury J, Chanan G, Killeen T, Lin Z, Gimelshein N, Antiga L, Desmaison A, Köpf A, Yang E, DeVito Z, Raison M, Tejani A, Chilamkurthy S, Steiner B, Fang L, Bai J, Chintala S (2019) Pytorch: an imperative style, high-performance deep learning library. In: Wallach HM, Larochelle H, Beygelzimer A, d'Alché-Buc F, Fox EB, Garnett R (eds) Advances in neural information processing systems 32: annual conference on neural information processing systems 2019, NeurIPS 2019, December 8–14, 2019, Vancouver, BC, Canada, pp 8024–8035

Popescu AC, Farid H (2005) Exposing digital forgeries in color filter array interpolated images. IEEE Trans Signal Proc 53(10):3948–3959

Rafi AM, Kamal U, Hoque R, Abrar A, Das S, Laganière R, Hasan K (2019) Application of densenet in camera model identification and post-processing detection. In: IEEE conference on computer vision and pattern recognition workshops, CVPR workshops 2019, Long Beach, CA, USA, June 16–20, 2019. Computer Vision Foundation/IEEE, pp 19–28

Rafi AM, Tonmoy TI, Kamal U, Wu QMJ, Hasan K (2021) Remnet: remnant convolutional neural network for camera model identification. Neural Comput Appl 33(8):3655–3670

Ramanath R, Snyder WE, Yoo Y, Drew MS (2005) Color image processing pipeline. IEEE Signal Process Mag 22(1):34–43

Sameer VU, Naskar R (2018) Eliminating the effects of illumination condition in feature based camera model identification. J Vis Commun Image Represent 52:24–32

San Choi K, Lam EY, Wong KKY (2006) Automatic source camera identification using the intrinsic lens radial distortion. Optics Express 14(24):11551–11565

Shaya OA, Yang P, Ni R, Zhao Y, Piva A (2018) A new dataset for source identification of high dynamic range images. Sensors 18(11):3801

Shullani D, Fontani M, Iuliani M, Shaya OA, Piva A (2017) VISION: a video and image dataset for source identification. EURASIP J Inf Secur 2017:15

Stamm MC, Bestagini P, Marcenaro L, Campisi P (2018) Forensic camera model identification: Highlights from the IEEE signal processing cup 2018 student competition [SP competitions]. IEEE Signal Process Mag 35(5):168–174

Swaminathan A, Wu M, Liu KJR (2007) Nonintrusive component forensics of visual sensors using output images. IEEE Trans Inf Forensics Secur 2(1):91–106

Swaminathan A, Wu M, Liu KJR (2009) Component forensics. IEEE Signal Proc Mag 26(2):38–48

Szegedy C, Ioffe S, Vanhoucke V, Alemi AA (2017) Inception-v4, inception-resnet and the impact of residual connections on learning. In: Singh SP, Markovitch S (eds) Proceedings of the thirty-first AAAI conference on artificial intelligence, February 4–9, 2017, San Francisco, California, USA. AAAI Press, pp 4278–4284

Takamatsu J, Matsushita Y, Ogasawara T, Ikeuchi K (2010) Estimating demosaicing algorithms using image noise variance. In: The twenty-third IEEE conference on computer vision and pattern recognition, CVPR 2010, San Francisco, CA, USA, 13–18 June 2010. IEEE Computer Society, pp 279–286

Tan M, Le QV (2019) Efficientnet: rethinking model scaling for convolutional neural networks. In: Chaudhuri K, Salakhutdinov R (eds) Proceedings of the 36th international conference on machine learning, ICML 2019, 9–15 June 2019, Long Beach, California, USA, vol 97 of Proceedings of machine learning research. PMLR, pp 6105–6114

Thai TH, Cogranne R, Retraint F (2014) Camera model identification based on the heteroscedastic noise model. IEEE Trans Image Proc 23(1):250–263

Thai TH, Retraint F, Cogranne R (2016) Camera model identification based on the generalized noise model in natural images. Digit Signal Proc 48:285–297

Tuama A, Comby F, Chaumont M (2016) Camera model identification with the use of deep convolutional neural networks. In: IEEE international workshop on information forensics and security, WIFS 2016, Abu Dhabi, United Arab Emirates, December 4–7, 2016. IEEE, pp 1–6

Wahab AWA, Ho ATS, Li S (2012) Inter-camera model image source identification with conditional probability features. In: Proceedings of IIEEJ 3rd image electronics and visual computing workshop (IEVC 2012)

Xu G, Gao S, Shi Y-Q, Hu RM, Su W (2009) Camera-model identification using markovian transition probability matrix. In: Ho ATS, Shi YQ, Kim HJ, Barni M (eds) Digital watermarking, 8th international workshop, IWDW 2009, Guildford, UK, August 24–26, 2009. Proceedings, vol 5703 of Lecture notes in computer science. Springer, pp 294–307

Xu G, Shi YQ (2012) Camera model identification using local binary patterns. In: Proceedings of the 2012 IEEE international conference on multimedia and Expo, ICME 2012, Melbourne, Australia, July 9–13, 2012. IEEE Computer Society, pp 392–397

Yao H, Qiao T, Ming X, Zheng N (2018) Robust multi-classifier for camera model identification based on convolution neural network. IEEE Access 6:24973–24982

Yu J, Craver S, Li E (2011) Toward the identification of DSLR lenses by chromatic aberration. In: Memon ND, Dittmann J, Alattar AM, Delp III EJ (eds) Media forensics and security III, San Francisco Airport, CA, USA, January 24–26, 2011, Proceedings, vol 7880 of SPIE Proceedings. SPIE, p 788010

Zhao X, Stamm MC (2016) Computationally efficient demosaicing filter estimation for forensic camera model identification. In: 2016 IEEE international conference on image processing, ICIP 2016, Phoenix, AZ, USA, September 25–28, 2016. IEEE, pp 151–155

Chapter 8
GAN Fingerprints in Face Image Synthesis

João C. Neves, Ruben Tolosana, Ruben Vera-Rodriguez, Vasco Lopes, Hugo Proença, and Julian Fierrez

The availability of large-scale facial databases, together with the remarkable progresses of deep learning technologies, in particular Generative Adversarial Networks (GANs), have led to the generation of extremely realistic fake facial content, raising obvious concerns about the potential for misuse. Such concerns have fostered the research on manipulation detection methods that, contrary to humans, have already achieved astonishing results in various scenarios. This chapter is focused on the analysis of GAN fingerprints in face image synthesis. In particular, it covers an in-depth literature analysis of state-of-the-art detection approaches for the entire face synthe-

---

[1] The present chapter is an adaptation from the following article: Neves et al. (2020). DOI: http://dx.doi.org/10.1109/JSTSP.2020.3007250.

J. C. Neves
NOVA LINCS, Universidade da Beira Interior, Covilha, Portugal
e-mail: jcneves@di.ubi.pt

R. Tolosana · R. Vera-Rodriguez · J. Fierrez (✉)
Biometrics and Data Pattern Analytics - BiDA Lab, Universidad Autonoma de Madrid, 28045 Madrid, Spain
e-mail: julian.fierrez@uam.es

R. Tolosana
e-mail: ruben.tolosana@uam.es

R. Vera-Rodriguez
e-mail: ruben.vera@uam.es

V. Lopes
University of Beira Interior, 6201-001 Covilhã, Portugal
e-mail: vasco.lopes@ubi.pt

H. Proença
IT - Instituto de Telecomunicações, 3810-193 Aveiro, Portugal
e-mail: hugomcp@di.ubi.pt

175
H. T. Sencar et al. (eds.), *Multimedia Forensics*, Advances in Computer Vision and Pattern Recognition, https://doi.org/10.1007/978-981-16-7621-5_8

sis manipulation. It also describes a recent approach to spoof fake detectors based on a GAN-fingerprint Removal autoencoder (GANprintR). A thorough experimental framework is included in the chapter, highlighting *(i)* the potential of GANprintR to spoof fake detectors, and *(ii)* the poor generalisation capability of current fake detectors.

8.1   Introduction

Images[1] and videos containing fake facial information obtained by digital manipulation have recently become a great public concern (Cellan-Jones 2019). Up until the advent of DeepFakes a few years ago, the number and realism of digitally manipulated fake facial contents were very limited by the lack of sophisticated editing tools, the high domain of expertise required, and the complex and time-consuming process involved to generate realistic fakes. The scientific communities of biometrics and security in the past decade paid some attention in understanding and protecting against those limited threats around face biometrics (Hadid et al. 2015), with special attention to presentation attacks conducted physically against the face sensor (camera) using various kinds of face spoofs (e.g. 2D or 3D printed, displayed, mask-based, etc.) (Hernandez-Ortega et al. 2019; Galbally et al. 2014).

However, nowadays it is becoming increasingly easy to automatically synthesise non-existent faces or even to manipulate the face of a real person in an image/video, thanks to the free access to large public databases and also to the advances on deep learning techniques that eliminate the requirements of manual editing. As a result, accessible open software and mobile applications such as *ZAO* and *FaceApp* have led to large amounts of synthetically generated fake content (ZAO 2019; FaceApp 2017).

The current methods to generate digital fake face content can be categorised into four different groups, regarding the level of manipulation (Tolosana et al. 2020c; Verdoliva 2020): *(i)* entire face synthesis, *(ii)* face identity swap, *(iii)* facial attribute manipulation and *(iv)* facial expression manipulation.

In this chapter, we focus on the entire face synthesis manipulation, where a machine learning model, typically based on Generative Adversarial Networks (GANs) (Goodfellow et al. 2014), learns the distribution of the human face data, allowing to generate non-existent faces by sampling this distribution. This type of facial manipulation provides astonishing results and is able to generate extremely realistic fakes. Nevertheless, contrary to humans, most state-of-the-art detection systems provide very good results against this type of facial manipulation, remarking how easy it is to detect the GAN "fingerprints" present in the synthetic images.

This chapter covers the following aspects in the topic of GAN Fingerprints:

- An in-depth literature analysis of the state-of-the-art detection approaches for the entire face synthesis manipulation, including the key aspects of the detection

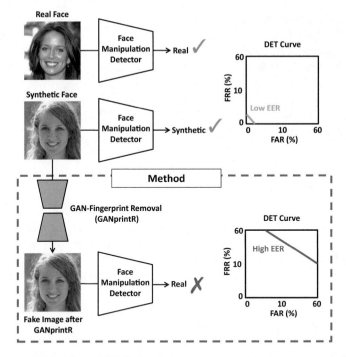

Fig. 8.1 Architecture of the GAN-fingerprint removal approach. In general, the state-of-the-art face manipulation detectors can easily distinguish between real and synthetic fake images. This usually happens due to the existence and exploitation by those detectors of GAN "fingerprints" produced during the generation of synthetic images. The GANprintR approach proposed in (Neves et al. 2020) aims to remove the GAN fingerprints from the synthetic images and spoof the facial manipulation detection systems, while keeping the visual quality of the resulting images

systems, the databases used for developing and evaluating these systems, and the main results achieved by them.

- An approach to spoof state-of-the-art facial manipulation detection systems, while keeping the visual quality of the resulting images. Figure 8.1 graphically summarises the approach presented in Neves et al. (2020) based on a GAN-fingerprint Removal autoencoder (GANprintR).
- A thorough experimental assessment of this type of facial manipulation considering fake detection (based on holistic deep networks, steganalysis, and local artifacts) and realistic GAN-generated fakes (with and without the proposed GANprintR) over different experimental conditions, i.e. controlled and in-the-wild scenarios.
- A recent database named iFakeFaceDB,[2] resulting from the application of the GANprintR approach to already very realistic synthetic images.

---

[2] https://github.com/socialabubi/iFakeFaceDB.

The remainder of the chapter is organised as follows. Section 8.2 summarises the state of the art on the exploitation of GAN fingerprints for the detection of entire face synthesis manipulation. Section 8.3 explains the GAN-fingerprint removal approach (GANprintR) presented in Neves et al. (2020). Section 8.4 summarises the key features of the real and fake databases considered in the experimental assessment of this type of facial manipulation. Sections 8.5 and 8.6 describe the experimental setup and results achieved, respectively. Finally, Sect. 8.7 draws the final conclusions and points out some lines for future work.

8.2 Related Work

Contrary to popular belief, image manipulation dates back to the dawn of photography. Nevertheless, image manipulation only became particularly important after the rise of digital photography, due to the use of image processing techniques or low-cost image editing software. As a consequence, in the last decades the research community devised several strategies for assuring authenticity of digital data. In addition, digital image tampering still required some level of expertise to deceive the humans' eye, and both factors helped reducing significantly the use of manipulated content for malicious purposes. However, after the proposal of Generative Adversarial Networks (Goodfellow et al. 2014), the possibility of synthesising realistic digital content became possible. Among the four possible levels of face manipulation, this chapter focuses on the entire face synthesis manipulation, particularly on the problem of distinguishing between real and fake facial images.

Typically, synthetic face detection methods rely on the "fingerprints" caused by the generation process. According to the type of fingerprints used, each approach can be broadly divided into three categories: *(i)* methods based on visual artifacts; *(ii)* methods based on frequency analysis; and *(iii)* learning-based approaches for automatic fingerprint estimation. Table 8.1 provides a comparison of the state-of-the-art synthetic face detection methods.

The following sections describe the state-of-the-art techniques for synthetic data generation and review the state-of-the-art methods capable of detecting synthetic face imagery according to the taxonomy described above.

8.2.1 Generative Adversarial Networks

Proposed by Goodfellow et al. (2014), GANs are a novel generative concept, composed of two neural networks contesting each other in the form of a competition. A generator learns to generate instances that resemble the training data, while a discriminator learns to distinguish between the real and the generated images, while serving the goal of penalising the generator. The goal is to have a generator that can learn how to generate plausible images that can fool the discriminator. While

Table 8.1 Comparison of the state-of-the-art synthetic face detection methods

Study	Features	Classifiers	Best performance	Databases
Visual artifacts				
McCloskey and Albright (2018)	Colour Histogram	SVM	AUC = 70%	NIST MFC2018
Matern et al. (2019)	Eye Colour	K-NN	AUC = 85.2%	Real: CelebA Fake: Own Database (PGGAN)
Yang et al. (2019)	Head Pose	SVM	AUC = 89%	Real: UADFV/DARPA MediFor Fake: UADFV/DARPA MediFor
He et al. (2019)	Colour-related	Random Forest	Acc. = 99%	Real: CelebA Fake: Own Database (PGGAN)
Li et al. (2020)	Correlation Between Adjacent Pixels in Multiple Colour Channels	–	Acc. = 91.87	Real: FFHQ/LFW/LSUN/FFHQ Fake: Own Database (ProGAN, StyGAN, BigGAN, CocoGAN, DCGAN and WGAN-GP)
Hu et al. (2020)	Difference Between the Two Corneal Specular Highlights	Rule-based	AUC = 94%	Real: FFHQ Fake: Own Database (StyleGan2)
High-frequency information				
Yu et al. (2018)	GAN Fingerprint	Rule-based	Acc. = 99.50%	Real: CelebA Fake: Own Database
Wang et al. (2020b)	CNN Neuron Behaviour	SVM	Acc. = 84.78%	Real: CelebA-HQ/FFHQ Fake: Own Database
Stehouwer et al. (2020)	Image-related	CNN + Attention	EER = 0.05%	Real: CelebA/FFHQ/FaceForensics++ Fake: Own Database
Marra et al. (2019a)	GAN Fingerprint	Rule-based	AUC = 99.9%	Real: RAISE Fake: Own Database (Cycle-GAN, ProGAN, and Star-GAN)
Albright et al. (2019)	GAN Fingerprint	Rule-based	Acc. = 98.33%	Real: MNIST/CelebA Fake: Own Database (ProGAN, SAGAN, SNGAN)
Guarnera et al. (2020)	Local Pixel Correlations	K-NN	Acc. = 99.81%	Real: CelebA Fake: Own Database (StarGAN, StyleGAN, StyleGAN2, GDWCT, AttGAN)

Table 8.1 (continued)

Study	Features	Classifiers	Best performance	Databases
High-frequency information				
Zhang et al. (2019)	Image Spectrum	CNN	Acc. = 97.2%	CycleGAN/AutoGAN
Durall et al. (2020)	Frequency features extracted from DFT	SVM	Acc. = 90%	Real: Own Database (CelebA, FFHQ) Fake: Own Database (100 K, StyleGan)
Frank et al. (2020)	Frequency features extracted from DCT	Ridge-regression	Acc. = 100%	Real: FFHQ Fake: Own Database (StyleGAN)
Bonettini et al. (2020)	Distribution of the quantized coefficients of the DCT	Random Forest	Acc. = 99.83%	GAN-generated (Marra et al. 2019b) (CycleGAN, ProGAN)
Learning-based				
Marra et al. (2018)	Image-related	CNN	Acc. = 95.07%	Real: Own Database(CycleGAN) Fake: Own Database(CycleGAN)
Hsu et al. (2020)	Raw Image	CNN	Precision = 88 Recall = 87.32	Real: CelebA Fake: Own Database (DCGAN, WGAP, WGAN-GP, LSGAN, PGGAN)
Marra et al. (2019c)	Raw Image Using Incremental Learning Strategy	CNN	Acc. = 99.37%	Real: CelebA-HQ Fake: DoGANS (CycleGAN, ProGAN, Glow, StarGAN)
Xuan et al. (2019)	Pre-processed Image Using Blur or Noise in Training	CNN	Acc. = 95.45%	Real: CelebA-HQ Fake: Own Database (DC-GAN, WGAN-GP, PGGAN)
Wang et al. (2020a)	Raw Image	CNN	mAP = 93	Own Database (using 11 synthesis models)
Hsu et al. (2020)	Raw Image	CNN	Precision = 96.76 Recall = 90.56	Real: CelebA Fake: Own Database (DCGAN, WGAP, WGAN-GP, LSGAN, PGGAN)
Nataraj et al. (2020)	Co-occurrence matrix of each colour channel (RGB)	CNN	Acc. = 87.96%	Own Database (ProGAN, StarGAN, GlowGAN, StyleGAN2)

Table 8.1 (continued)

Study	Features	Classifiers	Best performance	Databases
Learning-based				
Goebel et al. (2020)	Co-occurrence matrix of each colour channel (RGB)	CNN	Acc. = 98.17%	Own Database (StarGAN, CycleGAN, ProGAN, Spade, StyleGAN)
Bani et al. (2020)	Co-occurrence matrix of each colour channel (RGB) and for each colour channels pairs	CNN	Acc. = 99.70%	Real: FFHQ Fake: Own Database (StyleGAN2)
Hulzebosch et al. (2020)	Pre-processed Image Using Colour Transformations, Co-occurrence Matrices or High-pass Filters	CNN	Acc. = 99.9%	Real: CelebA-HQ/FFHQ Fake: Own Database (StarGAN, GLOW, ProGAN, StyleGAN)
Liu et al. (2020)	Global Texture Features captured by "Gram-Block" (extra layer)	CNN	Acc. = 95.51%	Real: CelebA-HQ/FFHQ Fake: Own Database (StyleGAN, PGGAN, DCGAN, DRAGAN, StarGAN)
Yu et al. (2020a)	Channel Differences, Image Spectrum	CNN	Acc. = 99.41%	Real: FFHQ Fake: Own Database (StyleGAN, StyleGAN2)

at the beginning, GANs were only capable of producing low-resolution images of faces with some notorious visual artifacts, in the last years several techniques have emerged for synthesising highly realistic content (including BigGAN Brock et al. 2019, CycleGAN Zhu et al. 2017, GauGAN Park et al. 2019, ProGAN Karras et al. 2018, StarGAN Choi et al. 2018, StyleGAN Karras et al. 2019, and StyleGAN2 Karras et al. 2020) that even humans cannot distinguish from the real ones. Next, we review the state-of-the-art approaches specifically devised for detecting a entire face synthesis manipulation.

8.2.2   GAN Detection Techniques

As denoted before, the images generated by the initial versions of GANs exhibited several visual artifacts, including distinct eye colour, holes in the face, deformed teeth, among others. For this reason, several approaches attempted to leverage these traits for detecting face manipulations (Matern et al. 2019; Yang et al. 2019; Hu et al. 2020). Matern et al. (2019) extracted several geometric facial features which were then fed to a Support Vector Machine (SVM) classifier to distinguish between real and synthetic face images. Yang et al. (2019) exploited the weakness of GANs in generating consistent head poses and trained a SVM to distinguish between real and synthetic faces based on the estimation of the 3D head pose. As the remaining artifacts became less noticeable, researchers focused on more subtle features of the face, as in Hu et al. (2020), where synthetic face detection was performed by analysing the difference between the two corneal specular highlights. Other visual artifact typically exploited is the probability distribution of colour channels. McCloskey and Albright (McCloskey and Albright 2018) hypothesised that the colour is markedly different between real camera images and fake synthesis images, and proposed a detection system based on the colour histogram and a linear SVM. He et al. (2019) exploited different colour channels (YCbCr, HSV and Lab) to extract from a CNN different deep representations, which were subsequently fed to a Random Forest classifier for distinguishing between real and synthetic data. Li et al. (2020) observed that it is easier to spot the differences between real and GAN-generated data in non-RGB colour spaces, since GANs are trained for producing content in RGB channels.

As the quality and realism of synthetic data improved, visual artifacts started to become ineffectual, which in turn fostered researchers to explore digital forensic techniques for the problem of synthetic data detection. Each camera sensor leaves a unique and stable mark on each acquired photo, denoted as the photo-response non-uniformity (PRNU) pattern (Lukás et al. 2006). This mark is usually denoted as the camera fingerprint, which inspired researchers to detect the presence of similar patterns in images synthesised by GANs. These approaches usually define the GAN fingerprint as a high-frequency signal available in the image. Marra et al. (2019a) defined GAN fingerprint as the high-level image information obtained by subtracting the image from its corresponding denoised version. Yu et al. (2018) improved (Marra et al. 2019a) by subtracting from the original image the corresponding reconstructed version obtained from an autoencoder, which was tuned based on the discriminability of the fingerprints inferred by this process. They learned a model fingerprint for each source (each GAN instance plus the real world), such that the correlation index between one image fingerprint and each model fingerprint gives the probability of the image being produced by a specific model. Their proposed approach was tested using real faces from CelebA database (Liu et al. 2015) and synthetic faces created through different GAN approaches (PGGAN Karras et al. 2018, SNGAN Miyato et al. 2018, CramerGAN Bellemare et al. 2017, and MMDGAN Binkowski et al. 2018), achieving a final accuracy of 99.50% for the best performance. Later, they extended their approach (Yu et al. 2020b) by proposing a novel strategy for the training of the generative model such that the fingerprints can be controlled by the user, and easily decoded from a synthetic image, allowing to solve the problem of source attribution, i.e. identifying the model that generated the image. In (Albright and McCloskey 2019), the authors proposed an alternative to (Yu et al. 2018) by replacing the autoencoder by an inverted GAN capable of reconstructing an image based on the attributes inferred from the original image. Zhang et al. (2019) proposed the use of the up-sampling artifact in the frequency domain as a discriminative feature for distinguishing veridical and synthetic data. Frank et al. (2020) reported similar conclusions regarding the discriminability of the frequency space of GAN-generated images. They relied on the Discrete Cosine Transform (DCT) for extracting features from either real and fake images, in order to train a linear classifier. Durall et al. (2020) found out that upconvolution or transposed convolution layers of GAN architectures are not capable of reproducing the spectral distribution of natural images. Based on this finding, they showed that generated face images can be easily identified by training a SVM with the features extracted with the Discrete Fourier Transform (DFT). Guarnera et al. (2020) used pixel correlation as a GAN fingerprint, since they noticed that the correlation of pixels in synthetic images are exclusively dependent on the operations performed by all the layers present in the GAN which generate it. Their proposed approach was tested using fake images generated by several GAN architectures (AttGAN, GDWCT, StarGAN, StyleGAN and StyleGAN2).

A distinct family of methods adopts a data-driven strategy for the problem of detecting GAN-generated imagery. In this strategy, a standard image classifier, typically a Convolutional Neural Network (CNN), is trained directly with raw images or through a modified version of them (Barni et al. 2020; Hsu et al. 2020). Marra et al. (2018) carried out a study about the classification accuracy of different CNN architectures when fed with raw images. It was observed that, in spite almost ideal performance was obtained, the performance decreased significantly when compressed images were used in the test set. Later, the authors proposed a strategy based on incremental learning for addressing this problem and the generalisation to unseen datasets (Marra et al. 2019c). Inspired by the forensic analysis of image manipulation (Cozzolino et al. 2014), Nataraj et al. (2019a) proposed a detection system based on a combination of pixel co-occurrence matrices and CNNs. Their proposed approach was initially tested in a database of various objects and scenes created through Cycle-GAN (Zhu et al. 2017). Besides, the authors performed an interesting analysis to see the robustness of the proposed approach against fake images created through different GAN architectures (CycleGAN vs. StarGAN), with good generalisation results. This idea was later improved in (Goebel et al. 2020) and (Barni et al. 2020).

The above studies show that a simple CNN is able to easily distinguish between real and synthetic data generated from specific GAN architectures, but is not capable of maintaining the same performance in data originated from GAN architectures not seen during training or even in data altered by image filtering operations. For this reason, Xuan et al. (2019) used an image pre-processing step in the training stage to remove artifacts of a specific GAN architecture. The same idea was exploited in (Hulzebosch et al. 2020) to improve the accuracy in real-world scenarios, where the particularities of the data (e.g. image compression) and the generator architecture are not known. Liu et al. (2020) observed that the texture of fake faces is substantially different from the real ones. Based on this observation, the authors devised a novel block to be added to the backbone of a CNN, the Gram-Block, which is capable of extracting global image texture features and improve the generalisation of the model against data generated by GAN architectures not used during training. Similarly, Yu et al. (2020a) introduced a novel convolution operator intended for separately processing the low- and high-frequency information of the image, improving the capability to detect the patterns of synthetic data available in the high-frequency band of the images. Finally, Wang et al. (2020a) studied the topic of generalisation to unseen datasets. For this, they collected a dataset consisting of fake images generated by 11 different CNN-based image generator models and concluded that the correct combination of pre-processing and data augmentation techniques allows a standard image classifier to generalise to unseen dataset even when trained with data obtained from a single GAN architecture.

To summarise this section, we conclude that state-of-the-art automatic detection systems against face synthesis manipulation have excellent performance, mostly because they are able to learn the GAN fingerprints present in the images. However, it is also clear that the dependence on the model fingerprint affects the generability and the reliability of the model, e.g. when presented with adversarial attacks (Gandhi and Jain 2020).

8.3   GAN Fingerprint Removal: GANprintR

GANprintR was originally presented in (Neves et al. 2020) and aims at transforming synthetic face images, such that their visual appearance is unaltered but the GAN fingerprints (the discriminative information that permits the distinction from real imagery) are removed. Considering that the fingerprints are high-frequency signals (Marra et al. 2019a), we hypothesised that their removal could be performed by an autoencoder, which acts as a non-linear low-pass filter. We claimed that by using this strategy, the detection capability of state-of-the-art facial manipulation detection methods significantly decreases, while at the same time humans still are not capable of perceiving that images were transformed.

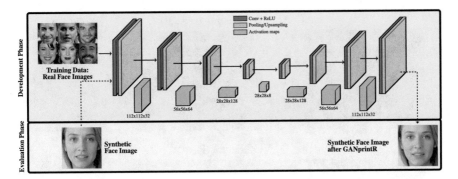

Fig. 8.2 GAN-fingerprint Removal module (GANprintR) based on a convolutional AutoEncoder (AE). The AE is trained using only real face images from the development dataset. In the evaluation stage, once the autoencoder is trained, we can pass synthetic face images through it to provide them with additional naturalness, in this way removing the GAN-fingerprint information that may be present in the initial fakes

In general, an autoencoder comprises two distinct networks, encoder ψ and decoder γ:

$$\psi : X \mapsto l$$
$$\gamma : l \mapsto X', \tag{8.1}$$

where X denotes the input image to the network, l is the latent feature representation of the input image after passing through the encoder ψ, and X' is the reconstructed image generated from l, after passing through the decoder γ. The networks ψ and γ can be learned by minimising the reconstruction loss $\mathcal{L}_{\psi,\gamma}(X, X') = ||X - X'||^2$ over a development dataset following an iterative learning strategy.

As result, when $\mathcal{L}$ is nearly 0, ψ is able to discard all redundant information from X and code it properly into l. However, for a reduced size of the latent feature representation vector, $\mathcal{L}$ will increase and ψ will be forced to encode in l only the most representative information of X. We claimed that this kind of autoencoder acts as a GAN-fingerprint removal system.

Figure 8.2 describes the GANprintR architecture based on a convolutional AutoEncoder (AE) composed of a sequence of 3×3 convolutional filters, coupled with ReLU activation functions. After each convolutional layer, a 2×2 max-pooling layer is used to progressively decrease the size of the activation map to $28 \times 28 \times 8$, which represents the bottleneck of the reconstruction model.

The AE is trained with images from a public dataset that comprises face imagery from real persons. In the evaluation phase, the AE is used to generate improved fakes from input fake faces where GAN "fingerprints", if present in the initial fakes, will be reduced. The main rationale of this strategy is that by training with real images the AE can learn the core structure of this type of natural data, which can then be exploited to improve existing fakes.

CASIA-WebFace (Real)

VGGFace2 (Real)

TPDNE (Synthetic)

100K-Faces (Synthetic)

PGGAN (Synthetic)

Fig. 8.3 Examples of the databases considered in the experiments of this chapter after applying the pre-processing stage described in Sect. 8.5.1

8.4 Databases

Four different public databases and one generated are considered in the experimental framework of this chapter. Figure 8.3 shows some examples of each database. We now summarise the most important features.

8.4.1 Real Face Images

- *CASIA-WebFace:* this database contains 494,414 face images from 10,575 actors and actresses of IMDb. Face images comprise random pose variations, illumination, facial expression and resolution.
- *VGGFace2:* this database contains 3,31 million images from 9,131 different subjects, with an average of 363 images per subject. Images were downloaded from the Internet and contain large variations in pose, age, illumination, ethnicity and profession (e.g. actors, athletes, and politicians).

8.4.2 Synthetic Face Images

- *TPDNE:* this database comprises 150,000 unique faces, collected from the website.[3] Synthetic images are based on the recent StyleGAN approach (Karras et al. 2019) trained with FFHQ database (Flickr-Faces-HQ 2019).
- *100K-Faces:* this database contains 100,000 synthetic images generated using StyleGAN (Karras et al. 2019). In this database the StyleGAN network was trained using around 29,000 photos of 69 different models, producing face images with a flat background.
- *PGGAN:* this database comprises 80,000 synthetic face images generated using the PGGAN network. In particular, we consider the publicly available model trained using the CelebA-HQ database.

8.5 Experimental Setup

This section describes the details of the experimental setup followed in the experimental framework of this chapter.

---

[3] https://thispersondoesnotexist.com.

8.5.1 Pre-processing

In order to ensure fairness in our experimental validation, we created a curated version of all the datasets where the confounding variables were removed. Two different factors were considered in this chapter:

- *Background*: this is a clearly distinctive aspect among real and synthetic face images as different acquisition conditions are considered in each database.
- *Head pose*: images generated by GANs hardly ever produce high variation from the frontal pose (Dang et al. 2020), contrasting with most popular real face databases such as CASIA-WebFace and VGGFace2. Therefore, this factor may falsely improve the performance of the detection systems since non-frontal images are more likely to be real faces.

To remove these factors from both the real and synthetic images, we extracted 68 face landmarks, using the method described in (Kazemi and Sullivan 2014). Given the landmarks of the eyes, an affine transformation was determined such that the location of the eyes appears in all images at the same distance from the borders. This step allowed to remove all the background information of the images while keeping the maximum amount of the facial regions. Regarding the head pose, landmarks were used to estimate the pose (*frontal* vs. *non-frontal*). In the experimental framework of this chapter, we kept only the frontal face images, in order to avoid biased results. After this pre-processing stage, we were able to provide images of constant size (224×224 pixels) as input to the systems. Figure 8.3 shows examples of the crop-out faces of each database after applying the pre-processing steps. The synthetic images obtained by this pre-processing stage are the ones used to create the database iFakeFaceDB after being processed by the GANprintR approach.

8.5.2 Facial Manipulation Detection Systems

Three different state-of-the-art manipulation detection approaches are considered in this chapter.

(1) *XceptionNet* (Chollet 2017): this network was selected, essentially because it provides the best detection results in the most recently published studies (Dang et al. 2020; Rössler et al. 2019; Dolhansky et al. 2019). We followed the same training approach considered in (Rössler et al. 2019): *(i)* the model was initialised with the weights obtained after training with the ImageNet dataset (Deng et al. 2009), *(ii)* we changed the last fully-connected layer of the ImageNet model by a new one (two classes, real or synthetic image), *(iii)* we fixed all weights up to the final layers and pre-trained the network for few epochs, and finally *(iv)* we trained the network for 20 more epochs and chose the best performing model based on validation accuracy.

(2) *Steganalysis* (Nataraj et al. 2019b): the method by Nataraj et al. was selected for providing an approach based on steganalysis, rather than directly extracting features from the images, as in the XceptionNet approach. In particular, this approach

calculates the co-occurrence matrices directly from the image pixels on each channel (red, green and blue), and passes this information through a custom CNN, which allows the network to extract non-linear robust features. Considering that the source code is not available from the authors, we replicated this technique to perform our experiments.

(3) *Local Artifacts* (Matern et al. 2019): we have chosen the method of Matern et al., because it provides an approach based on the direct analysis of the visual facial artifacts, in opposition to the remaining approaches that follow holistic strategies. In particular, the authors of that work claim that some parts of the face (e.g. eyes, teeth, facial contours) provide useful information about the authenticity of the image, and thus train a classifier to distinguish between real and synthetic face images using features extracted from these facial regions.

All our experiments were implemented under a PyTorch framework, with a NVIDIA Titan X GPU. The training of the Xception network was performed using the Adam optimiser with a learning rate of 10^{-3}, dropout for model regularisation with a rate of 0.5, and a binary cross-entropy loss function. Regarding the steganalysis approach, we reused the parameters adopted for Xception network, since the authors of (Nataraj et al. 2019b) did not detail the training strategy adopted. Regarding the local artifacts approach, we adopted the strategy for detecting "generated faces", where a k-nearest neighbour classifier was used to distinguish between real and synthetic face images based on eye colour features.

8.5.3 Protocol

The experimental protocol designed in this chapter aims at performing an exhaustive analysis of the state-of-the-art facial manipulation detection systems. As such, three different experiments were considered: *(i)* controlled scenarios, *(ii)* in-the-wild scenarios, and *(iii)* GAN-fingerprint removal.

Each database was divided into two disjoint datasets, one for the development of the systems (70%) and the other one for evaluation purposes (30%). Additionally, the development dataset was divided into two disjoint subsets, training (75%) and validation (25%). The same number of real and synthetic images were considered in the experimental framework. In addition, for real face images, different users were considered in the development and evaluation datasets, in order to avoid biased results.

The GANprintR approach was trained during 100 epochs, using the Adam optimizer with a learning rate of 10^{-3}, and a mean square error (MSE) to obtain the reconstruction loss. To ensure an unbiased evaluation, GANprintR was trained with images from the MS-Celeb dataset (Guo et al. 2016), since it is disjoint from the datasets used in the development and evaluation of all the fake detection systems used in our experiments.

8.6 Experimental Results

This section describes the results achieved in the experimental framework of this chapter.

8.6.1 Controlled Scenarios

In this section, we report the results of the detection of entire face synthesis in controlled scenarios, i.e. when samples from the same databases were considered for both development and final evaluation of the detection systems. This is the strategy commonly used in most studies, typically resulting in very good performance (see Sect. 8.2).

A total of six experiments were carried out: A.1 to A.6. Table 8.2 describes the development and evaluation databases considered in each experiment together with the corresponding final evaluation results in terms of EER. Additionally, we represent in Fig. 8.4 the evolution of the loss/accuracy of the XceptionNet and Steganalysis detection systems for Exp. A.1.

The analysis of Fig. 8.4 shows that both XceptionNet and Steganalysis approaches were able to learn discriminative features to detect between real and synthetic face images. The training process was faster for the XceptionNet detection system compared with Steganalysis, converging to a lower loss value in fewer epochs (close to zero after 20 epochs). The best validation accuracy achieved in Exp. A.1 for the XceptionNet and Steganalysis approaches were 99% and 95%, respectively. Similar trends were observed for the other experiments.

We now analyse the results included in Table 8.2 for experiments A.1 to A.6. Analysing the results obtained by the XceptionNet system, almost ideal performance

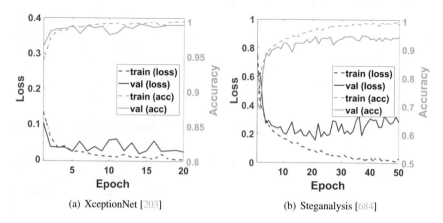

(a) XceptionNet [203]   (b) Steganalysis [684]

Fig. 8.4 **Exp. A.1:** Evolution of the loss/accuracy with the number of epochs

Table 8.2 Controlled and in-the-wild scenarios: manipulation detection performance in terms of EER (%) for different development and evaluation setups. R_{real} and R_{fake} denote the Recall of the real and fake classes, respectively. Controlled (Exp. A.1–A.6). In-the-wild (Exp. B.1–B.24). VF2 = VGGFace2. CASIA = CASIA-WebFace. All metrics are given in (%)

Experiment	Development		Evaluation		XceptionNet (Chollet 2017)			Steganalysis (Nataraj et al. 2019b)			Local artifacts (Matern et al. 2019)		
	Real	Synthetic	Real	Synthetic	EER	R_{real}	R_{fake}	EER	R_{real}	R_{fake}	EER	R_{real}	R_{fake}
A.1	**VF2**	**TPDNE**	**VF2**	**TPDNE**	**0.22**	**99.77**	**99.80**	**10.92**	**89.07**	**89.10**	**38.53**	**60.72**	**62.20**
B.1	VF2	TPDNE	VF2	100F	0.45	99.30	99.80	23.07	71.66	85.59	35.86	64.13	64.16
B.2	VF2	TPDNE	VF2	PGGAN	13.82	78.44	99.73	27.12	67.28	83.87	40.10	59.05	60.80
B.3	VF2	TPDNE	CASIA	100F	0.35	99.30	100.00	24.00	71.23	83.53	35.61	64.05	64.69
B.4	VF2	TPDNE	CASIA	PGGAN	13.72	78.47	100.00	28.05	66.81	81.61	39.87	59.0	61.4
A.2	**VF2**	**100F**	**VF2**	**100F**	**0.28**	**99.70**	**99.73**	**12.28**	**87.70**	**87.73**	**31.45**	**67.83**	**69.26**
B.5	VF2	100F	VF2	TPDNE	21.18	70.32	99.54	28.02	66.72	82.09	42.89	55.17	60.16
B.6	VF2	100F	VF2	PGGAN	44.43	52.96	97.71	32.62	62.35	79.31	48.70	50.53	52.87
B.7	VF2	100F	CASIA	TPDNE	21.07	70.37	99.94	28.85	66.29	80.14	46.04	52.50	55.98
B.8	VF2	100F	CASIA	PGGAN	44.32	53.01	99.71	33.45	61.90	77.15	51.89	47.8	48.6
A.3	**VF2**	**PGGAN**	**VF2**	**PGGAN**	**0.02**	**99.97**	**100.00**	**3.32**	**96.67**	**96.70**	**35.13**	**64.33**	**65.41**
B.9	VF2	PGGAN	VF2	TPDNE	16.85	74.79	100.00	33.32	60.42	91.74	40.84	57.55	61.17
B.10	VF2	PGGAN	VF2	100F	5.85	89.53	100.00	25.60	66.87	94.04	44.47	53.99	57.77
B.11	VF2	PGGAN	CASIA	TPDNE	16.85	74.79	100.00	35.73	59.19	81.85	39.89	58.02	62.82
B.12	VF2	PGGAN	CASIA	100F	5.85	89.53	100.00	28.02	65.73	86.50	43.53	54.5	59.5
A.4	**CASIA**	**TPDNE**	**CASIA**	**TPDNE**	**0.02**	**99.97**	**100.00**	**12.08**	**87.90**	**87.93**	**39.36**	**59.62**	**61.65**
B.13	CASIA	TPDNE	VF2	100F	1.75	99.35	97.20	36.68	59.58	71.82	39.03	60.67	61.25
B.14	CASIA	TPDNE	VF2	PGGAN	4.42	94.21	97.04	30.77	65.13	76.40	38.94	61.02	61.10
B.15	CASIA	TPDNE	CASIA	100F	0.32	99.37	100.00	34.12	61.02	78.41	38.05	61.20	62.67

(continued)

Table 8.2 (continued)

Experiment	Development		Evaluation		XceptionNet (Chollet 2017)			Steganalysis (Nataraj et al. 2019b)			Local artifacts (Matern et al. 2019)		
	Real	Synthetic	Real	Synthetic	EER	R_{real}	R_{fake}	EER	R_{real}	R_{fake}	EER	R_{real}	R_{fake}
B.16	CASIA	TPDNE	CASIA	PGGAN	2.98	94.37	100.00	28.20	66.48	82.19	37.96	61.5	62.5
A.5	**CASIA**	**100F**	**CASIA**	**100F**	**0.08**	**99.90**	**99.93**	**16.05**	**83.94**	**83.96**	**33.96**	**65.04**	**67.03**
B.17	CASIA	100F	VF2	TPDNE	5.93	97.69	90.95	34.00	62.64	71.80	43.11	55.00	59.83
B.18	CASIA	100F	VF2	PGGAN	10.08	89.64	90.20	45.63	52.91	58.71	46.36	52.37	55.92
B.19	CASIA	100F	CASIA	TPDNE	1.10	97.91	99.93	31.67	63.97	76.67	44.22	53.94	58.54
B.20	CASIA	100F	CASIA	PGGAN	5.25	90.55	99.93	43.30	54.34	64.74	47.49	51.3	54.6
A.6	**CASIA**	**PGGAN**	**CASIA**	**PGGAN**	**0.05**	**99.93**	**99.97**	**4.62**	**95.37**	**95.40**	**34.79**	**64.42**	**66.00**
B.21	CASIA	PGGAN	VF2	TPDNE	4.90	99.96	91.10	31.73	61.93	88.92	43.52	55.25	57.94
B.22	CASIA	PGGAN	VF2	100F	4.88	100.00	91.10	41.97	54.63	80.35	44.69	54.05	56.89
B.23	CASIA	PGGAN	CASIA	TPDNE	0.03	99.97	99.97	31.43	62.08	90.07	41.46	56.64	61.00
B.24	CASIA	PGGAN	CASIA	100F	0.02	100.00	99.97	41.67	54.79	82.22	42.63	55.5	60.0

is achieved with EER values less than 0.5%. These results are in agreement to previous studies in the topic (see Sect. 8.2), pointing for the potential of the XceptionNet model in controlled scenarios. Regarding the Steganalysis approach, a higher degradation of the system performance is observed, when compared with the XceptionNet approach, especially for the 100K-Face database, e.g. a 16% EER is obtained in Exp. A.5. Finally, it can be observed that the approach based on local artifacts was the least efficient to spot the differences between real and synthetic data, with an average 35.5% EER over all experiments.

In summary, for controlled scenarios XceptionNet has excellent manipulation detection accuracies, then Steganalysis provides good accuracies, and finally Local Artifacts have poor accuracy. In the next section we will see the limitations of these techniques in-the-wild.

8.6.2 In-the-Wild Scenarios

This section evaluates the performance of the facial manipulation detection systems in more realistic scenarios, i.e. in-the-wild. The following aspects are considered: *(i)* different development and evaluation databases, and *(ii)* different image resolution/blur among the development and evaluation of the models. This last point is particularly important, as the quality of raw images/videos is usually modified when, e.g. they are uploaded to social media. The effect of image resolution has been preliminary analysed in previous studies (Rössler et al. 2019; Korshunov and Marcel 2018), but for different facial manipulation groups, i.e. face swapping/identity swap and facial expression manipulation. The main goal of this section is to analyse the generalisation capability of state-of-the-art entire face synthesis detection in unconstrained scenarios.

First, we focus on the scenario of considering the same real but different synthetic databases in development and evaluation (Exp. B.1, B.2, B.5, B.6, and so on, provided in Table 8.2). In general, the results achieved in the experiments evidence a high degradation of the detection performance regardless of the facial manipulation detection approach. For the XceptionNet, the average EER is 11.2%, i.e. over 20 times higher than the results achieved in Exp. A.1–A.6 (<0.5% average EER). Regarding the Steganalysis approach, the average EER is 32.5%, i.e. more than 3 times higher than the results achieved in Exp. A.1–A.6 (9.8% average EER). For Local Artifacts, the observed average EER was 42.4%, with an average worsening of 19%. The large degradation of the first two detectors suggests that they might rely heavily on the GAN fingerprints of the training data. This result confirms the hypothesis that different GAN models produce different fingerprints, as also mentioned in previous studies (Yu et al. 2018). Moreover, these results suggest that these GAN fingerprints are the information used by the detectors to distinguish between real and synthetic data.

Table 8.2 also considers the case of using different real and synthetic databases for both development and evaluation (Exp. B.3, B.4, B.7, B.8, etc.). In this scenario, an average EERs of 9.3%, 32.3% and 42.3% in fake detection were obtained for XceptionNet, Steganalysis and Local Artifacts, respectively. When comparing these results with the EERs of the previous experiments (where only the synthetic evaluation set was changed), no significant gap in performance was found, which points that the change of synthetic data might be the main cause for performance degradation.

Finally, we also analyse how different image transformations affect facial manipulation detection systems. In this analysis, we focus only on the XceptionNet model as it provides much better results when compared with the remaining detection systems. For each baseline experiment (A.1 to A.6), the evaluation set (both real and fake images) was transformed by: *(i)* resolution downsizing (1/3 of the original resolution), *(ii)* a low-pass filter (9×9 Gaussian kernel, $\sigma = 1.7$), and *(iii)* jpeg image compression using a quality level of 60. The resulting EER together with the Recall, PSRN and SSIM values are provided in Table 8.3, together with the performance of the original images. The results suggest a high performance degradation in all experiments, proving the vulnerability of the fake detection system to unseen conditions, even if they result from simple image transformations.

To further understand the impact of these transformations, we evaluated an increasing downsize ratio in the performance of the fake detection system. Figure 8.5 depicts the detection performance results in terms of EER (%), from lower to higher modifications of the image resolution. In general, we can observe increasingly higher degradation of the fake detection performance for decreasing resolution. For example, when the image resolution is reduced by 1/4, the average EER increases 6% when compared with the raw image resolution (raw equals to 1/1). This performance degradation is even higher when we further reduce the image resolution, with EERs (%) higher than 15%. These results support the conclusion about a poor generalisation capacity of state-of-the-art facial manipulation detection systems to unseen conditions.

8.6.3 GAN-Fingerprint Removal

This section analyses the results of the strategy for GAN-fingerprint Removal (GANprintR). We evaluated to what extent our method is capable of spoofing state-of-the-art facial manipulation detection systems by improving fake images already obtained with some of the best and most realistic known methods for entire face synthesis. For this, the experiments A.1 to A.6 were repeated for the XceptionNet detection system, but the fake images of the evaluation set were transformed after passing through GANprintR.

Table 8.3 provides the results achieved for both the original fake data and after GANprintR. The analysis of the results shows that GANprintR obtains higher fake detection error than the remaining attacks, while maintaining a similar or even better visual quality. In all the experiments, the EER of the manipulation detection increases when using GANprintR to transform the synthetic face images. Also, the detection degradation is higher than other types of attacks for similar PSNR values and slightly

Table 8.3 Comparison between the GANprintR approach and typical image manipulations. The detection performance is provided in terms of EER (%) for experiments A.1 to A.6, when using different versions of the evaluation set. TDE stands for transformation of the evaluation data and details the technique used to modify the test set before fake detection. R_{real} and R_{fake} denote the Recall of the real and fake classes, respectively,

Experiment	TDE	EER (%)	R_{real} (%)	XceptionNet		
				R_{fake} (%)	PSNR (db)	SSIM
A.1	**Original**	**0.22**	**99.77**	**99.80**	–	–
	Downsize	1.17	98.83	98.87	35.55	0.93
	Low-pass filter	0.83	99.17	99.20	34.63	0.92
	jpeg compression	1.53	98.47	98.50	36.02	0.96
	GANprintR	10.63	89.37	89.40	35.01	0.96
A.2	**Original**	**0.28**	**99.70**	**99.73**	–	–
	Downsize	0.87	99.13	99.17	36.24	0.95
	Low-pass filter	2.87	97.10	97.13	35.22	0.93
	jpeg compression	1.83	98.17	98.20	36.76	0.97
	GANprintR	6.37	93.64	93.66	35.59	0.96
A.3	**Original**	**0.02**	**99.97**	**100.00**	–	–
	Downsize	3.70	96.27	96.30	34.85	0.91
	Low-pass filter	1.53	98.43	98.47	34.10	0.90
	jpeg compression	30.93	69.04	69.06	35.85	0.96
	GANprintR	17.27	82.71	82.73	34.82	0.95
A.4	**Original**	**0.02**	**99.97**	**100.00**	–	–
	Downsize	1.00	98.97	99.00	35.55	0.93
	Low-pass filter	0.07	99.90	99.93	34.63	0.92
	jpeg compression	2.50	97.47	97.50	36.02	0.96
	GANprintR	4.47	95.50	95.53	35.01	0.96
A.5	**Original**	**0.08**	**99.90**	**99.93**	–	–
	Downsize	6.27	93.70	93.73	36.24	0.95
	Low-pass filter	11.53	88.44	88.46	35.22	0.93
	jpeg compression	3.27	96.73	96.77	36.76	0.97
	GANprintR	11.47	88.50	88.53	35.59	0.96
A.6	**Original**	**0.05**	**99.93**	**99.97**	–	–
	Downsize	7.77	92.24	92.26	34.85	0.91
	Low-pass filter	2.10	97.90	97.93	34.10	0.90
	jpeg compression	5.37	94.64	94.66	35.85	0.96
	GANprintR	8.37	91.64	91.66	34.82	0.95

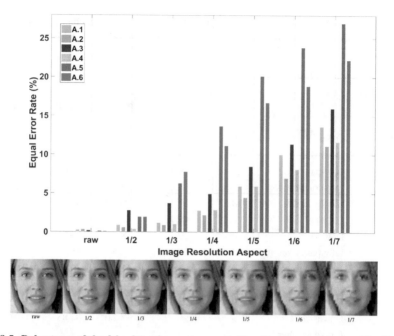

Fig. 8.5 Robustness of the fake detection system regarding the image resolution. The Xcep-tionNet model is trained with the raw image resolution and evaluated with lower image resolutions. Note how the EER increases significantly while reducing the image resolution

higher values of SSIM. In particular, the average EER when considering GANprintR is 9.8%, i.e. over 20 times higher than the results achieved when using the original fakes (<0.5% average EER). This suggests that our method is not simply removing high-frequency information (evidenced by the comparison with the low-pass filter and downsize) but it is also removing the GAN fingerprints from the fakes improving their naturalness. It is important to remark that different real face databases were considered for training the face manipulation detectors and our GANprintR module.

In addition, we provide in Fig. 8.6 an analysis of the impact of the latent feature representation of the autoencoder in terms of EER and PSNR. In particular, we follow the experimental protocol considered in Exp. A.3, and calculate the EER of Xcep-tionNet for detecting fakes improved with various configurations of GANprintR. Moreover, the PSNR for each set of transformed images is also included in Fig. 8.6 together with a face example of each configuration to visualise the image quality. The face examples included in Fig. 8.6 show no substantial differences between the original fake and the resulting fakes after GANprintR for the different latent fea-ture representation size of the GANprintR, which is confirmed by the tight range of PSNR values obtained along the different latent feature representations. The EER values of fake detection significantly increase as the size of latent feature represen-tations diminish, evidencing that GANprintR is capable of spoofing state-of-the-art detectors without significantly degrading the visual aspect of the image.

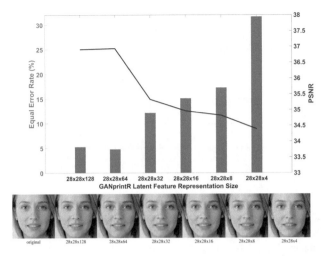

Fig. 8.6 Robustness of the fake detection system after GAN-fingerprint Removal (GAN-printR). The latent feature representation size of the AE is varied to analyse the impact on both system performance and visual aspect of the reconstructed images. Note how the EER increases significantly when considering GANprintR spoof approach, while maintaining a high visual similarity with the original image

Finally, to confirm that GANprintR is actually removing the GAN-fingerprint information and not just reducing the image resolution of the images, we performed a final experiment where we trained the XceptionNet for fake detection considering different levels of image resolution, and then tested it using fakes improved with GANprintR. Figure 8.7 shows the fake detection performance in terms of EER for different sizes of the latent feature representation of GANprintR. Five different GANprintR configurations are tested per image resolution. The obtained results point for the stability of EER values with respect to downsized synthetic images in training, concluding that GANprintR is actually removing the GAN-fingerprint information.

8.6.4   Impact of GANprintR on Other Fake Detectors

For completeness, we provide in this section a comparative analysis between the impact of the GANprintR approach on the three state-of-the-art manipulation detection approaches considered in this chapter. Table 8.4 reports the EER and Recall observed when using the original images and when using the modified version of the same images.

In Sect. 8.6.1 it has been concluded that XceptionNet stands out as the most reliable approach at recognising synthetic faces. The analysis of Table 8.4 evidences that this conclusion also holds when using images transformed by GANprintR. Nevertheless, it is also interesting to analyse the performance degradation caused by the GANprintR

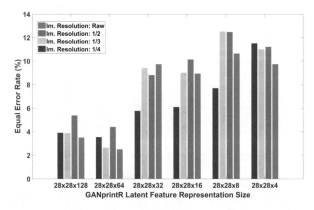

Fig. 8.7 Robustness of the fake detection system trained with different resolutions and then tested with fakes improved with GANprintR under various configurations (representation sizes). Five different GANprintR configurations are tested per image resolution level. The results observed point for the stability of EER values with respect to using downsized synthetic images in training. This observation supports the conclusion that GANprintR is actually removing the GAN fingerprints.

approach. The average number of percentage points that the EER has increased for XceptionNet, Steganalysis and Local Artifacts is 9.65, 14.68 and 4.91, respectively. Even though, in this case, the work of Matern et al. (2019) stands out for having the lowest performance degradation, we believe that this is primarily due to the high EER achieved in the original set of images.

8.7 Conclusions and Outlook

This chapter has covered the topic of GAN fingerprints in face image synthesis. We have first provided an in-depth literature analysis of the most popular GAN synthesis architectures and fake detection techniques, highlighting the good fake detection results achieved by most approaches due to the "fingerprints" inserted in the GAN generation process.

In addition, we have reviewed a recent approach to improve the naturalness of facial fake images and spoof state-of-the-art fake detectors: GAN-fingerprint Removal (GANprintR). GANprintR was originally presented in Neves et al. (2020) and is based on a convolutional autoencoder. The autoencoder is trained using only real face images from the development dataset. In the evaluation stage, once the autoencoder is trained, we can pass synthetic face images through it to provide them with additional naturalness, in this way removing the GAN-fingerprint information that may be present in the initial fakes.

A thorough experimental assessment of this type of facial manipulation has been carried out considering fake detection (based on holistic deep networks, steganalysis,

Table 8.4 Impact of the GANprintR approach on three state-of-the-art manipulation detection approaches. A significant performance degradation is observed in all manipulation detection approaches when exposed to images transformed by GANprintR. The detection performance is provided in terms of EER (%), while R_{real} and R_{fake} denote the Recall of the real and fake classes, respectively

Experiment	Data	XceptionNet			Steganalysis (Nataraj et al. 2019b)			Local artifacts (Matern et al. 2019)		
		EER (%)	R_{real} (%)	R_{fake} (%)	EER (%)	R_{real} (%)	R_{fake} (%)	EER (%)	R_{real} (%)	R_{fake} (%)
A.1	Original	0.22	99.77	99.80	10.92	89.07	89.10	38.53	60.72	62.20
	GANprintR	10.63	89.37	89.40	22.37	77.61	77.63	44.06	55.16	56.67
A.2	Original	0.28	99.70	99.73	12.28	87.70	87.73	31.45	67.83	69.26
	GANprintR	6.37	93.64	93.66	17.30	82.71	82.73	36.35	62.93	64.41
A.3	Original	0.02	99.97	100.00	3.32	96.67	96.70	35.13	64.33	65.41
	GANprintR	17.27	82.71	82.73	35.13	64.85	64.85	42.24	57.28	58.29
A.4	Original	0.02	99.97	100.00	12.08	87.90	87.93	39.36	59.62	61.65
	GANprintR	4.47	95.50	95.53	24.97	75.04	75.06	42.75	56.16	58.37
A.5	Original	0.08	99.90	99.93	16.05	83.94	83.96	33.96	65.04	67.03
	GANprintR	11.47	98.50	98.53	19.80	80.17	80.19	38.14	60.77	62.97
A.6	Original	0.05	99.93	99.97	4.62	95.37	95.40	34.79	64.42	66.00
	GANprintR	8.37	93.64	93.66	27.77	72.21	72.22	39.15	60.02	61.70

and local artifacts) and realistic GAN-generated fakes (with and without GANprintR) over different experimental conditions, i.e. controlled and in-the-wild scenarios. We highlight three major conclusions about the performance of the state-of-the-art fake detection methods: *(i)* the existing fake systems attain almost perfect performance when the evaluation data is derived from the same source used in the training phase, which suggests that these systems have actually learned the GAN "fingerprints" from the training fakes generated with GANs; *(ii)* the observed fake detection performance decreases substantially (over one order of magnitude) when the fake detection is exposed to data from unseen databases, and over seven times in case of substantially reduced image resolution; and *(iii)* the accuracy of the existing fake detection methods also drops significantly when analysing synthetic data manipulated by GANprintR.

In summary, our experiments suggest that the existing facial fake detection methods still have a poor generalisation capability and are highly susceptible to—even simple—image transformation manipulations, such as downsizing, image compression or others similar to the one proposed in this work. While loss of resolution may not be particularly concerning in terms of the potential misuse of the data, it is important to note that approaches such as GANprintR are capable of confounding detection methods, while maintaining a high visual similarity with the original image.

Having shown some of the limitations of the state-of-the-art in face manipulation detection, future work should research about strategies to harden such face manipulation detectors by exploiting databases such as iFakeFaceDBiFakeFaceDB.[4] Additionally, further works should study: *(i)* how improved fakes obtained in similar ways as GANprintR can jeopardise other kinds of sensitive data (e.g. other popular biometrics like fingerprint (Tolosana et al. 2020a), iris (Proença and Neves 2019), or behavioural traits (Tolosana et al. 2020b)), *(ii)* how to improve the security of systems dealing with other kinds of sensitive data (Hernandez-Ortega et al. 2021), and finally *(iii)* best ways to combine multiple manipulation detectors (Tolosana et al. 2021) in a proper way (Fiérrez et al. 2018) to deal with the growing sophistication of fakes.

Acknowledgements This work has been supported by projects: PRIMA (H2020-MSCA-ITN-2019-860315), TRESPASS-ETN (H2020-MSCA-ITN-2019-860813), BIBECA (RTI2018-101248-B-I00 MINECO/FEDER), Bio-Guard (Ayudas Fundación BBVA a Equipos de Investigación Científica 2017), by NOVA LINCS (UIDB/04516/2020) with the financial support of FCT—Fundação para a Ciência e a Tecnologia, through national funds, by FCT/MCTES through national funds and co-funded by EU under the project UIDB/EEA/50008/2020, and by FCT—Fundação para a Ciência e a Tecnologia through the research grant '2020.04588.BD'. We gratefully acknowledge the donation of the NVIDIA Titan X GPU used for this research made by NVIDIA Corporation. Ruben Tolosana is supported by Consejería de Educación, Juventud y Deporte de la Comunidad de Madrid y Fondo Social Europeo.

---

[4] https://github.com/socialabubi/iFakeFaceDB.

References

Albright M, McCloskey S (2019) Source generator attribution via inversion. In: IEEE Conference on computer vision and pattern recognition workshops, CVPR workshops 2019, Long Beach, CA, USA, June 16–20, 2019. Computer Vision Foundation/IEEE, pp 96–103

Barni M, Kallas K, Nowroozi E, Tondi B (2020) CNN detection of GAN-generated face images based on cross-band co-occurrences analysis. arXiv:abs/2007.12909

Bellemare MG, Danihelka I, Dabney W, Mohamed S, Lakshminarayanan B, Hoyer S, Munos R (2017) The cramer distance as a solution to biased wasserstein gradients. arXiv:abs/1705.10743

Binkowski M, Sutherland DJ, Arbel M, Gretton A (2018) Demystifying MMD GANs. In: 6th international conference on learning representations, ICLR 2018, Vancouver, BC, Canada, April 30 - May 3, 2018, conference track proceedings. OpenReview.net

Bonettini N, Bestagini P, Milani S, Tubaro S (2020) On the use of benford's law to detect GAN-generated images. arXiv:abs/2004.07682

Brock A, Donahue J, Simonyan K (2019) Large scale GAN training for high fidelity natural image synthesis. In: 7th international conference on learning representations, ICLR 2019, New Orleans, LA, USA, May 6–9, 2019. OpenReview.net

Cellan-Jones R (2019) Deepfake videos double in nine months. https://www.bbc.com/news/technology-49961089

Choi Y, Choi M-J, Kim M, Ha J-W, Kim S, Choo J (2018) StarGAN: unified generative adversarial networks for multi-domain image-to-image translation. In: 2018 IEEE conference on computer vision and pattern recognition, CVPR 2018, Salt Lake City, UT, USA, June 18–22, 2018. IEEE Computer Society, pp 8789–8797

Chollet F (2017) Xception: deep learning with depthwise separable convolutions. In: 2017 IEEE conference on computer vision and pattern recognition, CVPR 2017, Honolulu, HI, USA, July 21–26, 2017. IEEE Computer Society, pp 1800–1807

Cozzolino D, Gragnaniello D, Verdoliva L (2014) Image forgery detection through residual-based local descriptors and block-matching. In: 2014 IEEE international conference on image processing, ICIP 2014, Paris, France, October 27–30, 2014. IEEE, pp 5297–5301

Dang H, Liu F, Stehouwer J, Liu X, Jain AK (2020) On the detection of digital face manipulation. In: 2020 IEEE/CVF conference on computer vision and pattern recognition, CVPR 2020, Seattle, WA, USA, June 13–19, 2020. IEEE, pp 5780–5789

Deng J, Dong W, Socher R, Li L-J, Li K, Li F-F (2009) Imagenet: a large-scale hierarchical image database. In: 2009 IEEE computer society conference on computer vision and pattern recognition (CVPR 2009), 20–25 June 2009, Miami, Florida, USA. IEEE Computer Society, pp 248–255

Dolhansky B, Howes R, Pflaum B, Baram N, Canton-Ferrer C (2019) The deepfake detection challenge (DFDC) preview dataset. arXiv:abs/1910.08854

Durall R, Keuper M, Keuper J (2020) Watch your up-convolution: CNN based generative deep neural networks are failing to reproduce spectral distributions. In: 2020 IEEE/CVF conference on computer vision and pattern recognition, CVPR 2020, Seattle, WA, USA, June 13–19, 2020. IEEE, pp 7887–7896

FaceApp (2017) https://apps.apple.com/gb/app/faceapp-ai-face-editor/id1180884341

Fiérrez J, Morales A, Vera-Rodríguez R, Camacho D (2018) Multiple classifiers in biometrics. Part 2: trends and challenges. Inf Fusion 44:103–112

Flickr-Faces-HQ Dataset (FFHQ) (2019)

Frank J, Eisenhofer T, Schönherr L, Fischer A, Kolossa D, Holz T (2020) Leveraging frequency analysis for deep fake image recognition. In: Proceedings of the 37th international conference on machine learning, ICML 2020, 13–18 July 2020, Virtual Event, volume 119 of Proceedings of machine learning research. PMLR, pp 3247–3258

Galbally J, Marcel S, Fierrez J (2014) Biometric anti-spoofing methods: a survey in face recognition. IEEE Access 2:1530–1552

Gandhi A, Jain S (2020) Adversarial perturbations fool deepfake detectors. In: 2020 international joint conference on neural networks, IJCNN 2020, Glasgow, United Kingdom, July 19–24, 2020. IEEE, pp 1–8

Goebel M, Nataraj L, Nanjundaswamy T, Mohammed TM, Chandrasekaran S, Manjunath BS (2020) Detection, attribution and localization of GAN generated images. arXiv:abs/2007.10466

Goodfellow IJ, Pouget-Abadie J, Mirza M, Xu B, Warde-Farley D, Ozair S, Courville AC, Bengio Y (2014) Generative adversarial nets. In: Ghahramani Z, Welling M, Cortes C, Lawrence ND, Weinberger KQ (eds) Advances in neural information processing systems 27: annual conference on neural information processing systems 2014, December 8–13 2014, Montreal, Quebec, Canada, pp 2672–2680

Guarnera L, Giudice O, Battiato S (2020) Deepfake detection by analyzing convolutional traces. In: 2020 IEEE/CVF conference on computer vision and pattern recognition, CVPR workshops 2020, Seattle, WA, USA, June 14–19, 2020. IEEE, pp 2841–2850

Guo Y, Zhang L, Hu Y, He X, Gao J (2016) MS-Celeb-1M: a dataset and benchmark for large-scale face recognition. In: Leibe B, Matas J, Sebe N, Welling M (eds) Computer vision - ECCV 2016 - 14th European conference, Amsterdam, The Netherlands, October 11–14, 2016, proceedings, Part III, vol 9907 of Lecture notes in computer science. Springer, pp 87–102

Hadid A, Evans NWD, Marcel S, Fiérrez J (2015) Biometrics systems under spoofing attack: an evaluation methodology and lessons learned. IEEE Signal Process Mag 32(5):20–30

He P, Li H, Wang H (2019) Detection of fake images via the ensemble of deep representations from multi color spaces. In: 2019 IEEE international conference on image processing, ICIP 2019, Taipei, Taiwan, September 22–25, 2019. IEEE, pp 2299–2303

Hernandez-Ortega J, Fiérrez J, Morales A, Galbally J (2019) Introduction to face presentation attack detection. In: Marcel S, Nixon MS, Fiérrez J, Evans NWD (eds) Handbook of biometric anti-spoofing - presentation attack detection, Second Edition, advances in computer vision and pattern recognition. Springer, pp 187–206

Hernandez-Ortega J, Tolosana R, Fierrez J, Morales A (2021) DeepFakesON-Phys: deepfakes detection based on heart rate estimation. In: Proceedings of the 35th AAAI conference on artificial intelligence workshops

Hsu C-C, Zhuang Y-X, Lee C-Y (2020) Deep fake image detection based on pairwise learning. Appl Sci 10(1):370

Hu S, Li Y, Lyu S (2020) Exposing GAN-generated faces using inconsistent corneal specular highlights. arXiv:abs/2009.11924

Hulzebosch N, Ibrahimi S, Worring M (2020) Detecting CNN-generated facial images in real-world scenarios. In: 2020 IEEE/CVF conference on computer vision and pattern recognition, CVPR workshops 2020, Seattle, WA, USA, June 14–19, 2020. IEEE, pp 2729–2738

Karras T, Aila T, Laine S, Lehtinen J (2018) Progressive growing of GANs for improved quality, stability, and variation. In: 6th international conference on learning representations, ICLR 2018, Vancouver, BC, Canada, April 30–May 3, 2018, conference track proceedings. OpenReview.net

Karras T, Laine S, Aila T (2019) A style-based generator architecture for generative adversarial networks. In: IEEE conference on computer vision and pattern recognition, CVPR 2019, Long Beach, CA, USA, June 16–20, 2019. Computer Vision Foundation/IEEE, pp 4401–4410

Karras T, Laine S, Aittala M, Hellsten J, Lehtinen J, Aila T (2020) Analyzing and improving the image quality of StyleGAN. In: 2020 IEEE/CVF conference on computer vision and pattern recognition, CVPR 2020, Seattle, WA, USA, June 13–19, 2020. IEEE, pp 8107–8116

Kazemi V, Sullivan J (2014) One millisecond face alignment with an ensemble of regression trees. In: 2014 IEEE conference on computer vision and pattern recognition, CVPR 2014, Columbus, OH, USA, June 23–28, 2014. IEEE Computer Society, pp 1867–1874

Korshunov P, Marcel S (2018) Deepfakes: a new threat to face recognition? Assessment and detection. arXiv:abs/1812.08685

Li H, Li B, Tan S, Huang J (2020) Identification of deep network generated images using disparities in color components. Signal Process 174:107616

Liu Z, Luo P, Wang X, Tang X (2015) Deep learning face attributes in the wild. In: 2015 IEEE international conference on computer vision, ICCV 2015, Santiago, Chile, December 7–13, 2015. IEEE Computer Society, pp 3730–3738

Liu Z, Qi X, Torr PHS (2020) Global texture enhancement for fake face detection in the wild. In: 2020 IEEE/CVF conference on computer vision and pattern recognition, CVPR 2020, Seattle, WA, USA, June 13–19, 2020. IEEE, pp 8057–8066

Lukás J, Fridrich J, Goljan M (2006) Digital camera identification from sensor pattern noise. IEEE Trans Inf Forensics Secur 1(2):205–214

Marra F, Gragnaniello D, Cozzolino D, Verdoliva L (2018) Detection of GAN-generated fake images over social networks. In: IEEE 1st conference on multimedia information processing and retrieval, MIPR 2018, Miami, FL, USA, April 10–12, 2018. IEEE, pp 384–389

Marra F, Gragnaniello D, Verdoliva L, Poggi G (2019a) Do GANs leave artificial fingerprints? In: 2nd IEEE conference on multimedia information processing and retrieval, MIPR 2019, San Jose, CA, USA, March 28–30, 2019. IEEE, pp 506–511

Marra F, Saltori C, Boato G, Verdoliva L (2019b) Incremental learning for the detection and classification of GAN-generated images. In: 2019 IEEE international workshop on information forensics and security (WIFS). IEEE, pp 1–6

Marra F, Saltori C, Boato G, Verdoliva L (2019c) Incremental learning for the detection and classification of GAN-generated images. In: IEEE international workshop on information forensics and security, WIFS 2019, Delft, The Netherlands, December 9–12, 2019. IEEE, pp 1–6

Matern F, Riess C, Stamminger M (2019) Exploiting visual artifacts to expose deepfakes and face manipulations. In: Proceedings of the IEEE winter applications of computer vision workshops

McCloskey S, Albright M (2018) Detecting GAN-generated imagery using color cues. arXiv:abs/1812.08247

Miyato T, Kataoka T, Koyama M, Yoshida Y (2018) Spectral normalization for generative adversarial networks. In: 6th international conference on learning representations, ICLR 2018, Vancouver, BC, Canada, April 30–May 3, 2018, conference track proceedings. OpenReview.net

Nataraj L, Mohammed TM, Manjunath BS, Chandrasekaran S, Flenner A, Bappy JH, Roy-Chowdhury AK (2019a) Detecting gan generated fake images using co-occurrence matrices. Electr Imag 2019(5):532–1

Nataraj L, Mohammed TM, Manjunath BS, Chandrasekaran S, Flenner A, Bappy JH, Roy-Chowdhury AK (2019b) Detecting GAN generated fake images using co-occurrence matrices. In: Alattar AM, Memon ND, Sharma G (eds) Media watermarking, security, and forensics 2019, Burlingame, CA, USA, 13–17 January 2019. Ingenta

Neves JC, Tolosana R, Vera-Rodriguez R, Lopes V, Proença H, Fierrez J (2020) GANprintR: improved fakes and evaluation of the state of the art in face manipulation detection. IEEE J Sel Top Signal Process 14(5):1038–1048

Park T, Liu M-Y, Wang T-C, Zhu J-Y (2019) Semantic image synthesis with spatially-adaptive normalization. In: IEEE conference on computer vision and pattern recognition, CVPR 2019, Long Beach, CA, USA, June 16–20, 2019. Computer Vision Foundation/IEEE, pp 2337–2346

Proença H, Neves JC (2019) Segmentation-less and non-holistic deep-learning frameworks for iris recognition. In: IEEE conference on computer vision and pattern recognition workshops, CVPR workshops 2019, Long Beach, CA, USA, June 16–20, 2019. Computer Vision Foundation/IEEE, pp 2296–2305

Rössler A, Cozzolino D, Verdoliva L, Riess C, Thies J, Nießner M (2019) Faceforensics++: learning to detect manipulated facial images. In: 2019 IEEE/CVF international conference on computer vision, ICCV 2019, Seoul, Korea (South), October 27–November 2, 2019. IEEE, pp 1–11

Tolosana R, Gomez-Barrero M, Busch C, Ortega-Garcia J (2020a) Biometric presentation attack detection: beyond the visible spectrum. IEEE Trans Inf Forensics Secur 15:1261–1275

Tolosana R, Vera-Rodriguez R, Fierrez J, Ortega-Garcia J (2020b) BioTouchPass2: touchscreen password biometrics using time-aligned recurrent neural networks. IEEE Trans Inf Forensics Secur 15:2616–2628

Tolosana R, Vera-Rodriguez R, Fierrez J, Morales A, Ortega-Garcia J (2020c) DeepFakes and beyond: a survey of face manipulation and fake detection. Inf Fusion 64:131–148

Tolosana R, Romero-Tapiador S, Fierrez J, Vera-Rodriguez R (2021) DeepFakes evolution: analysis of facial regions and fake detection performance. In: Proceedings of the international conference on pattern recognition workshops

Verdoliva L (2020) Media forensics and deepfakes: an overview. IEEE J Sel Top Signal Process 14(5):910–932

Wang S-Y, Wang O, Zhang R, Owens A, Efros AA (2020a) CNN-generated images are surprisingly easy to spot... for now. In: 2020 IEEE/CVF conference on computer vision and pattern recognition, CVPR 2020, Seattle, WA, USA, June 13–19, 2020. IEEE, pp 8692–8701

Wang R, Juefei-Xu F, Ma L, Xie X, Huang Y, Wang J, Liu Y (2020b) Fakespotter: a simple yet robust baseline for spotting ai-synthesized fake faces. In: Bessiere C (ed) Proceedings of the twenty-ninth international joint conference on artificial intelligence, IJCAI 2020. ijcai.org, pp 3444–3451

Xuan X, Peng B, Wang W, Dong J (2019) On the generalization of GAN image forensics. In: Sun Z, He R, Feng J, Shan S, Guo Z (eds) Biometric recognition - 14th Chinese conference, CCBR 2019, Zhuzhou, China, October 12–13, 2019, Proceedings, vol 11818 of Lecture notes in computer science. Springer, pp 134–141

Yang X, Li Y, Lyu S (2019) Exposing deep fakes using inconsistent head poses. In: IEEE international conference on acoustics, speech and signal processing, ICASSP 2019, Brighton, United Kingdom, May 12–17, 2019. IEEE, pp 8261–8265

Yu N, Davis L, Fritz M (2018) Attributing fake images to GANs: analyzing fingerprints in generated images. arXiv:abs/1811.08180

Yu Y, Ni R, Zhao Y (2020a) Mining generalized features for detecting ai-manipulated fake faces. arXiv:abs/2010.14129

Yu N, Skripniuk V, Abdelnabi S, Fritz M (2020b) Artificial GAN fingerprints: rooting deepfake attribution in training data, pp arXiv–2007

ZAO (2019) https://apps.apple.com/cn/app/id1465199127

Zhang X, Karaman S, Chang S-F (2019) Detecting and simulating artifacts in GAN fake images. In: IEEE international workshop on information forensics and security, WIFS 2019, Delft, The Netherlands, December 9–12, 2019. IEEE, pp 1–6

Zhu J-Y, Park T, Isola P, Efros AA (2017) Unpaired image-to-image translation using cycle-consistent adversarial networks. In: IEEE international conference on computer vision, ICCV 2017, Venice, Italy, October 22–29, 2017. IEEE Computer Society, pp 2242–2251

Part III
Integrity and Authenticity

Chapter 9
Physical Integrity

Christian Riess

Physics-based methods anchor the forensic analysis in the physical laws of image and video formation. The analysis is typically based on simplifying assumptions to make the forensic analysis tractable. In scenes that satisfy such assumptions, different types of forensic analysis can be performed. The two most widely used applications are the detection of content repurposing and content splicing. Physics-based methods expose such cases with assumptions about the interaction of light and objects, and about the geometric mapping of light and objects onto the image sensor.

In this chapter, we review the major lines of research on physics-based methods. The approaches are categorized as geometric and photometric, and combinations of both. We also discuss the strengths and limitations of these methods, including an interesting unique property: most physics-based methods are quite robust to low-quality material, and can even be applied to analog photographs. The chapter closes with an outlook and with links to related forensic techniques such as the analysis of physiological signals.

9.1 Introduction

Consider a story that is on the internet. Let us also assume that an important part of that story is a picture or video to document the story. If this story goes viral, it becomes potentially relevant also to classical journalists. Relevance may either inherently be given, because the story reports a controversial event of public interest, or relevance may emerge from the sole fact that a significant number of people are discussing a

C. Riess (✉)
IT Security Infrastructures Lab, Friedrich-Alexander University Erlangen-Nürnberg,
Erlangen, Germany
e-mail: christian.riess@fau.de

© The Author(s) 2022

H. T. Sencar et al. (eds.), *Multimedia Forensics*, Advances in Computer Vision and Pattern
Recognition, https://doi.org/10.1007/978-981-16-7621-5_9

particular topic. Additionally, quality journalism also requires confidence in the truth value that is associated with a news story. Hence, classical journalistic work involves the work of eye witnesses and other trusted sources of documentation. However, stories that emerge from internet sources are oftentimes considerably more difficult to verify for a number of reasons. One reason may be because the origin of the story is difficult to determine. Another reason may be that reports are difficult to verify because involved locations and people are inaccessible, as it is oftentimes the case for reports from civil war zones. In these cases, a new specialization of journalistic research formed in the past years, namely journalistic fact checking. The techniques of fact checkers will be introduced in Sect. 9.1.1. We will note that the class of physics-based methods in multimedia forensics is closely related to these techniques. More precisely, physics-based approaches can be seen as computational support tools that smoothly integrate into the journalistic fact checking toolbox, as highlighted in Sect. 9.1.2. We close this section with an outline for the remaining chapter in Sect. 9.1.3.

9.1.1 Journalistic Fact Checking

Classical journalistic research about a potential news story aims to answer the so-called "Five W" questions who, what, why, when, and where about an event, i.e., who participated in the event, what happened, why, and at what time and place it happened. The answers to these questions are typically derived and verified from contextual information, witnesses, and supporting documents. Investigative journalists foster a broad network of contacts and sources to verify such events. However, for stories that emerge on the internet, this classical approach can be of limited effectivity, particularly if the sources are anonymized or covered by social media filters. Journalistic fact checkers still heavily rely on classical techniques, and, for example, investigate the surroundings in social networks to learn about possible political alignments and actors from where a message may have originated. For images or videos, context is also helpful.

One common issue with such multimedia content is repurposing. This means that the content itself is authentic, but it has been acquired at a different time or a different place than what is claimed in the associated story. To find cases of repurposing, the first step is a reverse image search as it is possible with specialized search engines like `tineye.com` or `images.google.com`.

Furthermore, a number of open-source intelligence tools are available for a further in-depth verification of the image content. Depending on the shown scene and the context of the news story, photogrammetric methods allow to validate time and place from the position of the sun and known or estimated shadow lengths of buildings, vehicles, or other known objects. Any landmarks in the scene, written text, or signs can be further used to constrain the possible location of acquisition, combined with, e.g., regionally annotated satellite maps. Another type of open-source information is monitoring tools for flights and shipping lines. Such investigations require a high manual work effort to collect and organize different pieces of evidence and to assess their individual credibility. In journalistic practice, however, these approaches are currently the gold standard to verify multimedia content from sources with very limited contextual information. Unfortunately, this manual effort also implies that an investigation may take several days or weeks. While it would oftentimes be desirable to quell wrong viral stories early on, this is in many cases not possible with current investigative tools and techniques.

9.1.2 Physics-Based Methods in Multimedia Forensics

In multimedia forensics, the group of physics-based methods follows the same spirit as journalistic fact checking. Like the journalistic methods, they operate on the content of the scene. They also use properties of known objects, and some amount of outside knowledge, such as the presence of a single dominant light source. However, there are two main differences. First, physics-based methods in multimedia forensics operate almost exclusively on the scene content. They use much less contextual knowledge, since this contextual information is strongly dependent on the actual case, which makes it difficult to address in general-purpose algorithms. Second, journalistic methods focus on validating time, place, and actors, while physics-based methods in multimedia forensics aim to verify that the image is authentic, i.e., that the scene content is consistent with the laws of physics.

The second difference makes physics-based forensic algorithms complementary to journalistic methods. They can add specific evidence whether an image is a composite from multiple sources. For example, if an image shows two persons on a free field who are only illuminated by the sun, then one can expect that both people are illuminated from the same direction. More subtle cues can be derived from the laws of perspective projection. This applies to all common acquisition devices, since it is the foundation of imaging with a pinhole camera. The way how light is geometrically mapped onto the sensor is tied to the camera and to the location of objects in the scene. In several cases, the parameters for this mapping can be calculated from specific objects. Two objects in an image with inconsistent parameters can indicate an image composition. Similarly, it is possible to compare the color of incident light on different objects, or the color of ambient light in shadow areas.

To achieve this type of analysis, physics-based algorithms use methods from related research fields, most notably from computer vision and photometry. Typically, the analysis analytically solves an assumed physical model for a quantity of interest. This has several important consequences, which distinguish physics-based approaches from statistical approaches in multimedia forensics. First, physics-based methods typically require an analyst to validate the model assumptions, and to perform a limited amount of annotations in the scene to access the known variables. In contrast, statistical approaches can in most cases work fully automatically, and are hence much better suitable for batch processing. Second, the applied physical models and analysis methods can be explicitly checked for their approximation error. This makes physics-based methods inherently explainable, which is an excellent property for defending a decision based on a physics-based analysis. The majority of physics-based approaches do not use advanced machine learning methods, since this would make the explainability of the results much more complicated. Third, physics-based algorithms require specific scene constellations in order to be applicable. For example, an analysis that assumes that the sun is the only light source in the scene is not applicable to indoor photographs or night time pictures. Conversely, the focus on scene content in conjunction with manual annotations by analysts makes physics-based methods by design extremely robust to the quality or processing history of an image. This is a clear benefit over statistical forensic algorithms, since their performance quickly deteriorates on images of reduced quality or complex post-processing. Some physics-based methods can even be applied to printouts, which is also not possible for statistical methods.

With these three properties, physics-based forensic algorithms can be seen as complementary to statistical approaches. Their applicability is limited to specific scenes and manual interactions, but they are inherently explainable and very robust to the processing history of an image.

Closely related to physics-based methods are behavioral cues of persons and physiological features as discussed in Chap. 11 of this book. These methods are not physics-based in the strict sense, as they do not use physical models. However, these methods share with physics-based methods the property that they operate on the scene content, and offer as such in many cases also resilience to various processing histories.

9.1.3   Outline of This Chapter

A defining property of physics-based methods is the underlying analytic models and their assumptions. The most widely used models will be introduced in Sect. 9.2. The models are divided into two parts: geometric and optical models are introduced in Sect. 9.2.1, and photometric and reflectance models are introduced in Sect. 9.2.2. Applications of these models in forensic algorithms are presented in Sect. 9.3. This part is subdivided into geometric methods, photometric methods, and combinations thereof. This chapter concludes with a discussion on the strength and weaknesses of physics-based methods and an outlook on emerging challenges in Sect. 9.4.

9.2   Physics-Based Models for Forensic Analysis

Consider a photograph of a building, for example, the Hagia Sophia in Fig. 9.1. The way how this physical object is converted to a digital image is called image formation. Three aspects of the image formation are particularly relevant for physics-based algorithms. These three aspects are illustrated below.

Fig. 9.1 Example photograph "Turkey-3019—Hagia Sophia" (full picture credits at the end of chapter)

First, landmark points like the tips of the towers are projected onto the camera. For all conventional cameras that operate with a lens, the laws of perspective projection determine the locations to which these landmarks are projected onto the sensor. Second, the dome of the building is bright where the sun is reflected into the camera, and darker at other locations. Brightness and color variations across pixels are represented with photometric models. Third, the vivid colors and, overall, the final representation of the image are obtained from the several camera-internal processing functions, which computationally perform linear and non-linear operations on the image.

The first two components of the image formation are typically used for a physics-based analysis. We introduce the foundation for the analysis of geometric properties of the scene in Sect. 9.2.1, and we introduce foundations for the analysis of reflectance properties of the scene in Sect. 9.2.2.

9.2.1 Geometry and Optics

The use of geometric laws in image analysis is a classical topic from the field of computer vision. Geometric algorithms typically examine the relation between the location of 3-D points in the scene and their 2-D projection onto the image. In computer vision, the goal is usually to infer a consistent 3-D structure from one or more 2-D images. One example is the stereo vision to calculate depth maps for an object or scene from two laterally shifted input images. In multimedia forensics, the

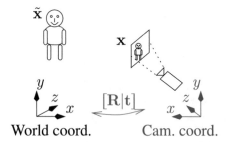
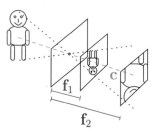

Fig. 9.2 The mapping between world- and camera coordinates is performed with the extrinsic camera parameters (left). The mapping from the 3-D scene onto the 2-D image plane is performed with the intrinsic camera parameters (right)

underlying assumption of a geometric analysis is that the geometric consistency of the scene is violated by an inserted object or by certain types of object editing.

For brevity of notation, let us denote a point in the image $\mathbf{I}$ at coordinate (y, x) as $\mathbf{x} = I(y, x)$. This point corresponds to a 3-D point in the world at the time when the image was taken. In the computer vision literature, this point is typically denoted as $\mathbf{X}$, but here it is denoted as $\tilde{\mathbf{x}}$ to avoid notational confusion with matrices. Vectors that are used in projective geometry are typically written in homogeneous coordinates. This also allows to conveniently express points at infinity. Homogeneous vectors have one extra dimension with an entry z_h, such that the usual Cartesian coordinates are obtained by dividing each other entry by z_h. Assume that we convert Cartesian coordinates to homogeneous coordinates, and choose $z_h = 1$. Then, the remaining vector elements x_i are identical to their corresponding Cartesian coordinates, since $x_i/1 = x_i$.

When taking a picture, the camera maps the 3-D point $\tilde{\mathbf{x}}$ from the 3-D world onto the 2-D image point $\mathbf{x}$. Mathematically, this is a perspective projection, which can be very generally expressed as

$$\mathbf{x} = \mathbf{K}[\mathbf{R}|\mathbf{t}]\tilde{\mathbf{x}} , \tag{9.1}$$

where $\mathbf{x}$ and $\tilde{\mathbf{x}}$ are written in homogeneous coordinates. This projection consists of two matrices: the matrix $[\mathbf{R}|\mathbf{t}] \in \mathbb{R}^{3 \times 4}$ contains the so-called extrinsic parameters. It transforms the point $\tilde{\mathbf{x}}$ from an arbitrary world coordinate system to the 3-D coordinates of the camera. The matrix $K \in \mathbb{R}^{3 \times 3}$ contains the so-called intrinsic parameters. It maps these 3-D coordinates onto the 2-D image plane of the camera.

Both mapping steps are illustrated in Fig. 9.2. The mapping of the world coordinate system to the camera coordinate system via the extrinsic camera parameters is shown on the left. The mapping of the 3-D scene onto the 2-D image plane is shown on the right.

Both matrices have a special form. The matrix of extrinsic parameters $[\mathbf{R}|\mathbf{t}]$ is a 3×4 matrix. It consists of a 3×3 rotation matrix in the first three columns and a 3×1 translation vector in the fourth column. The matrix of intrinsic parameters K

is an upper triangular 3×3 matrix

$$K = \begin{pmatrix} f_x & s & c_x \\ 0 & f_y & c_y \\ 0 & 0 & 1 \end{pmatrix} , \qquad (9.2)$$

where f_x and f_y are the focal length in pixels along x- and y-direction, (c_x, c_y) is the camera principal point, and s denotes pixel skewness. The camera principal point (oftentimes also called camera center) is particularly important for forensic applications, as it marks the center of projection. The focal length is also of importance, as it changes the size of the projection cone from world coordinates to image coordinates. This is illustrated in the right part of Fig. 9.2 with two example focal lengths: the smaller focal length f_1 maps the whole person onto the image plane, whereas the larger focal length f_2 only maps the torso around the center of projection (green) onto the image plane.

Forensic algorithms that build on top of these equations typically do not use the full projection model. If only a single image is available, and no further external knowledge on the scene geometry, then the world coordinate system can be aligned with the camera coordinate system. In this case, the x- and y-axes of the world coordinate system correspond to the x- and y-axes of the camera coordinate system, and the z-axis points with the camera direction into the scene. In this case, the matrix $[\mathbf{R}|\mathbf{t}]$ can be omitted. Additionally, it is a common assumption to assume that the pixels are square, the lens is sufficiently homogeneous, and that the camera skew is negligible. These assumptions simplify the projection model to only three unknown intrinsic parameters, namely

$$\mathbf{x} = \begin{pmatrix} f & 0 & c_x \\ 0 & f & c_y \\ 0 & 0 & 1 \end{pmatrix} \tilde{\mathbf{x}} . \qquad (9.3)$$

These parameters are the focal length f and the center of projection (c_x, c_y), which are used in several forensic algorithms.

One interesting property of the perspective projection is the so-called vanishing points: in the projected image, lines that are parallel in the 3-D world converge to a vanishing point on the image. Vanishing points are not necessarily visible within the shown scene but can also be outside of the scene. For example, a 3-D building typically consists of parallel lines in three mutually orthogonal directions. In this special case, the three vanishing points span an orthogonal coordinate system with the camera principal point at the center.

One standard operation in projective geometry is homography. It maps points from one plane onto another. One example application is shown in Fig. 9.3. On the left, a perspectively distorted package of printer paper is shown. The paper is in A4 format, and has hence a known ratio between height and width. After annotation of the corner points (red squares in the left picture), the package can be rectified to obtain a virtual frontal view (right picture). In forensic applications, we usually use

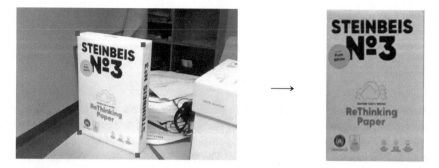

Fig. 9.3 Example homography. Left: rectification of an object with known aspect ratio (the A4 paper) using four annotated corners (red dots). Right: rectified object after calculation of the homography from the distorted paper plane onto the world coordinate plane

a mapping from a plane in the 3-D scene onto the camera image plane or vice versa as presented in Sect. 9.3.1. For this reason, we review the homography calculation in the following paragraphs in greater detail.

All points on a plane lie on a 2-D manifold. Hence, the mapping of a 2-D plane onto another 2-D plane is actually a transform of 2-D points onto 2-D points. We denote by $\hat{\mathbf{x}}$ a point on a plane in homogeneous 2-D world coordinates, and a point in homogeneous 2-D image coordinates as $\mathbf{x}$. Then, the homography is performed by multiplication with a 3×3 projection matrix $\mathbf{H}$,

$$\mathbf{x} = \mathbf{H}\hat{\mathbf{x}} \ . \tag{9.4}$$

Here, the equality sign holds only up to the scale of the projection matrix. This scale ambiguity does not affect the equation, but it implies that the homography matrix $\mathbf{H}$ has only 8 degrees of freedom instead of 9. Hence, to estimate the homography matrix, a total of 8 constraints are required. These constraints are obtained from point correspondences. A point correspondence is a pair of two matching points, where one point lies on one plane and the second point on the other plane, and both points mark the same location on the object. The x-coordinates and the y-coordinates of a correspondence, each contribute one constraint, such that a total of four point correspondences suffices to estimate the homography matrix.

The actual estimation is performed by solving Eq. 9.4 for the elements $\mathbf{h} = (h_{11}, \ldots, h_{33})^\mathrm{T}$ of $\mathbf{H}$. This estimation is known as Direct Linear Transformation, and can be found in computer vision textbooks, e.g., by Hartley and Zisserman (2013, Chap. 4.1). The solution has the form

$$\mathbf{A}\mathbf{h} = \mathbf{0} \ , \tag{9.5}$$

where $\mathbf{A}$ is a $2 \cdot N \times 9$ matrix. Each of the N available point correspondences contributes two rows to $\mathbf{A}$ of the form

Fig. 9.4 Distortions and aberrations from the camera optics. From left to right: image of a rectangular grid, and its image under barrel distortion and pincushion distortion, axial chromatic aberration and lateral chromatic aberration

$$\mathbf{a}_{2k} = (-\hat{x}_1, -\hat{y}_1, -1, 0, 0, 0, x_2\hat{x}_1, x_2\hat{x}_2, x_2) \tag{9.6}$$

and

$$\mathbf{a}_{2k+1} = (0, 0, 0, -\hat{x}_1, -\hat{y}_1, -1, y_2\hat{x}_1, y_2\hat{x}_2, y_2) . \tag{9.7}$$

Equation 9.5 can be solved via singular value decomposition (SVD). After the factorization into $A = U \Sigma V^\mathsf{T}$, the unit singular vector that corresponds to the smallest singular value is the solution for $\mathbf{h}$.

The homography matrix implicitly contains the intrinsic and extrinsic camera parameters. More specifically, the RQ-decomposition factorizes a matrix into a product of an upper triangular matrix and an orthogonal matrix (Hartley and Zisserman 2013, Appendix 4.1). Applied to the 3×3 matrix **H**, the 3×3 upper triangular matrix corresponds then to the intrinsic camera parameter matrix **K**. The orthogonal matrix corresponds to a 3×3 matrix $[\mathbf{r}_1\mathbf{r}_2\mathbf{t}]$, where the missing third rotation vector of the extrinsic parameters is calculated as the cross-product $\mathbf{r}_3 = \mathbf{r}_1 \times \mathbf{r}_2$.

The equations above implicitly assume a perfect mapping from points in the scene to pixels in the image. However, in reality, the camera lens is not perfect, which can slightly alter the pixel representation. We briefly state notable deviations from a perfect mapping without going further into detail. Figure 9.4 illustrates these deviations. The three grids on the left illustrate geometric lens distortions, which change the mapping of lines from the scene onto the image. Probably, the most well-known types are barrel distortion and pincushion distortion. Both types of distortions either stretch or shrink the projection radially around a center point. Straight lines appear then either stretched away from the center (barrel distortion) or contracted toward the center (pincushion distortion). While lens distortions affect the mapping of macroscopic scene elements, another form of lens imperfection is chromatic aberration, which is illustrated on the right of Fig. 9.4. Here, the mapping of a world point between the lens and image plane is shown. However, different wavelengths (and hence colors) are not mapped onto the same location on the sensor. The first illustration shows axial chromatic aberration, where individual color channels focus at different distances. In the picture, only the green color channel converges at the image plane and is in focus; the other two color channels are slightly defocused. The second illustration shows lateral chromatic aberration, which displaces different wavelengths. In extreme cases, this effect can also be seen when zooming into a high-resolution photograph as a slight color seam at object boundaries that are far from the image center.

9.2.2 *Photometry and Reflectance*

The use of photometric and reflectance models in image analysis is also a classical topic from the field of computer vision. These models operate on brightness or color distributions of pixels. If a pixel records the reflectance from the surface of a solid object, its brightness and color are a function of the object material, and the relative orientation of the surface patch, the light sources, and the camera in the scene. In computer vision, classical applications are, for example, intrinsic image decomposition for factorizing an object into geometry and material color (called albedo), photometric stereo and shape from shading for reconstructing the geometry of an object from the brightness distribution of its pixels, and color constancy for obtaining a canonical color representation that is independent of the color of the illumination. In multimedia forensics, the underlying assumption is that the light source and the camera induce global consistency constraints that are violated when an object is inserted or otherwise edited.

The color and brightness of a single pixel $I(y, x)$ are oftentimes modeled as irradiance of light that is reflected from a surface patch onto the camera lens. To this end, the surface patch itself must be illuminated by one or more light sources. Each light source is assumed to emit photons with a particular energy distribution, which is the spectrum of the light source. In human vision, the visible spectrum of a light source is perceived as the color of the light. The surface patch may reflect this light as diffuse or specular reflectance, or a combination of both.

Diffuse reflectance, also called Lambertian reflectance, is by far the most commonly assumed model. The light-object interaction is illustrated in the left part of Fig. 9.5. Here, photons entering the object surface are scattered within the object, and then emitted in a random direction. Photons are much more likely to enter an object when they perpendicularly hit the surface than when they hit at a flatter angle. This is mathematically expressed as a cosine between the angle of incidence and the surface normal. The scattering within the object changes the spectrum toward the albedo of the object, such that the spectrum after leaving the object corresponds to the object color given the color of the light source. Since the photons are emitted in random directions, the perceived brightness is identical from all viewing directions.

Specular reflectance occurs when photons do not enter the object, but instead are reflected at the surface. The light-object interaction is illustrated in the right part

Fig. 9.5 Light-object interaction for diffuse (Lambertian) and specular reflectance

of Fig. 9.5. Specular reflectance has two additional properties. First, the spectrum of the light is barely affected by this minimal interaction with the surface. Second, specular reflectance is not scattered in all directions, but only in a very narrow angle of exitance that is opposite to the angle of incidence. This can be best observed on mirrors, which exhibit almost perfect specular reflectance.

The photometric formation of a pixel can be described by a physical model. Let $\mathbf{i} = I(y, x)$ be the usual RGB color representation at pixel $I(y, x)$. We assume that this pixel shows an opaque surface object. Then, the irradiance of that pixel into the camera is quite generally described as

$$\mathbf{i}(\boldsymbol{\theta}_e, v) = f_{cam} \left(\int_{\boldsymbol{\theta}_i} \int_{\lambda} r(\boldsymbol{\theta}_i, \boldsymbol{\theta}_e, v, \lambda) \cdot e(\boldsymbol{\theta}_i, \lambda) \cdot c_{rgb}(\lambda) \, \mathrm{d}\boldsymbol{\theta}_i \, \mathrm{d}\lambda \right) . \qquad (9.8)$$

Here, $\boldsymbol{\theta}_i$ and $\boldsymbol{\theta}_e$ denote the 3-D angles under which a ray of light falls onto the surface and exits the surface, λ denotes the wavelength of light, v is the normal vector of the surface point, $e(\boldsymbol{\theta}_i, \lambda)$ denotes the amount of light coming from angle $\boldsymbol{\theta}_i$ with wavelength λ, $r(\boldsymbol{\theta}_i, \boldsymbol{\theta}_e, v, \lambda)$ is the reflectance function indicating the fraction of light that is reflected in direction $\boldsymbol{\theta}_e$ after it arrived with wavelength λ from angle $\boldsymbol{\theta}_i$, and $c_{rgb}(\lambda)$ models the camera's color sensitivity in the three RGB channels to wavelength λ. The inner expression integrates over all angles of incidence and all wavelengths to model all light that potentially leaves this surface patch in the direction of the camera. The function f_{cam} captures all further processing in the camera. For example, consumer cameras always apply a non-linear scaling called gamma factor to each individual color channel to make the colors appear more vivid. Theoretically, f_{cam} could also include additional processing that involves multiple pixels, such as demosaicking, white balancing, and lens distortion correction, although this is of minor concern for the algorithms in this chapter.

The individual terms of Eq. 9.8 are illustrated in Fig. 9.6. Rays of light are emitted by the light source and arrive at a surface with term $e(\boldsymbol{\theta}_i, \lambda)$. Reflections from the surface with function $r(\boldsymbol{\theta}_i, \boldsymbol{\theta}_e, v, \lambda)$ are filtered for their colors on the camera sensor with c_{rgb}, and further processed in the camera with f_{cam}.

For algorithm development, Eq. 9.8 is in most cases too detailed to be useful. Hence, most components are typically neutralized by additional assumptions. For example, algorithms that focus on geometric arguments, such as the analysis of lighting environments, typically assume to operate on a grayscale image (or just a single-color channel), and hence remove all influences from the wavelength λ and the color sensitivity c_{rgb} from the model. Conversely, algorithms that focus on color typically ignore the integral over $\boldsymbol{\theta}_i$ and all other influences of $\boldsymbol{\theta}_i$ and $\boldsymbol{\theta}_e$, and also assume a greatly simplified color sensitivity function c_{rgb}. Both types of algorithms oftentimes assume that the camera response function f_{cam} is linear. In this case, it is particularly important to invert the gamma factor as a pre-processing step when analyzing pictures from consumer cameras. The exact inversion formula depends on

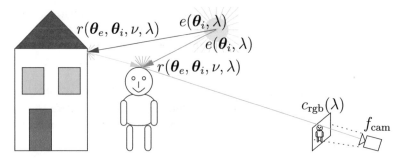

Fig. 9.6 Photometric image formation. Light sources emit light rays, which arrive as $e(\theta_i, \lambda)$ with wavelength λ at an angle of incidence θ_i on a surface. The surface reflectance $r(\theta_i, \theta_e, \nu, \lambda)$ models the reflection. Several of these rays may mix and eventually reach the camera sensor with color filters c_{rgb}. Further in-camera processing is denoted by f_{cam}

the color space. For the widely used sRGB color space, the inversion formula for a single RGB-color channel i_{sRGB} of a pixel is

$$
i_{\text{linearRGB}} =
\begin{cases}
\frac{25 i_{sRGB}}{323} & \text{if } i_{sRGB} < 0.04045 \\
\left(\frac{200 i_{sRGB} + 11}{211}\right)^{\frac{12}{5}} & \text{otherwise}
\end{cases}
\tag{9.9}
$$

where we assume that the intensities are in a range from 0 to 1.

The reflectance $r(\theta_i, \theta_e, \lambda)$ is oftentimes assumed to be purely diffuse, i.e., without any specular highlights. This reflectance model is called Lambertian reflectance. In this simple model, the angle of exitance θ_e is ignored. The amount of light that is reflected from an object only depends on the angle of incidence θ_i and the surface normal ν. More specifically, the amount of reflected light is the cosine between the angle of incidence and the surface normal ν of the object at that point. The full reflectance function also encodes the color of the object, written here as $s_d(\lambda)$, which yields a product of the cosine due to the ray geometry and the color,

$$
r_{\text{Lambertian}}(\theta_i, \nu, \lambda) = \cos(\theta_i, \nu) s_d(\lambda) .
\tag{9.10}
$$

In many use cases of this equation, it is common to pull $s_d(\lambda)$ out of the equation (or to set it to unity, thereby ignoring the impact of color) and to only consider the geometric term $\cos(\theta_i, \nu)$.

The dichromatic reflectance model is a linear combination of purely diffuse and specular reflectance, i.e.,

$$
r_{\text{dichromatic}}(\theta_i, \theta_e, \nu, \lambda) = \cos(\theta_i, \nu) s_d(\lambda) + w_s(\theta_i, \nu, \theta_e) s_s(\lambda) .
\tag{9.11}
$$

Here, the purely diffuse term is identical to the Lambertian reflectance Eq. 9.10. The specular term again decomposes into a geometric part w_s and a color part s_s. Both are to my knowledge not explicitly used in forensics, and can hence be superficially treated. However, the geometric part w_s essentially contains the mirror equation, i.e., the angle of incidence θ_i and the angle of exitance θ_e have to be mirrored around the surface normal. The object color s_s is typically set to unity with an additional assumption, which is called the neutral interface reflectance function. This has the effect that the color of specular highlights is equal to the color of the light source when inserting the dichromatic reflectance function $r_{\text{dichromatic}}$ into the full photometric model in Eq. 9.8.

9.3 Algorithms for Physics-Based Forensic Analysis

This section reviews several forensic approaches that exploit the introduced physical models. We introduce in Sect. 9.3.1 geometric methods that directly operate on the physical models from Sect. 9.2.1. We introduce in Sect. 9.3.2 photometric methods that exploit the physical models from Sect. 9.2.2. Then, we present in Sect. 9.3.3 methods that slightly relax assumptions from both geometric and photometric approaches.

9.3.1 Principal Points and Homographies

We start with a method that extracts relatively strong constraints from the scene. Consider a picture that contains written text, for example, signs with political messages during a rally. If a picture of that sign is not taken exactly frontally, but at an angle, then the letters are distorted according to the laws of perspective projection. Conotter et al. proposed a direct application of homographies to validate a proper perspective mapping of the text (Conotter et al. 2010): An attacker who aims to manipulate the text on the sign must also distort the text. However, if the attacker fits the text only in a way that is visually plausible, the laws of perspective projection are still likely to be violated.

To calculate the homography, it is necessary to obtain a minimum of four point correspondences between the image coordinates and a reference picture in world coordinates. However, if only a single picture is available, there is no outside reference for the point correspondences. Alternatively, if the text is printed with a commonly used font, the reference can be synthetically created by rendering the text in that font. From this reference and the text in the image are SIFT keypoints extracted to obtain point correspondences for homography calculation. The matrix **A** from Eq. 9.5 is composed with at least four corresponding keypoints points and solved. To avoid degenerate solutions, the corresponding points must be selected such that there are always at least two corresponding points that do not lie on a line with other corresponding points. The homography matrix **H** can then be used to perform the inverse homography, i.e., from image coordinates to world coordinates, and to calculate the root mean square error (RMSE) between the reference and the transformed image.

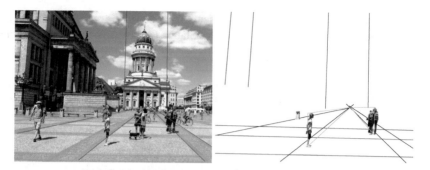

Fig. 9.7 Example scene for height measurements ("berlin square" by zoetnet; full picture credits at the end of the chapter). Lines that converge to the three vanishing points are shown in red, green, and blue. With an object or person with known height, the heights of other objects or persons can be calculated

This method benefits from its relatively strong scene constraints. If it is plausible to assume that the written text is indeed on a planar surface, and if a sufficiently similar font is available (or even the original font), then it suffices to calculate the reprojection error of a homography onto the reference. A similar idea can be used if two images show the same scene, and one of the images has been edited (Zhang et al. 2009a). In this case, the second image serves as a reference for the mapping of the first image. However, many scenes do not provide an exact reference like the exact shape of the written text or a second picture. Nevertheless, if there is other knowledge about the scene, slightly more complex variations of this idea can be used.

For example, Yao et al. show that the vanishing line of a ground plane perpendicular to the optical axis can be used to calculate the height ratio of two persons or objects at an identical distance to the camera (Yao et al. 2012). This height ratio may be sufficient when additional prior knowledge is available like the actual body height of the persons. Iuliani et al. generalize this approach to ground planes with non-zero tilt angles (Iuliani et al. 2015). This is illustrated in Fig. 9.7. The left side shows in red, green, and blue reference lines from the rectangular pattern on the floor and vertical building structures to estimate the vanishing points. The right side shows that the height of the persons on the reference plane can then be geometrically related. Note two potential pitfalls in this scene: first, when the height of persons is related, it may still be challenging to compensate for different body poses (Thakkar and Farid 2021). Second, the roadblock on the left could in principle provide the required reference height. However, it is not exactly on the same reference plane, as the shown public square is not entirely even. Hence, measurements with that block may be wrong.

Another example is to assume some prior information about the geometry of objects in the scene. Then, a simplified form of camera calibration can be performed on each object. In a second step, all objects can be checked for their agreement on the calibration parameters. If two objects disagree, it is assumed that one of these objects has been inserted from another picture with different calibration parameters.

For example, Johnson and Farid proposed to use the eyes of persons for camera calibration (Johnson and Farid 2007a). The underlying assumption is that a specific part, namely the eyes' limbi lie on a plane in 3-D space. The limbus is the boundary between the iris and the white of the eye. The limbi are assumed to be perfect circles when facing the camera directly. When viewed at an angle, the appearance of the limbi is elliptical. Johnson and Farid estimate the homography from these circles instead of isolated points. Since the homography also contains information about the camera intrinsic parameters, the authors calculate the principal point under the additional assumptions that the pixel skew is zero, and that the focal length is known from metadata or contextual knowledge of the analyst. In authentic photographs, the principal point can be assumed to be near the center of the image. If the principal point largely deviates from the image center, it can be plausibly concluded that the image is cropped (Xianzhe et al. 2013; Fanfani et al. 2020). Moreover, if there are multiple persons in the scene with largely different principal points, it can be concluded that the image is spliced from two sources. In this case, it is likely that the relative position of one person was changed from the source image to the target image, e.g., by copying that person from the left side of the image to the right side. Another useful type of additional knowledge can be structures with orthogonal lines like man-made buildings. These orthogonal lines can be used to estimate their associated vanishing points, which also provides the camera principal point (Iuliani et al. 2017).

An inventive variation of these ideas has been used by Conotter et al. (2012). They investigate ballistic motions in videos, as it occurs, e.g., for a video of a thrown basketball. Here, the required geometric constraint does not come from an object per se, but instead from the motion pattern of an object. The assumption of a ballistic motion pattern includes a linear parabolic motion without external forces except for initial acceleration and gravity. This also excludes drift from wind. Additionally, the object is assumed to be rigid and compact with a well-defined center of mass. Under these assumptions, the authors show that the consistency of the motion can be determined by inserting the analytic motion equation into the perspective projection Eq. 9.1. This model holds not only for a still camera, but also for a moving camera if the camera motion can be derived from additional static objects in the surrounding. The validation of true physical motion is additionally supported by the cue that the projection of the object size becomes smaller when the object moves away from the camera and vice versa.

Kee and Farid (2009) and Peng et al. (2017a) use the additional assumption that 3-D models are known for the heads of persons in a scene. This is a relatively strong assumption, but such a 3-D model could, for example, be calculated for people of public interest from which multiple photographs from different perspectives exist, or a 3-D head model could be captured as part of a court case. With this assumption, Kee and Farid show that approximately co-planar landmarks from such a head model can also be used to estimate a homography when the focal length can be retrieved from the metadata (Kee and Farid 2009). Peng et al. use the full 3-D set of facial landmarks and additional face contours to jointly estimate the principal point and the focal length (Peng et al. 2017a). Their main contribution is to show that spliced images of faces can be exposed when the original faces have been captured with different focal length settings.

9.3.2 Photometric Methods

Pixel colors and brightness offer several cues to physics-based forensic analysis. In contrast to the approaches of the previous section, the photometric methods do not model perspective projections, but instead argue about local intensity distributions of objects. We distinguish two main directions of investigation, namely the distribution of the amount of light that illuminates an object from various directions and the color of light.

9.3.2.1 Lighting Environments

Johnson and Farid proposed the first method to calculate the distribution of incident light on an object or person (Johnson and Farid 2005, 2007b). We discuss this method in greater detail due to its importance in the field of physics-based methods. The method operates on a simplified form of the generalized irradiance model from Eq. 9.8. One simplifying assumption is that the camera function is linear, or that a non-linear camera response has been inverted in a pre-processing step. Another simplifying assumption is that the algorithm is only used on objects of a single color, hence all dependencies of Eq. 9.8 on wavelength λ can be ignored. Additionally, Lambertian reflectance is assumed, which removes the dependency on θ_e. These assumptions lead to the irradiance model

$$\mathbf{i}(v) = \int_{\theta_i} \cos(\theta_i, v) \cdot e(\theta_i) \, d\theta_i \tag{9.12}$$

for a single image pixel showing object surface normal v. Here, we directly inserted Eq. 9.10 for the Lambertian reflectance model without the wavelength λ, i.e., the color term in $s_d(\lambda)$ in Eq. 9.10 is set to unity.

Johnson and Farid present two coupled ideas to estimate the distribution of incident light: First, the integral over the angles of incident rays can be summarized by only 9 parameters in the orthonormal spherical harmonics basis. Second, these nine basis coefficients can be easily regressed when a minimum of 9 surface points are available where both the intensity $\mathbf{i}$ and their associated surface normal vectors v are known.

Spherical harmonics are a frequency representation of intensities on a sphere. In our case, they are used to model the half dome of directions from which light might fall onto a surface point of an opaque object. We denote the spherical harmonics basis functions as $h_{i,j}(x, y, z)$ where $j \leq 2i - 1$ and the parameters x, y, and z are points on a unit sphere, i.e., $\|(x, y, z)^{\mathrm{T}}\|_2^2 = 1$. Analogous to other frequency transforms like the Fourier transform or the DCT, the zeroth basis function $h_{0,0}$ is the DC component that contains the overall offset of the values. Higher orders, i.e., where $i > 0$, contain increasingly higher frequencies. However, it suffices to consider only the basis functions up to second order (i.e., $i \leq 2$), to model all possible intensity distributions that can be observed under Lambertian reflectance. These orders $i = \{0, 1, 2\}$ consist of a total of 9 basis functions. For these 9 basis functions, the coefficients are estimated in the proposed algorithm.

The solution for the lighting environment requires knowledge about the surface normals from image locations where the brightness is measured. Obtaining such surface normals from only a 2-D image can be difficult. Johnson and Farid propose two options to address this challenge. First, if the 3-D structure of an object is known, a 3-D model can be created and fitted to the 2-D image. Then, the normals from the fitted 3-D model can be used. This approach has been demonstrated by Kee and Farid for faces, for which reliable 3-D model fitting methods are available (Kee and Farid 2010). Second, if no explicit knowledge about the 3-D structure of an object exists, then it is possible to estimate surface normals from occluding contours of an object. At an occluding contour, the surface normal is approximately coplanar with the image plane. Hence, the z-component is zero, and the x- and y-components can be estimated as lines that are orthogonal to the curvature of the contour. However, without the z-component, also the estimated lighting environment can only be estimated as a projection onto the 2-D image plane. On the other hand, the spherical harmonics model also becomes simpler. By setting all coefficients that contain z-components to zero, only 5 unknown coefficients remain (Johnson and Farid 2005, 2007b).

The required intensities can be directly read from the pixel grid. Johnson and Farid use the green color channel, since this color channel is usually most densely sampled by the Bayer pattern, and it has a high sensitivity to brightness differences. When 2-D normals from occluding contours are used, the intensities in the actual pixel location at the edge of an object might be inaccurate due to sampling errors and in-camera processing. Hence, in this case the intensity is extrapolated from the nearest pixels within the object along the line of the normal vector.

The normals and intensities from several object locations are the known factors in a linear system of equations

$$\mathbf{Al} = \mathbf{i} \, , \tag{9.13}$$

where $\mathbf{i} \in \mathbb{R}^{N \times 1}$ are the observed intensities at N pixel locations on an object. For these N locations, matching surface normals must be available and all constraints must be satisfied. In particular, the N locations must be selected from the same surface material, and they must be directly illuminated. $\mathbf{A}$ is the matrix of the spherical harmonics basis functions. In the 3-D case, its shape is $\mathbb{R}^{N \times 9}$, and in the 2-D case it is $\mathbb{R}^{N \times 5}$. Each of the N rows in this matrix evaluates the basis functions for the surface normal of the associated pixel. The vectors $\mathbf{l}$ are the unknown coefficients of the basis functions, with dimension $\mathbb{R}^{9 \times 1}$ in the 3-D case and $\mathbb{R}^{5 \times 1}$ in the 2-D case. An additional Tikhonov regularizer dampens higher frequency spherical harmonics coefficients, which yields the objective function

$$E(\mathbf{l}) = \|\mathbf{Al} - \mathbf{i}\|_2^2 + \mu\|\mathbf{Rl}\|_2^2 \; , \tag{9.14}$$

with the Tikhonov regularizer $\mathbf{R} = \mathrm{diag}(1\ 2\ 2\ 2\ 3\ 3\ 3\ 3\ 3)$ for the 3-D case, and $\mathbf{R} = \mathrm{diag}(1\ 2\ 2\ 3\ 3)$ for the 2-D case. A minimum is found after differentiation with respect to $\mathbf{l}$ and solving for $\mathbf{l}$, i.e.,

$$\mathbf{l} = (\mathbf{A}^{\mathrm{T}}\mathbf{A} + \mu\mathbf{R}^{\mathrm{T}}\mathbf{R})^{-1}\mathbf{A}^{\mathrm{T}}\mathbf{i} \; . \tag{9.15}$$

The lighting environments of two objects can be directly compared via correlation. A simple approach is to render two spheres from the coefficients of each object, and to correlate the rendered intensities in the 3-D case, or just the intensities along the boundary of the sphere in the 2-D case. Alternatively, the coefficients can also be directly compared after a suitable transform (Johnson and Farid 2005, 2007b).

This first work on the analysis of lighting environments inspired many follow-up works to relax the relatively restrictive assumptions of the method. Fan et al. (2012) explored shape-from-shading as an intermediate step to estimate 3-D surface normals when an explicit 3-D object model is not available. Riess et al. added a reflectance term to the 2-D variant of the algorithm, such that also contours from different materials can be used (Riess et al. 2017). Carvalho et al. investigate human annotations of 3-D surface normals on objects other than faces (Carvalho et al. 2015). Peng et al. automate the 3-D variant by Kee and Farid (2010) via automated landmark detection (Peng et al. 2016). Seuffert et al. show that high-quality 3-D lighting estimation requires a good-fitting geometric model (Seuffert et al. 2018). Peng et al. further include a texture term for 3-D lighting estimation on faces (Peng et al. 2015, 2017b).

Some methods also provide alternatives to the presented models. The assumption of an orthographic projection has also been relaxed, and will be separately discussed in Sect. 9.3.3. Matern et al. integrate over the 2-D image gradients, which provides a slightly less accurate, but very robust 2-D lighting estimation (Matern et al. 2020). Huang et al. propose to calculate 3-D lighting environments from general objects using surface normals from shape-from shading algorithms (Huang and Smith 2011). However, this classic computer vision approach requires relatively simple objects such as umbrellas to be robustly applicable. Zhou et al. train a neural network to learn the lighting estimation from faces in a fully data-driven manner (Zhou et al. 2018).

An example 2-D lighting estimation is shown on top of Fig. 9.8. The person is wearing a T-shirt, which makes it difficult to find surface normals that are of the same material and at the same time point in a representative number of directions. Contours for the T-shirt and the skin are annotated in green and red, together with their calculated normals. The x-axes of the scatterplots contain the possible angles of the surface normals between $-\pi$ and $+\pi$. The y-axes contain the image intensities. The left scatterplot naively combines the intensities of the black T-shirt (green) and the light skin (red). Naively fitting the spherical harmonics to the distribution of both materials leads to a failure case: the light skin dominates the estimation, such that the dominant light source (maximum of the red line) appears to be located below the person. On the other hand, using only the skin pixels or only the T-shirt pixels leads to

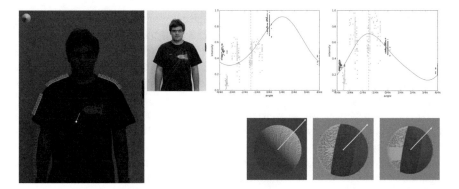

Fig. 9.8 2-D lighting estimation on objects with multiple materials. Top: Example 2-D lighting estimation according to Johnson and Farid (2005, 2007b). From left to right: contour annotations and estimated normals for two materials (green and red lines) on a person. The scatterplots show the angle of the normal along the x-axis with the associated pixel intensities along the y-axis. The left scatterplot mixes both materials, which leads to a wrong solution. The right scatterplot performs an additional reflectance normalization (Riess et al. 2017), such that both materials can be used simultaneously. Bottom: 2-D lighting environments can also be estimated from image gradients, which are overall less accurate, but considerably more robust to material transitions and wrong or incomplete segmentations

a very narrow range of surface normals, which also makes the estimation unreliable. The right plot shows the same distribution, but with the reflectance normalization by Riess et al. (2017), which correctly estimates the location of the dominant light source above the person. The three spheres on the bottom illustrate the estimation of the light source from image gradients according to Matern et al. (2020). This method only assumes that objects are mostly convex, and that the majority of local neighborhoods for gradient computation consists of the same material. With these modest assumptions, the method can oftentimes still operate on objects with large albedo differences, and on object masks with major errors in the segmentation.

9.3.2.2 Color of Illumination

The spectral distribution of light determines the color formation in an image. For example, the spectrum of sunlight is impacted by the light path through the atmosphere, such that sunlight in the morning and in the evening is more reddish than at noon. As another example, camera flashlight oftentimes exhibits a strong blue component. Digital cameras typically normalize the colors in an image with a manufacturer-specific white-balancing function that oftentimes uses relatively simple heuristics. Post-processing software such as Adobe Lightroom provides more sophisticated functions for color adjustment.

These many influencing factors on the colors of an image make their forensic analysis interesting. It is a reasonable assumption that spliced images exhibit differences in the color distribution if the spliced image components had different lighting conditions upon acquisition or if they had undergone different post-processing.

Forensics analysis methods assume that the camera white-balancing is a global image transformation that does not introduce local color inconsistencies in an original image. Hence, local inconsistencies in the color formation are attributed to potential splicing manipulations.

There are several possibilities to locally analyze the consistency of the lighting color. One key requirement is the need to estimate the illuminant color locally on individual objects to expose inconsistencies. Particularly well-suited for local illumination color estimation is the dichromatic reflectance model from Eq. 9.11 in combination with the neutral interface reflectance function. This model includes diffuse and specular reflectance, with the additional assumption that the specular portion exhibits the color of the light source.

To our knowledge, the earliest method that analyzes reflectance has been presented by Gholap and Bora (2008). This work proposes to calculate the intersection of so-called dichromatic lines of multiple objects to expose splicing, using a classical result by Tominaga and Wandell (1989): on a monochromatic object, the distribution of diffuse and specular pixels forms a 2-D plane in the 3-D RGB space. This plane becomes a line when projecting it onto the 2-D r-g chromaticity space, where $r = i_R/(i_R + i_G + i_B)$ and $g = i_G/(i_R + i_G + i_B)$ denote the red and green color channels, normalized via division by the sum of all color channels of a pixel. If the scene is illuminated by a single, global illuminant, the dichromatic lines of differently colored objects all intersect at one point. Hence, Gholap and Bora propose to check for three or more scene objects whether their dichromatic lines intersect in a single point. This approach has the advantage that it is very simple. However, to our knowledge it barely provides possibilities to validate the model assumptions, i.e., ways to check whether the assumption of dichromatic reflectance and a single global illuminant holds for the objects under investigation.

Riess and Angelopoulou (2010) propose the inverse-intensity chromaticity (IIC) space by Tan et al. (2004) to directly estimate the color of the illuminant. The IIC space is simple to calculate, but provides very convenient properties for forensic analysis. It also operates on surfaces of dichromatic reflectance, and requires only a few pixels of identical material that exhibit a mixture of specular and diffuse reflectance. For each color channel, the IIC space is calculated as a 2-D chart for related pixels. Each pixel is transformed to a tuple

$$i_{R,G,B} \rightarrow \left(\frac{1}{i_{R,G,B}}, \frac{i_{R,G,B}}{i_R + i_G + i_B} \right) , \tag{9.16}$$

where again i_R, i_G, and i_B denote the red, green, and blue color channels. On pixels with different portions of specular and diffuse reflectance, the distribution forms a triangle or a line that indicates the chromaticity of the light source at the y-axis intercept. This is illustrated in Fig. 9.9. For persons, the nose region is oftentimes a well-suited location to observe partially specular pixels. Close-ups of the manual annotations are shown together with the distributions in IIC space. In this original

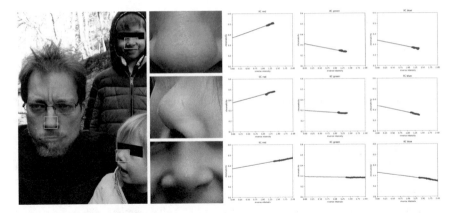

Fig. 9.9 Example application of illuminant color estimation via IIC color charts. For each person in the scene, a few pixels are selected that are of the same material, and that exhibit a mixture of specular and diffuse reflectance. The associated IIC diagrams are shown on the right. The pixel distributions (blue) point toward the estimated color of the light source at the y-axis intercept (red lines)

image, the pixel distributions for the red, green, and blue IIC diagrams point to almost identical illuminant chromaticities. In real scenes, the illuminant colors are oftentimes achromatic. However, the IIC diagrams are very sensitive to post-processing. Thus, if an image is spliced from sources with different post-processing, major differences in the IIC diagrams may be observed.

A convenient property of IIC diagrams is their interpretability. Pixel distributions that do not match the assumptions do not form a straight line. For example, purely diffuse pixels tend to cluster in circular shapes, and pixels of different materials form several clusters, which indicates to an analyst the need to revise the segmentation.

One disadvantage of the proposed method is the need for manual annotations, since the automated detection of specularities is a severely underconstrained problem, and hence not reliable. To mitigate this issue, Riess and Angelopoulou propose to estimate the color of the illuminant on small, automatically segmented image patches of approximately uniform chromaticity. They introduce the notion of "illuminant map" for a picture where each image patch is colored with the estimated illuminant chromaticity. An analyst can then consider only those regions that match the assumptions of the approach, i.e., that exhibit partially specular and partially diffuse reflectance on dichromatic materials. These regions can be considered reliable, and their estimated illuminant colors can be compared to expose spliced images. However, for practical use, we consider illuminant maps oftentimes inferior to a careful, fully manual analysis.

Another analytic approach based on the dichromatic reflectance model has been proposed by Francis et al. (2014). It is conceptually similar, but separates specular and diffuse pixels directly in RGB space. Several works use machine learning to automate the processing and to increase the robustness of illuminant descriptors. For example, Carvalho et al. use illuminant maps that are calculated from the IIC space, and also from the statistical gray edge illuminant estimator, to train a classifier for the authenticity assessment of human faces (de Carvalho et al. 2013). By constraining the application of the illuminant descriptors only to faces, the variety of possible surface materials is greatly restricted, which can increase the robustness of the estimator. However, this approach also performs further processing on the illuminant maps, which slightly leaves the ground of purely physics-based methods and enters the domain of statistical feature engineering. Hadwiger et al. investigate a machine learning approach to learn the color formation of digital cameras (Hadwiger et al. 2019). They use images with Macbeth color charts to learn the relationship between ground truth colors and their representations of different color camera pipelines. In different lines of work on image colors, Guo et al. investigate fake colorized image detection via hue and saturation statistics (Guo et al. 2018). Liu et al. investigate methods to assess the ambient illumination in shadow regions for their consistency by estimating the shadow matte (Liu et al. 2011).

9.3.3 Point Light Sources and Line Constraints in the Projective Space

Outdoor scenes that are only illuminated by the sun provide a special set of constraints. The sun can be approximated as a point light source, i.e., a source where all light is emitted from a single point in 3-D space. Also, indoor scenes may in special cases contain light sources that can be approximated as a single point, like a single candle that illuminates the scene (Stork and Johnson 2006).

Such a single-point light source makes the modeling of shadows particularly simple: the line that connects the tip of an object with the tip of its cast shadow also intersects the light source. Multiple such lines can be used to constrain the position of the light source. Conversely, shadows that have been artificially inserted may violate these laws of projection.

Several works make use of this relationship. Zhang et al. (2009b) and Wu et al. (2012) show that cast shadows of an object can be measured and validated against the length relationships of the persons or objects that cast the shadow. One assumption of these works is that the shadow is cast on a ground plane, such that the measurements are well comparable. Conversely, Stork and Johnson use cast shadows to verify that a scene is illuminated by a specific light source, by connecting occluding contours that likely stem from that light source (Stork and Johnson 2006).

However, these approaches require a clear correspondence between the location on the object that casts the shadow and the location where the shadow is cast to. This is not an issue for poles and other thin cylindrical objects, as the tip of the object can be easily identified. However, it is oftentimes difficult to find such correspondences on objects of more complex shapes.

O'Brien and Farid hence propose a geometric approach with much more relaxed requirements to mark specific locations for object-shadow correspondences (O'Brien and Farid 2012). The method is applicable not only to cast shadows, but also to a mixture of cast shadows and attached shadows as they occur on smooth surfaces. The idea is that if the 2-D image plane would be infinitely large, there would be a location onto which the light source is projected, even if it is outside of the actual image. Object-shadow correspondences can constrain the projected location of the light source. The more accurate an object-shadow correspondence can be identified, the stronger is the constraint. However, it is also possible to incorporate very weak constraints like attached shadows. If the image has been edited, and, for example, an object has been inserted with incorrect illumination, then the constraints from that inserted object are likely to violate the remaining constraints in the image.

The constraints are formed from half-planes in the 2-D image plane. An attached shadow separates the 3-D space into two parts, namely one side that is illuminated and one side that is in shadows. The projection of the illuminated side onto the 2-D image plane corresponds to one half-plane with a boundary normal that corresponds to the attached shadow edge. An object-shadow correspondence with some uncertainty about the exact location at the object and the shadow separates the 3-D space into a 3-D wedge of possible light source locations. Mapping this wedge onto the 2-D image plane leads to a 2-D wedge. Such a wedge can be constructed from two opposing half-planes.

An application of this approach is illustrated in Fig. 9.10. Here, the pen and the spoon exhibit two different difficulties for analyzing the shadows, which is shown in the close-ups on the top right: the shadow of the pen is very unsharp, such that it is difficult to locate its tip. Conversely, the shadow of the spoon is very sharp, but the round shape of the spoon makes it difficult to determine from which location on the spoon surface the shadow tip originates. However, both uncertainties can be modeled as half-plane constraints shown at the bottom. These constraints have a common intersection in the image plane outside of the image, shown in pink. Hence, the shadows of both objects are consistent. Further objects could now be included in the analysis, and their half-plane constraints have to analogously intersect the pink area.

In follow-up work, this approach has been extended by Kee et al. to also include constraints from object shading (Kee et al. 2013). This approach is conceptually highly similar to the estimation of lighting environments as discussed in Sect. 9.3.2.1. However, instead of assuming an orthographic projection, the lighting environments are formulated here within the framework of perspective projection to smoothly integrate with the shadow constraints.

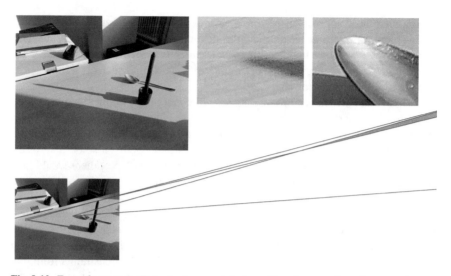

Fig. 9.10 Example scene for forensics from cast shadows. The close-ups show the uncertain areas of shadow formation. Left: the shadow of the tip of the pen is very unsharp. Right: the round shape of the spoon makes it difficult to exactly localize the location that casts the tip of the shadow. Perspective constraints according to O'Brien and Farid allow to nevertheless use these uncertainty regions for forensic analysis (O'Brien and Farid 2012): the areas of the line constraints overlap in the pink region of the image plane (although outside of the actual image), indicating that the shadows of both objects are consistent

9.4 Discussion and Outlook

Physics-based methods for forensic analysis are based on simplified physical models to validate the authenticity of a scene. Journalistic verification uses physics-based methods mostly to answer questions about the time and place of an acquisition, the academic literature mostly focuses on the detection of inconsistencies within an image or video.

Conceptually, physics-based methods are quite different from statistical approaches. Each physics-based approach requires certain scene elements to perform an analysis, while statistical methods can operate on almost arbitrary scenes. Also, most physics-based methods require the manual interaction of an analyst to provide "world knowledge", for example, to annotate occluding contours, or to select partially specular pixels. On the other hand, physics-based methods are mostly independent of the image or video quality, which makes them particularly attractive for analyzing low-quality content or even analog content. Also, physics-based methods are inherently explainable by verification of their underlying models. This would in principle also make it well possible to perform a rigorous analysis of the impact of estimation errors from various error sources, which is much more difficult for statistical approaches. Surprisingly, such robustness investigations have until now only been performed to a limited extent, e.g., for the estimation of vanishing points (Iuliani et al. 2017), perspective distortions (Peng et al. 2017a), or height measurements (Thakkar and Farid 2021).

The future of physics-based methods is challenged by the technical progress in two directions: First, the advent of computational images in modern smartphones. Second, physically plausible computer-generated scene elements from modern methods in computer graphics and computer vision.

Computational images compensate for the limited camera optics in smartphones, which are due to space constraints not competitive with high-quality cameras. Hence, when a modern smartphone captures "an image", it actually captures a short video, and calculates a high-quality single image from that frame sequence. However, if the image itself is not the result of a physical image formation process, the validity of physics-based models for forensic analysis is inherently questioned. It is currently an open question to which extent these computations affect the presented physics-based algorithms.

Recent computer-generated scene elements are created with an increasing contribution of learned physics-based models for realistically looking virtual reality or augmented reality (VR/AR) applications. These approaches are a direct competition to physics-based forensic algorithms, as they use very similar models to minimize their representation error. This emphasizes the need for multiple complementary forensic tools to expose manipulations from cues that are not relevant for the specific tasks of such VR/AR applications, and are hence not considered in their optimization.

9.5 Picture Credits

- Figure 9.1 is published by Dennis Jarvis (archer10, https://www.flickr.com/photos/archer10/ with an Attribution-ShareAlike 2.0 Generic (CC BY-SA 2.0) License. Full text to the license is available at https://creativecommons.org/licenses/by-sa/2.0/; link to the original picture is https://www.flickr.com/photos/archer10/2216460729/. The original image is downsampled for reproduction.
- Figure 9.7 is published by zoetnet, https://flickr.com/photos/zoetnet/ with an Attributed 2.0 Generic (CC BY 2.0) License. Full text to the license is available at https://creativecommons.org/licenses/by/2.0/; link to the original picture is https://flickr.com/photos/zoetnet/9527389096/. The original image is downsampled for reproduction, and annotations of perspective lines are added.

References

Carvalho T, Farid H, Kee E (2015) Exposing photo manipulation from user-guided 3d lighting analysis. In: Alattar AM, Memon ND, Heitzenrater C (eds) Media watermarking, security, and forensics 2015, San Francisco, CA, USA, February 9–11, 2015, Proceedings, vol 9409 of SPIE Proceedings. SPIE, p 940902

Conotter V, O'Brien James F, Farid H (2012) Exposing digital forgeries in ballistic motion. IEEE Trans Inf Forensics Secur 7(1):283–296

Conotter V, Boato G, Farid H (2010) Detecting photo manipulation on signs and billboards. In: Proceedings of the international conference on image processing, ICIP 2010, September 26–29, Hong Kong, China. IEEE, pp 1741–1744

de Carvalho TJ, Riess C, Angelopoulou E, Pedrini H, de Rezende Rocha A (2013) Exposing digital image forgeries by illumination color classification. IEEE Trans Inf Forensics Secur 8(7):1182–1194

Fanfani M, Iuliani M, Bellavia F, Colombo C, Piva A (2020) A vision-based fully automated approach to robust image cropping detection. Signal Proc Image Commun 80

Fan W, Wang K, Cayre F, Xiong Z (2012) 3d lighting-based image forgery detection using shape-from-shading. In: Proceedings of the 20th European signal processing conference, EUSIPCO 2012, Bucharest, Romania, August 27–31, 2012. IEEE, pp 1777–1781

Francis K, Gholap S, Bora PK (2014) Illuminant colour based image forensics using mismatch in human skin highlights. In: Twentieth national conference on communications, NCC 2014, Kanpur, India, February 28–March 2, 2014. IEEE, pp 1–6

Gholap S, Bora PK (2008) Illuminant colour based image forensics. In: IEEE region 10 conference, TENCON 2008, Hyderabad, India, November 19–21 2008

Guo Y, Cao X, Zhang W, Wang R (2018) Fake colorized image detection. IEEE Trans Inf Forensics Secur 13(8):1932–1944

Hadwiger B, Baracchi D, Piva A, Riess C (2019) Towards learned color representations for image splicing detection. In: IEEE international conference on acoustics, speech and signal processing, ICASSP 2019, Brighton, United Kingdom, May 12–17, 2019. IEEE, pp 8281–8285

Hartley R, Zisserman A (2004) Multiple view geometry in computer vision. Cambridge University Press, Cambridge

Huang R, Smith WAP (2011) Shape-from-shading under complex natural illumination. In: Macq B, Schelkens P (eds) 18th IEEE international conference on image processing, ICIP 2011, Brussels, Belgium, September 11–14, 2011. IEEE, pp 13–16

Iuliani M, Fanfani M, Colombo C, Piva A (2017) Reliability assessment of principal point estimates for forensic applications. J Vis Commun Image Represent 42:65–77

Iuliani M, Fabbri G, Piva A (2015) Image splicing detection based on general perspective constraints. In: 2015 IEEE international workshop on information forensics and security, WIFS 2015, Roma, Italy, November 16–19, 2015. IEEE, pp 1–6

Johnson MK, Farid H (2006) Exposing digital forgeries by detecting inconsistencies in lighting. In: Eskicioglu AM, Fridrich JJ, Dittmann J (eds) Proceedings of the 7th workshop on Multimedia & Security, MM&Sec 2005, New York, NY, USA, August 1–2, 2005. ACM, pp 1–10

Johnson MK, Farid H (2007a) Detecting photographic composites of people. In: Shi YQ, Kim H-J, Katzenbeisser S (eds) Digital watermarking, 6th international workshop, IWDW 2007, Guangzhou, China, December 3–5, 2007, Proceedings, vol 5041 of Lecture notes in computer science. Springer, pp 19–33

Johnson MK, Farid H (2007b) Exposing digital forgeries in complex lighting environments. IEEE Trans Inf Forensics Secur 2(3–1):450–461

Kee E, Farid H (2009) Detecting photographic composites of famous people. Technical Report Computer Science Technical Report TR2009-656, Department of Computer Science, Dartmouth College

Kee E, Farid H (2010) Exposing digital forgeries from 3-d lighting environments. In: 2010 IEEE international workshop on information forensics and security, WIFS 2010, Seattle, WA, USA, December 12–15, 2010. IEEE, pp 1–6

Kee E, O'Brien JF, Farid H (2013) Exposing photo manipulation with inconsistent shadows. ACM Trans Graph 32(3):28:1–28:12

Liu Q, Cao X, Deng C, Guo X (2011) Identifying image composites through shadow matte consistency. IEEE Trans Inf Forensics Secur 6(3–2):1111–1122

Matern F, Riess C, Stamminger M (2020) Gradient-based illumination description for image forgery detection. IEEE Trans Inf Forensics Secur 15:1303–1317

O'Brien JF, Farid H (2012) Exposing photo manipulation with inconsistent reflections. ACM Trans Graph 31(1):4:1–4:11

Peng B, Wang W, Dong J, Tan T (2015) Improved 3d lighting environment estimation for image forgery detection. In: 2015 IEEE international workshop on information forensics and security, WIFS 2015, Roma, Italy, November 16–19, 2015. IEEE, pp 1–6

Peng B, Wang W, Dong J, Tan T (2016) Automatic detection of 3d lighting inconsistencies via a facial landmark based morphable model. In: 2016 IEEE international conference on image processing, ICIP 2016, Phoenix, AZ, USA, September 25–28, 2016. IEEE, pp 3932–3936

Peng B, Wang W, Dong J, Tan T (2017a) Position determines perspective: investigating perspective distortion for image forensics of faces. In: 2017 IEEE conference on computer vision and pattern recognition workshops, CVPR workshops 2017, Honolulu, HI, USA, July 21–26, 2017. IEEE Computer Society, pp 1813–1821

Peng B, Wang W, Dong J, Tan T (2017b) Optimized 3d lighting environment estimation for image forgery detection. IEEE Trans Inf Forensics Secur 12(2):479–494

Riess C, Angelopoulou E (2010) Scene illumination as an indicator of image manipulation. In: Böhme R, Fong PWL, Safavi-Naini R (eds) Information hiding - 12th international conference, IH 2010, Calgary, AB, Canada, June 28–30, 2010, Revised Selected Papers, vol 6387 of Lecture notes in computer science. Springer, pp 66–80

Riess C, Unberath M, Naderi F, Pfaller S, Stamminger M, Angelopoulou E (2017) Handling multiple materials for exposure of digital forgeries using 2-d lighting environments. Multim Tools Appl 76(4):4747–4764

Seuffert J, Stamminger M, Riess C (2018) Towards forensic exploitation of 3-d lighting environments in practice. In: Langweg H, Meier M, Witt BC, Reinhardt D (eds) Sicherheit 2018, Beiträge der 9. Jahrestagung des Fachbereichs Sicherheit der Gesellschaft für Informatik e.V. (GI), 25.–27.4.2018, Konstanz, volume P-281 of LNI. Gesellschaft für Informatik e.V., pp 159–169

Stork DG, Johnson MK (2006) Estimating the location of illuminants in realist master paintings computer image analysis addresses a debate in art history of the baroque. In: 18th international conference on pattern recognition (ICPR 2006), 20–24 August 2006, Hong Kong, China. IEEE Computer Society, pp 255–258

Tan RT, Nishino K, Ikeuchi K (2004) Color constancy through inverse-intensity chromaticity space. J Opt Soc Amer A 21(3):321–334

Thakkar N, Farid H (2021) On the feasibility of 3D model-based forensic height and weight estimation. In: Workshop on media forensics (in conjunction with CVPR)

Tominaga S, Wandell Brian A (1989) Standard surface-reflectance model and illuminant estimation. J Opt Soc Am A 6(4):576–584

Wu L, Cao X, Zhang W, Wang Y (2012) Detecting image forgeries using metrology. Mach Vis Appl 23(2):363–373

Xianzhe M, Ru SN, Yan Li Y (2013) Detecting photographic cropping based on vanishing points. Chinese J Electron 22(2):369–372

Yao H, Wang S, Zhao Y, Zhang X (2012) Detecting image forgery using perspective constraints. IEEE Signal Proc Lett 19(3):123–126

Zhang W, Cao X, Feng Z, Zhang J, Wang P (2009a) Detecting photographic composites using two-view geometrical constraints. In: Proceedings of the 2009 IEEE international conference on multimedia and Expo, ICME 2009, June 28–July 2, 2009, New York City, NY, USA. IEEE, pp 1078–1081

Zhang W, Cao X, Zhang J, Zhu J, Wang P (2009b) Detecting photographic composites using shadows. In: Proceedings of the 2009 IEEE international conference on multimedia and Expo, ICME 2009, June 28–July 2, 2009, New York City, NY, USA. IEEE, pp 1042–1045

Zhou H, Sun J, Yacoob Y, Jacobs DW (2018) Label denoising adversarial network (LDAN) for inverse lighting of faces. In: 2018 IEEE conference on computer vision and pattern recognition, CVPR 2018, Salt Lake City, UT, USA, June 18–22, 2018. IEEE Computer Society, pp 6238–6247

Chapter 10
Power Signature for Multimedia Forensics

Adi Hajj-Ahmad, Chau-Wai Wong, Jisoo Choi, and Min Wu

There has been an increasing amount of work surrounding the Electric Network Frequency (ENF) signal, an environmental signature captured by audio and video recordings made in locations where there is electrical activity. ENF is the frequency of power distribution networks, 60 Hz in most of the Americas and 50 Hz in most other parts of the world. The ubiquity of this power signature and the appearance of its traces in media recordings motivated its early application toward time–location authentication of audio recordings. Since then, more work has been done toward utilizing this signature for other forensic applications, such as inferring the grid in which a recording was made, as well as applications beyond forensics, such as temporally synchronizing media pieces. The goal of this chapter is to provide an overview of the research work that has been done on the ENF signal and to provide an outlook for the future.

A. Hajj-Ahmad
Amazon, Inc, Seattle, WA, USA

C.-W. Wong · J. Choi
Electrical and Computer Engineering and Forensic Sciences Cluster, North Carolina State University, Raleigh, NC, USA

M. Wu (✉)
Electrical and Computer Engineering and Institute for Advanced Computer Studies, University of Maryland, College Park, MD, USA
e-mail: minwu@umd.edu

© The Author(s) 2022
H. T. Sencar et al. (eds.), *Multimedia Forensics*, Advances in Computer Vision and Pattern Recognition, https://doi.org/10.1007/978-981-16-7621-5_10

10.1 Electric Network Frequency (ENF): An Environmental Signature for Multimedia Recordings

In this chapter, we discuss the Electric Network Frequency (ENF) signal, an environmental signature that has been under increased study since 2005 and has been shown to be a useful tool for a number of information forensics and security applications. The ENF signal is such a versatile tool that later work has also shown that it can have applications beyond security, such as for digital archiving purposes and multimedia synchronization.

The ENF signal is a signal influenced by the electric power grid. Most of the power provided by the power grid comes from turbines that work as generators of alternating current. The rotational velocity of these turbines determines the ENF, which usually has a nominal value 60 Hz in most of the Americas and 50 Hz in most other parts of the world. The ENF fluctuates around its nominal value as a result of power-frequency control systems that maintain the balance between the generation and consumption of electric energy across the grid (Bollen and Gu 2006). These fluctuations can be seen as random, unique in time, and typically very similar in all locations of the same power grid. The changing instantaneous value of the ENF over time is what we define as the *ENF signal*.

What makes the ENF particularly relevant to multimedia forensics is that the ENF can be embedded in audio or video recordings made in areas where there is electrical activity. It is in this way that the ENF serves as an environmental signature that is intrinsically embedded in media recordings. Once we can extract this invisible signature well, we can answer many questions about the recording it was embedded in. For example: When was the recording made? Where was it made? Has it been tampered with? A recent study has shown that ENF traces can even be extracted from images captured by digital cameras with rolling shutters, which can allow researchers to determine the nominal frequency of the area in which an image was captured.

Government agencies and research institutes of many countries have conducted ENF-related research and developmental work. This includes academia and government in Romania (Grigoras 2005, 2007, 2009), Poland (Kajstura et al. 2005), Denmark (Brixen 2007a, b, 2008), the United Kingdom (Cooper 2008, 2009a, b, 2011), the United States (Richard and Peter 2008; Sanders 2008; Liu et al. 2011, 2012; Ojowu et al. 2012; Garg et al. 2012, 2013; Hajj-Ahmad et al. 2013), the Netherlands (Huijbregtse and Geradts 2009), Brazil (Nicolalde and Apolinario 2009; Rodríguez et al. 2010; Rodriguez 2013; Esquef et al. 2014, Egypt (Eissa et al. 2012; Elmesalawy and Eissa 2014), Israel (Bykhovsky and Cohen 2013), Germany (Fechner and Kirchner 2014), Singapore (Hua 2014), Korea (Kim et al. 2017; Jeon et al. 2018), and Turkey (Vatansever et al. 2017, 2019). Among the work studied, we can see two major groups of study. The first group addresses challenges in accurately estimating an ENF signal from media signals. The second group focuses on possible applications of the ENF signal once it is extracted properly. In this chapter, we conduct a comprehensive literature study on the work done so far and outline avenues for future work.

The rest of this chapter is organized as follows. Section 10.2 describes the methods for ENF signal extraction. Section 10.3 discusses the findings of works studying the presence of ENF traces and providing statistical models for ENF behavior. Section 10.4 explains how ENF traces are embedded in videos and images and how they can be extracted. Section 10.5 delves into the main forensic and security ENF applications proposed in the literature. Section 10.6 describes an anti-forensics framework for understanding the interplay between an attacker and an ENF analyst. Section 10.7 extends the conversation toward ENF applications beyond security. Section 10.8 summarizes the chapter and provides an outlook for the future.

10.2   Technical Foundations of ENF-Based Forensics

As will be discussed in this chapter, the ENF signal can have a number of useful real-world applications, both in multimedia forensics-related fields and beyond. A major first step to these applications, however, is to properly extract and estimate the ENF signal from an ENF-containing media signal. But, how can we check that what we have extracted is indeed the ENF signal? For this purpose, we introduce the notion of *power reference signals* in Sect. 10.2.1. Afterward, we discuss various proposed methods to estimate the ENF signal from a media signal. We focus here on ENF extraction from digital one-dimensional (1-D) signals, typically audio signals, and delay the explanation of extracting ENF from video and image signals to Sect. 10.4.

10.2.1   Reference Signal Acquisition

Power reference recordings can be very useful to ENF analysis. The ENF traces found in the power recordings are typically much stronger than the ENF traces found in audio recordings. This is why they can be used as a reference and a guide for ENF signals extracted from audio recordings, especially in cases where one has access to a pair of simultaneously recorded signals, one a power signal and one an audio signal; the ENF in both should be very similar at the same instants of time. In this section, we describe different methods used in the ENF literature to acquire power reference recordings.

The Power Information Technology Laboratory at the University of Tennessee, Knoxville (UTK), operates the North American Power Grid Frequency Monitoring Network System (FNET), or GridEye. The FNET/GridEye is a power grid situational awareness tool that collects real-time, Global Position System (GPS) timestamped measurements of grid reference data at the distribution level (Liu et al. 2012). A framework for FNET/GridEye is shown in Fig. 10.1. The FNET/GridEye system consists of two major components, which are the frequency disturbance recorders (FDRs) and the information management system (IMS). The FDRs are the sensors of the system; each FDR is an embedded microprocessor system that performs local

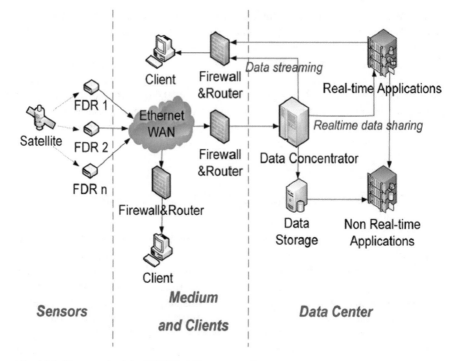

Fig. 10.1 Framework of the FNET/GridEye system (Zhang et al. 2010)

GPS-synchronized measurements, such as computing the instantaneous ENF values over time. In this setup, the FDR estimates the power frequency values at a rate of 10 records/s using phasor techniques (Phadke et al. 1983). The measured data is sent to the server through the Internet, where the IMS collects the data, stores it, and provides a platform for the visualization and analysis of power system phenomena. More information on the FNET/GridEye system can be found in FNET Server Web Display (2021), Zhong et al. (2005), Tsai et al. (2007), and Zhang et al. (2010).

A system similar to the FNET/GridEye system, named the wide area management systems (WAMS), has been set up in Egypt, where the center providing the information management functions is at Helwan University (Eissa et al. 2012; Elmesalawy and Eissa 2014). The researchers operating this system have noted that during system disturbances, the instantaneous ENF value is not the same across all points of the grid. It follows that in such cases, the ENF value from a single point in the grid may not be a reliable reference. For this purpose, in Elmesalawy and Eissa (2014), they propose a method for establishing ENF references from a number of FDRs deployed in multiple locations of the grid, rather than from a single location.

Recently, the authors of Kim et al. (2017) presented an ENF map that takes a different route: Instead of relying on installing specialized hardware, they built their ENF map by extracting ENF signals from the audio tracks of open-source online streaming multimedia data obtained from such sources as "Ustream", "Earthcam", and "Skyline

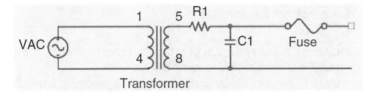

Fig. 10.2 Sample generic schematic of sensoring hardware (Top et al. 2012)

webcams". Most microphones used in such streaming services are mains-powered, which makes the ENF traces captured in the recordings stronger than those that would be captured in recordings made using battery-powered recorders. Kim et al. (2017) addressed in detail the challenges that come with the proposed approach, including accounting for packet loss, aligning different ENF signals temporally, and interpolating ENF signals geographically to account for locations that are not covered by the streaming services.

Systems such as the ones discussed offer tremendous benefits in power frequency monitoring and coverage, yet one does not need access to them in order to acquire ENF references locally. An inexpensive hardware circuit can be built to record a power signal or measure ENF variations, given access to an electric wall outlet. Typically, a transformer is used to convert the voltage from the wall outlet voltage levels down to a level that an analog-to-digital converter can capture. Figure 10.2 shows a sample generic circuit that can be built to record the power reference signal (Top et al. 2012).

There is more than one design to build the sensor hardware. In the example of Fig. 10.2, an anti-aliasing filter is placed in the circuit along with a fuse for safety purposes. In some implementations, such as in Hajj-Ahmad et al. (2013), a step-down circuit is connected to a digital audio recorder that records the raw power signal, whereas in other implementations, such as in Fechner and Kirchner (2014), the step-down circuit is connected to a BeagleBone Black board, via a Schmitt trigger that computes an estimate of the ENF signal spontaneously. In the former case, the recorded digital signal is processed later using ENF estimation techniques, which will be discussed in the next section, to extract the reference ENF signal while in the latter case, the ENF signal is ready to be used as a reference for the analysis.

As can be seen, there are several ways to acquire ENF references depending on the resources one has access to. An ENF signal extracted through these measurements typically has a high signal-to-noise ratio (SNR) and can be used as a reference in ENF research and applications.

10.2.2 ENF Signal Estimation

In this section, we discuss several approaches that have been proposed in the literature to extract the ENF signal embedded in audio signals.

A necessary stage before estimating the changing instantaneous ENF value over time is preprocessing the audio signal. Typically, since the ENF component is in a low-frequency band, a lowpass filter with proper anti-aliasing can be applied to the ENF-containing audio signal to make the computations of the estimation algorithms easier. For some estimation approaches, it also helps to bandpass the ENF-containing audio signal around the frequency band of interest, i.e., frequency band surrounding the nominal ENF value. Besides that, the ENF-containing signal is then divided into consecutive overlapping or nonoverlapping *frames*. The aim of the ENF extraction process would be to apply a frequency estimation approach on each frame to estimate its most dominant frequency around the nominal ENF value. This frequency estimate would be the estimated instantaneous ENF value for the frame. Concatenating the frequency estimates of all the frames together would form the extracted ENF signal. The length of the frame, typically on the order of seconds, determines the resolution of the extracted ENF signal. Typically, a trade-off exists here. Smaller frame size would better capture the ENF variations but may result in poorer performance of the frequency estimation approach, and vice versa.

Generally speaking, an ENF estimation approach can be one of three types: (1) a time-domain approach, (2) a nonparametric frequency-domain approach, and (3) a parametric frequency-domain approach.

10.2.2.1 Time-Domain Approach

The time-domain zero-crossing approach is fairly straightforward, and it is one of the few ENF estimation approaches that is not preceded by dividing the recording into consecutive frames for individual processing. As described in Grigoras (2009), a bandpass filter with 49–51 Hz or 59–61Hz cutoff is first applied to the ENF-containing signal without downsampling initially. This is done to separate the ENF waveform from the rest of the recording. Afterward, the zero-crossings of the remaining ENF signal are computed, and the time differences between consecutive zero values are computed and used to obtain the instantaneous ENF estimates.

10.2.2.2 Nonparametric Frequency-Domain Approach

Nonparametric approaches do not assume any explicit model for the data. Most of these approaches are based on the Fourier analysis of the signal.

Most nonparametric frequency-domain approaches are based on the periodogram-based or spectrogram-based approach utilizing the short-time Fourier transform (STFT). STFT is often used for signals with a time-varying spectrum, such as speech signals. After the signal is segmented into overlapping frames, each frame undergoes Fourier analysis to determine the frequencies present. A spectrogram is then defined as the squared magnitude of the STFT and is usually displayed as a two-dimensional (2-D) intensity plot, with the two axes being time and frequency, respectively (Hajj-Ahmad et al. 2012).

Because of the slowly varying nature of the ENF signal, it is reasonable to consider the instantaneous frequency within the duration of a frame approximately constant for analysis. Given a sinusoid of a fixed frequency embedded in noise, the power spectral density (PSD) estimated by the STFT should ideally exhibit a peak at the frequency of the sinusoidal signal. Estimating this frequency well gives a good estimate for the ENF value of this frame.

A straightforward approach to estimating this frequency would be finding the frequency that has the maximum power spectral component. Directly choosing this frequency as the ENF value, however, typically leads to a loss in accuracy, because the spectrum is computed for discretized frequency values and the actual frequency of the maximum energy may not be aligned with these discretized frequency values. For this reason, typically, the STFT-based ENF estimation approach carries out further computations to obtain a more refined estimate. Examples of such operations are quadratic or spline interpolations being done about the detected spectral peak, or a weighted approach where the ENF estimate is found by weighing the frequency bins around the nominal value based on their spectrum intensities (Hajj-Ahmad et al. 2012; Cooper 2008, 2009b; Grigoras 2009; Liu et al. 2012).

In addition to the STFT-based approach that is most commonly used, the authors in Ojowu et al. (2012) advocate the use of a nonparametric, adaptive, and high-resolution technique known as the time-recursive iterative adaptive approach (TR-IAA). This algorithm reaches the spectral estimates of a given frame by minimizing a quadratic cost function using a weighted least squares formulation. It is an iterative technique that takes 10–15 iterations to converge, where the spectral estimate is initialized to be either the spectrogram or the final spectral estimate of the preceding frame (Glentis and Jakobsson 2011). As compared to STFT-based techniques, this approach is more computationally extensive. In Ojowu et al. (2012), the authors report that the STFT-based approach gives slightly better estimates of the network frequency when the SNR is high, yet the adaptive TR-IAA approach achieves a higher ENF estimation accuracy in the presence of interference from other signals.

To enhance the ENF estimation accuracy, Ojowu et al. (2012) propose a frequency tracking method based on dynamic programming that finds a minimum cost path. For each frame, the method generates a set of candidate frequency peak locations from which the minimum cost path is generated. The cost function selected in this proposed extension takes into account the slowly varying nature of the ENF and penalizes significant jumps in frequency from frame to frame. The minimum path found is the estimated ENF signal.

In low SNR scenarios, the sets of candidate peak frequency locations used by Ojowu et al. (2012) may not be precisely estimated. To avoid using imprecise peak locations, the Zhu et al. (2018, 2020) apply dynamic programming directly to a 2-D time–frequency representation such as the spectrogram. The authors also propose to extract multiple ENF traces of unequal strengths by iteratively estimating and erasing the strongest one. A near-real-time variant of the multiple ENF extraction algorithm was proposed in Zhu et al. (2020) to facilitate efficient online frequency tracking.

There have been other proposed modifications to nonparametric approaches in the literature to improve the final ENF signal estimate. In Ling Fu et al. (2013), the authors propose a discrete Fourier transform (DFT)-based binary search algorithm to lower the computational complexity. Instead of calculating the full Fourier spectrum, the proposed method uses the DFT to calculate a spectral line at the midpoint of a frequency interval at each iteration. At the end of the iteration, the frequency interval will be replaced by its left or right half based on the relative strength of the calculated spectral line and that of the two ends of the interval. The search stops once the frequency interval is narrow enough, and the estimated frequency of the current frame will be used to initialize the candidate frequency interval of the next frame.

In Dosiek (2015), the author considers the ENF estimation problem as a frequency demodulation problem. By considering the captured power signal to be a carrier sinusoid of nominal ENF value modulated by a weak stochastic signal, a 0-Hz intermediate frequency signal can be created and analyzed instead of using the higher frequency modulated signal. This allows the ENF to be estimated through the use of FM algorithms.

In Georgios Karantaidis and Constantine Kotropoulos (2018), the authors suggested using refined periodograms as the basis for the nonparametric frequency estimation approach, including Welch, Blackman–Tukey, and Daniel, as well as the Capon method, which is a filter bank approach based on a data-dependent filter.

In Xiaodan Lin and Xiangui Kang (2018), the authors propose to improve the extraction of ENF estimates in cases of low SNR by exploiting the low-rank structure of the ENF signal in an approach that uses robust principal component analysis to remove interference from speech content and background noise. Weighted linear prediction is then involved in extracting the ENF estimates.

10.2.2.3 Parametric Frequency-Domain Approach

Parametric frequency-domain ENF estimation approaches assume an explicit model for the signal and the underlying noise. Due to such an explicit assumption about the model, the estimates obtained using parametric approaches are expected to be more accurate than those obtained using nonparametric approaches if the modeling assumption is correct (Manolakis et al. 2000). Two of the most widely used parametric frequency estimation methods are based on the subspace analysis of a signal–noise model, namely the MUltiple SIgnal Classification (MUSIC) (Schmidt 1986) and Estimation of Signal Parameters via Rotational Invariance Techniques (ESPRIT) (Roy and Kailath 1989). These methods can be used to estimate the frequency of a signal composed of P complex frequency sinusoids embedded in white noise. As ENF signals consist of a real sinusoid, the value of P for ENF signals is 2 when no harmonic signals exist.

The MUSIC algorithm is a subspace-based approach to frequency estimation that relies on eigendecomposition and the properties between the signal and noise subspaces for sinusoidal signals with additive white noise. MUSIC makes use of the orthogonality between the signal and noise subspaces to compute a pseudo-spectrum for a signal. This pseudo-spectrum should have P frequency peaks corresponding to the dominant frequency components in the signal, i.e., the ENF estimates.

The ESPRIT algorithm makes use of the rotational property between staggered subspaces that are invoked to produce the frequency estimates. In our case, this property relies on observations of the signal over two intervals of the same length staggered in time. ESPRIT is similar to MUSIC in the sense that they are both subspace-based approaches, but it is different in that it works with the signal subspace rather than the noise subspace.

10.2.3 Higher Order Harmonics for ENF Estimation

The electric power signal is subject to waveform distortion due to the presence of nonlinear elements in the power system. A nonlinear element takes a non-sinusoidal current for a sinusoidal voltage. Thus, even for a nondistorted voltage waveform, the current through a nonlinear element is distorted. This distorted current waveform, in turn, leads to a distorted voltage waveform. Though most elements of the power network are linear, transformers are not, especially during sustained over-voltages and power electronic components. Besides that, a main reason behind the distortion is due to nonlinear load, mainly power-electronic converters (Bollen and Gu 2006).

A significant way in which the waveform distortion of the power signal manifests itself is in harmonic distortion, where the power signal waveform can be decomposed into a sum of harmonic components with the fundamental frequency being close to the 50/60 Hz nominal ENF value. It follows that scaled versions of almost the same variations appear in many of the harmonic bands, although the strength of the traces may differ at different harmonics with varying recording environments and devices used. An example of this can be seen in Fig. 10.3. In extracting the ENF signal, we can take advantage of the presence of ENF traces at multiple harmonics of the nominal frequency in order to make our ENF signal estimate more robust.

In Bykhovsky and Cohen (2013), the authors extend the ENF model from a single-tone signal to a multi-tone harmonic one. Under this model, they use the Cramer–Rao bound for the estimation model that shows that the harmonic model can lead to a theoretical $O\left(M^3\right)$ factor improvement in the ENF estimation accuracy, with M being the number of harmonics. The authors then derive a maximum likelihood estimator for the ENF signal, and their results show a significant gain as compared with the results on the single-tone model.

In Hajj-Ahmad et al. (2013), the authors propose a *spectrum combining* approach to ENF signal estimation, which exploits the different ENF components appearing in a signal and strategically combines them based on the local SNR values. A hypothesis

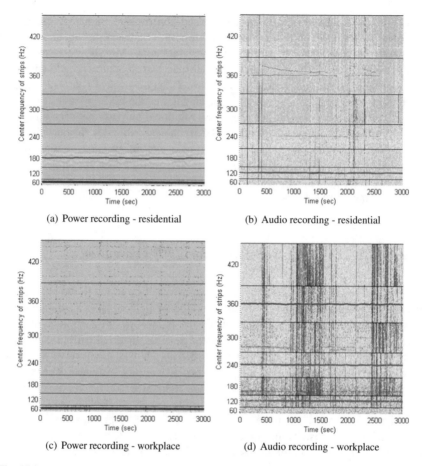

(a) Power recording - residential

(b) Audio recording - residential

(c) Power recording - workplace

(d) Audio recording - workplace

Fig. 10.3 Spectrogram strips about the harmonics 60 Hz for two sets of simultaneously recorded power and audio measurements (Hajj-Ahmad et al. 2013)

testing performance of an ENF-based timestamp verification application is examined, which validates that the proposed approach achieves a more robust and accurate estimate than conventional ENF approaches that rely solely on a single tone.

10.3 ENF Characteristics and Embedding Conditions

Despite that there has been an increasing amount of work recently geared toward extracting the ENF signal from media recordings and then using it for innovative applications, there has been relatively less work done toward understanding the conditions that promote or hinder the embedding of ENF in media recordings. Further work also can be done toward understanding the way the ENF behaves. In this section, we will first go over the existing literature on the conditions that affect ENF capture and then discuss recent statistical modeling efforts on the ENF signal.

10.3.1 Establishing Presence of ENF Traces

Understanding the conditions that promote or hinder the capture of ENF traces in media recordings would go a long way in helping us benefit from the ENF signal in our applications. It would also help us understand better the situations in which ENF analysis is applicable.

If a recording is made with a recorder connected to the electric mains power, it is generally accepted that ENF traces will be present in the resultant recording (Fechner and Kirchner 2014; Kajstura et al. 2005; Brixen 2007a; Huijbregtse and Geradts 2009; Cooper 2009b, 2011; Garg et al. 2012). The strength and presence of the ENF traces depend on the recording device's internal circuitry and electromagnetic compatibility characteristics (Brixen 2008).

If the recording is made with a battery-powered recorder, the question of whether or not the ENF will be captured in the recording becomes more complex. Broadly speaking, ENF capturing can be affected by several factors that can be divided into two groups: factors related to the environment in which the recording was made and factors related to the recording device used to make the recording. Interactions between different factors may as well lead to different results. For instance, electromagnetic fields in the place of recording promote ENF capturing if the recording microphone is *dynamic* but not in the case where the recording microphone is *electret*. Table 10.1 shows a sample of factors that have been studied in the literature for their effect on ENF capture in audio recordings (Fechner and Kirchner 2014; Brixen 2007a; Jidong Chai et al. 2013).

Overall, the most common cause of ENF capture in audio recordings is the acoustic mains hum, which can be produced by mains-powered equipment in the place of recording. The hypothesis that this background noise is a carrier of ENF traces was confirmed in Fechner and Kirchner (2014). Experiments carried out in an indoor setting suggested high robustness of ENF traces where the ENF traces were present in a recording made 10 m away from a noise source located in a different room. Future work will still need to conduct large-scale empirical studies to infer how likely real-world audio recordings will contain distinctive ENF traces.

Hajj-Ahmad et al. (2019) conducted studies exploring factors that affect the capture of ENF traces. They demonstrated that moving a recorder while making a recording will likely compromise the quality of the ENF being captured, due to the Doppler effect, possibly in conjunction with other factors such as air pressure and other vibrations. They also showed that using different recorders in the same recording setting can lead to different strengths of ENF traces captured and at different harmonics. Further studies along this line will help understand better the applicability of ENF research and inform the design of scalable ENF-based applications.

Table 10.1 Sample of factors affecting ENF capture in audio recordings made by battery-powered recorders (Hajj-Ahmad et al. 2019)

	Factors	Effect
Environmental	Electromagnetic (EM) fields	Promote ENF capture in recordings made by dynamic microphones but not in those made by electret microphones
	Acoustic mains hum	Promotes ENF capture; sources include fans, power adaptors, lights, and fridges
	Electric cables in vicinity	Not sufficient for ENF capture
Device-related	Type of microphone	Different types have different reactions to the same sources, e.g., to EM fields
	Frequency band of recorders	Recorders may be incapable of recording low frequencies, e.g., around 50/60 Hz
	Internal compression by recorder	Strong compression, e.g., Adaptive Multi-Rate, can limit ENF capturing

In Zhu et al. (2018), Zhu et al. (2020), the authors propose an ENF traces presence test for 1-D signals. The presence of the ENF is first tested for each frame using a test statistic quantifying the relative energy within a small neighborhood of the frequency of the spectral peak. A frame can be classified as "voiced" or "unvoiced" by thresholding the test statistic. The algorithm merges frames of the same decision type into a segment while allowing frames of the other type to be sparsely presented in a long segment. This refinement produces a final decision result containing two types of segments that are sparsely interleaved over time.

In Vatansever et al. (2017), the authors address the issue of the presence of ENF traces in videos. In particular, the problem they address is that ENF analysis on videos can be computationally expensive, and typically, there is no guarantee that a video would contain ENF traces prior to analysis. In their paper, they propose an approach to assess a video before further analysis to understand whether it contains ENF traces. Their ENF detection approach is based on using *superpixels*, which are steady object regions having very close reflectance properties, in a representative video frame. They show that their algorithm can work on video clips as short as 2 min and can operate independently of the camera image sensor type, i.e., complementary metal-oxide semiconductor (CMOS) or charge-coupled device (CCD).

10.3.2 Modeling ENF Behavior

Understanding how the ENF behaves over time can be important toward understanding if a detected frequency component corresponds to the ENF or not. Several studies have carried out statistical modeling on the ENF variations. In Fig. 10.4, we can see that ENF values collected from the UK grid over one month follow a Gaussian-like distribution (Cooper 2009b).

Other studies have been carried out on the North American grids, where there are four interconnections, namely Eastern Interconnection (EI), Western Interconnection (WECC), Texas Interconnection (ERCOT), and a Quebec Interconnection. The studies show that the ENF generally follows Gaussian distributions in the Eastern interconnection, the Western interconnection, and the Quebec interconnection, yet the mean and standard deviation values are different across interconnections. For instance, the Western interconnection shows a smaller standard deviation than the Eastern interconnection indicating that it maintains a slightly tighter control over the ENF variations. The differences in density are reflective of the control strategies employed on the grids and the size of the grids (Liu et al. 2011, 2012; Garg et al. 2012; Top et al. 2012). A further indication of the varying behaviors of ENF variations across different grids is that a study has shown that the ENF data collected in Singapore do not strictly follow a Gaussian distribution (Hua 2014).

In Garg et al. (2012), the authors have shown that the ENF signal in the North American Eastern interconnection can be modeled as a piecewise wide sense stationary (WSS) signal. Following that, they model the ENF signal as a piecewise autoregressive (AR) process, and show how this added understanding of the ENF signal behavior can be an asset toward improving performance in ENF applications, such as with time–location authentication, which will be discussed in Sect. 10.5.1.

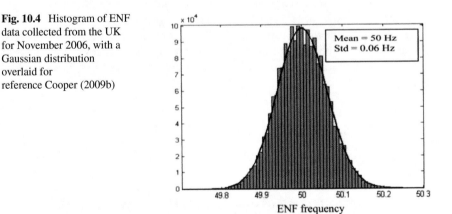

Fig. 10.4 Histogram of ENF data collected from the UK for November 2006, with a Gaussian distribution overlaid for reference Cooper (2009b)

10.4   ENF Traces in the Visual Track

In Garg et al. (2011), the authors showed for the first time that the ENF signal can be sensed and extracted from the light illumination using a customized photodiode circuit. In this section, we explain how the ENF signal can be extracted from video and image signals captured by digital cameras. First, we describe each stage of the pipeline that an ENF signal goes through and is processed. Different types of imaging sensors, namely CCD and CMOS, will be discussed. Second, we explain the steps to extract the ENF signal. We start from simple videos with white-wall scenes and then move to more complex videos with camera motion, foreground motion, and brightness change.

10.4.1   Mechanism of ENF Embedding in Videos and Images

The intensity fluctuation in visual recordings is caused by the alternating pattern of the supply current/voltage. We define *ENF embedding* as the process of adding intensity fluctuations caused by the ENF to visual recordings. The steps of the ENF embedding are illustrated in Fig. 10.5. The process starts by converting the AC voltage/current into a slightly flickering light signal. The light then travels through the air with its energy attenuated. When it arrives at an object, the flickering light interacts with the surface of the object, producing a reflected light that flickers at the same speed. The light continues to travel through the air and the lens and goes through another round of attenuation before it arrives at the imaging sensor of a camera or the retina of a human. Through a short temporal accumulation process that is lowpass in nature, a final sensed signal will contain the flickering component in addition to the pure visual signal. We describe each stage of ENF embedding in Sect. 10.4.1.1.

Depending on whether a camera's image sensor type is CCD or CMOS, the flickering due to the ENF will be captured at the frame level or the row level, respectively. The latter scenario has an equivalent sampling rate that is hundreds or thousands of times that of the former scenario. We provide more details about CCD and CMOS in Sect. 10.4.1.2.

10.4.1.1   Physical Embedding Processes

Given the supply voltage

$$v(t) = A(t) \cos \left(2\pi \int_{-\infty}^{t} f(\tau)d\tau + \phi_0 \right) \tag{10.1}$$

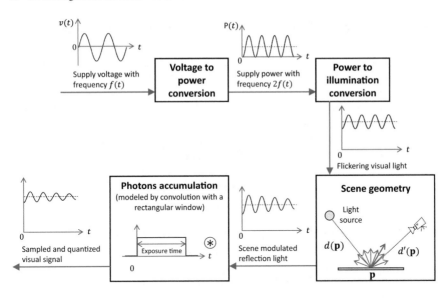

Fig. 10.5 Illustration of the ENF embedding process for visual signals. The supply voltage $v(t)$ with frequency $f(t)$ is first converted into the visual light signal, which has a DC component and an AC component flickering at $2f(t)$ Hz. The visual light then reaches an object and interacts with the object's surface. The reflected light emitted from the object arrives at a camera's imaging sensor at which photons are collected for a duration of the exposure time, resulting in a sampled and quantized temporal visual signal whose intensity fluctuates at $2f(t)$ Hz

with a time-varying voltage amplitude $A(t)$, a time-varying frequency $f(t)$ Hz, and a random initial phase ϕ_0, it follows from the power law that the frequency of supply power is double that the supply voltage,[1] namely

$$P(t) = v^2(t)/R = A^2(t) \cos^2 \left(2\pi \int_{-\infty}^{t} f(\tau)d\tau + \phi_0 \right) / R \qquad (10.2a)$$

$$= \frac{A^2(t)}{2R} \left\{ \cos \left(2\pi \int_{-\infty}^{t} [2f(\tau)]d\tau + 2\phi_0 \right) + 1 \right\}. \qquad (10.2b)$$

Note that for the nonnegative supply power $P(t)$, the ratio of the amplitude of the AC component to the strength of the DC is equal to one. The voltage to power conversion is illustrated by the first block of Fig. 10.5.

Next, the supply power in the electrical form is converted into the electromagnetic wave by a lighting device. The conversion process equivalently applies a lowpass filter to attenuate the relative strength of the AC component based on the optoelectronic principles of the fluorescent or incandescent light. The resulting visual light contains a stable DC component and a relatively smaller flickering AC component with a

---

[1] We simplify the above analysis by assuming that the load is purely resistive but the result applies to inductive and capacitive loads as well.

time-varying frequency $2f(t)$. The power to light conversion is illustrated by the second block of Fig. 10.5.

A ray of the visual light then travels to an object and interacts with the object's surface to produce reflected or reemitted light that can be picked up by the eyes of an observer or a camera's imaging system. The overall effect is a linear scaling to the magnitude of the visual light signal, which preserves its flickering component for the light arriving at the lens. The third block of Fig. 10.5 illustrates a geometric setup where a light source shines on an object, and a camera acquires the reflected light. Given a point **p** on the object, the intensity of reflected light arriving at the camera is dependent on the light–objective distance $d(\mathbf{p})$ and the camera–object distance $d'(\mathbf{p})$ per the *inverse-square law* that light intensity is inversely proportional to the squared distance. The intensity of the reflected light is also determined by the reflection characteristics of the object at **p**. The reflection characteristics can be broadly summarized into the *diffuse reflection* and the *specular reflection*, which are widely adopted in computer vision and computer graphics for understanding and modeling everyday vision tasks (Szeliski 2010). The impact of reflection on the intensity of the reflected light is a combined effect of albedo, the direction of the incident light, the orientation of the object's surface, and the direction of the camera. From the camera's perspective, the light intensities at different locations of the object form an image of the object (Szeliski 2010). Adding a time dimension, the intensities of all locations fluctuate synchronously at the speed of $2f(t)$ Hz due to the changing appearance caused by the flickering light.

The incoming light is further sampled and quantized by the camera's sensing unit to create digital images or videos. Through the imaging sensor, the photons of the incoming light reflected from the scene are converted into electrons and subsequently into voltage levels that represent the intensity levels of pixels of a digital image. Photons are collected for a duration of the *exposure time* to accumulate a sizable mass to clearly depict the scene. Hajj-Ahmad et al. (2016) show that the accumulation of photons can be viewed as convolving a rectangular window to the perceived light signal arriving at the lens as illustrated in the last block of Fig. 10.5. This is a second lowpass filtering process that further reduces the strength of the AC component relative to the DC.

10.4.1.2 Rolling Versus Global Shutter

When images and videos are digitized by cameras, depending on whether the camera uses a *global-shutter* or *rolling-shutter* sensing mechanism, the ENF signal will be embedded into the visual data in different ways. CCD cameras usually capture images using global shutters. It exposes and reads out all pixels of an image/frame simultaneously, hence the ENF is embedded by capturing the global intensity of the image. In contrast, CMOS cameras usually capture images using rolling shutters (Jinwei Gu et al. 2010). A rolling shutter exposes and reads out only one row of pixels at a time, hence the ENF is embedded by sequentially capturing the intensity of every row of an image/frame. The rolling shutter digitizes rows of each image/frame

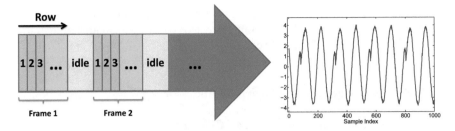

Fig. 10.6 (Left) Rolling-shutter sampling time diagram of a CMOS camera: Rows of each frame are sequentially exposed and read out, followed by an idle period before proceeding to the next frame (Su et al. 2014a). (Right) The row signal of a white-wall video generated by averaging the pixel intensities of each row and concatenating the averaged values for all frames (Su et al. 2014b). The discontinuities are caused by idle periods

sequentially, making it possible to sense the ENF signal much faster than using the global shutter—the effective sampling rate is scaled up by a multiplicative factor that equals the number of rows of sensed images, which is usually on the order of hundreds or thousands. The left half of Fig. 10.6 shows a timing diagram of when each row and each frame are acquired using a rolling shutter. Rows of each frame are sampled uniformly in time, followed by an idle period before proceeding to the next frame (Su et al. 2014a). The right half of Fig. 10.6 shows a *row signal* of a white-wall video (Su et al. 2014b), which can be used for frequency estimation/tracking. The row signal is generated by averaging the pixel intensities of each row and concatenating the averaged values for all frames (Su et al. 2014b). Note that the discontinuities are caused by the missing values during the idle periods.

10.4.2   ENF Extraction from the Visual Track

ENF signals can be embedded in both global-shutter and rolling-shutter videos as explained in Sect. 10.4.1.2. However, being able to successfully extract an ENF signal from a global-shutter video needs a careful selection of a camera's frame rate. This is because global-shutter videos may suffer from frequency contamination due to aliasing caused by an insufficient sampling rate normally ranging from 24 to 30 fps. Two typical failure examples are using a 25-fps camera in an environment with light flickering 100 Hz or using a 30-fps camera with 120 Hz light. We will detail this challenge in Sect. 10.4.2.1.

The rolling shutter, traditionally considered to be detrimental to image qualities (Jinwei Gu et al. 2010), can sample the flickering of the light signal hundreds or thousands of times faster than the global shutter. A faster sampling rate eliminates the need to worry about the ENF's potential contamination due to aliasing.

We will explain the steps to extract the ENF signal starting from simple videos with no visual content in Sect. 10.4.2.2 to complex videos with camera motion, foreground motion, and brightness change in Sect. 10.4.2.3.

10.4.2.1   Challenges of Using Global-Shutter Videos

In Garg et al. (2013), the authors used a CCD camera to capture white-wall videos under indoor lighting to demonstrate the feasibility of extracting ENF signals from the visual track. The white-wall scene can be considered to contain no visual content except for an intensity bias, hence it can be used to demonstrate the essential steps to extract ENF from videos. The authors took the average of the pixel values in each H-by-W frame of the video and obtained a 1-D sinusoid-like time signal $s_{\text{frame}}(t)$ for frequency estimation, namely

$$s_{\text{frame}}(t) = \frac{1}{HW} \sum_{x=0}^{H-1} \sum_{y=0}^{W-1} I(x, y, t), \quad t = 0, 1, 2, \ldots, \quad (10.3)$$

where H is the number of rows, W is the number of columns, I is the video intensity, and x, y, and t are its row, column, and time indices, respectively. Here, the subscript "frame" of the signal symbol implies that its sampling is conducted at the frame level. Figure 10.7(a) and (b) shows the spectrograms calculated from the power signal and the time signal $s_{\text{frame}}(t)$ of frame averaged intensity. It is revealed that the ENF estimated from the video has the same trend as that from the power mains and has doubled dynamic range, confirming the feasibility of ENF extraction from videos.

One major challenge in using a global-shutter-based CCD camera for ENF estimation is the aliasing effect caused by the insufficient sampling rate. Most consumer digital cameras adopt a frame rate of around 30 fps, while the ENF signal and its harmonics appear at integer multiples of 100 120 Hz. The ENF signal, therefore, suffers from a severe aliasing effect due to an insufficient sampling rate. In Garg et al. (2013),

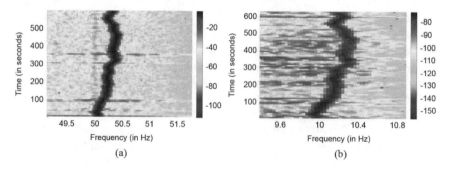

Fig. 10.7 Spectrograms of fluctuating ENF measured from **a** power mains, and **b** frames of a white-wall video recording (Garg et al. 2013)

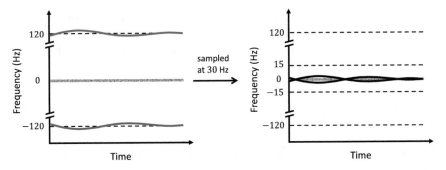

Fig. 10.8 Issue of mixed frequency components caused by aliasing when global-shutter CCD cameras are used. (Left) Time-frequency representation of the original signal containing substantial signal contents around ± 120 Hz and DC (0 Hz); (Right) Mixed/aliased frequency components around DC after being sampled at 30 fps. Two mirrored components, in general, cannot be separated once mixed. Their overlap with the DC component further hinders the estimation of the desired frequency component

the illumination was 100 Hz and the camera's sampling rate was at 30 fps, resulting in an aliased component centered 10 Hz as illustrated in Fig. 10.7(b). Such a combination of ENF nominal frequency and the CCD camera's frame rate does not affect the ENF extraction. However, the 30-fps sampling rate can cause major difficulties in ENF estimation for global-shutter videos captured in 120-Hz ENF countries such as the US. We point out two issues using the illustrations in Fig. 10.8. First, two mirrored frequency components cannot be easily separated once they are mixed. Since the power signal is real-valued, a minored -120 Hz component also exists in its frequency domain as illustrated in the left half of Fig. 10.8. When the frame rate is 30 Hz, both ± 120 Hz components will be aliased to 0 Hz upon sampling, creating a symmetrically overlapping pattern at 0 Hz as shown in the right half of Fig. 10.8. Once the two desired frequency components are mixed, they cannot, in general, be separated without ambiguity hence making the ENF extraction impossible in most cases. Second, to make the matter worse, the native DC content around 0 Hz may further distort the mirrored and aliased ENF components (Garg et al. 2013), which may further hinder the estimation of the desired frequency component.

10.4.2.2 Rolling-Shutter Videos with No Visual Content

To address the challenge of insufficient sampling rate, Garg et al. (2013); Su et al. (2014a); Choi and Wong (2019) exploited the fact that a rolling shutter acquires rows of each frame in a sequential manner, which effectively upscales the sampling rate by a multiplicative factor of the number of rows in a frame. In the rolling-shutter scenario, the authors again use a white-wall video to demonstrate the essential steps of extracting ENF traces. As discussed in Sect. 1.4.1.2, the ENF is embedded by sequentially affecting the row intensities. To extract the ENF traces, it is intuitive to average the similarly affected intensity values within each row to produce one value indicating the impact of ENF at the timestamp at which the row is exposed. More precisely, after averaging, the resulting "video" data is indexed only by row x and time t:

$$I_{\text{row}}(x, t) = \frac{1}{W} \sum_{y=0}^{W-1} I(x, y, t), \quad x = 0, \ldots, H-1, \ t = 0, 1, \ldots. \quad (10.4)$$

Su et al. (2014a) define a 1-D row signal by concatenating $I_{\text{row}}(x, t)$ along time, namely

$$s_{\text{row}}(n) = I_{\text{row}}(n \bmod H, \ \text{floor}(n/H)), \quad n = 0, 1, \ldots \quad (10.5)$$

Using a similar naming convention as in (10.3), the subscript in $s_{\text{row}}(n)$ implies that its sampling is equivalently conducted at the row level, which is hundreds or thousands of times faster than the frame rate. This allows the frequency estimation of ENF to be conducted on a signal of a much higher rate without suffering from potentially mixed signals due to aliasing.

The multi-rate signal analysis (Parishwad 2006) is used in Su et al. (2014a) to analyze the concatenated signal (10.5) and shows that such direct concatenation by ignoring a frame's idle period can result in slight distortion to the estimated ENF traces. To avoid such distortion, Choi and Wong (2019) show that the signal must be concatenated as if it were sampled uniformly in time. That is, the missing sample points due to the idle period need to be filled in with zeros before the concatenation. This zero-padding approach can produce undistorted ENF traces but requires the knowledge of the duration of the idle period ahead of time. The idle period is related to a camera-model specific parameter named *read-out time* (Hajj-Ahmad et al. 2016), which will be discussed in Sect. 10.5.4. The authors in Vatansever et al. (2019) further illustrate how the frequency of the main ENF harmonic is replaced with new ENF components depending on the length of the idle period. Their model reveals that the power of the captured ENF signal is inversely proportional to the idle period length.

10.4.2.3 Rolling-Shutter Videos with Complex Visual Content

In practical scenarios, an ENF signal in the form of light intensity variation coexists with the nontrivial video content reflecting camera motion, foreground motion, and brightness change. To be able to extract an ENF signal from such videos, the general principle is to construct a reasonable estimator for the video content $V(x, y, t)$ and then subtract it from the original video $I(x, y, t)$ in order to single out the light intensity variation due to ENF (Su et al. 2014a, b). The resulting ENF-only residual video $\hat{E}(x, y, t) = I(x, y, t) - \hat{V}(x, y, t)$ can then be used for ENF extraction by following the procedure laid out in Sect. 10.4.2.2. A conceptual diagram is shown in Fig. 10.9. Below, we formalize the procedure to generate an ENF-only residual video using a static-scene example. The procedure is also applicable to videos with more

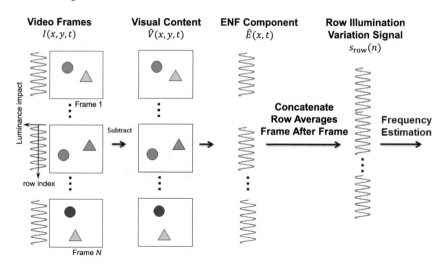

Fig. 10.9 A schematic of how ENF can be extracted from rolling-shutter videos

complex scenes once motion estimation and compensation are conducted to estimate the visual content $V(x, y, t)$. We will also explain how camera motion, foreground motion, and brightness change can be addressed.

ENF-Only Residual Video Generation
Based on the ENF embedding mechanism discussed in Sect. 10.4.1.2, we formulate an additive ENF embedding model[2] for rolling-shutter captured video as follows:

$$I(x, y, t) = V(x, y, t) + E(x, t), \tag{10.6}$$

where $V(x, y, t)$ is the visual content and $E(x, t)$ is the ENF component that depends on row index x and time index t. For a fixed t, the ENF component $E(x, t)$ is a sinusoid-like signal.

Our goal is to estimate $E(x, t)$ given a rolling-shutter captured video $I(x, y, t)$ modeled by (10.6). Once we obtain the estimate $\hat{E}(x, t)$, one can choose to use the direct concatenation (Su et al. 2014a) or the periodic zero padding (Choi and Wong 2019) to generate a 1-D time signal for frequency estimation.

Dangling modifier. An intuitive approach to estimate $E(x, t)$, is to first obtain a reasonable estimate $\hat{V}(x, y, t)$ of the visual content and subtract it from the raw video $I(x, y, t)$, namely

$$\hat{E}(x, y, t) = I(x, y, t) - \hat{V}(x, y, t). \tag{10.7}$$

---

[2] Although a multiplicative model may better describe the optophysical process of how ENF is modulated into the reflected light, the multiplicative model is not very analytically tractable for ENF analysis. The additive model is also more correct when the AC component in the digitized image is relatively weaker than the DC component as pointed out in Sect. 10.4.1.1.

For a static scene video, since the visual contents of every frame are identical, it is intuitive to obtain an estimator by taking the average of a random subset $\mathcal{T} \subset \{0, 1, 2, \ldots\}$ of video frames

$$\hat{V}(x, y, t) = \frac{1}{|\mathcal{T}|} \sum_{t \in \mathcal{T}} I(x, y, t). \tag{10.8}$$

Such a random average can help cancel out the ENF component in (10.6). Once $\hat{E}(x, y, t)$ is obtained, the same procedure of averaging over the column index (Su et al. 2014a) as in (10.4) is conducted to estimate $E(x, t)$, as the pixels of the same row in $\hat{E}(x, y, t)$ are perturbed by the ENF in the same way:

$$\hat{E}(x, t) = \frac{1}{W} \sum_{y=0}^{W-1} \hat{E}(x, y, t) \tag{10.9a}$$

$$= E(x, t) + \frac{1}{W} \sum_{y=0}^{W-1} [V(x, y, t) - \hat{V}(x, y, t)]. \tag{10.9b}$$

When $\sum_{y=0}^{W-1} \hat{V}(x, y, t)$ is a good estimator for $\sum_{y=0}^{W-1} V(x, y, t)$ for all (x, t), the second term is close to zero, making $\hat{E}(x, t)$ a good estimator for $E(x, t)$. To extract the ENF signal, $\hat{E}(x, t)$ is then vectorized into a 1-D time signal as in (10.5), namely

$$s_{\text{row}}(n) = \hat{E}(n \bmod H, \text{floor}(n/H)), \quad n = 0, 1, \ldots \tag{10.10}$$

Camera Motion

For videos with more complex scenes, appropriate video processing tools need to be used to generate the estimated visual content $\hat{V}(x, y, t)$. We first consider the class of videos containing merely camera motions such as panning, rotation, zooming, and shaking. Two frames that are not far apart in time can be regarded as being generated by almost identical scenes projected to the image plane with some offset in the image coordinates. The relationship of the two frames can be established by finding each pixel of the frame of interest at t_0 and its new location in a frame at t_1, namely

$$I\left(x + dx(x, y, t_1), \ y + dy(x, y, t_1), \ t_0\right) \approx I(x, y, t_1), \tag{10.11}$$

where (dx, dy) is the motion vector associated with the frame of interest pointing to the new location in the frame at t_1. Motion estimation and compensation are carried out in Su et al. (2014a), Su et al. (2014b) using optical flow to obtain a set of estimated frames $\hat{I}_i(x, y, t_0)$ for the frame of interest $I(x, y, t_0)$. As the ENF signal has a relatively small magnitude compared to the visual content and the motion compensation procedure often introduces noise, an average over all these motion-compensated frames, namely

$$\hat{V}(x, y, t) = \frac{1}{n} \sum_{i=1}^{n} \hat{I}_i(x, y, t), \tag{10.12}$$

should lead to a good estimation of the visual content per the law of large numbers.

Foreground Motion
We now consider the class of videos with foreground motions only. In this case, the authors in Su et al. (2014a), Su et al. (2014b) use a video's static regions to estimate the ENF signal. Given two image frames, the regions that are not affected by foreground motion in both frames are detected by thresholding the pixel-wise differences. The row averages of the static regions are then calculated to generate a 1-D time signal for ENF estimation.

Brightness Compensation
Many cameras are equipped with an automatic brightness control unit that would adjust a camera's sensitivity in response to the changes of the global or local light illumination. For example, as a person in a bright background moves closer to the camera, the background part of the image can appear brighter when the control unit relies on the overall intensity of the frame to adjust the sensitivity (Su et al. 2014b). Such a brightness change can bias the estimated visual content and lead to a failure in the content elimination process.

Su et al. (2014b) found that the intensity values of two consecutive frames roughly follow the linear relationship

$$I(x, y, t) \approx \beta_1(t)I(x, y, t + 1) + \beta_0(t), \quad \forall(x, y), \tag{10.13}$$

where $\beta_1(t)$ is a slope and $\beta_0(t)$ is an intercept for the frame pair $(I(x, y, t), I(x, y, t + 1))$. By estimating the parameters, background intensity values of different frames can be matched to the same level, which allows a precise visual content estimation and elimination.

10.4.3   ENF Extraction from a Single Image

In an extension to ENF extraction from videos, Wong et al. (2018) showed that ENF can even be extracted from a single image taken by a camera with a rolling shutter. As with rolling-shutter videos, each row of an image is sequentially acquired and contains the temporal flickering component due to the ENF. The acquisition of all rows within an image happens during the frame period that is usually 1/30 or 1/25 s based on the common frame rates, which is too short to extract a meaningful temporal ENF signal for most ENF-based analysis tasks. However, an easier binary classification problem can be answered (Wong et al. 2018): Is it possible to tell whether the capturing geographic region of an image has 50 Hz or 60 Hz supply frequency?

There are several unsolved research challenges in ENF from images. For example, the proposed ENF classifier in Wong et al. (2018) works well for images with synthetically added ENF, but its performance on real images may be further improved by incorporating a more sophisticated physical embedding model. Moreover, instead of making a binary decision, one may construct a ternary classifier to test whether ENF is present in a given image. Lastly, one may take advantage of a few frames captured in a continuous shooting mode to estimate the rough shape of the ENF signal for authentication purposes.

10.5 Key Applications in Forensics and Security

The ENF signal has been shown to be a useful tool in solving several problems that are faced in digital forensics and security. In this section, we highlight major ENF-based forensic applications discussed in the ENF literature in recent years. We start with joint time–location authentication, and proceed to discuss ENF-based tampering detection, ENF-based localization of media signals, and ENF-based approaches for camera forensics.

10.5.1 Joint Time–Location Authentication

The earliest works on ENF-based forensic applications have focused on ENF-based time-of-recording authentication of audio signals (Grigoras 2005, 2007; Cooper 2008; Kajstura et al. 2005; Brixen 2007a; Sanders 2008). The ENF pattern extracted from an audio signal should be very similar to the ENF pattern extracted from a power reference recording simultaneously recorded. So, if a power reference recording is available from a claimed time-of-recording of an audio signal, the ENF pattern can be extracted from the audio signal and compared against the ENF pattern from the power recording. If the extracted audio ENF pattern is similar to the reference pattern, then the audio recording's claimed timestamp is deemed authentic.

To discern the measure of similarity between the two ENF patterns, the minimum mean squared error (MMSE) or Pearson's correlation coefficient can be used. Certain modifications to these matching criteria have been proposed in the literature.

One such modification to the matching criteria was proposed in Garg et al. (2012) where the authors exploit their findings that the US ENF signal can be modeled as a piecewise linear AR process to achieve better matching in ENF-based time-of-recording estimation. In this work, the authors extract the *innovation signals* resulting from the AR modeling of the ENF signals, and use these innovation signals for matching instead of the original ENF signals. Experiments done under a hypothesis detection framework show that this approach provides higher confidence in time-of-recording estimation and verification.

Other matching approaches have been proposed as well. For instance, in Hua (2014), the authors proposed a new threshold-based dynamic matching algorithm, termed the error correction matching algorithm (ECM), to carry out the matching. Their approach accounts for noise affecting ENF estimates due to limited frequency resolution. In their paper, they illustrate the advantages of their proposed approach using both synthetic and real signals. The performance of this ECM approach, along with a simplified version of it termed the bitwise similarity matching (BSM) approach, is compared against the conventional MMSE matching criterion in Hua (2018). The authors find that due to the complexity of practical situations, the performance of the ECM and BSM approaches over the benchmark MMSE approach may not be guaranteed. However, in the situations examined in the paper, the finding is that ECM results in the most accurate matching results while BSM sacrifices matching accuracy for processing speed.

In Vatansever et al. (2019), the authors propose a time-of-recording authentication approach tailored to ENF signals extracted from videos taken with cameras using rolling shutters. As mentioned in Sect. 10.4.1.2, such cameras typically have an idle period between each frame, thus resulting in missed ENF samples in the resultant video. In Vatansever et al. (2019), and based on multiple idle period assumptions, missing illumination periods in each frame are interpolated to compensate for the missing samples. Each interpolated time series then yields an ENF signal that can be matched to the ground-truth ENF reference through correlation coefficients to find or verify the time-of-recording.

An example of a real-life case where ENF-based time-of-recording authentication was used was described in Kajstura et al. (2005). In 2003, the Institute of Forensic Research in Cracow, Poland, was asked to investigate a 55-min long recording of a conversation between two businessmen made using a Sony portable digital recorder (model: ICD-MS515). The time of the creation of the recording indicated by the evidential recorder differed by close to 196 days from the time reported by witnesses of the conversation. Analysis of the audio recording revealed the presence of ENF traces. The ENF signal extracted from the audio recording was compared against reference ENF signals provided by the power grid operator, and it was revealed that the true time of the creation of the recording was the one reported by the witnesses. In this case, the date/time setting on the evidential recorder at the time of recording was incorrect.

A broader way to look at this particular ENF forensics application is not just as a time-of-recording authentication but as a joint time–location authentication. Consider the case where a forensic analyst is asked to verify, or discover, the correct time-of-recording and the location-of-recording of a media recording. In addition to authenticating the time-of-recording, the analyst may as well be able to identify the location-of-recording on a power grid level. Provided the analyst has access to ENF reference data from candidate times-of-recording and candidate grids-of-origin, comparing the ENF pattern extracted from the media recording to the reference ENF patterns extracted from candidate time/location ENF patterns should point to the correct time-of-recording and grid-of-origin of the media recording.

This application was the first that piqued the interest of the forensics community in the ENF signal. Yet, wide application in practical settings still faces several challenges.

First, the previous discussion has assumed that the signal has not been tampered with. If it had been tampered with, the previously described approach may not be successful. Section 10.5.2 describes approaches to verify the integrity of a media recording using its embedded ENF.

Second, depending on the initial information available about the recording, exhaustively matching the extracted media ENF pattern against all possible ENF patterns may be too expensive. An example of a recent work addressing this challenge is Pop et al. (2017), where the authors proposed computing a bit sequence encoding the trend of an ENF signal when it is added into the ENF database. When presented with a query ENF signal to be timestamped, its trend bit sequence is computed and compared to the reference trend bit sequences as a way to prune its candidate timestamps, and reduce the final number of reference ENF signals it would need to be compared against using conventional approaches such as MMSE and correlation coefficient.

Third, this application assumes the availability of reference ENF data from the time and location of the media recording under study. In the case where reference data is incomplete or unavailable, a forensic analyst will need to resort to other measures to carry out the analysis. In Sect. 10.5.3, one such approach is described, which infers the grid-of-origin of a media recording without the need for concurrent reference ENF data.

10.5.2 Integrity Authentication

Audio forgery techniques can be used to conduct piracy over the Internet, falsify court evidence, or modify security device recordings or recordings of events taking place in different parts of the world (Gupta et al. 2012). This is especially relevant today with the prevalent use of social media by people who add on to news coverage of global events by sharing their own videos and recordings of what they experienced.

In some cases, it is instrumental to be able to ascertain whether recordings are authentic or not. Transactions involving large sums of money, for example, can take place over the phone. In such cases, it is possible to change the numbers spoken in a previously recorded conversation and then replay the message. If the transaction is later challenged in court, the forged phone conversation can be used as an alibi by the accused to argue that the account owner did authorize the transaction (Gupta et al. 2012). This would be a case where methods to detect manipulation in audio recordings are necessary. In this section, we discuss the efforts made to use the captured ENF signal in audio recordings for integrity authentication.

The basic idea of ENF-based authentication is that if an ENF-containing recording has been tampered with, the changes made will affect the extracted ENF signal as well. Examples of tampering include removing parts of the recording and inserting parts that do not belong. Examining the ENF of a tampered signal would reveal

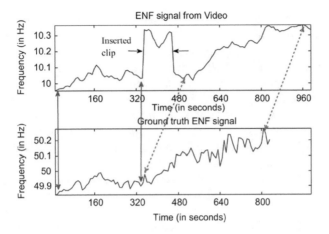

Fig. 10.10 ENF matching result demonstrating video tampering detection based on ENF traces (Cooper 2009b)

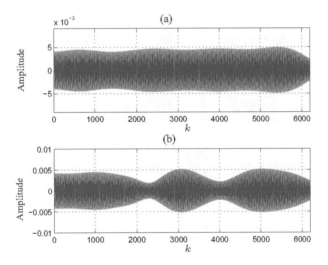

Fig. 10.11 Audio signals from Brazil bandpassed around the 60 Hz value, **a** filtered original signal, and **b** filtered edited signal (Nicolalde and Apolinario 2009)

discontinuities in the extracted ENF signal that raise the question of tampering. The easiest scenario in which to check for tampering would be the case where we have reference ENF available. In such a case, the ENF signal can be compared to the reference ENF signal from the recording's claimed time and location. This comparison would either support or question the integrity of the recording. Figure 10.10 shows an example where comparing an ENF signal extracted from a video recording with its reference ENF signal reveals the insertion of a foreign video piece.

The case of identifying tampering in ENF-containing recordings with no reference ENF data becomes more complicated. The approaches studied in the literature have focused on detecting ENF tampering based on ENF phase and/or magnitude changes. Generally speaking, when an ENF-containing recording has been tampered with, it is highly likely to find discontinuities in the phase of the ENF at regions where the tampering took place.

In Nicolalde and Apolinario (2009), the authors noticed that phase changes in the embedded ENF signal result in a *modulation effect* when the recording containing the ENF traces is bandpassed about the nominal ENF band. An example of this is shown in Fig. 10.11, where there are visible differences in the relative amplitude in an edited signal versus in its original version. Here, the decrease in amplitude occurs at locations where the authors had introduced edits into the audio signal.

Plotting the phase of an ENF recording will also reveal changes indicating tampering. In Rodríguez et al. (2010), the authors describe their approach to compute the phase that goes as follows. First, a signal is downsampled and then passed through a bandpass filter centered around the nominal ENF value. If no ENF traces are found at the nominal ENF, due to attacks aimed at avoiding forensics analysis, for instance (Chuang et al. 2012; Chuang 2013), bands around the higher harmonics of the nominal ENF can be used instead (Rodriguez 2013). Afterward, the filtered signal is divided into overlapping frames of N_C cycles of the nominal ENF, and the phase of each segmented frame is computed using DFT or using a high-precision Fourier analysis method called DFT[1] (Desainte-Catherine and Marchand 2000). When the phase of the signal is plotted, one can not only visually inspect for tampering but can also glean insights as to whether the editing was the result of a fragment deletion or of a fragment insertion, as can be seen in Fig. 10.12.

Rodríguez et al. (2010) also proposed an automatic approach for discriminating between an edited signal and an unedited one, which follows from the phase estimation just described. The approach depends on a statistic F defined as

$$F = 100 \log \left(\frac{1}{N-1} \sum_{n=2}^{N} \left[\phi(n) - m_\phi \right]^2 \right), \qquad (10.14)$$

where N is the number of frames used for phase estimation, $\phi(n)$ denotes the estimate phase of frame n, and m_ϕ is the average of the computed phases. The process of determining whether or not an ENF-containing recording is authentic is formulated under a detection framework with the null hypothesis H_0 and the alternative hypothesis H_1 denoting an original signal and an edited signal, respectively. If F is greater than a threshold γ, then the alternative hypothesis H_1 is favored; otherwise, the null hypothesis H_0 is favored.

For optimal detection, the goal is to obtain a value for the threshold γ that maximizes the value of the probability of detection $P_D = \Pr (F > \gamma \mid H_1)$. To do so, the authors in Rodríguez et al. (2010) prepare a *corpus* of audio signals, including original and edited signals, and evaluate the database with the proposed automatic approach with a range of γ values. The value chosen for γ is the one it takes at the

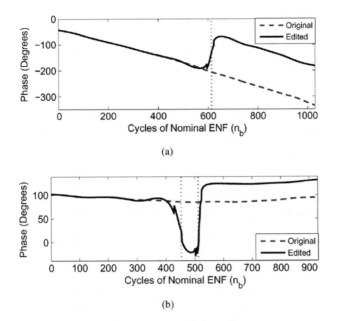

Fig. 10.12 Phase estimated using DFT from edited Spanish audio signals where the edits were **a** fragment deletion and **b** fragment insertion (Rodríguez et al. 2010)

equal error rate (EER) point where the probability of miss $P_M = 1 - P_D$ is equal to the probability of false alarm $P_{FA} = \Pr(F > \gamma \mid H_0)$. Experiments carried out on the Spanish databases AHUMADA and GAUDI resulted in an EER of 6%, and experiments carried out on Brazilian databases Carioca 1 and Carioca 2 resulted in an EER of 7%.

Esquef et al. (2014), the authors proposed a novel edit detection method for forensic audio analysis that includes modification on their approach in Rodríguez et al. (2010). In this version of the approach, the detection criterion is based on unlikely variations of the ENF magnitude. A data-driven magnitude threshold is chosen in a hypothesis framework similar to the one in Rodríguez et al. (2010). The authors conduct a qualitative evaluation of the influences of the edit duration and location as well as noise contamination on the detection ability. This new approach achieved a 4% EER on the Carioca 1 database versus 7% EER on the same database for the previous approach of Rodríguez et al. (2010). The authors report that their results indicate that amplitude clipping and additive broadband noise severely affect the performance of the proposed method, and further research is needed to improve the detection performance in more challenging scenarios.

The authors further extended their work in Esquef et al. (2015). Here, they modified the detection criteria by taking advantage of the typical pattern of ENF variations elicited by audio edits. In addition to the threshold-based detection strategy of Esquef et al. (2014), a verification of the pattern of the anomalous ENF variations is carried

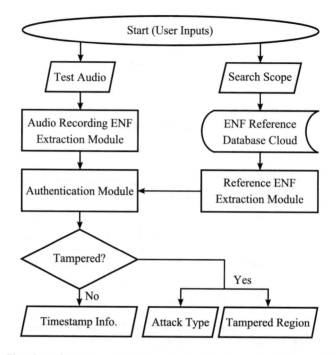

Fig. 10.13 Flowchart of the proposed ENF-based authentication system in Hua et al. (2016)

out, making the edit detector less prone to false positives. The authors confronted this newly proposed approach with that of Esquef et al. (2014) and demonstrated experimentally that the new approach is more reliable as it tends to yield lower EERs such as a reduction from 4% EER to 2% EER on the Carioca 1 database. Experiments carried out on speech databases degraded with broadband noise also support the claim that the modified approach can achieve better results than the original one.

Other groups have also addressed and extended the explorations into ENF-based integrity authentication and tampering detection. We mention some highlights in what follows.

In Fuentes et al. (2016), the authors proposed a phase locked loop (PLL)-based method for determining audio authenticity. They use a voltage-controlled oscillator (VCO) in a PLL configuration that produces a synthetic signal similar to that of a preprocessed ENF-containing signal. Some corrections are made to the VCO signal to make it closer to the ENF, but if the ENF has strong phase variations, there will remain large differences between the ENF signal and the VCO signal, signifying tampering. An automatic threshold-based decision on the audio authenticity is made by quantifying the frequency variations of the VCO signal. The experimental results shown in this work show that the performance of the proposed approach is on par with that of previous works (Rodríguez et al. 2010; Esquef et al. 2014), achieving, for instance, 2% EER on the Carioca 1 database.

In Hua et al. (2016), the authors present a solution to the ENF-based audio authentication system that jointly performs timestamp verification (if tampering is not detected) and detection of tampering type and tampering region (if tampering is detected). A high-level description of this system is shown in Fig. 10.13. The authentication tool here is an *absolute error map* (AEM) between the ENF signal under study and reference ENF signals from a database. The AEM is quantitatively a matrix containing all the raw information of absolute errors where the extracted ENF signal is matched to each possible shift of a reference signal. The AEM-based solutions rely coherently on the authentic and trustworthy portions of the ENF signal under study to make their decisions on authenticity. If the signal is deemed authentic, the system returns the timestamp at which a match is found. Otherwise, the authors propose two variant approaches that analyze the AEM to find the tampering regions and characterize the tampering type (insertion, deletion, or splicing). The authors frame their work here as a proof-of-concept study, demonstrating the effectiveness of their proposal by synthetic performance analysis and experimental results.

In Reis et al. (2016), Reis et al. (2017), the authors propose their ESPRIT-Hilbert-based tampering detection scheme with SVM classifier (SPHINS) framework for tampering detection, which uses an ESPRIT-Hilbert ENF estimator in conjunction with an outlier detector based on the sample kurtosis of the estimated ENF. The computed kurtosis values are vectorized and applied to an SVM classifier to indicate the presence of tampering. They report a 4% EER performance on the clean Carioca 1 database and that their proposed approach gives improved results, as compared to those in Rodríguez et al. (2010); Esquef et al. (2014), for low SNR regimes and in scenarios with nonlinear digital saturation, when applied to the Carioca 1 database.

In Lin and Kang (2017), the authors propose an approach where they apply a wavelet filter to an extracted ENF signal to reveal the detailed ENF fluctuations and then carry out autoregressive (AR) modeling on the resulting signal. The AR coefficients are used as input features for an SVM system for tampering detection. They report that, as compared to Esquef et al. (2015), their approach can achieve improved performance in noisy conditions and can provide robustness against MP3 compression.

In the same vein as ENF-based integrity authentication, Su et al. (2013); Lin et al. (2016) address the issue of recaptured audio signals. The authors in Su et al. (2013) demonstrate that recaptured recordings may contain two sets of ENF traces, one from the original time of the recording and the other due to the recapturing process. They tackle the more challenging scenario that the two traces overlap in frequency by proposing a decorrelation-based algorithm to extract the ENF traces with the help of the power reference signal. Lin et al. (2016) employ a convolutional neural network (CNN) to decide if a certain recording is recaptured or original. Their deep neural network relies on spectral features based on the nominal ENF and its harmonics. Their paper goes into further depth, examining the effect of the analysis window on the approach's performance and visualizing the intermediate feature maps to gain insight on what the CNN learns and how it makes its decisions.

10.5.3 ENF-Based Localization

The ENF traces captured by audio and video signals can be used to infer information about the location in which the media recording was taken. In this section, we discuss approaches to use the ENF to do *inter-grid localization*, which entails inferring the grid in which the media recording was made, and *intra-grid localization*, which entails pinpointing the location-of-recording of the media signal within a grid.

10.5.3.1 Inter-Grid Localization

This section describes a system that seeks to identify the grid in which an ENF-containing recording was made, without the use of concurrent power references Hajj-Ahmad et al. (2013), Hajj-Ahmad et al. (2015). Such a system can be very important for multimedia forensics and security. It can pave the way to identify the origins of videos such as those of terrorist attacks, ransom demands, and child exploitation. It can also reduce the computational complexity and facilitate other ENF-based forensics applications, such as the time-of-recording authentication application of Sect. 10.5.1. For instance, if a forensic analyst is given a media recording of unknown time and location information, inferring the grid-of-origin of the recording would help in narrowing down the set of power reference recordings that need to be used to compare the media ENF pattern.

Upon examining ENF signals from different power grids, it can be noticed that there are differences between them in the nature and the manner of the ENF variations (Hajj-Ahmad et al. 2015). These differences are generally attributed to the control mechanisms used to regulate the frequency around the nominal value and the size of the power grid. Generally speaking, the larger the power grid is, the smaller the frequency variations are. Figure 10.14 shows the 1-min average frequency during a 48-hour period from five different locations in five different grids. As can be seen in this figure, Spain, which is part of the large continental European grid, shows a small range of frequency values as well as fast variations. The smallest grids of Singapore and Great Britain show relatively large variations in the frequency values (Bollen and Gu 2006).

Given an ENF-containing media recording, a forensic analyst can process the captured ENF signal to extract its statistical features to facilitate the identification of the grid in which the media recording was made. A machine learning system can be built that learns the characteristics of ENF signals from different grids and uses it to classify ENF signals in terms of their grids-of-origin. Hajj-Ahmad et al. (2015) have developed such a machine learning system. In that implementation, ENF signal segments are extracted from recordings 8 min long each. Statistical, wavelet-based, and linear-predictive-related features are extracted from these segments and used to train an SVM multiclass system.

The authors make a distinction between "clean" ENF data extracted from power recordings and "noisy" ENF data extracted from audio recordings, and various com-

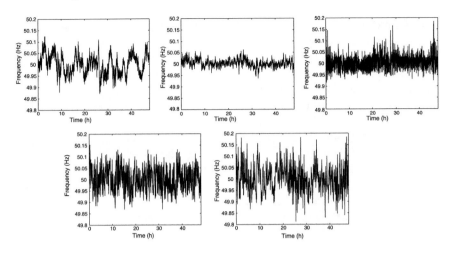

Fig. 10.14 Frequency variations measured in Sweden (top left), in Spain (top center), on the Chinese east coast (top right), in Singapore (bottom left), and in Great Britain (bottom right) (Bollen and Gu 2006)

binations of these two sets of data are used in different training scenarios. Overall, the authors were able to achieve an average accuracy of 88.4% on identifying ENF signals extracted from power recordings from eleven candidate power grids, and an average accuracy of 84.3% on identifying ENF signals extracted from audio recordings from eight candidate grids. In addition, the authors explored using multi-conditional systems that can adapt to cases where the noise conditions of the training and testing data are different. This approach was able to improve the identification accuracy of noisy ENF signals extracted from audio recordings by 28% when the training dataset is limited to clean ENF signals extracted from power recordings.

This ENF-based application was the focus of the 2016 Signal Processing Cup (SP Cup), an international undergraduate competition overseen by the IEEE Signal Processing Society. The competition engaged participants from nearly 30 countries. 334 students formed 52 teams that registered for the competition; among them, more than 200 students in 33 teams turned in the required submissions by the open competition deadline in January 2016. The top three teams from the open competition attended the final stage of the competition at ICASSP 2016 in Shanghai, China, to present their final work (Wu et al. 2016).

Most student participants from the SP Cup have uploaded their final reports to SigPort (Signal Processing 2021). A number of them have also further developed their works and published them externally. A common improvement on the SVM approach originally proposed in Hajj-Ahmad et al. (2015) was to make the classification system a multi-stage one, with earlier stages distinguishing between 50 Hz grid versus 60 Hz grid, or an audio signal versus a power signal. An example of such a system is proposed in Suresha et al. (2017).

In Šarić et al. (2016), a team from Serbia examined different machine learning algorithms that can be used to achieve inter-grid localization, including K-nearest neighbors, random forests, SVM, linear perceptron, and neural networks. In their setup, the classifier that achieved the highest accuracy overall was random forest. They also explored adding additional features to those of Hajj-Ahmad et al. (2015), related to the extrema and rising edges in the ENF signal, which showed performance improvements between 3% and 19%.

Aside from SP-Cup-related works, Jeon et al. (2018) demonstrated how their ENF map proposed in Jeon et al. (2018) can be utilized for inter-grid location identification as part of their LISTEN framework. Their experimental setup included identifying grid-of-origin of audio streams extracted from Skype and Torfan from seven power grids, achieving classification accuracies in the 85–90% range for audio segments of length 10–40 mins. Their proposed system also tackled intra-grid localization, which we will discuss next.

10.5.3.2   Intra-Grid Localization

Though the ENF variations at different points in the same grid are very similar, research has shown that there can be discernible differences between these variations. The differences can be due to the local load characteristics of a given city and the time needed to propagate a response to demand and supply in the load to other parts of the grid (Bollen and Gu 2006). System disturbances, such as short circuits, line switching, and generator disconnections, can be contributing causes to these different ENF values (Elmesalawy and Eissa 2014). When a significant change in the load occurs somewhere in the grid, it has a localized effect on the ENF in the given area. This change in the ENF then propagates across the grid at, typically, a speed of approximately 500 ms (Tsai et al. 2007).

In Hajj-Ahmad et al. (2012), the authors conjecture that small and large changes in the load may cause location-specific signatures in local ENF patterns. Following this conjecture, such differences may be exploited to pinpoint the location-of-recording of a media signal within a grid. Due to the finite propagation speed of frequency disturbances across the grid, the ENF signal is anticipated to be more similar for locations close to each other as compared to locations that are farther apart. As an experiment, concurrent recordings of power signals were done in three cities in the Eastern North American grid shown in Fig. 10.15a.

Examining the ENF signals extracted from the different power recordings, the authors found that they have a similar general trend but possible microscopic differences. To better capture these differences, the extracted ENF signals are passed through a highpass filter, and the correlation coefficient between the highpassed ENF signals is used as a metric to examine the location similarity between recordings. Figure 10.15b shows the correlation coefficients between various 500 s ENF segments from different locations at the same time. We can see that the closer the two city-of-origins are, the more similar their ENF signals are, and the farther they are apart, the less similar their ENF signals are. From here, one can start to think about

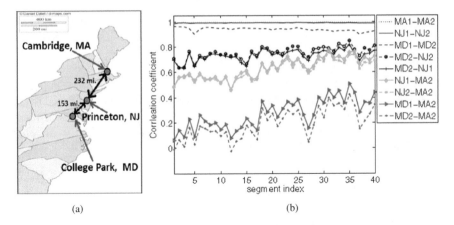

Fig. 10.15 **a** Locations of power recordings. **b** Pairwise correlation coefficients between 500 s long, highpass-filtered ENF segments extracted from power recordings made in the locations shown in **a** (Hajj-Ahmad et al. 2012)

using the ENF signal as a location stamp and exploiting the relation between pairwise ENF similarity and distance between locations of recording. A highpass-filtered ENF query can be compared to highpass-filtered ENF signals from known city anchors in a grid as a means toward pinpointing the location-of-recording of an ENF query.

The authors further examined the use of the ENF signal as an intra-grid location stamp in Garg et al. (2013), where they proposed a half-plane intersection method to estimate the location of ENF-containing recordings and found that the localization accuracy can be improved with the increase in the number of locations used as anchor nodes. This study conducted experiments on power ENF signals. With audio and video signals, the situation is more challenging because of the noisy nature of the embedded ENF traces. The location-specific signatures are best captured using instantaneous frequencies estimated at 1 s temporal resolution, and reliable ENF signal extraction at such a high temporal resolution is an ongoing research problem. Nevertheless, there is a potential of being able to use ENF signals as a location stamp.

Jeon et al. (2018) have included intra-grid localization as part of the capabilities of their LISTEN framework for inferring the location-of-recording of a target recording, which relies on an ENF map built using ENF traces collected from online streaming sources. After narrowing down the power grid to which a recording belongs through inter-grid localization, their intra-grid localization approach relies on calculating the Euclidean distance between a time-series sequence of interpolated signals in the chosen power grid and the target recording's ENF signal. The approach then narrows down the location of recording to a specific region within the grid.

In Chai et al. (2016); Yao et al. (2017); Cui et al. (2018), the authors rely on the FNET/GridEye system, described in Sect. 10.2.1, to achieve intra-grid localization. In these works, the authors work on ENF-based intra-grid localization for ENF signals extracted from clean power signals without using concurrent ENF power references. Such intra-grid localization is made possible by exploiting geolocation-specific temporal variations induced by electromechanical propagation, nonlinear load, and recurrent local disturbance (Yao et al. 2017).

In Cui et al. (2018), the authors apply features similar to those proposed in intra-grid localization approaches and rely on a random forest classifier for training. In Yao et al. (2017), the authors apply wavelet analysis to ENF signals and feed detail signals into a neural network for training. The results show that the system works better at larger geographical scales and when the training and testing ENF-containing recordings were recorded closer together in time.

10.5.4 ENF-Based Camera Forensics

An additional ENF-based application that has been proposed uses the ENF traces captured in a video recording to characterize the camera that was used to produce the video (Hajj-Ahmad et al. 2016). This is done within a nonintrusive framework that only analyzes the video at hand to extract a characterizing inner parameter of the camera used. The focus was on CMOS cameras equipped with rolling shutters, and the parameter estimated was the read-out time T_{ro}, which is the time it takes for the camera to acquire the rows of a single frame. T_{ro} is typically not listed in a camera's user manual and is usually less than the frame period equal to the reciprocal of a video's frame rate.

This work was inspired by prior work on flicker-based video forensics that addresses issues in the entertainment industry pertaining to movie piracy-related investigations (Baudry et al. 2014; Hajj-Ahmad 2015). The focus of that work was on pirated videos that are produced by camcording video content displayed on an LCD screen. Such pirated videos commonly exhibit an artifact called the flicker signal, which results from the interplay between the backlight of the LCD screen and the recording mechanism of the video camera. In Hajj-Ahmad (2015), this flicker signal is exploited to characterize the LCD screen and camera producing the pirated video by estimating the frequency of the screen's backlight signal and the camera's read-out time T_{ro} value.

Both the flicker signal and the ENF signal are signatures that can be intrinsically embedded in a video due to the camera's recording mechanism and the presence of a signal in the recording environment. For the flicker signal, the environmental signal is the backight signal of the LCD screen, while for the ENF signal, it is the electric lighting signal in the recording environment. In Hajj-Ahmad et al. (2016), the authors leverage the similarities between the two signals to adapt the flicker-based approach to an ENF-based approach targeted at characterizing the camera producing the ENF-containing video. The authors carried out an experiment involving ENF-containing videos produced using five different cameras, and the results showed that the proposed approach achieves high accuracy in estimating the discriminating T_{ro} parameter with a relative estimation error within 1.5%.

Vatansever et al. (2019) further extend the work on the application by proposing an approach that can operate on videos that the approach in Hajj-Ahmad et al. (2016) cannot address, which are cases where the main ENF component is aliased 0 Hz when the nominal ENF is a multiple of the video camera's frame rate, e.g., a video camera with 25 fps capturing 100 Hz ENF. The approach examines the two strongest frequency components at which the ENF component appears, following a model proposed in the same paper, and deduces the read-out time T_{ro} based on the ratio of strength between these two components.

10.6 Anti-Forensics and Countermeasures

ENF signal-based forensic investigations such as time–location authentication, integrity authentication, and recording localization discussed in Sect. 10.5 rely on the assumption that the ENF signal buried in a hosting signal is not maliciously altered. However, there may exist adversaries performing anti-forensic operations to mislead forensic investigations. In this section, we examine the interplay between forensic analysts and adversaries to better understand the strengths and limitations of ENF-based forensics.

10.6.1 Anti-Forensics and Detection of Anti-Forensics

Anti-forensic operations by adversaries aim at altering the ENF traces to invalidate or mislead the ENF analysis. One intuitive approach is to *superimpose* an alternative ENF signal without removing the native ENF signal. This approach can confuse the forensic analyst when the two ENF traces are overlapping and have comparable strengths, under which it is impossible to separate the native ENF signal without using side information (Su et al. 2013; Zhu et al. 2020). A more sophisticated approach is to *remove* the ENF traces either in the frequency domain by using a bandstop filter around the nominal frequency (Chuang et al. 2012; Chuang 2013) or in the time domain by subtracting a time signal synthesized via frequency modulation based on some accurate estimates of the ENF signal such as those from the power mains (Su et al. 2013). The most malicious approach is to mislead the forensic analyst by *replacing* the native ENF signal with some alternative ENF signal (Chuang et al. 2012; Chuang 2013). This can be done by first removing the native signal and then adding an alternative ENF signal.

Forensic analysts are motivated to devise ways to detect the use of anti-forensic operations (Chuang et al. 2012; Chuang 2013), so that the forged signal can be rejected or be recovered to allow forensic analysis. In case an anti-forensic operation for ENF is carried out merely around the fundamental frequency, the consistency of the ENF components from multiple harmonic frequency bands can be used to detect such operation. The abrupt discontinuity in a spectrogram may be another indicator for an anti-forensic operation. In Su et al. (2013), the authors formulated a composite hypothesis testing problem to detect the operation of superimposing an alternative ENF signal.

Adversaries may respond to forensic analysts by improving their anti-forensic operations (Chuang et al. 2012; Chuang 2013). Instead of replacing the ENF component at the fundamental frequency, adversaries can attack all harmonic bands as well. The abrupt discontinuity check for spectrogram can be addressed by envelope adjustment via the Hilbert transform. Such dynamic interplay continues by evolving their actions in response to each other's actions (Chuang et al. 2012; Chuang 2013), and actions tend to become more complex as the interplay goes on.

10.6.2 Game-Theoretic Analysis on ENF-Based Forensics

Chuang et al. (2012), Chuang (2013) quantitatively analyzed the dynamic interplay between forensic analysts and adversaries using the game-theoretic framework. A representative but simplified scenario that involves different actions was quantitatively evaluated. The optimal strategies in terms of Nash equilibrium, namely the status in which no player can increase his/her own benefit (or formally, the *utility*) via unilateral strategy changes, were derived. The optimal strategies and the resulting forensic and anti-forensic performances at the equilibrium can lead to a comprehensive understanding of the capability of ENF-based forensics in a specific scenario.

10.7 Applications Beyond Forensics and Security

The ENF signal as an intrinsic signature of multimedia recordings gives rise to applications beyond forensics and security. For example, an ENF signal's unique fluctuations can be used as a time–location signature to allow the synchronization of multiple pieces of media signals that overlap in time. This allows the synchronized signal pieces to jointly reveal more information than the individual pieces (Su et al. 2014c, b; Douglas et al. 2014). In another example, the slowly varying trend of the ENF signal can serve as an anchor for the detection of abnormally recorded segments within a recording (Chang and Huang 2010; Feng-Cheng Chang and Hsiang-Cheh Huang 2011) and can thus allow for the restoration of tape recordings suffering from irregular motor speed (Su 2014).

10.7.1 Multimedia Synchronization

Conventional multimedia synchronization approaches rely on passive/active calibration of the timestamps or identifying common *contextual information* from two or more recordings. Voice and music are the contextual information that can be used for audio synchronization. Overlapping visual scenes, even recorded from different viewing angles, can be exploited for video synchronization. ENF signals naturally embedded in the audio and visual tracks of recordings can complement conventional synchronization approaches as they would not rely on common contextual information.

ENF-based synchronization can be categorized into two major scenarios, one where multiple recordings for synchronization overlap in time, and one where they do not. If multiple recordings from the same power grid are recorded with time overlap, then they can be readily synchronized by the ENF without using other information. In Su et al. (2014c), Douglas et al. (2014), Su et al. (2014b), the authors use the ENF signals extracted from audio tracks for synchronization. For videos containing both the visual and audio tracks, it is possible to use either modality as the source of ENF traces for synchronization. Su et al. (2014b) demonstrate using the ENF signals from the visual track for synchronization. Using the visual track for multimedia synchronization is important for such scenarios as video surveillance cases where an audio track may not be available.

In Golokolenko and Schuller (2017), the authors show that the idea can be extended to synchronizing the audio streams from wireless/low-cost USB sound cards in a microphone array. Sound cards have their own respective sample buffers, which fill up at different rates, thus leading to synchronization mismatches of streams by different sound cards. Conventional approaches to address the synchronization require specialized hardware, which is not a requirement for a synchronization approach based on using the captured ENF traces.

If recordings were not captured with time overlap, then reference ENF databases are needed to help determine the absolute recording time of each piece. This falls back to the time-of-recording authentication scenario discussed in Sect. 10.5.1—Each recording will need to determine its timestamp by matching against one or multiple reference databases, depending on whether the source location of the recording is known.

10.7.2   Time-Stamping Historical Recordings

ENF traces can help archivists by providing a time-synchronized exhibit of multimedia files. Many twentieth century recordings are important cultural heritage records, but some may lack necessary metadata, such as the date and the exact time of recording. Su et al. (2013), Su et al. (2014c), Douglas et al. (2014) found ENF traces in the 1960s phone conversation recordings of President Kennedy in the White House, and in the multi-channel recordings of the 1970 NASA Apollo 13 mission. In Su et al. (2014c), the ENF was used to align two recordings of around 4 hours in length from the Apollo 13 mission, and the result was confirmed by the audio contents of the two recordings.

A set of time-synchronized exhibits of multimedia files with reliable metadata of recording time and high-SNR ENF traces can be used to reconstruct a reference ENF database for various purposes (Douglas et al. 2014). Such a reconstructed ENF database from a set of recordings can be valuable for applying ENF analysis to sensing data collected in the past before power signals recorded for reference purposes is available.

10.7.3 Audio Restoration

Audio signal digitized from an analog tape has its content frequency "drifted" off the original value due to the inconsistent rolling speed between the recorder and the digitizer. The ENF signal, although varying along time, has a general trend that remains around the nominal frequency. When ENF traces are embedded in a tape recording, this consistent trend can serve as the anchor for correcting the abnormal speed of the recording (Su 2014). Figure 10.16a shows the drift of the general trend of the ENF followed by an abrupt jump in a recording from the NASA Apollo 11 Mission. The abnormal speed can be corrected by temporally stretching or compressing the frame segments of an audio signal using techniques such as multi-rate conversion and interpolation (Su 2014). The spectrogram of the resulting corrected signal is shown in Fig. 10.16b in which the trend of the ENF is constant without an abrupt jump.

10.8 Conclusions and Outlook

This chapter has provided an overview of the ENF signal, an environment signature that can be captured by multimedia recordings made in locations where there is elec-

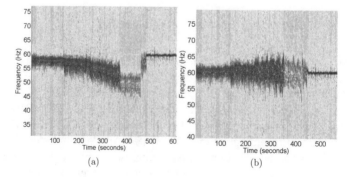

Fig. 10.16 The spectrogram of an Apollo mission control recording **a** before and **b** after speed correction on the digitized signal (Su 2014)

trical activity. We first examined the embedding mechanism of ENF traces, followed by how ENF signals, can be extracted from audio and video recordings. We noted that in order to reliably extract ENF signals from visual recordings, it is often helpful to exploit the rolling-shutter mechanism commonly used by cameras with CMOS image sensor arrays. We then systematically reviewed how the ENF signature can be used for media recordings' time and/or location authentication, integrity authentication, and localization. We also touched on anti-forensics in which adversaries try to mislead forensic analysts and discussed countermeasures. Applications beyond forensics, including multimedia synchronization and audio restoration, were also discussed.

Looking into the future, we notice many new problems have naturally arisen as technology continues to evolve. For example, given the popularity of short video clips that last for less than 30 s, ENF analysis tools need to accommodate short recordings. The technical challenge lies in how to efficiently make use of the ENF information than traditional ENF analysis tools that focus on recordings of minutes or even hours long.

Another technical direction is to push forward the limit of ENF extraction algorithms from working on videos to images. Is it possible to extract enough ENF information from a sequence of fewer than ten images captured over a few seconds/minutes to infer the time and/or location? For the case of ENF extraction from a single rolling-shutter image, is it possible to know more beyond whether the image is captured at a 50 or 60 Hz location?

Exploring new use cases of ENF analysis based on its intrinsic ENF embedding mechanism is also a natural extension. For example, deepfake video detection may be a potential avenue for ENF analysis. ENF can be naturally embedded through the flickering of indoor lighting into the facial region and the background of a video recording. A consistency test may be designed to complement detection algorithms based on other visual/statistical cues.

References

Baudry S, Chupeau B, Vito MD, Doërr G (2014) Modeling the flicker effect in camcorded videos to improve watermark robustness. In: IEEE international workshop on information forensics and security (WIFS), pp 42–47

Bollen M, Gu I (2006) Signal processing of power quality disturbances. Wiley-IEEE Press

Brixen EB (2007a) Techniques for the authentication of digital audio recordings. In: Audio engineering society convention 122

Brixen EB (2007b) Further investigation into the ENF criterion for forensic authentication. In: Audio engineering society convention 123

Brixen EB (2008) ENF; quantification of the magnetic field. In: AES international conference: audio forensics-theory and practice

Bykhovsky D, Cohen A (2013) Electrical network frequency (ENF) maximum-likelihood estimation via a multitone harmonic model. IEEE Trans Inf Forensics Secur 8(5):744–753

Catalin G (2009) Applications of ENF analysis in forensic authentication of digital audio and video recordings. J Audio Eng Soc 57(9):643–661

Chai J, Liu F, Yuan Z, Conners RW, Liu Y (2013) Source of ENF in battery-powered digital recordings. In: Audio engineering society convention 135

Chai J, Zhao J, Guo J, Liu Y (2016) Application of wide area power system measurement for digital authentication. In: IEEE/PES transmission and distribution conference and exposition, pp 1–5

Chang F-C, Huang H-C (2010) Electrical network frequency as a tool for audio concealment process. In: International conference on intelligent information hiding and multimedia signal processing, IIH-MSP '10, Washington, DC, USA. IEEE Computer Society, pp 175–178

Chang F-C, Huang H-C (2011) A study on ENF discontinuity detection techniques. In: International conference on intelligent information hiding and multimedia signal processing, IIH-MSP '11, Washington, DC, USA. IEEE Computer Society, pp 9–12

Choi J, Wong C-W (2019) ENF signal extraction for rolling-shutter videos using periodic zero-padding. In: IEEE international conference on acoustics, speech and signal processing (ICASSP), pp 2667–2671

Chuang W-H, Garg R, Wu M (2012) How secure are power network signature based time stamps? In: ACM conference on computer and communications security, CCS '12, New York, NY, USA, 2012. ACM, pp 428–438

Chuang W-H, Garg R, Wu M (2013) Anti-forensics and countermeasures of electrical network frequency analysis. IEEE Trans Inf Forensics Sec 8(12):2073–2088

Cooper AJ (2008) The electric network frequency (ENF) as an aid to authenticating forensic digital audio recordings an automated approach. In: AES international conference: audio forensics-theory and practice

Cooper Alan J (2011) Further considerations for the analysis of ENF data for forensic audio and video applications. Int J Speech Lang Law 18(1):99–120

Cooper AJ (2009a) Digital audio recordings analysis: the electric network frequency (ENF) criterion. Int J Speech Lang Law 16(2):193–218

Cooper AJ (2009b) An automated approach to the electric network frequency (ENF) criterion: theory and practice. Int J Speech Lang Law 16(2):193–218

Cui Y, Liu Y, Fuhr P, Morales-Rodriguez M (2018) Exploiting spatial signatures of power ENF signal for measurement source authentication. In: IEEE international symposium on technologies for homeland security (HST), pp 1–6

Dosiek L (2015) Extracting electrical network frequency from digital recordings using frequency demodulation. IEEE Signal Process Lett 22(6):691–695

Eissa MM, Elmesalawy MM, Liu Y, Gabbar H (2012) Wide area synchronized frequency measurement system architecture with secure communication for 500kV/220kV Egyptian grid. In: IEEE international conference on smart grid engineering (SGE), p 1

Elmesalawy MM, Eissa MM (2014) New forensic ENF reference database for media recording authentication based on harmony search technique using GIS and wide area frequency measurements. IEEE Trans Inf Forensics Secur 9(4):633–644

Esquef PA, Apolinario JA, Biscainho LWP (2014) Edit detection in speech recordings via instantaneous electric network frequency variations. IEEE Trans Inf Forensics Secur 9(12):2314–2326

Esquef PAA, Apolinário JA, Biscainho LWP (2015) Improved edit detection in speech via ENF patterns. In: IEEE international workshop on information forensics and security (WIFS), pp 1–6

Fechner N, Kirchner M (2014) The humming hum: Background noise as a carrier of ENF artifacts in mobile device audio recordings. In: Eighth international conference on IT security incident management IT forensics (IMF), pp 3–13

FNET Server Web Display (2021) http://fnetpublic.utk.edu, Accessed Jun. 2021

Fuentes M, Zinemanas P, Cancela P, Apolinário JA (2016) Detection of ENF discontinuities using PLL for audio authenticity. In: IEEE latin American symposium on circuits systems (LASCAS), pp 79–82

Fu L, Markham PN, Conners RW, Liu Y (2013) An improved discrete Fourier transform-based algorithm for electric network frequency extraction. IEEE Trans Inf Forensics Secur 8(7):1173–1181

Garg R, Hajj-Ahmad A, Wu M (2013) Geo-location estimation from electrical network frequency signals. In: IEEE international conference on acoustics, speech and signal processing (ICASSP), pp 2862–2866

Garg R, Varna AL, Hajj-Ahmad A, Wu M (2013) 'Seeing' ENF: Power-signature-based timestamp for digital multimedia via optical sensing and signal processing. IEEE Trans Inf Forensics Secur 8(9):1417–1432

Garg R, Varna AL, Wu M (2011) 'Seeing' ENF: Natural time stamp for digital video via optical sensing and signal processing. In: ACM international conference on multimedia, MM '11, New York, NY, USA. ACM, pp 23–32

Garg R, Varna AL, Wu M (2012) Modeling and analysis of electric network frequency signal for timestamp verification. In: IEEE international workshop on information forensics and security (WIFS), pp 67–72

Glentis GO, Jakobsson A (2011) Time-recursive IAA spectral estimation. IEEE Signal Process Lett 18(2):111–114

Golokolenko O, Schuller G (2017) Investigation of electric network frequency for synchronization of low cost and wireless sound cards. In: European signal processing conference (EUSIPCO), pp 693–697

Grigoras C (2005) Digital audio recordings analysis: the electric network frequency (ENF) criterion. Int J Speech Lang Law 12(1):63–76

Grigoras C (2007) Applications of ENF criterion in forensic audio, video, computer and telecommunication analysis. Forensic Sci Int 167(2–3):136–145

Gu J, Hitomi Y, Mitsunaga T, Nayar S (2010) Coded rolling shutter photography: Flexible space-time sampling. In: IEEE international conference on computational photography (ICCP), pp 1–8, Cambridge, MA

Gupta S, Cho S, Kuo JCC (2012) Current developments and future trends in audio authentication. IEEE MultiMed 19(1):50–59

Hajj-Ahmad A, Berkovich A, Wu M (2016) Exploiting power signatures for camera forensics. IEEE Signal Process Lett 23(5):713–717

Hajj-Ahmad A, Wong C-W, Gambino S, Zhu Q, Yu M, Wu M (2019) Factors affecting ENF capture in audio. IEEE Trans Inf Forensics Secur 14(2):277–288

Hajj-Ahmad A, Baudry S, Chupeau B, Doërr G (2015) Flicker forensics for pirate device identification. In: ACM workshop on information hiding and multimedia security, IH&MMSec '15, New York, NY, USA. ACM, pp 75–84

Hajj-Ahmad A, Garg R, Wu M (2012) Instantaneous frequency estimation and localization for ENF signals. In: APSIPA annual summit and conference, pp 1–10

Hajj-Ahmad A, Garg R, Wu M (2013) ENF based location classification of sensor recordings. In: IEEE international workshop on information forensics and security (WIFS), pp 138–143

Hajj-Ahmad A, Garg R, Wu M (2013) Spectrum combining for ENF signal estimation. IEEE Signal Process Lett 20(9):885–888

Hajj-Ahmad A, Garg R, Wu M (2015) ENF-based region-of-recording identification for media signals. IEEE Trans Inf Forensics Secur 10(6):1125–1136

Hua G (2018) Error analysis of forensic ENF matching. In: IEEE international workshop on information forensics and security (WIFS), pp 1–7

Hua G, Zhang Y, Goh J, Thing VLL (2016) Audio authentication by exploring the absolute-error-map of ENF signals. IEEE Trans Inf Forensics Secur 11(5):1003–1016

Hua G, Goh J, Thing VLL (2014) A dynamic matching algorithm for audio timestamp identification using the ENF criterion. IEEE Trans Inf Forensics Secur 9(7):1045–1055

Huijbregtse M, Geradts Z (2009) Using the ENF criterion for determining the time of recording of short digital audio recordings. In: Geradts ZJMH, Franke KY, Veenman CJ (eds) Computational Forensics, *Lecture Notes in Computer Science*, vol 5718. Springer Berlin Heidelberg, pp 116–124

Huijbregtse M, Geradts Z (2009) Using the ENF criterion for determining the time of recording of short digital audio recordings. In: International workshop on computational forensics (IWCF), pp 116–124

Jeon Y, Kim M, Kim H, Kim H, Huh JH, Yoon JW (2018) I'm listening to your location! inferring user location with acoustic side channels. In: World Wide Web conference, WWW '18, pages 339–348, Republic and Canton of Geneva, Switzerland. International World Wide Web Conferences Steering Committee

Kajstura M, Trawinska A, Hebenstreit J (2005) Application of the electric network frequency (ENF) criterion: a case of a digital recording. Forensic Sci Int 155(2–3):165–171

Karantaidis G, Kotropoulos C (2018) Assessing spectral estimation methods for electric network frequency extraction. In: Pan-Hellenic conference on informatics, PCI '18, New York, NY, USA. ACM, pp 202–207

Kim H, Jeon Y, Yoon JW (2017) Construction of a national scale ENF map using online multimedia data. In: ACM on conference on information and knowledge management, CIKM '17, New York, NY, USA. ACM, pp 19–28

Lin X, Liu J, Kang X (2016) Audio recapture detection with convolutional neural networks. IEEE Trans Multimed 18(8):1480–1487

Lin X, Kang X (2017) Supervised audio tampering detection using an autoregressive model. In: IEEE international conference on acoustics, speech and signal processing (ICASSP), pp 2142–2146

Lin X, Kang X (2018) Robust electric network frequency estimation with rank reduction and linear prediction. ACM Trans Multimed Comput Commun Appl 14(4):84:1–84:13

Liu Y, Yuan Z, Markham PN, Conners RW, Liu Y (2011) Wide-area frequency as a criterion for digital audio recording authentication. In: Power and energy society general meeting, pp 1–7

Liu Y, Yuan Z, Markham PN, Conners RW, Liu Y (2012) Application of power system frequency for digital audio authentication. IEEE Trans Power Deliv 27(4):1820–1828

Manolakis DG, Ingle VK, Kogon SM(2000) Statistical and adaptive signal processing. McGraw-Hill, Inc

Myriam D-C, Sylvain M (2000) High-precision fourier analysis of sounds using signal derivatives. J Audio Eng Soc 48(7/8):654–667

Nicolalde DP, Apolinario J (2009) Evaluating digital audio authenticity with spectral distances and ENF phase change. In: IEEE international conference on acoustics, speech and signal processing (ICASSP), Washington, DC, USA. IEEE Computer Society, pp 1417–1420

Oard DW, Wu M, Kraus K, Hajj-Ahmad A, Su H, Garg R (2014) It's about time: Projecting temporal metadata for historically significant recordings. In: iConference, Berlin, Germany, pp 635–642

Ojowu O, Karlsson J, Li J, Liu Y (2012) ENF extraction from digital recordings using adaptive techniques and frequency tracking. IEEE Trans Inf Forensics Secur 7(4):1330–1338

Phadke AG, Thorp JS, Adamiak MG (1983) A new measurement technique for tracking voltage phasors, local system frequency, and rate of change of frequency. IEEE Trans Power Appar Syst PAS-102(5):1025–1038

Pop G, Draghicescu D, Burileanu D, Cucu H, Burileanu C (2017) Fast method for ENF database build and search. In: International conference on speech technology and human-computer dialogue (SpeD), pp 1–6

Ralph S (1986) Multiple emitter location and signal parameter estimation. IEEE Trans Antennas Propag 34(3):276–280

Reis PMGI, da Costa JPCL, Miranda RK, Del Galdo G (2016) Audio authentication using the kurtosis of ESPRIT based ENF estimates. In: International conference on signal processing and communication systems (ICSPCS), pp 1–6

Reis PMGI, da Costa JPCL, Miranda RK, Del Galdo G (2017) ESPRIT-Hilbert-based audio tampering detection with SVM classifier for forensic analysis via electrical network frequency. IEEE Trans Inf Forensics Secur 12(4):853–864

Rodríguez DPN, Antonio Apolinário J, Wagner Pereira Biscainho L (2010) Audio authenticity: Detecting ENF discontinuity with high precision phase analysis. IEEE Trans Inf Forensics and Secur 5(3):534–543

Rodriguez DP, Apolinario JA, Biscainho LWP (2013) Audio authenticity based on the discontinuity of ENF higher harmonics. In: Signal processing conference (EUSIPCO), pp 1–5

Roy R, Kailath T (1989) ESPRIT-estimation of signal parameters via rotational invariance techniques. IEEE Trans Acoust Speech Signal Process 37(7):984–995

Sanders RW (2008) Digital audio authenticity using the electric network frequency. In: AES international conference: audio forensics-theory and practice

Sanders R, Popolo PS (2008) Extraction of electric network frequency signals from recordings made in a controlled magnetic field. In: Audio engineering society convention 125

Šarić Z, Žunić A, Zrniéc T, Knežević M, Despotović D, Delić T (2016) Improving location of recording classification using electric network frequency (ENF) analysis. In: International symposium on intelligent systems and informatics (SISY), pp 51–56

Signal Processing (SP) Cup Project Reports (2021) https://sigport.org/events/sp-cup-project-reports, Accessed Jun. 2021

Su H (2014) Temporal and spatial alignment of multimedia signals. PhD thesis, University of Maryland, College Park

Su H, Garg R, Hajj-Ahmad A, Wu M (2013) ENF analysis on recaptured audio recordings. In: IEEE international conference on acoustics, speech and signal processing (ICASSP), pp 3018–3022

Su H, Hajj-Ahmad A, Garg R, Wu M (2014a) Exploiting rolling shutter for ENF signal extraction from video. In: IEEE international conference on image processing (ICIP), pp 5367–5371

Su H, Hajj-Ahmad A, Wong C-W, Garg R, Wu M (2014b) ENF signal induced by power grid: A new modality for video synchronization. In: ACM international workshop on immersive media experiences, ImmersiveMe '14, New York, NY, USA, 2014. ACM, pp 13–18

Su H, Hajj-Ahmad A, Wu M, Oard DW (2014c) Exploring the use of ENF for multimedia synchronization. In: IEEE international conference on acoustics, speech and signal processing (ICASSP), pp 4613–4617

Suresha PB, Nagesh S, Roshan PS, Gaonkar PA, Meenakshi GN, Ghosh PK (2017) A high resolution ENF based multi-stage classifier for location forensics of media recordings. In: National conference on communications (NCC), pp 1–6

Szeliski R (2010)Computer vision: algorithms and applications, chapter 2.2. Springer

Top P, Bell MR, Coyle E, Wasynczuk O (2012) Observing the power grid: Working toward a more intelligent, efficient, and reliable smart grid with increasing user visibility. IEEE Signal Process Mag 29(5):24–32

Tsai S-J, Zhang L, Phadke AG, Liu Y, Ingram MR, Bell SC, Grant IS, Bradshaw DT, Lubkeman D, Tang L (2007) Frequency sensitivity and electromechanical propagation simulation study in large power systems. IEEE Trans Circuits Syst I: Regular Papers 54(8):1819–1828

Vaidyanathan PP (2006) Multirate systems and filter banks. Pearson Education India

Vatansever S, Dirik AE, Memon N (2017) Detecting the presence of ENF signal in digital videos: a superpixel-based approach. IEEE Signal Process Lett 24(10):1463–1467

Vatansever S, Dirik AE, Memon N (2019) Analysis of rolling shutter effect on ENF based video forensics. IEEE Trans Inf Forensics Secur 1

Wong C-W, Hajj-Ahmad A, Wu M (2018) Invisible geo-location signature in a single image. In: IEEE international conference on acoustics, speech and signal processing (ICASSP), pp 1987–1991

Wu M, Hajj-Ahmad A, Kirchner M, Ren Y, Zhang C, Campisi P (2016) Location signatures that you don't see: highlights from the IEEE Signal Processing Cup 2016 student competition [SP Education]. IEEE Signal Process Mag 33(5):149–156

Yao W, Zhao J, Till MJ, You S, Liu Y, Cui Y, Liu Y (2017) Source location identification of distribution-level electric network frequency signals at multiple geographic scales. IEEE Access 5:11166–11175

Zhang Y, Markham P, Xia T, Chen L, Ye Y, Wu Z, Yuan Z, Wang L, Bank J, Burgett J, Conners RW, Liu Y (2010) Wide-area frequency monitoring network (FNET) architecture and applications. IEEE Trans Smart Grid 1(2):159–167

Zhong Z, Xu C, Billian BJ, Zhang L, Tsai SS, Conners RW, Centeno VA, Phadke AG, Liu Y (2005) Power system frequency monitoring network (FNET) implementation. IEEE Trans Power Syst 20(4):1914–1921

Zhu Q, Chen M, Wong C-W, Wu M (2018) Adaptive multi-trace carving based on dynamic pro-
    gramming. In: Asilomar conference on signals, systems, and computers, pp 1716–1720
Zhu Q, Chen M, Wong C-W, Wu M (2020) Adaptive multi-trace carving for robust frequency
    tracking in forensic applications. IEEE Trans Inf Forensics Secur 16:1174–1189

Chapter 11
Data-Driven Digital Integrity Verification

Davide Cozzolino, Giovanni Poggi, and Luisa Verdoliva

Images and videos are by now a dominant part of the information flowing on the Internet and the preferred communication means for younger generations. Besides providing information, they elicit emotional responses, much stronger than text does. It is probably for these reasons that the advent of AI-powered deepfakes, realistic and relatively easy to generate, has raised great concern among governments and ordinary people alike. Yet, the art of image and video manipulation is much older than that. Armed with conventional media editing tools, a skilled attacker can create highly realistic fake images and videos, so-called "cheapfakes". However, despite the nickname, there is nothing cheap in cheapfakes. They require significant domain knowledge to be crafted and their detection, can be much more challenging than the detection of AI-generated material. In this chapter, we focus on the detection of cheapfakes, that is, image and video manipulations carried out with conventional means. However, we skip conventional detection methods, already thoroughly described in many reviews, and focus on the data-driven deep learning-based methods proposed in recent years. We look at manipulation detection methods under various perspectives. First, we analyze in detail the forensics traces they rely on, and then the main architectural solutions proposed for detection and localization, together with the associated training strategies. Finally, major challenges and future directions are highlighted.

11.1 Introduction

In the last few years, multimedia forensics has been drawing ever-increasing attention in the scientific community and beyond. Indeed, with the editing software tools

D. Cozzolino (✉) · G. Poggi · L. Verdoliva
University Federico II of Naples, Naples, Italy
e-mail: luisa.verdoliva@unina.it

© The Author(s) 2022
H. T. Sencar et al. (eds.), *Multimedia Forensics*, Advances in Computer Vision and Pattern Recognition, https://doi.org/10.1007/978-981-16-7621-5_11

available nowadays, modifying images or videos has become extremely easy. In the wrong hands, and used for the wrong purposes, this capability may represent major threats for both individuals and the society as a whole. In this chapter, we focus on conventional manipulations, also known as cheapfakes, in contrast to deepfakes that are based on the use of deep learning tools for their generation (Paris and Donovan 2019). Generating a cheapfake does not require artificial-intelligence-based technology, however a skilled attacker can carry out very realistic fakes, which can have disruptive consequences. In Fig. 11.1, we show some examples of very common manipulations: inserting an object copied from a different image (splicing), replicating an object coming from the same image (copy-move), or removing an object by extending the background (inpainting). Usually, to better fit the object in the scene, some suitable post-processing operations are also applied, like resizing, rotation, boundary smoothing, or color adjustment.

A large number of methods have been proposed for forgery detection and localization, both for images and videos (Farid 2016; Korus 2017; Verdoliva 2020). Here, we focus on methods that verify the digital (as opposed to physical or semantic) integrity of the media asset, namely, discover the occurrence of a manipulation by detecting the pixel-level inconsistencies it caused. It is worth emphasizing that even well-crafted manipulations, which do not leave visible artifacts on the image, always modify its statistics, leaving traces that can be exploited by pixel-level analysis tools. In fact, the image formation process inside a camera comprises a certain number of operations, both hardware and software, specific of each camera, which leave distinctive marks on each acquired image (Fig. 11.2). For example, camera models use widely different demosaicing algorithms, as well as different quantization tables

Fig. 11.1 Examples of image manipulations carried out using conventional media editing tools. From left to right: splicing (alien material inserted in the image), copy-move (an object has been cloned), inpainting (objects removed using background patches)

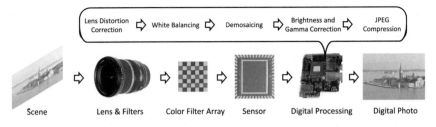

| Lens Distortion Correction ⇨ | White Balancing ⇨ | Demosaicing ⇨ | Brightness and Gamma Correction ⇨ | JPEG Compression |

| Scene | Lens & Filters | Color Filter Array | Sensor | Digital Processing | Digital Photo |

Fig. 11.2 In-camera processing pipeline. An image is captured using a complex acquisition system. The optical filter reduces undesired light components and the lenses focus the light on the sensor, where the red-green-blue (RGB) components are extracted by means of a color filter array (CFA). Then, a sequence of internal processing steps follow, including lens distortion correction, enhancement, demosaicing, and finally compression

for JPEG compression. Their traces allow one to identify the camera model that acquired a given image. So, when parts of different images are spliced together, a spatial anomaly in these traces can be detected, revealing the manipulation. Likewise, the post-camera editing process may introduce some specific artifacts, as well as disrupt camera-specific patterns. Therefore, by emphasizing possible deviations with respect to the expected behavior, one can establish with good confidence if the digital integrity has been violated.

In the multimedia forensics literature, there has been intense work on detecting any possible type of manipulation, global or local, irrespective of its purpose. Several methods detect a large array of image manipulations, like resampling, median filtering, and contrast enhancement. However, these operations can also be carried out only to improve the image appearance or for other legitimate purposes, with no malicious intent. For this reason, in this chapter, we will focus on detecting local (semantically) relevant manipulations, like copy-moves, inpainting, and compositions, that can modify the meaning of the image.

Early digital integrity method followed a model-based approach, relying on some mathematical or statistical models of data and attacks. More recent methods, instead, are for the most part data-driven, based on deep learning: they exploit the availability of large amounts of data to learn how to detect specific or generic forensic clues. These can be further classified based on whether they give as output an image-level detection score that indicates the probability that the image has been manipulated or not, or a pixel-level localization mask that provides a score for each pixel. In addition, we will distinguish methods based on the training strategy they apply:

- two-class learning, where training is carried out on datasets of labeled pristine and manipulated data;
- one-class learning, where only real data are used during training and manipulations are regarded as anomalies.

In the rest of the chapter, we first analyze in detail the forensics clues exploited in the various approaches, then review the main architectural solutions proposed for detection and localization, together with their training strategies and the most popular datasets proposed in this field. Finally, we draw conclusions and highlight major challenges and future directions.

11.2 Forensics Clues

Learning-based detection methods are typically designed on the basis of the specific forensic artifacts they look for. For example, some proposed convolutional neural networks (CNNs) base their decision on low-level features (e.g., camera-based), while others focus on higher level features (e.g., edge boundaries in a composition), and still others on features of both types. Therefore, this section will review the most interesting clues that can be used for digital integrity verification.

11.2.1 Camera-Based Artifacts

One powerful approach is to exploit the many artifacts that are introduced during the image acquisition process (Fig. 11.2). Such artifacts are specific of the individual camera (the device) or the camera model, and therefore represent some form of signature superimposed on the image, which can be exploited for forensic purposes. As an example, if part of the image is replaced with content taken from another source, this added material will lack the original camera artifacts, allowing detection of the attack. Some of the most relevant artifacts are related to the sensor, the lens, the color filter array, the demosaicing algorithm, and the JPEG quantization tables. In all cases, with the aim of analyzing weak artifacts, the image content (the scene) represents an undesired strong disturbance. Therefore, it is customary to work on the so-called noise residuals, obtained by suppressing the high-level content. To this end, one can estimate the "true" image by means of a denoising algorithm, and subtract it from the original image, or use some high-pass filters in the spatial or in the transform domain (Lyu and Farid 2005; He et al. 2012).

A powerful camera-related artifact is the photo-response non-uniformity noise (PRNU), successfully used not only for forgery detection but also for source identification (Chen et al. 2008). The PRNU pattern is due to unavoidable imperfections in the sensor manufacturing process. It is unique for each individual camera, stable in time, and present in all acquired images, therefore it can be considered as a device fingerprint. Using such a fingerprint, manipulations of any type can be detected. In fact, whenever the image is modified, the PRNU pattern is corrupted or even fully removed, a strong clue of a possible manipulation (Lukàš et al. 2006). The main drawbacks of PRNU-based methods are

(i) the need of images taken from the same camera under test to obtain a good estimate of the reference pattern and
(ii) the need to have a PRNU pattern spatially aligned with the image under test.

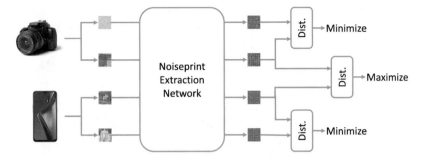

Fig. 11.3 Training a noiseprint extraction network. Pairs of patches feed the two branches of the Siamese network that extract the corresponding noiseprint patches. The output distance must be minimized for patches coming from the same camera and spatial location, maximized for the others

In Cozzolino and Verdoliva (2020), a learning-based strategy is proposed to extract a stronger camera fingerprint, the image noiseprint. Like the PRNU pattern, it behaves as a fingerprint embedded in all acquired images. Contrary to it, it depends on *all* camera-model artifacts which contribute to the digital history of an image. Noiseprints are extracted by means of a CNN, trained to capture, and emphasize all camera-model artifacts, while suppressing the high-level semantic content. To this end, the CNN is trained in a Siamese configuration, with two replicas of the network, identical for architecture and weights, on two parallel branches. The network is fed with pairs of patches drawn from a large number of pristine images of various camera models. When the patches on the two branches are aligned (same camera model and same spatial position, see Fig. 11.3) they can be expected to contain much the same artifacts. Therefore, the output of each branch can be used as reference for the input of the other one, by-passing the need for clean examples. Eventually, the network learns to extract the desired noiseprint, and can be used with no further supervision on images captured by any camera model, both inside and outside the training set.

Of course, the extracted noiseprints will always contain traces of the imperfectly rejected high-level image, acting as noise for the forensic purposes. However, by exploiting all sorts of camera-related artifacts, noiseprints enable more powerful and reliable forgery detection procedures than PRNU patterns. On the down side, the noiseprint does not allow one to distinguish between two devices of the same camera model. Therefore, it should not be regarded as a substitute of the PRNU, but rather as a complementary tool.

In Fig. 11.4, we show some examples of noiseprints extracted from different fake images, with the corresponding heat maps obtained by feature clustering. In the first case, the noiseprint clearly shows the 8×8 grid of the JPEG format. Sometimes, traces left by image manipulations on the extracted noiseprint are so strong to allow easy localization even by direct inspection. The approach can be also easily extended to videos (Cozzolino et al. 2019), but larger datasets are necessary to make up for the reduced quality of the sources, due to heavy compression. Interestingly, noiseprints can be also used in a supervised setting (Cozzolino and Verdoliva 2018). Given a

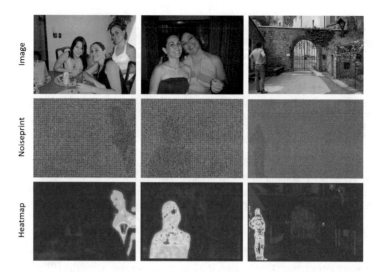

Fig. 11.4 Examples of manipulated images (top) with their corresponding noiseprints (middle) and heat maps obtained by a proper clustering (bottom)

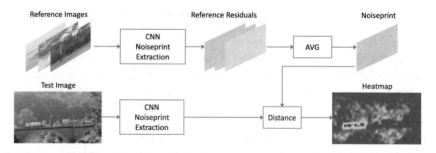

Fig. 11.5 Noiseprint in supervised modality. The localization procedure is the classic pipeline used with PRNU-based methods: the noiseprint is extracted from a set of pristine images taken by the same camera of the image under analysis, their average represents a clean reference, namely, a reliable estimate of the camera noiseprint that can be used for image forgery localization

suitable training set of images, a reliable estimate of the noiseprint is built, where high-level scene leakages are mostly removed, and used in a PRNU-like fashion to discover anomalies in the image under analysis (see Fig. 11.5).

As we mentioned before, many different artifacts arise from the in-camera processing pipeline, related to sensor, lens, and especially the internal processing algorithms. Of course, one can also focus on just one of these artifacts to develop a manipulation detector. Indeed, this has been the dominant approach in the past, with many model-based methods proposed to exploit very specific features, leveraging a deep domain knowledge. In particular, there has been intense research on the CFA-demosaicing combination. Most digital cameras use a periodic color filter array, so that each individual sensor element records light only in a certain range of wavelengths (i.e., red,

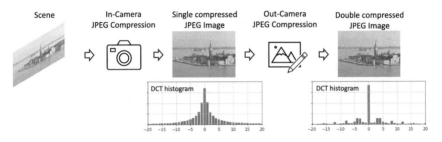

Fig. 11.6 Histograms of the DCT coefficients clearly show the statistical differences between images subject to a single or double JPEG compression

green, blue). The missing color information is then interpolated from surrounding pixels in the demosaicing process, which introduces a subtle periodic correlation pattern in all acquired images. Both CFA configuration and demosaicing algorithms are specific of each camera model, leading to different correlation patterns which can be exploited for forensic means. For example, when a region is spliced in a photo taken by another camera model, its periodic pattern will appear anomalous. Most of the methods proposed in this context are model-based (Popescu and Farid 2005; Cao and Kot 2009; Ferrara et al. 2012), however recently a convolutional neural network specifically tailored to the detection of mosaic artifacts has been proposed (Bammey et al. 2020). The network is trained only on authentic images and includes 3×3 convolutions with a dilation of 2, so as to examine pixels that all belong to the same color channel and detect possible inconsistencies. In fact, a change of the mosaic can lead to forged blocks being detected at incorrect positions modulo $(2, 2)$.

11.2.2 JPEG Artifacts

Many algorithms in image forensics exploit traces left by the compression process. Some methods look for anomalies in the statistical distribution of original DCT samples, assumed to comply with the Benford law (Fu et al. 2007). However, the most popular approach is to detect double compression traces (Bianchi and Piva 2012). In fact, when a JPEG-compressed image undergoes a local manipulation and then is compressed again, double compression (or double quantization, DQ) artifacts appear all over the image (with very high probability) in the forged area. This phenomenon is exploited also in the well-known Error Level Analysis (ELA), widespread among practitioners for its simplicity.

The effects of double compression are especially visible in the distributions of the DCT coefficients which, after the second compression, show characteristic periodic peaks and valleys (Fig. 11.6). Based on this observation, Wang and Zhang (2016) performs splicing detection by means of a CNN trained with histograms of DCT coefficients. The histograms are computed either from single-compressed patches (forged areas) or double-compressed ones (pristine areas), with quality factors in the range 60–95. To reduce the dimensionality of the feature vector, only a small interval near the peak of each histogram is chosen to represent the whole histogram. Following this seminal paper, also Amerini et al. (2017) and Park et al. (2018) use histograms of DCT coefficients as input, while (Kwon et al. 2021) uses one-hot coded DCT coefficients as already proposed in Yousfi and Fridrich (2020) for steganalysis. Also the quantization tables can bring useful information, and are used as additional input in Park et al. (2018); Kwon et al. (2021). Leveraging domain-specific prior knowledge, the approach proposed in Barni et al. (2017) works on noise residuals rather than on image pixels, and uses the first layers of the CNN to extract histogram-related features. In Park et al. (2018), instead, the quantization tables from the JPEG header are used as input together with the histogram features. It is worth observing that methods based on deep learning provide good results also in cases where conventional methods largely fail, such as when test images are compressed with a quality factor (QF) never seen in training or when the second QF is larger than the first one. Double compression detectors have been also proposed for H.264 video analysis, with a two-stream neural network that analyzes separately intra-coded and predictive frames (Nam et al. 2019).

11.2.3 Editing Artifacts

The editing process often generates a trail of precious traces, besides artifacts related to re-compression. Indeed, when a new object is inserted in an image, it typically requires several post-processing steps to fit the new context smoothly. These include geometric transformations, like rotation and scaling, contrast adjustment, and blurring, to smooth the object-background boundaries. Both rotation and resampling, for example, require some forms of interpolation, which in turn generates periodic artifacts. Therefore, some methods (Kirchner 2008) look for traces of resampling as a proxy for possible forgeries. In this section, we describe the most used and discriminative editing-based traces exploited in forensics algorithms.

Copy-moves

A very common manipulation consists in replicating or hiding objects. Of course, the presence of identical regions is a strong hint of forgery, but clones are often modified to disguise traces, and near-identical natural objects also exist, which complicates the forensic analysis. Studies on copy-move detection date back to 2003, with the seminal work of Fridrich (2003). Since then, a large literature has grown on this topic proposing solutions that allow for copy-move detection even in the presence of rotation, resizing, and other geometric distortions (Christlein et al. 2012). Methods based on keypoints are very efficient, while block-based dense methods (Cozzolino et al. 2015) are more accurate and deal also with removals.

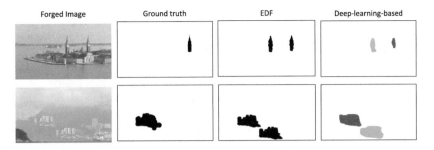

Fig. 11.7 Examples of additive copy-moves. From left to right: manipulated images and corresponding ground truths, binary localization maps obtained using a dense approach (Cozzolino et al. 2015) and those obtained by means of a deep-learning-based one (Chen et al. 2020). The latter method allows telling apart the source and copied areas

A first solution that relies on deep learning has been proposed in Wu et al. (2018), where an end-to-end CNN-based solution is able to jointly optimize the three main steps of a classic copy-move solution, that is, feature extraction, matching, and post-processing. To improve performance across different scenarios, multiscale feature analysis and hierarchical feature mapping are employed in Zhong and Pun (2019). Overall, deep-learning-based methods appear to outperform conventional methods on low-resolution images (i.e., small-size images) and can be designed to distinguish the copy from the original region. In fact, typically, copy-move detection methods generate a map where both the original object and its clone are highlighted, but do not establish which is which. The problem of source-target disambiguation is considered in Wu et al. (2018); Barni et al. (2019); Chen et al. (2020). The main idea originally proposed in Wu et al. (2018) is to use a CNN for manipulation detection and another one for similarity computation. Then a suitable fusion of these two outputs helps to distinguish the source region from its copies (Fig. 11.7).

Inpainting

Inpainting is a very effective image manipulation approach that can be used to hide objects/people. The main classic approaches for their detection rely on methods inspired by copy-move detection, given that in the widespread exemplar-based inpainting, multiple small regions are copied from all over the image and combined together to cover the object. Patch-based inpainting is addressed in Zhu et al. (2018) with an encoder/decoder architecture which provides a localization map as output. In Li and Huang (2019), instead, a fully convolutional network is used to localize the inpainted area in the residual domain. The CNN does not look any specific editing artifact caused by inpainting, but only trained on a large number of examples of real and manipulated images, making the detection very effective for the analyzed methods.

Splicings

Other approaches focus on anomalies which appear at the boundaries of objects when a composition is performed, due to the inconsistencies between regions drawn from different sources. For example, Salloum et al. (2018) uses a multi-task fully convolutional network comprising a specific branch for detecting boundaries between inserted regions and background and another one for analyzing the surface of the manipulation. Both in this work and in Chen et al. (2021), training is carried out also using the edge map extracted from the ground truth. In other papers (Liu et al. 2018; Shi et al. 2018; Zhang et al. 2018; Phan-Xuan et al. 2019), instead, training is performed on patches that are either authentic or composed of pixels belonging to the two classes. Therefore, the CNN is forced to look for the boundaries of the manipulation.

Face Warping

In Wang et al. (2019), a CNN-based solution is proposed to detect artifacts introduced by a specific Photoshop tool, Face-Aware Liquify, which performs image warping of human faces. The approach is very successful for this task because the network is trained on manipulated images automatically generated by the very same tool. To deal with more challenging situations and increase robustness, data augmentation is applied by including resizing, JPEG compression, and various types of histogram editing.

11.3 Localization Versus Detection

Integrity verification can be carried out at two different levels: image-level (detection) or pixel-level (localization). In the first case, a global integrity score is provided, which indicates the likelihood that a manipulation occurred anywhere in the image. The score can be a real number, typically an estimate of the probability of manipulation, or a binary decision (yes/no). Localization instead builds upon the hypothesis that a detector has already decided on the presence of a manipulation and tries to establish for each individual pixel if it has been manipulated. Therefore, the output has the same size of the image and, again, can be real valued, corresponding to a probability map or more in general a heat map, or binary, corresponding to a decision map (see Fig. 11.8). It is worth noting that most localization methods in practice also try to detect the presence of a manipulation and do not make any assumptions on the fake/pristine nature of the image. Many methods focus on localization, others on detection or on both tasks. A brief summary of the main characteristics of these approaches is presented in Table 11.1. In this section, we analyze the major directions adopted in the literature to address these problems by using deep-learning-based solutions.

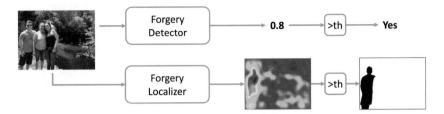

Fig. 11.8 The output of a method for digital integrity verification can be either a (binary or continuous) global score related to the probability that the image has been tampered or a localization map, where the score is computed at pixel level

11.3.1   Patch-Based Localization

The images to analyze are usually much larger than the input layer of neural networks. To avoid resizing the input image, which may destroy precious evidence, several approaches propose to work on small patches, with size typically spanning from 64×64 to 256×256 pixels, and analyze the whole image patch by patch. Localization can be performed in different ways based on the learning strategy, as described in the following.

Two-class Learning

Supervised learning relies on the availability of a large dataset of both real and manipulated patches. Once the patch is classified at test time, localization is carried out by performing a sliding-window analysis of the whole image. Most of the early methods are designed in this way. For example, to detect traces of double compression, in Park et al. (2018) a dataset of single-compressed and double-compressed JPEG blocks was generated using a very large number of quantization tables. Clearly, the more diverse and varied the dataset is, the more effective the training will be, with strong impact on the final performance. Beyond double compression, one can be interested in detecting traces of other types of processing, like blurring or resampling. In this case, one has only to generate a large number of patches of such kind. The main issue is the variety of parameters one can consider and that does not allow to easily cover the whole space. Patch-based learning is also used to detect object insertion in an image, based on the presence of boundary artifacts. In this case, manipulated patches include pixels coming from two different sources (the host image and the alien object), with a simple edge separating the two regions (Zhang et al. 2018; Liu et al. 2018).

Table 11.1 An overview of the main approaches proposed in the literature together with their main features: the type of artifact they aim to detect, the training and test input, i.e., patches, resized images or original images. In addition, we also indicate if they have been designed for localization (L), detection (D), or both

Acronym [ref]	Artifact	Training	Testing	L	D
(1) Localization, Patch-based, Two-class					
Wang and Zhang (2016), Barni et al. (2017), Amerini et al. (2017), Park et al. (2018)	JPEG	Patch at 64-256	Patch at 64-256	✓	
Liu et al. (2018), Shi et al. (2018), Phan-Xuan et al. (2019)	Splicing	Patch at 32-64	Patch at 32-64	✓	
(2) Localization, Patch-based, One-class					
Cozzolino and Verdoliva (2016)	Camera-based	None	Original size	✓	
Bondi et al. (2017)	Camera-based	Patch at 64	Original size	✓	✓
Self-consistency Huh et al. (2018)	Camera-based	Patch at 128	Original size	✓	✓
RFM Yao et al. (2020)	Camera-based	Patch at 64	Original size	✓	✓
Adaptive-CFA Bammey et al. (2020)	Demosaicing	Patch at 648	Original size	✓	✓
Forensicgraph Mayer and Stamm (2019)	Camera-based	Patch at 128-256	Original size	✓	✓
Noiseprint Cozzolino and Verdoliva (2020)	Camera-based	Patch at 48	Original size	✓	
Niu et al. (2021)	JPEG	Patch at 64	Original size	✓	✓
(3) Localization, Image-based, Segmentation-like					
Wu et al. (2018)	Copy-move	Resized to 256	Resized to 256	✓	✓
MFCN Salloum et al. (2018)	Splicing	Patch	Original size	✓	
BusterNet Wu et al. (2018)	Copy-move	Patch at 256	Original size	✓	✓
Zhu et al. (2018)	Inpainting	Patch at 256	Patch at 256	✓	
Li and Huang (2019)	Inpainting	Original size	Original size	✓	
ManTra-Net Wu et al. (2019)	Editing	Patch at 256	Original size	✓	✓
RRU-Net Bi et al. (2019)	Splicing	Resized to 384	Original size	✓	✓
D-Unet Bi et al. (2020)	Splicing	Resized to 512	Resized to 512	✓	✓

(continued)

Table 11.1 (continued)

Acronym [ref]	Artifact	Training	Testing	L	D
H-LSTM Bappy et al. (2019)	Splicing	Resized to 256	Resized to 256	✓	
LSTM-Skip Mazaheri et al. (2019)	Splicing	Resized to 256	Resized to 256	✓	
MAGritte Kniaz et al. (2019)	Splicing	Resized to 512	Resized to 512	✓	
CMSD/STRD Chen et al. (2020)	Copy-move	Resized to 256	Resized to 256	✓	✓
SPAN Hu et al. (2020)	Editing	Resized to 521	Resized to 512	✓	
GSCNet Shi et al. (2020)	Editing	Resized to 256	Resized to 256	✓	
DU-DC-EC Net Zhang and Ni (2020)	Splicing	Patch at 128	Original size	✓	✓
DOA-GAN Islam et al. (2020)	Copy-move	Resized to 320	Resized to 320	✓	✓
CAT-Net Kwon et al. (2021)	JPEG	Patch at 512	Original size	✓	
MVSS-Net Chen et al. (2021)	Splicing	Resized to 512	Original size	✓	✓
PSCC-Net Liu et al. (2021)	Editing	Patch at 256	Original size	✓	✓
MS-CRF-Att Rao et al. (2021)	Splicing	Patch at 256	Original size	✓	✓
(4) Localization, Image-based, Obj.Detection-like					
RGB-N Zhou et al. (2018)	Editing	Resized to 600	Resized to 600	✓	✓
Const. R-CNN Yang et al. (2020)	Editing	Resized to 600	Resized to 600	✓	✓
(5) Detection					
SRM-CNN Rao and Ni (2016)	Splicing	Patch at 128	Patch at 128		✓
E2E Marra et al. (2019)	Editing	Original size	Original size		✓

One-class learning

Several methods in forensics look at the problem using an anomaly detection perspective, developing a model of the pristine data and looking for anomalies with respect to this model, which can suggest the presence of a manipulation. The idea is to extract a specific type of forensic feature during training, whose distribution depends strongly on the source image or class of images. Then, at test time, these features are extracted from all patches of the image (with suitable stride) and clustered based on their statistics. If two distinct clusters emerge, the smallest one is regarded as anomalous, likely due to a local manipulation, and the corresponding patches provide a map of the tampered area. The most successful features used for this task are low level, like those related to the CFA pattern (Bammey et al. 2020), compression traces (Niu et al. 2021), or all the in-camera processing operations (Cozzolino and Verdoliva 2016; Bondi et al. 2017; Cozzolino and Verdoliva 2020). For example, some features allow one to classify different camera models with high reliability. In the presence of a splicing, the features observed in the pristine and the forged regions will be markedly different, indicating that different image parts were acquired by different camera models (Bondi et al. 2017; Yao et al. 2020).

Autoencoders can be very useful to implement a one-class approach. In fact, a network trained on a large number of patches extracted from untampered images eventually learns to reproduce pristine images with good accuracy. On the contrary, anomalies will give rise to large reconstruction errors, pointing to a possible manipulation. Noise inconsistencies are used in Cozzolino and Verdoliva (2016) to localize splicings without requiring a training step. At test time, handcrafted features are extracted from the patches of the image and an iterative algorithm based on feature labeling is used to detect the anomaly region. The same approach has been extended to videos in D'Avino et al. (2017), using an LSTM recurrent network to account for temporal dependencies.

In Bammey et al. (2020), instead, the idea is to learn locally the possible configurations of a CFA mosaic. Whenever such pattern is not found in the sliding-window analysis at test time, an anomaly is identified, and then a possible manipulation. A similar approach can be followed with reference to the JPEG quality factor: in this case, if an inconsistency is found in the image, this means that the anomalous region was compressed with a different compression level than the rest of the image, a strong evidence of local modifications (Niu et al. 2021).

11.3.2   *Image-Based Localization*

Majority of approaches that carry out localization rely on fully convolutional networks and leverage architectures developed originally for image segmentation, regarding the problem itself as a special form of segmentation between pristine and manipulated regions. A few methods, instead, take inspiration from object detection and make decisions on regions extracted from a preliminary region proposal phase. In the following, we review the main ideas.

Segmentation-like approach

These methods take as input the whole image and produce a localization map of the same size. Therefore, they avoid the problem of post-processing a large number of patch-level results in some sensible way. On the down side, they need a large collection of fake and real images for training, as well as their associated pixel-level ground truth, two rather challenging requirements. Actually, many methods keep being trained on patches and keep working on patches at testing time, but all the pieces of information are processed jointly in the network to provide an image-level output, just as it happens with denoising networks. Therefore, for these methods, training can be performed at patch level by considering many different types of local manipulations (Wu et al. 2019; Liu et al. 2021) or including boundaries between the pristine and forged area (Salloum et al. 2018; Rao et al. 2021; Zhang and Ni 2020). Many other solutions, instead, use the information from the whole image by resizing it to a fixed dimension (Wu et al. 2018; Bappy et al. 2019; Mazaheri et al. 2019; Bi et al. 2019, 2020; Kniaz et al. 2019; Shi et al. 2020; Chen et al. 2020; Hu et al. 2020; Islam et al. 2020; Chen et al. 2021). The image is reduced to the size of the network input, and therefore fine-grained pixel-level dependencies typical of in-camera and out-camera artifacts are destroyed. Moreover, local manipulations may become extremely small after resizing and easily neglected in the overall loss. This approach is conceptually simple but causes a significant loss of information. To increase the number of training samples and generalize across a large variety of possible manipulations, in Zhou et al. (2019) an adversarial strategy is also adopted. Even the loss functions used in these works are typically inspired by semantic segmentation, such as pixel-wise cross-entropy (Bi et al. 2019, 2020; Hu et al. 2020), weighted cross-entropy (Bappy et al. 2019; Rao et al. 2021; Salloum et al. 2018; Zhang and Ni 2020), and dice (Chen et al. 2021). Since often the manipulation region is much smaller than the whole image in Li and Huang (2019), it is proposed to adopt the focal loss, which assigns a modulating factor to the cross-entropy term and then can address the class imbalance problem.

Object Detection-Like Approach

A different strategy is to identify only the region box that includes the manipulation adopting approaches that are applied for the object detection task. This path is first followed in Zhou et al. (2018) relying on the Region Proposal Network (RPN) (Ren et al. 2017) very popular for object detection. This is adapted to provide regions with potential manipulations, by using features extracted both from the RGB channels and from the noise residual as input. The approach proposed in Yang et al. (2020), instead, adopts Mask R-CNN (He et al. 2020) and includes an attention region proposal network to identify the manipulated region.

11.3.3 Detection

Despite the obvious importance of detection, which should take place before local-ization is attempted, in the literature there has been limited attention to this specific problem. A large number of localization methods have been used to perform detection through a suitable post-processing of the localization heatmap aimed at extracting a global score. In some cases, the average or the maximum value of the heatmap is used as decision statistics (Huh et al. 2018; Wu et al. 2019; Rao et al. 2021). However, localization methods often perform clustering or segmentation in the target image and therefore they tend to find a forged area also in pristine images, generating a large number of false alarms.

Other methods propose ad hoc information fusion strategies. Specifically, in Rao and Ni (2016); Boroumand and Fridrich (2018), training is carried out on a patch-level analysis then statistics-based information is aggregated for the whole image to form a feature vector as the input of a classifier, either a support vector machine or a neural network. Unfortunately, a patch-level analysis does not allow taking into account both local information (through textural analyses) and global information over the whole image (through contextual analyses) at the same time. That is, by focusing on a local analysis, the big picture may go lost.

Based on these considerations, a different direction is followed in Marra et al. (2019). The network takes as input the whole image without any resizing, so as not to lose precious traces of manipulation hidden in its fine-grain structure. All patches are analyzed jointly, through a suitable aggregation, to make the final image-level decision on the presence of a manipulation. More important, also in the training phase, only image-level labels are used to update the network weights, and only image-level information back-propagates until the early patch-level feature extraction layers. To this end, a gradient checkpointing technique is used, which allows to keep in memory all relevant variables at the cost of a limited increase in computation. This approach allows for the joint optimization of patch-level feature extraction and image-level decision. The first layers extract features that are instrumental to carry out a global analysis and the last layers exploit this information to highlight local anomalies that

Fig. 11.9 Example of localization analysis based on the detection approach proposed in Marra et al. (2019). From left to right: test image, corresponding activation map, and ROI-based localization result with ground-truth box (green) and automatic ones (magenta). Detection scores are shown on the corner of each box

would not appear at the patch level. In Marra et al. (2019), this end-to-end approach is shown to also allow a good interpretation of results by means of activation maps and a good localization of forgeries (see Fig. 11.9).

11.4   Architectural Solutions

In this section, we describe the most popular and interesting deep-learning-based architectural solutions proposed in digital forensics.

11.4.1   Constrained Networks

To exploit low-level features related to camera-based artifacts, one needs to suppress the scene content. This can be done by pre-processing the input image, through a preliminary denoising step, or by means of suitable high-pass filters, like the popular spatial rich models (SRM) initially proposed in the context of steganalysis (Fridrich and Kodovsky 2012) and applied successfully in image forensics (Cozzolino et al. 2014). Several CNN architectures try to include this pre-processing in their architecture by means of a constrained first layer which performs the desired high-pass filtering. For example, in Rao and Ni (2016), a set of fixed high-pass filters inspired to the spatial rich model are used to compute residuals of the input patches, while in Bayar and Stamm (2016) the high-pass filters of the first convolutional layer are learnt using the following constraints:

$$w_k(0,0) = -1$$
$$\sum_{i,j} w_k(i,j) = 0 \tag{11.1}$$

where w_k (i, j) is the weight of k-th filter at the (i, j) position and $(0, 0)$ indicates the central position. This solution is pursued also in other works, such as (Wu et al. 2019; Yang et al. 2020; Chen et al. 2021), while many others use a fixed filtering (Li and Huang 2019; Barni et al. 2017; Zhou et al. 2018) or adopt a combination of fixed high-pass filters and trainable ones (Wu et al. 2019; Zhou et al. 2018; Bi et al. 2020; Zhang and Ni 2020).

A slightly different perspective is considered in Cozzolino et al. (2017) where a CNN architecture is built to replicate exactly the co-occurrence-based methods of Fridrich and Kodovsky (2012). All the phases of this procedure, i.e., extraction of noise residuals, scalar quantization, computation of co-occurrences, and histogram are implemented using standard convolutional or pooling layers. Once established the equivalence, the network is modified to enable its fine-tuning and further improve performance. Since the resulting network has a lightweight structure, fine-tuning can be carried out using a small training set, limiting computation time.

11.4.2 Two-Branch Networks

In order to exploit different types of clues at the same time, it is possible to adopt a CNN architecture that accepts multiple inputs (Zhou et al., 2018; Shi et al., 2018; Bi et al., 2020; Kwon et al., 2021). For example, a two-branch approach is proposed in Zhou et al. (2018) in order to take into account both low-level features extracted from noise residuals and high-level features extracted from RGB images. In Amerini et al. (2017); Kwon et al. (2021), instead, information extracted from the spatial and DCT domain is analyzed so as to take explicitly into account also compression artifacts. Two-branch solutions can also be used to produce multiple outputs, as done in Salloum et al. (2018), where the fully convolutional network has two output branches: one provides the localization map of splicing while the other provides a map containing only the edges of the forged region. In the context of copy-move detection, the two-branch structure is used to address separately the problems of detecting duplicates and disambiguating between the original object an its clones (Wu et al. 2018; Barni et al. 2019).

11.4.3 Fully Convolutional Networks

Fully convolution networks (FCNs) can be extremely useful since they preserve spatial dependencies and generate an output of the same size of the input image. Therefore, the network can be trained to output a heat map which allows to perform tampering localization. Various architectures have shown to be particularly useful for forensics applications.

Some methods (Bi et al. 2019; Shi et al. 2020; Zhang and Ni 2020) rely on a plain U-Net architecture (Ronneberger et al. 2015), one of the most successful network for semantic segmentation, initially proposed for biomedical applications. A more innovative variation of FCNs is proposed in Bappy et al. (2019); Mazaheri et al. (2019) where the architecture includes a long short-term memory (LSTM) module. The blocks of the image are sequentially processed by the LSTM. In this way, the network can model the spatial relationships between neighboring blocks, which facilitates detecting manipulated blocks as those that break the natural statistics of authentic ones. Other papers include spatial pyramid mechanisms in the architectures, in order to capture relationships at different scales of the image (Bi et al. 2020; Wu et al. 2019; Hu et al. 2020; Liu et al. 2021; Chen et al. 2021).

Both LSTM modules and spatial pyramids aim at overcoming the intrinsic limits of networks based only on convolutions where, due to the limited overall receptive field, the decision on a pixel depends only on the information in a small region around the pixel. With the same aim, it is also possible to use attention mechanisms (Chen et al. 2008, 2021; Liu et al. 2021; Rao et al. 2021), as in Shi et al. (2020), where channel attention is used in the Gram matrix to measure the correlation between any two feature maps.

11.4.4 Siamese Networks

Siamese networks can be extremely useful for several forensic purposes. In fact, by leveraging the Siamese configuration, with two identical networks working in parallel, one can focus on the similarity and differences among patches, making up for the absence of ground-truth data.

This approach is applied in Mayer and Stamm (2018), where a constrained network is used to extract camera-model features, while another network is trained to learn the similarity between pairs of such features. This strategy is further developed in Mayer and Stamm (2019) where a graph-based representation is introduced to better identify the forensic relationships among all patches within an image. A Siamese network is also used in Huh et al. (2018) to establish if image patches were captured with different imaging pipelines. The proposed approach is self-supervised and localizes image splicings by predicting the consistency of EXIF attributes between pairs of patches in order to establish whether they came from a single coherent image. Once trained on pristine images, featured by their EXIF header, the network can be used on any new image without further supervision. As already seen in Sect. 1.2.1, a Siamese approach can help to extract a camera-model fingerprint, where artifacts related to camera model are emphasized (Cozzolino and Verdoliva 2020, 2018; Cozzolino et al. 2019).

11.5   Datasets

For two-class learning-based approaches, having good data for training is of paramount importance. By "good", we mean abundant (for data-hungry modern CNNs), representative (of the many possible manipulations encountered in the wild) and well curated (e.g., balanced, free from biases). Moreover, to assess the performance of new proposals, it is important to compare results on multiple datasets with different features. The research community has made considerable efforts through the years to release a number of datasets with a wide array of image manipulations. However, not all of them possess the right properties to support, by themselves, the development of learning-based methods. In fact, a way too popular approach is to split a single dataset in training, validation, and test set, carrying out training and experiments on this single source, a practice that may easily induce some forms of polarization or overfitting, if the dataset is not built with great care.

In Fig. 11.10, we show a list of the most widespread datasets that have been proposed since 2011. During this short time lapse, the size of the datasets has rapidly increased, by roughly two orders of magnitude, together with the variety of manipulations (see Table 11.2). although the capacity to fool an observer has not grown at a similar pace. In this section, the most widespread datasets are described, and their features are briefly discussed.

The Columbia dataset (Hsu and Chang 2006) is one of the first dataset proposed in the literature. It has been extensively used, however it includes only unrealistic forgeries without any semantic meaning. A more realistic dataset with splicings is DSO-1 (Carvalho et al. 2013) that in turn has the limitation that all images have the

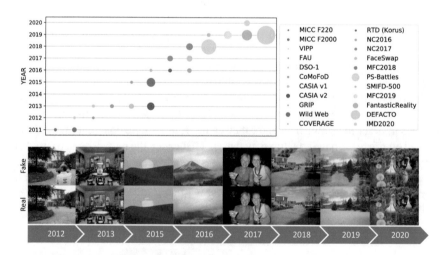

Fig. 11.10 An overview of the most used datasets in the current literature for image manipulation detection and localization. We can observe that more recent ones are much larger, while the level of realism has not really improved upon time. Note that the size of the circles corresponds to the number of samples in the dataset

Table 11.2 List of datasets including generic image manipulations. Some of them are targeted for specific manipulations, while others are more various and contain different types of forgeries

Dataset	Refs.	manipulations	# prist.	# forged
Columbia gray	Ng and Chang (2004)	Splicing (unrealistic)	933	912
Columbia color	Hsu and Chang (2006)	Splicing (unrealistic)	182	180
CASIA v1	Dong et al. (2013)	Splicing, copy-move	800	921
CASIA v2	Dong et al. (2013)	Splicing, copy-move	7,200	5,123
MICC F220	Amerini et al. (2011)	Copy-move	110	110
MICC F2000	Amerini et al. (2011)	Copy-move	1,300	700
VIPP	Bianchi and Piva (2012)	Splicing	68	69
FAU	Christlein et al. (2012)	Copy-move	48	48
DSO-1	Carvalho et al. (2013)	Splicing	100	100
CoMoFoD	Tralic et al. (2013)	Copy-move	260	260
Wild Web	Zampoglou et al. (2015)	Real-world cases	90	9,657
GRIP	Cozzolino et al. (2015)	Copy-move	80	80
RTD (Korus)	Korus and Huang (2016)	Various	220	220
COVERAGE	Wen et al. (2016)	Copy-move	100	100
NC2016	Guan et al. (2019)	Various	560	564
NC2017	Guan et al. (2019)	Various	2,667	1,410
FaceSwap	Zhou et al. (2017)	Face swapping	1,758	1,927
MFC2018	Guan et al. (2019)	Various	14,156	3,265
PS-Battles	Heller et al. (2018)	Various	11,142	102,028
MFC2019	MFC (2019)	Various	10,279	5,750
DEFACTO	Mahfoudi et al. (2019)	Various	–	229,000
FantasticReality	Kniaz et al. (2019)	Splicing	16,592	19,423
SMIFD-500	Rahman et al. (2019)	Various	250	250
IMD2020	Novozámský et al. (2020)	Various	414	2,010

same resolution, they are in uncompressed format and there is no information on how many cameras were used. A very large dataset of face swapping has been introduced in Zhou et al. (2017), which contains around 4,000 real and manipulated images using two different algorithms. Several datasets have been designed for copy-move forgery detection (Christlein et al. 2012; Amerini et al. 2011; Tralic et al. 2013; Cozzolino et al. 2015; Wen et al. 2016), one of the most studied forms of manipulation. Some of them modify the copied object in multiple ways, including rotation, resizing, and change of illumination, to stress the capabilities of copy-move methods.

Many other datasets include various types of manipulations so as to present a more realistic scenario. The CASIA dataset (Dong et al. 2013) includes both splicings and copy-moves and inserted objects are post-processed to better fit the scene. Nonetheless, it exhibits a strong polarization, as highlighted in Cattaneo and G. Roscigno (2014): authentic images can be easily separated from tampered ones since they were compressed with different quality factors. Forgeries of various nature are present also in the Realistic Tampering Dataset proposed in Korus and Huang (2016) that comprises all uncompressed, but very realistic manipulations. The Wild Web Dataset is instead a collection of cases from the Internet (Zampoglou et al. 2017).

The U.S. National Institute of Standards and Technology (NIST) has released several large datasets (Guan et al. 2019) aimed at testing forensic algorithms in increasingly realistic and challenging conditions. The first one, NC2016, is quite limited and present some redundancies, such as same images repeated multiple times in different conditions. While this can be of interest to study the behavior of some algorithms, it prevents the dataset to be split for training, validation, and test to avoid overfitting. Much more challenging and large datasets have been proposed by NIST in subsequent years: NC2017, MFC2018, and MFC2019. These datasets are very large and present very different types of manipulations, resolutions, formats, compression levels, and acquisition devices.

Some very large datasets have been released recently. DEFACTO (Mahfoudi et al. 2019) comprises over 200,000 images with a wide variety of realistic manipulations, both conventional, like splicings, copy-moves, and removals, and more advanced, like face morphing. The PS-Battles Dataset, instead, provides for each original image a varying number of manipulated versions (Heller et al. 2018), for a grand total of 102,028 images. However, it does not provide ground truths and the original themselves are not always pristine. SMIFD-500 (Social Media Image Forgery Detection Database) Rahman et al. (2019) is a set of 500 images collected from popular social media, while FantasticReality (Kniaz et al. 2019) is a dataset of splicing manipulations split into two parts: "Rough" and "Realistic". The "Rough" part contains 8k splicings with obvious artifacts while the "Realistic" part provides 8k splicings that were retouched manually to obtain a realistic output.

A dataset of manipulated images has been also built in Novozámský et al. (2020). It features 35,000 images collected from 2,322 different camera models, and manipulated in various ways, e.g., copy-paste, splicing, retouching, and inpainting. Moreover, it also includes a set of 2,000 realistic fake images downloaded from the Internet along with the corresponding real images.

Since some methods work on anomaly detection and are trained only on pristine data, well-curated datasets comprising a large number of real images are also of interest. The Dresden image database (Gloe and Böhme 2010) contains over 14,000 JPEG images from 73 digital cameras of 25 different models. To allow studying the peculiarities of different models and devices, the same scene is captured by a large number of cameras. Raw images can be found both in Dresden and in the RAISE dataset (Dang-Nguyen et al. 2015), composed of 8,156 images taken by 3 different camera models. A dataset comprising both SDR (standard dynamic range) and HDR (high dynamic range) images has been presented in Shaya et al. (2018).

A total of 5,415 images were captured by 23 mobile devices in different scenarios under controlled acquisition conditions. The VISION dataset, instead, Shullani et al. (2017) includes 34,427 images and 1,914 videos from 35 devices of 11 major brands. They appear both in their original format and after uploading/downloading on various platforms (Facebook, YouTube, WhatsApp) so as to allow studies on data retrieved from social networks. Another mixed dataset, SOCRATES, is proposed in Galdi et al. (2017). It contains 6,200 images and 680 videos captured using 67 smartphones of 14 brands and 42 models.

11.6 Major Challenges

In this section, we want to highlight some major challenges of current deep-learning-based approaches.

- Localization versus Detection. To limit computational complexity and memory storage, deep networks work on relatively small input patches. Therefore, to process large images, one must either resize them or analyze them patch-wise. The first solution does not fit forensic needs, as it destroys precious statistical evidence related to local textures. On the other hand, a patch-wise analysis may miss image-level phenomena. A feature extractor trained on patches can only learn good features for local decisions, which are not necessarily good to make image-level decisions. Most deep-learning-based methods sweep under the rug the detection problem and focus on localization. Then, detection is addressed as an afterthought through some forms of processing of the localization map. Not surprisingly, the resulting detection performance is often very poor.

  Table 11.3 compares the localization and detection performances of some patch-wise methods. We present results in terms of Area Under the Curve (AUC) for detection and the Matthews Correlation Coefficient (MCC) for localization. MCC is robust to unbalanced classes (the forgery is often small compared to the whole image) its absolute value belongs to the range $[0, 1]$. It represents the cross-correlation coefficient between the decision map and the ground truth, computed as

  $$\text{MCC} = \frac{TP \times TN - FP \times FN}{\sqrt{(TP + FP)(TP + FN)(TN + FP)(TN + FN)}}$$

  where

  - TP (true positive): # positive pixels declared positive;
  - TN (true negative): # negative pixels declared negative;
  - FP (false positive): # negative pixels declared positive;
  - FN (false negative): # positive pixels declared negative;

We can note that, while the localization results are relatively good, detection results are much worse and often no better than coin flipping. A more intense effort on the detection problem seems necessary.

Table 11.3 Localization by means of MCC and detection by means of AUC performance compared. With a few exceptions, the detection performance is very poor (AUC = 0.5 amounts to coin flipping)

MCC/AUC	DS0-1	VIPP	RTD	PS-Battles
	100/100	62/62	220/220	80/80
RRU-Net Bi et al. (2019)	0.250/0.522	0.288/0.538	0.240/0.524	0.303/0.583
Const. R-CNN Yang et al. (2020)	0.355/0.575	0.382/0.480	0.370/0.522	0.423/0.517
EXIF-SC Huh et al. (2018)	0.529/0.763	0.402/0.633	0.278/0.541	0.345/0.567
SPAN Hu et al. (2020)	0.318/0.697	0.336/0.556	0.260/0.564	0.320/0.583
Adaptive-CFA Bammey et al. (2020)	0.121/0.380	0.130/0.498	0.127/0.472	0.094/0.526
Noiseprint Cozzolino and Verdoliva (2020)	0.758/0.865	0.532/0.568	0.345/0.531	0.476/0.495

- Robustness. Forensic analyses rely typically on the weak traces associated with the digital history of the target image/video. Such traces are further weakened and tend to disappear altogether in the presence of other quality-impairing processing steps. As an example, consider the image of Fig. 11.11, with a large splicing, localized with high accuracy. After compression or resizing, the localization map becomes more fuzzy, and decision unreliable. Likewise, in the images of Fig. 11.12 with copy-moves, the attribution of the two copies becomes incorrect after heavy resizing. On the other hand, most images and videos of interest are found on social networks, where they are routinely heavily resized and recompressed for storage efficiency, with non-deterministic parameters. Therefore, robustness is a central problem that all multimedia forensic methods should confront with from the beginning.
- Generalization. Carrying out a meaningful training is a hard task in image forensics, in fact, it is very easy to introduce some sort of bias. For example, one may train and test the network on images acquired only by a limited number of devices, identifying the provenance source instead of the artifact trace, or may use different JPEG compression pipelines for real and fake images, giving rise to different digital histories. A network will no doubt learn all these features if they help discriminating between pristine and tampered images, and may neglect more general but weaker traces of manipulation. This will lead to a very good performance on a training-aligned test set, and very bad performance on independent tests. Therefore, it is important to avoid biases to the extent possible, and even more important to test the trained networks on multiple and independent test sets.
- Semantic Detection. Developing further on detection, a major problem is telling apart innocent manipulations from malicious ones. An image may be manipulated for many reasons, very frequently only for improving its appearance or to better fit it in the context of interest. These images should not be detected, as they are not really of interest and their number would easily overwhelm any subsequent analysis system. A good forensic algorithm should be able to single out only malicious attacks, based on the type of manipulation, often concerning only a small part of the image, and especially based on the context, including all other related media and textual information.

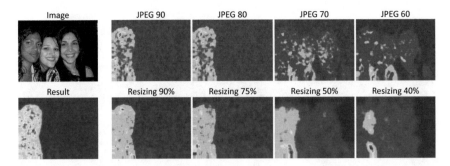

Fig. 11.11 Results of splicing localization in the presence of different levels of compression and resizing. MCC is 0.998, it decreases to 0.985, 0.679, 0.412, and 0.419 by varying the compression level: JPEG quality equal to 90, 80, 70, and 60, respectively. MCC decreases to 0.994, 0.991, 0.948, and 0.517 by varying the resizing factor: 0.90, 0.75, 0.50, and 0.40, respectively

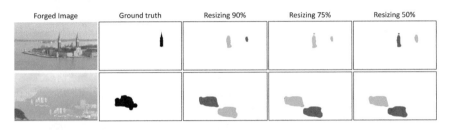

Fig. 11.12 Results of copy-move detection and attribution in the presence of resizing

11.7   Conclusions and Future Directions

Manipulating images and videos is becoming simpler and simpler, and fake media are becoming widespread, with the well-known associated risks. Therefore, the development of reliable multimedia forensic tools is more urgent than ever. In recent years, the main focus of research has been on deepfakes, images, and videos (often of persons) fully generated by means of AI tools. Yet, detecting computer-generated material seems to be relatively simple, for the time being. These images and videos lack some characteristic features of real media assets (think of device and model fingerprints) and exhibit instead traces of their own generation process (e.g., GAN fingerprints). Moreover, they can be generated in large number, providing for a virtually unlimited training set for deep learning methods to learn from.

Cheapfakes, instead, though crafted with conventional methods, keep evading more easily the scrutiny of forensic tools. Also for them, detectors based on deep learning can be expected to provide improved performance. However, creating large unbiased datasets with manipulated images representative of the wide variety of attacks is not simple. In addition, the manipulations themselves rely on real (as opposed to generated) data, hardly detected as anomalous. Finally, manipulated images are often resized and recompressed, further weakening the feeble forensic

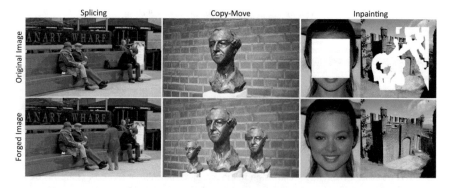

Fig. 11.13 Examples of conventional manipulations carried out using deep learning methods: splicing (Tan et al. 2018), copy-move (Thies et al. 2019), inpainting (Zhu et al. 2021)

traces a detector can rely on. All this said, the current state of deep-learning-based methods for cheapfake detection can be considered promising, and further improvements will certainly come.

However, new challenges loom on the horizon. The acquisition and processing technology evolves rapidly, deleting or weakening well-known forensic traces, just think of the effects of video stabilization and new shooting modes on PRNU-based methods (Mandelli et al. 2020; Baracchi et al. 2020). Of course, new traces will appear, but research may struggle keeping this fast pace. Likewise, also manipulation methods evolve, see Fig. 11.13. For example, the inpainting process has been significantly improved thanks to deep learning, with recent methods which fill image regions using newly generated semantically consistent content. In addition, methods have been already proposed to conceal the traces left by various types of attacks. Along this line, another peculiar problem of deep-learning-based methods is their vulnerability to adversarial attacks. The class (pristine/fake) of an asset can be easily changed by means of a number of simple methods, without significantly impairing its visual quality.

As a final remark, we underline the need for interpretable tools. Deep learning methods, despite their excellent performance, act as black boxes, providing little or no hints on why certain decisions are made. Of course, such decisions may be easily questioned and may hardly stand in a court of justice. Improving explainability, though largely unexplored, is therefore a major topic for current and future work in this field.

Acknowledgements This material is based on research sponsored by the Defense Advanced Research Projects Agency (DARPA) and the Air Force Research Laboratory (AFRL) under agreement number FA8750-20-2-1004. The U.S. Government is authorized to reproduce and distribute reprints for Governmental purposes notwithstanding any copyright notation thereon. The views and conclusions contained herein are those of the authors and should not be interpreted as necessarily representing the official policies or endorsements, either expressed or implied, of DARPA and AFRL or the U.S. Government. This work is also supported by the PREMIER project, funded by the Italian Ministry of Education, University, and Research within the PRIN 2017 program and by a Google gift.

References

Al Shaya O, Yang P, Ni R, Zhao Y, Piva A (2018) A new dataset for source identification of high dynamic range images. Sensors 18

Amerini I, Ballan L, Caldelli R, Del Bimbo A, Serra G (2011) A SIFT-Based forensic method for copy-move attack detection and transformation recovery. IEEE Trans Inf Forensics Secur 6(3):1099–1110

Amerini I, Uricchio T, Ballan L, Caldelli R (2017) Localization of JPEG double compression through multi-domain convolutional neural networks. In: IEEE computer vision and pattern recognition (CVPR) workshops

Bammey Q, von Gioi R, Morel J-M (2020) An adaptive neural network for unsupervised mosaic consistency analysis in image forensics. In: IEEE conference on computer vision and pattern recognition (CVPR), pp 14182–14192

Bappy JH, Simons C, Nataraj L, Manjunath BS, Roy-Chowdhury AK (2019) Hybrid LSTM and encoder-decoder architecture for detection of image forgeries. IEEE Trans Image Process 28(7):3286–3300

Baracchi D, Iuliani M, Nencini A, Piva A (2020) Facing image source attribution on iPhone X. In: International workshop on digital forensics and watermarking (IWDW)

Barni M, Bondi L, Bonettini N, Bestagini P, Costanzo A, Maggini M, Tondi B, Tubaro S (2017) Aligned and non-aligned double JPEG detection using convolutional neural networks. J Vis Commun Image Represent 49:153–163

Barni M, Phan Q-T, Tondi B (2021) Copy move source-target disambiguation through multi-branch CNNS. IEEE Trans Inf Forensics Secur 16:1825–1840

Bayar B, Stamm MC (2016) A deep learning approach to universal image manipulation detection using a new convolutional layer. In: ACM workshop on information hiding and multimedia security

Bianchi T, Piva A (2012) Image forgery localization via block-grained analysis of JPEG artifacts. IEEE Trans. Inf. Forensics Secur 7(3):1003–1017

Bi X, Wei Y, Xiao B, Li W (2019) RRU-Net: the ringed residual U-Net for image splicing forgery detection. In: IEEE computer vision and pattern recognition (CVPR) workshops

Bi X, Liu Y, Xiao B, Li W, Pun C-M, Wang G, Gao X (2020) D-Unet: a dual-encoder u-net for image splicing forgery detection and localization. arXiv:2012.01821

Bondi L, Lameri S, Güera D, Bestagini P, Delp EJ, Tubaro S (2017) Tampering detection and localization through clustering of camera-based CNN features. In: IEEE computer vision and pattern recognition (CVPR) workshops

Boroumand M, Fridrich J (2018) Deep learning for detecting processing history of images. In: IS&T electronic imaging: media watermarking, security, and forensics

Cao H, Kot AC (2009) Accurate detection of demosaicing regularity for digital image forensics. IEEE Trans Inf Forensics Secur 5:899–910

Cattaneo G, Roscigno G (2014) A possible pitfall in the experimental analysis of tampering detection algorithms. In: International conference on network-based information systems, pp 279–286

Chen M, Fridrich J, Goljan M, Lukàš J (2008) Determining image origin and integrity using sensor noise. IEEE Trans Inf Forensics Secury 3(4):74–90

Chen X, Dong C, Ji J, Cao J, Li X (2021) Image manipulation detection by multi-view multi-scale supervision. arXiv:2104.06832

Chen B, Tan W, Coatrieux G, Zheng Y, Shi YQ (2020) A serial image copy-move forgery localization scheme with source/target distinguishment. IEEE Trans. Multimed

Christlein V, Riess C, Jordan J, Angelopoulou E (2012) An evaluation of popular copy-move forgery detection approaches. IEEE Trans Inf Forensics Secur 7(6):1841–1854

Cozzolino D, Verdoliva L (2020) Noiseprint: a CNN-based camera model fingerprint. IEEE Trans Inf Forensics Secur 15(1):14–27

Cozzolino D, Poggi G, Verdoliva L (2015) Efficient dense-field copy-move forgery detection. IEEE Trans Inf Forensics Secur 10(11):2284–2297

Cozzolino D, Gragnaniello D, Verdoliva L (2014) Image forgery detection through residual-based local descriptors and block-matching. In: IEEE International Conference on Image Processing (ICIP), pp 5297–5301

Cozzolino D, Poggi G, Verdoliva L (2017) Recasting residual-based local descriptors as convolutional neural networks: an application to image forgery detection. In: ACM workshop on information hiding and multimedia security, pp 1–6

Cozzolino D, Poggi G, Verdoliva L (2019) Extracting camera-based fingerprints for video forensics. In: IEEE computer vision and pattern recognition (CVPR) workshops, pp 130–137

Cozzolino D, Verdoliva L (2016) Single-image splicing localization through autoencoder-based anomaly detection. In: IEEE workshop on information forensics and security (WIFS), pp 1–6

Cozzolino D, Verdoliva L (2018) Camera-based image forgery localization using convolutional neural networks. In: European signal processing conference (EUSIPCO), Sep 2018

Dang-Nguyen DT, Pasquini C, Conotter V, Boato G (2015) RAISE: a raw images dataset for digital image forensics. In: 6th ACM multimedia systems conference, pp 219–1224

D'Avino D, Cozzolino D, Poggi G, Verdoliva L (2017) Autoencoder with recurrent neural networks for video forgery detection. In: IS&T international symposium on electronic imaging: media watermarking, security, and forensics

de Carvalho T, Riess C, Angelopoulou E, Pedrini H, Rocha A (2013) Exposing digital image forgeries by illumination color classification. IEEE Trans Inf Forensics Secur 8(7):1182–1194

Dong J, Wang W, Tan T (2013) CASIA image tampering detection evaluation database. In: IEEE China summit and international conference on signal and information processing, pp 422–426

Farid H (2016) Photo forensics. The MIT Press

Ferrara P, Bianchi T, De Rosa A, Piva A (2012) Image forgery localization via fine-grained analysis of CFA artifacts. IEEE Trans Inf Forensics Secur 7(5):1566–1577

Fridrich J, Kodovsky J (2012) Rich models for steganalysis of digital images. IEEE Trans Inf Forensics Secur 7:868–882

Fridrich J, Soukal D, Lukáš J (2003) Detection of copy-move forgery in digital images. In: Proceedings of the 3rd digital forensic research workshop

Fu D, Shi YQ, Su W (2007) A generalized Benford's law for JPEG coefficients and its applications in image forensics. In: Proceedings of the SPIE, Security, Steganography, and Watermarking of Multimedia Contents IX

Galdi C, Hartung F, Dugelay J-L (2017) Videos versus still images: asymmetric sensor pattern noise comparison on mobile phones. Media watermarking, security and forensics. In: IS&T EI

Gloe T, Böhme R (2010) The 'Dresden Image Database' for benchmarking digital image forensics. In: Proceedings of the 25th annual ACM symposium on applied computing, vol 2, pp 1585–1591, Mar 2010

Grgic S, Tralic D, Zupancic I, Grgic M (2013) CoMoFoD-New database for copy-move forgery detection. In: Proceedings of the 55th international symposium ELMAR, pp 49–54

Guan H, Kozak M, Robertson E, Lee Y, Yates AN, Delgado A, Zhou D, Kheyrkhah T, Smith J, Fiscus J (2019) MFC datasets: large-scale benchmark datasets for media forensic challenge evaluation. In: IEEE winter conference on applications of computer vision (WACV) workshops, pp 63–72

He Z, Lu W, Sun W, Huang J (2012) Digital image splicing detection based on Markov features in DCT and DWT domain. Pattern Recogn 45:4292–4299

He K, Gkioxari G, Dollár P, Girshick R (2020) Mask R-CNN. IEEE Trans Pattern Anal Mach Intell 42(2):386–397

Heller S, Rossetto L, Schuldt H (2018) The ps-battles dataset-an image collection for image manipulation detection. arXiv:1804.04866

Hsu Y-F, Chang S-F (2006) Detecting image splicing using geometry invariants and camera characteristics consistency. In: IEEE international conference on multimedia and expo (ICME), pp 549–552

Huh M, Liu A, Owens A, Efros AA (2018) Fighting fake news: image splice detection via learned self-consistency. In: European conference on computer vision (ECCV)

Hu X, Zhang Z, Jiang Z, Chaudhuri S, Yang Z, Nevatia R (2020) SPAN: spatial pyramid attention network for image manipulation localization. In: European conference on computer vision (ECCV), pp 312–328

Islam A, Long C, Basharat A, Hoogs A (2020) DOA-GAN: dual-order attentive generative adversarial network for image copy-move forgery detection and localization. In: IEEE conference on computer vision and pattern recognition (CVPR), pp 4675–4684

Kirchner M (2008) Fast and reliable resampling detection by spectral analysis of fixed linear predictor residue. In: 10th ACM workshop on multimedia and security, pp 11–20

Kniaz VV, Knyaz V, Remondino F (2019) The point where reality meets fantasy: Mixed adversarial generators for image splice detection. Adv Neural Inf Process Syst (NIPS) 32:215–226

Korus P (2017) Digital image integrity—a survey of protection and verification techniques. Digit Signal Process 71:1–26

Korus P, Huang J (2016) Evaluation of random field models in multi-modal unsupervised tampering localization. In: IEEE international workshop on information forensics and security, pp 1–6, Dec 2016

Kwon M-J, Yu I-J, Nam S-H, Lee H-K (2021) CAT-Net: compression artifact tracing network for detection and localization of image splicing. In: IEEE winter conference on applications of computer vision (WACV), pp 375–384, Jan 2021

Li H, Huang J (2019) Localization of deep inpainting using high-pass fully convolutional network. In: IEEE international conference on computer Vision (ICCV), pp 8301–8310

Liu Y, Guan Q, Zhao X, Cao Y (2018) Image forgery localization based on multi-scale convolutional neural networks. In: ACM workshop on information hiding and multimedia security

Liu X, Liu Y, Chen J, Liu X (2021) PSCC-Net: progressive spatio-channel correlation network for image manipulation detection and localization. arXiv:2103.10596

Lukàš J, Fridrich J, Goljan M (2006) Detecting digital image forgeries using sensor pattern noise. In: Proceedings of the SPIE, pp 362–372

Lyu S, Farid H (2005) How realistic is photorealistic? IEEE Trans Signal Process 53(2):845–850

Mahfoudi G, Tajini B, Retraint F, Morain-Nicolier F, Dugelay JL, France B, Pic M (2019) DEFACTO: image and face manipulation dataset. In: European signal processing conference

Mandelli S, Bestagini P, Verdoliva L, Tubaro S (2020) Facing device attribution problem for stabilized video sequences. IEEE Trans Inf Forensics Secur 15(1):14–27

Marra F, Gragnaniello D, Verdoliva L, Poggi G (2020) A full-image full-resolution end-to-end-trainable CNN framework for image forgery detection. IEEE Access 8:133488–133502

Mayer O, Stamm MC (2020) Exposing fake images with forensic similarity graphs. IEEE J Sel Top Signal Process 14(5):1049–1064

Mayer O, Stamm MC (2018) Learned forensic source similarity for unknown camera models. In: IEEE international conference on acoustics, speech and signal processing (ICASSP), pp 2012–2016, Apr 2018

Mazaheri G, Chowdhury Mithun N, Bappy JH, Roy-Chowdhury AK (2019) A skip connection architecture for localization of image manipulations. In: IEEE computer vision and pattern recognition (CVPR) workshops

MFC2019. https://www.nist.gov/itl/iad/mig/media-forensics-challenge-2019-0

Nam S-H, Park J, Kim D, Yu I-J, Kim T-Y, Lee H-K (2019) Two-Stream network for detecting double compression of H.264 videos. In: IEEE international conference on image processing (ICCV)

Ng T-T, Chang S-F (2004) A data set of authentic and spliced image blocks. Tech. Rep. TR203-2004-3, Columbia University

Niu Y, Tondi B, Zhao Y, Ni R, Barni M (2021) Image splicing detection, localization and attribution via JPEG primary quantization matrix estimation and clustering. arXiv:2102.01439

Novozámský A, Mahdian B, Saic S (2020) IMD2020: a large-scale annotated dataset tailored for detecting manipulated images. In: IEEE winter conference on applications of computer vision (WACV) workshops

Paris B, Donovan J (2019) Deepfakes and cheap fakes. Data & Society, USA

Park J, Cho D, Ahn W, Lee H-K (2018) Double JPEG detection in mixed JPEG quality factors using deep convolutional neural network. In: European conference on computer vision (ECCV)

Phan-Xuan H, Le-Tien T, Nguyen-Chinh T, Do-Tieu T, Nguyen-Van Q, Nguyen-Thanh T (2019) Preserving spatial information to enhance performance of image forgery classification. In: International conference on advanced technologies for communications (ATC), pp 50–55

Popescu AC, Farid H (2005) Exposing digital forgeries in color filter array interpolated images. IEEE Trans Signal Process 53(10):3948–3959

Rahman M, Tajrin J, Hasnat A, Uzzaman N, Atiqur Rahaman R (2019) SMIFD: novel social media image forgery detection database. In: International conference on computer and information technology (ICCIT), pp 1–6

Rao Y, Ni J (2016) A deep learning approach to detection of splicing and copy-move forgeries in images. In: IEEE international workshop on information forensics and security (WIFS), pp 1–6

Rao Y, Ni J, Xie H (2021) Multi-semantic CRF-based attention model for image forgery detection and localization. Signal Process 183:108051

Ren S, He K, Girshick R, Sun J (2017) Faster R-CNN: towards real-time object detection with region proposal networks. IEEE Trans Pattern Anal Mach Intell 39(6):1137–1149

Ronneberger O, Fischer P, Brox T (2015) U-net: Convolutional networks for biomedical image segmentation. In: Medical image computing and computer-assisted intervention (MICCAI), pp 234–241

Salloum R, Ren Y, Jay Kuo CC (2018) Image splicing localization using a multi-task fully convolutional network (MFCN). J Vis Commun Image Represent 201–209

Shi Z, Shen X, Kang H, Lv Y (2018) Image manipulation detection and localization based on the dual-domain convolutional neural networks. IEEE Access 6:76437–76453

Shi Z, Shen X, Chen H, Lyu Y (2020) Global semantic consistency network for image manipulation detection. IEEE Signal Process Lett 27:1755–1759

Shullani D, Fontani M, Iuliani M, Al Shaya O, Piva A (2017) VISION: a video and image dataset for source identification. EURASIP J Inf Secur 1–16

Tan F, Bernier C, Cohen B, Ordonez V, Barnes C (2018) Where and who? Automatic semantic-aware person composition. In: IEEE winter conference on applications of computer vision (WACV), pp 1519–1528

Thies J, Zollhöfer M, Nießner M (2019) Deferred neural rendering: image synthesis using neural textures. ACM Trans Grap (TOG) 38(4):1–12

Verdoliva L (2020) Media forensics and DeepFakes: an overview. IEEE J Sel Top Signal Process 14(5):910–932

Wang S-Y, Wang O, Owens A, Zhang R, Efros AA (2019) Detecting photoshopped faces by scripting photoshop. In: International conference on computer vision (ICCV)

Wang Q, Zhang R (2016) Double JPEG compression forensics based on a convolutional neural network. EURASIP J Inf Secur 1–12

Wen B, Zhu Y, Subramanian R, Ng T-T, Shen X, Winkler S (2016) COVERAGE—a novel database for copy-move forgery detection. In: IEEE international conference on image processing (ICIP), pp 161–165

Wu Y, Abd-Almageed W, Natarajan P (2018) BusterNet: detecting copy-move image forgery with source/target localization. In: European conference on computer vision (ECCV), pp 170–186

Wu Y, Abd-Almageed W, Natarajan P (2018) Image copy-move forgery detection via an end-to-end deep neural network. In: IEEE winter conference on applications of computer vision (WACV)

Wu Y, AbdAlmageed W, Natarajan P (2019) ManTra-Net: manipulation tracing network for detection and localization of image forgeries with anomalous features. In: IEEE Conference on Computer Vision and Pattern Recognition (CVPR)

Yang C, Li H, Lin F, Jiang B, Zhao H (2020) Constrained R-CNN: a general image manipulation detection model. In: IEEE international conference on multimedia and expo (ICME), pp 1–6

Yao H, Xu M, Qiao T, Wu Y, Zheng N (2020) Image forgery detection and localization via a reliability fusion map. Sensors 20(22):6668

Yousfi Y, Fridrich J (2020) An intriguing struggle of CNNs in JPEG Steganalysis and the OneHot solution. IEEE Signal Process. Lett. 27:830–834

Zampoglou M, Papadopoulos S, Kompatsiaris Y (2017) Large-scale evaluation of splicing localization algorithms for web images. Multimed Tools Appl 76(4):4801–4834

Zampoglou M, Papadopoulos S, Kompatsiaris Y (2015) Detecting image splicing in the wild (web). In: IEEE international conference on multimedia and expo (ICME) workshops

Zhang R, Ni J (2020) A dense u-net with cross-layer intersection for detection and localization of image forgery. In: IEEE conference on acoustics, speech and signal processing (ICASSP), pp 2982–2986

Zhang Z, Zhang Y, Zhou Z, Luo J (2018) Boundary-based Image Forgery Detection by Fast Shallow CNN. In: IEEE international conference on pattern recognition (ICPR)

Zhong J-L, Pun C-M (2019) An end-to-end dense-inceptionnet for image copy-move forgery detection. IEEE Trans Inf Forensics Secur 15:2134–2146

Zhou P, Chen B-C, Han X, Najibi M, Shrivastava A, Lim S-N, Davis L (2020) Generate, segment, and refine: towards generic manipulation segmentation 34:13058–13065, Apr 2020

Zhou P, Han X, Morariu V, Davis L (2017) Two-stream neural networks for tampered face detection. In: IEEE computer vision and pattern recognition (CVPR) workshops, pp 1831–1839

Zhou P, Han X, Morariu V, Davis L (2018) Learning rich features for image manipulation detection. In: IEEE conference on computer vision and pattern recognition (CVPR)

Zhu X, Qian Y, Zhao X, Sun B, Sun Y (2018) A deep learning approach to patch-based image inpainting forensics. Signal Process: Image Commun 67:90–99

Zhu M, He D, Li X, Li C, Li F, Liu X, Ding E, Zhang Z (2021) Image inpainting by end-to-end cascaded refinement with mask awareness. IEEE Trans Image Process 30:4855–4866

Chapter 12
DeepFake Detection

Siwei Lyu

One particular disconcerting form of disinformation are the impersonating audios/ videos backed by advanced AI technologies, in particular, deep neural networks (DNNs). These media forgeries are commonly known as the DeepFakes. The AI-based tools are making it easier and faster than ever to create compelling fakes that are challenging to spot. While there are interesting and creative applications of this technology, it can be weaponized to cause negative consequences. In this chapter, we survey the state-of-the-art DeepFake detection methods. We introduce the technical challenges in DeepFake detection and how researchers formulate solutions to tackle this problem. We discuss the pros and cons, as well as the potential pitfalls and drawbacks of each types of the solutions. Notwithstanding this progress, there are a number of critical problems that are yet to be resolved for existing DeepFake detection methods. We will also highlight a few of these challenges and discuss the research opportunities in this direction.

12.1 Introduction

Falsified images and videos created by AI algorithms, more commonly known as *DeepFakes*, are a recent twist to the disconcerting problem of online disinformation. Although fabrication and manipulation of digital images and videos are not new (Farid 2012), the rapid developments of deep neural networks (DNNs) in recent years have made the process of creating convincing fake images/videos increasingly easier and faster. DeepFake videos first caught the public's attention in late 2017, when

S. Lyu (✉)
Department of Computer Science and Engineering, University at Buffalo,
State University of New York, Buffalo, NY 14260, USA
e-mail: siweilyu@buffalo.edu

© The Author(s) 2022
H. T. Sencar et al. (eds.), *Multimedia Forensics*, Advances in Computer Vision and Pattern
Recognition, https://doi.org/10.1007/978-981-16-7621-5_12

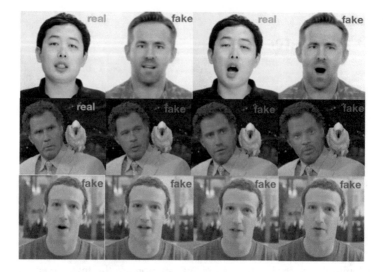

Fig. 12.1 Examples of DeepFake videos: (top) Head puppetry, (middle) face swapping, and (bottom) lip syncing

a Reddit account with the same name began posting synthetic pornographic videos generated using a DNN-based face-swapping algorithm. Subsequently, technologies that make DeepFakes have been mainstreamed through readily available software freely available on GitHub.[1] There are also emerging online services[2] and start-up companies also commercialized tools that that can generate DeepFake videos on demand.[3]

Currently, there are three major types of DeepFake videos.

- Head puppetry entails synthesizing a video of a target person's whole head and upper shoulder using a video of a source person's head, so the synthesized target appears to behave the same way as the source.
- Face swapping involves generating a video of the target with the faces replaced by synthesized faces of the source while keeping the same facial expressions.
- Lip syncing is to create a falsified video by only manipulating the lip region so that the target appears to speak something that s/he does not speak in reality.

Figure 12.1 shows some example frames of each type of DeepFake videos aforementioned.

While there are interesting and creative applications of DeepFake videos, due to the strong association of faces to the identity of an individual, they can also be

[1] E.g., FakeApp (FakeApp 2020), DFaker (DFaker github 2019), faceswap-GAN (faceswap-GAN github 2019), faceswap (faceswap github 2019), and DeepFaceLab (DeepFaceLab github 2020).

[2] E.g., https://deepfakesweb.com.

[3] E.g., Synthesia (https://www.synthesia.io/) and Canny AI https://www.cannyai.com/.

weaponized. Well-crafted DeepFake videos can create illusions of a person's presence and activities that do not occur in reality, which can lead to serious political, social, financial, and legal consequences (Chesney and Citron 2019). The potential threats range from revenge pornographic videos of a victim whose face is synthesized and spliced in, to realistically looking videos of state leaders seeming to make inflammatory comments they never actually made, a high-level executive commenting about her company's performance to influence the global stock market, or an online sex predator masquerades visually as a family member or a friend in a video chat.

Since the first known case of DeepFake videos were reported in December 2017,[4] the mounting concerns over the negative impacts of DeepFakes have spawned an increasing interest in DeepFake detection in the Multimedia Forensics research community, and the first dedicated DeepFake detection algorithm was developed by my group in June 2018 (Li et al. 2018). Subsequently, there are avid developments with many DeepFake detection methods developed in the past few years, e.g., Li et al. (2018, 2019a, b), Li and Lyu (2019) Yang et al. (2019), Matern et al. (2019), Ciftci et al. (2020), Afchar et al. (2018), Güera and Delp (2018a, b), McCloskey and Albright (2018), Sabir et al. (2019), Rössler et al. (2019), Nguyen et al. (2019a, b, c), Nataraj et al. (2019), Xuan et al. (2019), Jeon et al. (2020), Mo et al. (2018), Liu et al. (2020), Fernando et al. (2019), Shruti et al. (2020), Koopman et al. (2018), Amerini et al. (2019), Chen and Yang (2020), Wang et al. (2019a), Guarnera et al. (2020), Durall et al. (2020), Frank et al. (2020), Ciftci and Demir (2019), Chen and Yang (2020), Stehouwer et al. (2019), Bonettini et al. (2020), Khalid and Woo (2020). Correspondingly, there have also been several large-scale benchmark datasets to evaluate DeepFake detection performance (Yang et al. 2019; Korshunov and Marcel 2018; Rössler et al. 2019; Dufour et al. 2019; Dolhansky et al. 2019; Ciftci and Demir 2019; Jiang et al. 2020; Wang et al. 2019b; Khodabakhsh et al. 2018; Stehouwer et al. 2019). DeepFake detection has also been supported by government funding agencies and private companies alike.[5]

A climax of these efforts is the first DeepFake Detection Challenge from late 2019 to early 2020 (Deepfake detection challenge 2021). Overall, the DeepFake Detection Challenge corresponds to the state-of-the-art, with the winning solutions being a tour de force of advanced DNNs (an average precision of 82.56% by the top performer). These provide us effective tools to expose DeepFakes that are automated and mass produced by AI algorithms. However, we need to be cautious in reading these results. Although the organizers have made their best effort to simulate situations that Deep-Fake videos are deployed in real life, there is still a significant discrepancy between the performance on the evaluation dataset and a more real dataset—when tested on unseen videos, the top performer's accuracy reduced to 65.18%. In addition, all solu-

---

[4] https://www.vice.com/en/article/gydydm/gal-gadot-fake-ai-porn.

[5] Detection of DeepFakes is one of the goals of the DARPA MediFor (2016–2020), which also sponsored the NIST MFC 2018 and 2020 Synthetic Data Detection Challenge (NIST MFC 2018), and SemaFor (2020–2024) programs. The Global DeepFake Detection Challenge (Deepfake detection challenge 2021) in 2020 was sponsored by leading tech companies including Facebook, Amazon, and Microsoft.

tions are based on clever designs of DNNs and data augmentations, but provide little insight beyond the "black box"-type classification algorithms. Furthermore, these detection results may not completely reflect the actual detection performance of the algorithm on a single DeepFake video, especially ones that have been manually processed and perfected after being generated from the AI algorithms. Such "crafted" DeepFake videos are more likely to cause real damages, and careful manual post processing can reduce or remove artifacts that the detection algorithms predicate on.

In this chapter, we survey the state-of-the-art DeepFake detection methods. We introduce the technical challenges in DeepFake detection and how researchers formulate solutions to tackle this problem. We discuss the pros and cons, as well as the potential pitfalls and drawbacks of each types of the solutions. We also provide an overview of research efforts for DeepFake detection and a systematic comparison of existing datasets, methods, and performances. Notwithstanding this progress, there are a number of critical problems that are yet to be resolved for existing DeepFake detection methods. We will also highlight a few of these challenges and discuss the research opportunities in this direction.

12.2 DeepFake Video Generation

Although in recent years there have been many sophisticated algorithms for generating realistic synthetic face videos (Bitouk et al. 2008; Dale et al. 2011; Suwajanakorn et al. 2015, 2017; Thies et al. 2016; Korshunova et al. 2017; Pham et al. 2018; Karras et al. 2018a, 2019; Kim et al. 2018; Chan et al. 2019), most of these have not been in mainstream as open-source software tools that anyone can use. It is a much simpler method based on the work of neural image style transfer (Liu et al. 2017) that becomes the *tool of choice* to create DeepFake videos in scale, with several independent open-source implementations. We refer to this method as the *basic DeepFake maker*, and it is underneath many DeepFake videos circulated on the Internet or in the existing datasets.

The overall pipeline of the basic DeepFake maker is shown in Fig. 12.2 (left). From an input video, faces of the target are detected, from which facial landmarks are further extracted. The landmarks are used to align the faces to a standard configuration (Kazemi and Sullivan 2014). The aligned faces are then cropped and fed to an auto-encoder (Kingma and Welling 2014) to synthesize faces of the donor with the same facial expressions as the original target's faces.

The auto-encoder is usually formed by two convoluntional neural networks (CNNs), i.e., the *encoder* and the *decoder*. The encoder E converts the input target's face to a vector known as the *code*. To ensure the encoder capture identity-independent attributes such as facial expressions, there is one single encoder regardless the identities of the subjects. On the other hand, each identity has a dedicated decoder D_i, which generates a face of the corresponding subject from the code. The encoder and decoder are trained in tandem using uncorresponded face sets of multiple subjects in an unsupervised manner, Fig. 12.2 (right). Specifically, an encoder-

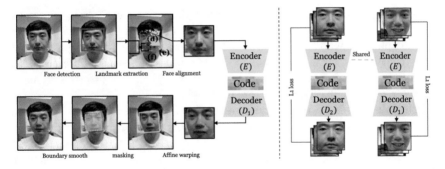

Fig. 12.2 Synthesis (left) and training (right) of the basic DeepFake maker algorithm. See texts for more details

decoder pair is formed alternatively using E and D_i for input face of each subject, and optimize their parameters to minimize the reconstruction errors (ℓ_1 difference between the input and reconstructed faces). The parameter update is performed with the back-propagation until convergence.

The synthesized faces are then warped back to the configuration of the original target's faces and trimmed with a *mask* from the facial landmarks. The last step involves smoothing the boundaries between the synthesized regions and the original video frames. The whole process is automatic and runs with little manual intervention.

12.3 Current DeepFake Detection Methods

In this section, we provide a brief overview of current DeepFake detection methods. As this is an actively researched area with the literature growing rapidly, we cannot promise comprehensive coverage of all existing methods. Furthermore, as the focus here is on DeepFake videos, we exclude detection methods for GAN-generated images. Our objective is to summarize the existing detection methods at a more abstract level, pointing out some common characteristics and challenges that can benefit future developments of more effective detection methods.

12.3.1 General Principles

As detection of DeepFakes is a problem in Digital Media Forensics, it complies with three general principles of Digital Media Forensics.

- **Principle 1**: *Manipulation operations leave traces in the falsified media.*

Digitally manipulated or synthesized media (including DeepFakes) are created by a process other than a capturing device recording an event actually occurs in the physical world. This fundamental difference in the creation process will be reflected in the resulting falsified media, albeit revealed with different scales depending on the amount of changes to the original media. Based on this principle, we can conclude that DeepFakes are detectable.

- **Principle 2***: A single forensic measure can be circumvented.*
  Any forensic detection method is based on differentiating the real and falsified media on certain characteristics in the media signals. However, the same characteristics can be exploited to evade the very forensic detections, either by hiding the traces or disrupting them. This principle is the premise for anti-forensic measures for DeepFake detections.
- **Principle 3***: There is an intention behind every falsified media.*
  A falsified media in the wild is made for a reason. It could be a satire or a prank but also a malicious attack to the victim's reputation and credibility. Understanding the motivation behind the DeepFake can provide richer information to the detection of DeepFakes, as well as to prevent and mitigate the damages.

12.3.2 Categorization Based on Methodology

We categorize existing DeepFake detection methods into three types. The first two work by seeking specific artifacts in DeepFake videos that differ them from the real videos (Principle 1 of Sect. 12.3.1).

12.3.2.1 Signal Feature-Based Methods

The **signal feature-based methods** (e.g., Afchar et al. 2018; Güera and Delp 2018b; McCloskey and Albright 2018; Li and Lyu 2019) look for abnormalities at the signal level, treating the videos as a sequence of frames $f(x, y, t)$, and a synchronized audio signal $a(t)$ if the audio track exists. Such abnormalities are often caused by various steps in the processing pipeline of DeepFake generation.

For instance, our recent work (Li and Lyu 2019) exploits the signal artifacts introduced by the resizing and interpolation operations during the post-processing of DeepFake generation. Similarly, the work in Li et al. (2019a) focuses on the boundary when the synthesized face regions are blended into the original frame. In Frank et al. (2020), the authors observe that synthesized faces often exhibit abnormalities in high frequencies, due to the up-sampling operation. Based on this observation, a simple detection method in the frequency domain is proposed. In Durall et al. (2020), the detection method uses a classical frequency domain analysis followed by an SVM classifier. The work in Guarnera et al. (2020) uses the EM algorithm to extract a set of local features in the frequency domain that are used to distinguish frames of DeepFakes from those of the real videos.

The main advantage of signal feature-based methods is that the abnormalities are usually fundamental to generation process, and fixing them may require significant changes to the underlying DNN model or post-processing steps. On the other hand, the reliance on signal features also means that signal feature-based DeepFake detection methods are susceptible to disturbance to the signals, such as interpolation (rotation, up-sizing), down-sampling (down-sizing), additive noise, blurring, and compression.

12.3.2.2   Physical/Physiological-Based Methods

The **physical/physiological-based methods** (e.g., Li et al. 2018; Yang et al. 2019; Matern et al. 2019; Ciftci et al. 2020; Hu et al. 2020) expose DeepFake videos based on their violations to the fundamental laws of physics or human physiology. The DNN model synthesizing human faces do not have direct knowledge about physiological traits of human faces or the physical laws of the surrounding environment, such information may be incorporated into the model indirectly (and inefficiently) through training data. This lack of knowledge can lead to detectable traits that are also intuitive to humans. For instance, the first dedicated DeepFake detection method (Li et al. 2018) works by detecting the inconsistency or lack of realistic eye blinking in DeepFake videos. This is due to the training data for the DeepFake generation model, which are often obtained from the Internet as the portrait images of a subject. As the portrait images are dominantly those with the subject's eyes open, the DNN generation models trained using them cannot reproduce closed eyes for a realistic eye blinking. The work of Yang et al. (2019) use the physical inconsistencies in the head poses of the DeepFake videos due to splicing the synthesized face region into the original frames. The work of Matern et al. (2019) summarizes various appearance artifacts that can be observed in DeepFake videos, such as the inconsistent eye colors, missing reflections, and fuzzy details in the eye and teeth areas. The authors then propose a set of simple features for for DeepFake detection. The work in Ciftci et al. (2020) introduces a DeepFake detection method based on biological signals extracted from facial regions that are invisible to human eyes in portrait videos, such as the slight color changes due to the blood flow caused by heart beats. These biological signals are used to reveal the spatial coherence and temporal consistency.

The main advantage of physical/physiological-based methods is its superior intuitiveness and explainability. This is especially important when the detection results are used by practitioners such as journalists. On the other hand, physical/physiological-based detection methods are limited by the effectiveness and robustness of the underlying Computer Vision algorithms, and their strong reliance on the semantic cues also limits their applicability.

12.3.2.3 Data-Driven Methods

The third category of detection methods are **data-driven methods** (e.g., Sabir et al. 2019; Rössler et al. 2019; Nguyen et al. 2019a, b, c; Nataraj et al. 2019; Xuan et al. 2019; Wang et al. 2019b; Jeon et al. 2020; Mo et al. 2018; Liu et al. 2020; Li et al. 2019a, b; Güera and Delp 2018a; Fernando et al. 2019; Shruti et al. 2020; Koopman et al. 2018; Amerini et al. 2019; Chen and Yang 2020; Wang et al. 2019a; Guarnera et al. 2020; Durall et al. 2020; Frank et al. 2020; Ciftci and Demir 2019; Chen and Yang 2020; Stehouwer et al. 2019; Bonettini et al. 2020; Khalid and Woo 2020), which do not target at specific features, but use videos labeled as real or DeepFake to train ML models (oftentimes DNNs) that can differentiate DeepFake videos from the real ones. Note that feature-based methods can also use DNN classifiers, so what makes the data-driven methods different is that the cues for classifying the two types of videos are implicit and found by the ML models. As such, the success of data-driven methods largely depends on the quality and diversity of training data, as well as the design of the ML models.

Early data-driven DeepFake detection methods reuse standard DNN models (e.g., VGG-Net Do et al. 2018, XceptionNet Rössler et al. 2019) designed for other Computer Vision tasks such as object detection and recognition. There also exist methods that use more specific or novel network architectures. MesoNet (Afchar et al. 2018) is an early detection method that uses a customized DNN model to detect DeepFakes, and interprets the detection results by correlating them with some mesoscopic level image features. The DeepFake detection method of Liu et al. (2020) uses the Gram-Net to capture differences in global image texture, which has the additional "Gram blocks" in the conventional CNN models that can calculate the Gram matrices at different levels of the network. The detection methods in Nguyen et al. (2019c, b) use capsule networks and show that it can achieve similar performance but with fewer parameters than the CNN models.

It should be mentioned that, to date, the state-of-the-art performance in DeepFake detection is obtained with the data-drive methods based on large-scale DeepFake datasets and innovative model design and training procedures. For instance, the top performer of the DeepFake Detection Challenge[6] is based on the use of an ensemble DNN models that entails seven different EfficientNet models Tan and Le (2019). The runner-up method[7] also uses ensemble models that includes EfficientNet and XceptionNet models but also introduce a new data augmentation methods (WSDAN) to anticipate the different configurations of faces.

---

[6] https://github.com/selimsef/dfdc_deepfake_challenge.

[7] https://github.com/cuihaoleo/kaggle-dfdc.

12.3.3 Categorization Based on Input Types

Another way to categorize existing DeepFake detection methods can be obtained by the type of the inputs. Most existing DeepFake detection methods are based on binary classification at the frame level, i.e., determining the likelihood of an individual frame as real or of DeepFake. Although simple and easy to implement, there are two issues related with the frame-based detection methods. First, the temporal consistency among frames are not explicitly considered, as (i) many DeepFake videos exhibit temporal artifacts and (ii) real or DeepFake frames tend to appear in continuous intervals. Second, it necessitates an extra step when video-level integrity score is needed: we have to aggregate the scores over individual frames to compute such a score using certain types of aggregation rules, common choices of which include the average, the maximum, or the average of top range. Temporal methods (e.g., Sabir et al. 2019; Amerini et al. 2019; Koopman et al. 2018), on the other hand, take the whole frame sequences as input, and use the temporal correlation between the frames as an intrinsic feature. Temporal methods often uses sequential models such as RNNs as basic model structure, and directly output the videos level prediction. The audio-visual detection methods also use the audio track as an input, and detect DeepFakes based on the asynchrony between the audios and frames.

12.3.4 Categorization Based on Output Types

Regardless of the underlying methodology or types of input, the majority of existing DeepFake detection methods formulate the problem as binary classification, which returns a label for each input video that signifies if the input is a real or a DeepFake. Often, the predicted labels are accompanied with a confidence score, a real value in [0, 1], which may be interpreted as the probability of the input to belong to one of the classes (i.e., real or DeepFake). A few methods extend this to a multi-class classification problem, where the labels also reflect different types of DeepFake generation models. There are also methods that solve the location problem (e.g., Li et al. 2019b; Huang et al. 2020), which further identifies the spatial area (in the form of bounding boxes or masked out region) and time interval of DeepFake treatments.

12.3.5 The DeepFake-o-Meter Platform

Unfortunately, the plethora of DeepFake detection algorithms were not taken full advantage of. On the one hand, differences in training datasets, hardware, and learning architectures across research publications make rigorous comparisons of different detection algorithms challenging. On the other hand, the cumbersome process of downloading, configuring, and installing of individual detection algorithms deny

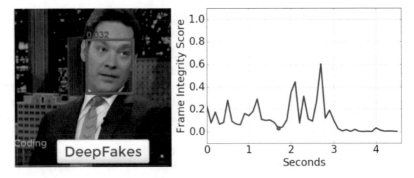

Fig. 12.3 Detection results on DeepFake-o-meter of a state-of-the-art DeepFake detection method (Li and Lyu 2019) over a fake video on youtube.com. The lower integrity score (range in [0, 1]) suggests a video frame more likely to be generated using DeepFake algorithms

the access of the state-of-the-art DeepFake detection methods to most users. To this end, we have developed `DeepFake-o-meter` http://zinc.cse.buffalo.edu/ubmdfl/deepfake-o-meter/, an open platform for DeepFake detections. For developers of DeepFake detection algorithms, it provides an API architecture to wrap individual algorithms and run on a third-party remote server. For researchers, it is an evaluation/benchmarking platform to compare multiple algorithms on the same input. For users, it provides a convenient portal to use multiple state-of-the-art detection algorithms. Currently, we have incorporated 11 state-of-the-art DeepFake image and video detection methods. A sample analysis result is shown in Fig. 12.3.

12.3.6 Datasets

The availability of large-scale datasets of DeepFake videos is an enabling factor to the development of DeepFake detection methods. The first DeepFake dataset UADFV (Yang et al. 2019) only has 49 DeepFake videos with visible artifacts when it was released in June 2018. Subsequently, more DeepFake datasets are proposed with increasing quantities and qualities. Several examples of synthesized DeepFake video frames are shown in Fig. 12.4.

- **UADFV** (Yang et al. 2019): This dataset contains 49 real videos downloaded from Youtube, which were used to create 49 DeepFake videos.
- **DeepfakeTIMIT** (Korshunov and Marcel 2018): The original videos in this dataset is from the VidTIMIT database, from which a total of 640 DeepFake videos were generated.
- **FaceForensics++** (Rössler et al. 2019): FaceForensics++ is a forensics dataset consisting of 1,000 original video sequences that have been manipulated with four automated face manipulation methods: DeepFakes, Face2Face, FaceSwap and NeuralTextures.

Fig. 12.4 Example frames of DeepFake videos from the Celeb-DF dataset (Li et al. 2020). The left most column corresponds to frames from the original videos, and the other columns are frame from DeepFake videos swapped with synthesized faces

- **DFD**: The Google/Jigsaw DeepFake detection dataset has 3,068 DeepFake videos generated based on 363 original videos of 28 consented individuals of various genders, ages, and ethnic groups.
- **DFDC** (Dolhansky et al. 2019): The Facebook DeepFake Detection Challenge (DFDC) Dataset is part of the DeepFake detection challenge, which has 4,113 DeepFake videos created based on 1,131 original videos of 66 consented individuals of various genders, ages and ethnic groups.
- **Celeb-DF** (Li et al. 2020): The Celeb-DF dataset contains real and DeepFake synthesized videos having similar visual quality on par with those circulated online. It includes 590 original videos collected from YouTube with subjects of different ages, ethic groups and genders, and 5,639 corresponding DeepFake videos.
- **DeeperForensics-1.0** (Jiang et al. 2020): DeeperForensics-1.0 consists of 10,000 DeepFake videos generated by a new DNN-based face swapping framework.
- **DFFD** (Stehouwer et al. 2019): The DFFD dataset contains 3,000 videos of four types of digital manipulations as identity swap, expression, swap, attribute manipulation, and entire synthesized faces.

12.3.7 Challenges

Albeit impressive progress has been made in the performance of detection of Deep-Fake videos, there are several concerns over the current detection methods that suggest caution.

Performance Evaluation. Currently, the problem of detecting DeepFake videos is commonly formulated, solved, and evaluated as a binary classification problem, where each video is categorized as real or a DeepFake. Such dichotomy is easy to set up in controlled experiments, where we develop and test DeepFake detection algorithms using videos that are either pristine or made with DeepFake generation algorithms. However, the picture is murkier when the detection method is deployed in real world. For instance, videos can be fabricated or manipulated in ways other than DeepFakes, so not being detected as a DeepFake video does not necessarily suggest the video is a real one. Also, a DeepFake video may be subject to other types of manipulations and a single label may not comprehensively reflect such. Furthermore, in a video with multiple subjects' faces only one or a few are generated with DeepFake for a fraction of the frames. So the binary classification scheme needs to be extended to multi-class, multi-label, and local classification/detection to fully handle the complexities of real-world media forgeries.

Explainability of Detection Results. Current DeepFake detection methods are mostly designed to perform batch analysis over a large collection videos. However, when the detection methods are used in the field by journalists or law enforcement, we usually need only to analyze a small number of videos. Numerical score corresponding to the likelihood of a video being generated using a synthesis algorithm is not as useful to the practitioners if it is not corroborated with proper reasoning of the score. In such scenarios, it is very typical to request a justification for the numerical score for the analysis to be acceptable for publishing or used in court. However, many data-driven Deepfake detection methods, especially those based on the use of deep neural networks, usually lack explainability due to the black box nature of the DNN models.

Generalization Across Datasets. As a DNN-based DeepFake detection method needs to be trained on a specific training dataset, a lack of generalization has been observed for DeepFake videos created using models not represented in the training dataset. This is different from overfitting, as the learned model may perform well on testing videos that are created with the same model but not used in training. The problem is caused by "domain shifting" when the trained detection method is applied to DeepFake videos generated using different generation models. A simple solution is to enlarge the training set to represent more diverse generation models, but a more flexible approach is needed to scale up to previously unseen models.

Social Media Laundering. A large fraction of online videos are now spread through social networks, e.g., FaceBook, Instagram, and Twitter. To save network bandwidth and also to protect the users' privacy, these videos are usually striped off meta-data, down-sized, and then heavy compressed before they are uploaded to the social platforms. These operations, commonly known as *social media laundering*, are detrimental to recover traces of underlying manipulation, and at the same time increase the false positive detections, i.e., classifying a real video as a DeepFake. So far, most data-driven DeepFake detection methods that use signal-level features are much affected by social media laundering. A practical measure to improve the robustness of DeepFake detection methods to social media laundering is to actively incorporate simulations of such effects in training data, and also enhance evaluation datasets to include performance on social media laundered videos, both real and synthesized.

Anti-forensics. With the increasing effectiveness of DeepFake detection methods, we also anticipate developments of corresponding anti-forensic measures, which take advantage of the vulnerabilities of current DeepFake detection methods to conceal revealing traces of DeepFake videos. The data-driven DNN-based DeepFake detection methods are particularly susceptible to anti-forensic attacks due to the known vulnerability of general deep neural network classification models (see Principle 2 of Sect. 12.3.1). Indeed, recent years have witnessed a rapid development of anti-forensic methods based on adversarial attacks to DNN models targeting DeepFake detectors (Huang et al. 2020; Carlini and Farid 2020; Gandhi and Jain 2020; Neekhara 2020). Anti-forensic measures can also be developed in the other aspect, to disguise a real video as a DeepFake video by adding simulated signal level features used by current detection algorithms, a situation we term as *fake DeepFake*. Further, DeepFake detection methods must improve to handle such intentional and adversarial attacks.

12.4 Future Directions

Besides continuing improving to solve the aforementioned limitations, we also envision a few important directions of DeepFake detection methods that will receive more attention in the coming years.

Other Forms of DeepFake Videos. Although face swapping is currently the most widely known form of DeepFake videos, it is by no means the most effective. In particular, for the purpose of impersonating someone, face swapping DeepFake videos have several limitations. Psychological studies (Sinha et al. 2009) show that human face recognition largely relied on information gleaned from face shape and hairstyle. As such, to create convincing impersonating effect, the person whose face is to be replaced (the target) has to have similar face shape and hairstyle to the person whose face is used for swapping (the donor). Second, as the synthesized faces need to be spliced into the original video frame, the inconsistencies between the synthesized region and the rest of the original frame can be severe and difficult to conceal. In these respects, other forms of DeepFake videos, namely, head puppetry and lip-syncing, are more effective and thus should become the focus of subsequent research in DeepFake detection. Methods studying whole face synthesis or reenactment have experienced fast development in recent years. Although there have not been as many easy-to-use and free open-source software tools generating these types of DeepFake videos as for the face-swapping videos, the continuing sophistication of the generation algorithms will change the situation in the near future. Because the synthesized region is different from face swapping DeepFake videos (the whole face in the former and lip area in the latter), detection methods designed based on artifacts specific to face swapping are unlikely to be effective for these videos. Correspondingly, we should develop detection methods that are effective to these types of DeepFake videos.

Audio DeepFakes. AI-based impersonation are not limited to imagery, recent AI-synthesized content-generation are leading to the creation of highly realistic audios (Ping et al. 2017; Gu and Kang 2018; AlBadawy and Lyu 2020). Using synthesized audios of the impersonating target can significantly make the DeepFake videos more convincing and compounds its negative impact. As audio signals are 1D signals and have very different nature from images and videos, different methods need to be developed to specifically target such forgeries. This problem has drawn attention in the speech processing community recently with part of the most recent Global ASVspoofing Challenge (https://www.asvspoof.org/). Dedicated to AI-driven voice conversion detection, and a few dedicated methods for audio DeepFake detection, e.g., AlBadawy et al. (2019), have also shown up recently. In the coming years, we expect more developments in these areas, in particular, those can leverage features in both visual and audio features of the fake videos.

Intent Inference. Even though the potential negative impacts of DeepFake videos are tremendous, in reality, the majority of DeepFake videos are not created not with a malicious intent. Many DeepFake videos currently circulated online are of a prank-some, humorous, or satirical nature. As such, it is important to expose the underlying intent of a DeepFake in the context of legal or journalistic investigation (Principle 3 in Sect. 12.3.1). Inferring intention may require more semantic and contextual understanding of the content, few forensic methods are designed to answer this question, but this is certainly a direction that future forensic methods will focus on.

Human Performance. Although the potential negative impacts of online DeepFake videos are widely recognized, currently there is a lack of formal and quantitative study of the perceptual and psychological factors underlying their deceptiveness. Interesting questions such as if there exist an *uncanny valley*[8] for DeepFake videos, what is the *just noticeable difference* between high-quality DeepFake videos and real videos to human eyes, or what type/aspects of DeepFake videos are more effective in deceiving the viewers, have yet to be answered. To pursue these questions, it calls for close collaboration among researchers in digital media forensics and in perceptual and social psychology. There is no doubt that such studies are invaluable to research in detection techniques as well as a better understanding of the social impact that DeepFake videos can cause.

Protection measures. However, given the speed and reach of the propagation of online media, even the currently best forensic techniques will largely operate in a postmortem fashion, applicable only after AI synthesized fake face images or videos emerge. We aim to develop *proactive* approaches to protect individuals from becoming the victims of such attacks, which complement to the forensic tools.

One such method we have recently studied (Li et al. 2019c; Sun et al. 2020) is to add specially designed patterns known as the *adversarial perturbations* that are imperceptible to human eyes but can result in detection failures. The rationale is as follows. High-quality AI face synthesis models need large number of, typically in the range of thousands, sometimes even millions, training face images collected using automatic face detection methods, i.e., the *face sets*. Adversarial perturbations "pollute" a face set to have few actual faces and many non-faces with low or no utility as training data for AI face synthesis models. The proposed adversarial perturbation generation method can be implemented as a service of photo/video sharing platforms before a user's personal images/videos are uploaded or as a standalone tool that the user can use, to process the images and videos before they are uploaded online.

12.5 Conclusion and Outlook

We predict that several future technological developments will further improve the visual quality and generation efficiency of the fake videos. Firstly, one critical disadvantage of the current DeepFake generation methods are that they cannot produce good details such as skin and facial hairs. This is due to the loss of information in the encoding step of generation. However, this can be improved by incorporating GAN models (Goodfellow et al. 2014) which have demonstrated performance in recov-

---

[8] The uncanny valley in this context refers to the phenomenon whereby a DeepFake generated face bearing a near-identical resemblance to a human being arouses a sense of unease or revulsion in the viewers.

ering facial details in recent works (Karras et al. 2018b, 2019, 2020). Secondly, the synthesized videos can be more realistic if they are accompanied with realistic voices, which combines video and audio synthesis together in one tool.

In the face of this, the overall running efficiency, detection accuracy, and more importantly, false positive rate, have to be improved for wide practical adoption. The detection methods also need to be more robust to real-life post-processing steps, social media laundering, and counter-forensic technologies. There is a perpetual competition of technology, know-hows, and skills between the forgery makers and digital media forensic researchers. The future will reckon the predictions we make in this work.

References

Afchar D, Nozick V, Yamagishi J, Echizen I (2018) Mesonet: a compact facial video forgery detection network. In: WIFS

AlBadawy E, Lyu S (2020) Voice conversion using speech-to-speech neuro-style transfer. In: Interspeech, Shanghai, China

AlBadawy E, Lyu S, Farid H (2019) Detecting ai-synthesized speech using bispectral analysis. In: Workshop on media forensics (in conjunction with CVPR), Long Beach, CA, United States

Amerini I, Galteri L, Caldelli R, Del Bimbo A (2019) Deepfake video detection through optical flow based cnn. In: Proceedings of the IEEE international conference on computer vision workshops, pp 0-0

Bitouk D, Kumar N, Dhillon S, Belhumeur P, Nayar SK (2008) Face swapping: automatically replacing faces in photographs. ACM Trans Graph (TOG)

Bonettini N, Cannas ED, Mandelli S, Bondi L, Bestagini P, Tubaro S (2020) Video face manipulation detection through ensemble of cnns. arXiv:2004.07676

Carlini N, Farid H (2020) Evading deepfake-image detectors with white- and black-box attacks. arXiv:2004.00622

Chan C, Ginosar S, Zhou T, Efros AA (2019) Everybody dance now. In: ICCV

Chen Z, Yang H (2020) Manipulated face detector: joint spatial and frequency domain attention network. arXiv:2005.02958

Chesney R, Citron DK (2019) Deep Fakes: a looming challenge for privacy, democracy, and national security. In: 107 California Law Review (2019, Forthcoming); U of Texas Law, Public Law Research Paper No. 692; U of Maryland Legal Studies Research Paper No. 2018-21

Ciftci UA, Demir I (2019) Fakecatcher: detection of synthetic portrait videos using biological signals. arXiv:1901.02212

Ciftci UA, Demir I, Yin L (2020) How do the hearts of deep fakes beat? Deep fake source detection via interpreting residuals with biological signals. In: IEEE/IAPR international joint conference on biometrics (IJCB)

Dale K, Sunkavalli K, Johnson MK, Vlasic D, Matusik W, Pfister H (2011) Video face replacement. ACM Trans Graph (TOG)

DeepFaceLab github. https://github.com/iperov/DeepFaceLab. Accessed 4 July 2020

Deepfake detection challenge. https://deepfakedetectionchallenge.ai

DFaker github. https://github.com/dfaker/df. Accessed 4 Nov 2019

Do N-T, Na I-S, Kim S-H (2018) Forensics face detection from gans using convolutional neural network

Dolhansky B, Howes R, Pflaum B, Baram N, Ferrer CC (2019) The deepfake detection challenge (DFDC) preview dataset. arXiv:1910.08854

Dufour N, Gully A, Karlsson P, Vorbyov AV, Leung T, Childs J, Bregler C (2019) Deepfakes detection dataset by google & jigsaw

Durall R, Keuper M, Keuper J (2020) Watch your up-convolution: Cnn based generative deep neural networks are failing to reproduce spectral distributions. arXiv:2003.01826

faceswap github. https://github.com/deepfakes/faceswap. Accessed 4 Nov 2019

faceswap-GAN github. https://github.com/shaoanlu/faceswap-GAN. Accessed 4 Nov 2019

FakeApp. https://www.malavida.com/en/soft/fakeapp/. Accessed 4 July 2020

Farid H (2012) Digital image forensics. MIT Press, Cambridge

Fernando T, Fookes C, Denman S, Sridharan S (2019) Exploiting human social cognition for the detection of fake and fraudulent faces via memory networks. Computer vision and pattern recognition. arXiv:1911.07844

Frank J, Eisenhofer T, Schönherr L, Fischer A, Kolossa D, Holz T (2020) Leveraging frequency analysis for deep fake image recognition. arXiv:2003.08685

Gandhi A, Jain S (2020) Adversarial perturbations fool deepfake detectors. arXiv:2003.10596

Goodfellow I, Pouget-Abadie J, Mirza M, Xu B, Warde-Farley D, Ozair S, Courville A, Bengio Y (2014) Generative adversarial nets. In: NeurIPS

Guarnera L, Battiato S, Giudice O (2020) Deepfake detection by analyzing convolutional traces. In: Proceedings of the IEEE conference on computer vision and pattern recognition workshops

Güera D, Delp EJ (2018a) Deepfake video detection using recurrent neural networks. In: 2018 15th IEEE international conference on advanced video and signal based surveillance (AVSS). IEEE, pp 1–6

Güera D, Delp EJ (2018b) Deepfake video detection using recurrent neural networks. In: AVSS

Gu Y, Yongguo K (2018) Multi-task WaveNet: a multi-task generative model for statistical parametric speech synthesis without fundamental frequency conditions. In: Interspeech, Hyderabad, India

Huang Y, Juefeixu F, Wang R, Xie X, Ma L, Li J, Miao W, Liu Y, Pu G (2020) Fakelocator: robust localization of gan-based face manipulations via semantic segmentation networks with bells and whistles. Computer vision and pattern recognition. arXiv:2001.09598

Hu S, Li Y, Lyu S (2009) Exposing GAN-generated faces using inconsistent corneal specular highlights. arXiv:11924:2020

Jeon H, Bang Y, Woo SS (2020) Fdftnet: facing off fake images using fake detection fine-tuning network. Computer vision and pattern recognition. arXiv:2001.01265

Jiang L, Wu W, Li R, Qian C, Loy CC (2020) Deeperforensics-1.0: a large-scale dataset for real-world face forgery detection. arXiv:2001.03024

Karras T, Aila T, Laine S, Lehtinen J (2018a) Progressive growing of GANs for improved quality, stability, and variation. In: ICLR

Karras T, Aila T, Laine S, Lehtinen J (2018b) Progressive growing of GANs for improved quality, stability, and variation. In: International conference on learning representations (ICLR)

Karras T, Laine S, Aila T (2019) A style-based generator architecture for generative adversarial networks. In: CVPR

Karras T, Laine S, Aittala M, Hellsten J, Lehtinen J, Aila T (2020) Analyzing and improving the image quality of stylegan. In: Proceedings of the IEEE/CVF conference on computer vision and pattern recognition, pp 8110–8119

Kazemi V, Sullivan J (2014) One millisecond face alignment with an ensemble of regression trees. In: CVPR

Khalid H, Woo SS (2020) Oc-fakedect: classifying deepfakes using one-class variational autoencoder. In: Proceedings of the IEEE/CVF conference on computer vision and pattern recognition (CVPR) workshops

Khodabakhsh A, Ramachandra R, Raja KB, Wasnik P, Busch C (2018) Fake face detection methods: can they be generalized?, pp 1–6

Kim H, Garrido P, Tewari A, Xu W, Thies J, Nießner N, Pérez P, Richardt C, Zollhöfer M, Theobalt C (2018) Deep video portraits. ACM Trans Graph (TOG)

Kingma DP, Welling M (2014) Auto-encoding variational bayes. In: ICLR

Koopman M, Rodriguez AM, Geradts Z (2018) Detection of deepfake video manipulation. In: The 20th Irish machine vision and image processing conference (IMVIP), pp 133–136

Korshunova I, Shi W, Dambre J, Theis L (2017) Fast face-swap using convolutional neural networks. In: ICCV

Korshunov P, Marcel S (2018) Deepfakes: a new threat to face recognition? Assessment and detection. arXiv:1812.08685

Li Y, Lyu S (2019) Exposing deepfake videos by detecting face warping artifacts. In: IEEE conference on computer vision and pattern recognition workshops (CVPRW)

Li Y, Chang M-C, Lyu S (2018) In Ictu Oculi: exposing AI generated fake face videos by detecting eye blinking. In: IEEE international workshop on information forensics and security (WIFS)

Li L, Bao J, Zhang T, Yang H, Chen D, Wen F, Guo B (2019a) Face x-ray for more general face forgery detection. arXiv:1912.13458

Li J, Shen T, Zhang W, Ren H, Zeng D, Mei T (2019b) Zooming into face forensics: a pixel-level analysis. Computer vision and pattern recognition. arXiv:1912.05790

Li Y, Yang X, Wu B, Lyu S (2019c) Hiding faces in plain sight: disrupting ai face synthesis with adversarial perturbations. arXiv:1906.09288

Li Y, Sun P, Qi H, Lyu S (2020) Celeb-DF: a Large-scale challenging dataset for DeepFake forensics. In: IEEE conference on computer vision and patten recognition (CVPR), Seattle, WA, United States

Liu M-Y, Breuel T, Kautz J (2017) Unsupervised image-to-image translation networks. In: NeurIPS

Liu Z, Qi X, Jia J, Torr P (2020) Global texture enhancement for fake face detection in the wild. arXiv:2002.00133

Matern F, Riess C, Stamminger M (2019) Exploiting visual artifacts to expose deepfakes and face manipulations. In: IEEE winter applications of computer vision workshops (WACVW)

McCloskey S, Albright M (2018) Detecting gan-generated imagery using color cues. arXiv:1812.08247

Mo H, Chen B, Luo W (2018) Fake faces identification via convolutional neural network. In: Proceedings of the 6th ACM workshop on information hiding and multimedia security, pp 43–47

Nataraj L, Mohammed TM, Manjunath BS, Chandrasekaran S, Flenner A, Bappy JH, Roy-Chowdhury AK (2019) Detecting gan generated fake images using co-occurrence matrices. Electron Imag (2019)5:532–1

Neekhara P (2020) Adversarial deepfakes: evaluating vulnerability of deepfake detectors to adversarial examples. arXiv:2002.12749

Nguyen HH, Fang F, Yamagishi J, Echizen I (2019a) Multi-task learning for detecting and segmenting manipulated facial images and videos. In: IEEE international conference on biometrics: theory, applications and systems (BTAS)

Nguyen HH, Yamagishi J, Echizen I (2019b) Capsule-forensics: using capsule networks to detect forged images and videos. In: ICASSP 2019-2019 IEEE international conference on acoustics, speech and signal processing (ICASSP). IEEE, pp 2307–2311

Nguyen HH, Yamagishi J, Echizen I (2019c) Use of a capsule network to detect fake images and videos. arXiv:1910.12467

NIST MFC 2018 Synthetic Data Detection Challenge. https://www.nist.gov/itl/iad/mig/media-forensics-challenge-2019-0

Pham HX, Wang Y, Pavlovic V (2018) Generative adversarial talking head: bringing portraits to life with a weakly supervised neural network. arXiv:1803.07716

Ping W, Peng K, Gibiansky A, Arik SO, Kannan A, Narang S, Raiman J, Miller J (2017) Deep voice 3: 2000-speaker neural text-to-speech. arXiv:1710.07654

Rössler A, Cozzolino D, Verdoliva L, Riess C, Thies J, Nießner M (2019) FaceForensics++: learning to detect manipulated facial images. In: ICCV

Sabir E, Cheng J, Jaiswal A, AbdAlmageed W, Masi I, Natarajan P (2019) Recurrent convolutional strategies for face manipulation detection in videos. Interfaces (GUI) 3:1

Shruti Agarwal HF, El-Gaaly T, Lim S-N (2020) Detecting deep-fake videos from appearance and behavior shruti. arXiv:2004.14491

Sinha P, Balas B, Ostrovsky Y, Russell R (2009) Face recognition by humans: 20 results all computer vision researchers should know about. https://www.cs.utexas.edu/users/grauman/courses/spring2007/395T/papers/sinha_20results.pdf

Stehouwer J, Dang H, Liu F, Liu X, Jain AK (2019) On the detection of digital face manipulation. Computer vision and pattern recognition. arXiv:1910.01717

Sun P, Li Y, Qi H, Lyu S (2020) Landmark breaker: obstructing deepfake by disturbing landmark extraction. In: IEEE workshop on information forensics and security (WIFS), New York, NY, United States

Suwajanakorn S, Seitz SM, Kemelmacher-Shlizerman I (2015) What makes tom hanks look like tom hanks. In: ICCV

Suwajanakorn S, Seitz SM, Kemelmachershlizerman I (2017) Synthesizing obama: learning lip sync from audio. ACM Trans Graph 36(4):95

Tan M, Le Q (2019) EfficientNet: rethinking model scaling for convolutional neural networks. In: Chaudhuri K, Salakhutdinov R (eds) Proceedings of the 36th international conference on machine learning, vol 97 of Proceedings of machine learning research, Long Beach, California, USA, 09–15 Jun 2019. PMLR, pp 6105–6114

Thies J, Zollhofer M, Stamminger M, Theobalt C, Niessner M (2016) Face2Face: real-time face capture and reenactment of rgb videos. In: IEEE conference on computer vision and pattern recognition (CVPR)

Wang R, Juefeixu F, Ma L, Xie X, Huang Y, Wang J, Liu Y (2019a) Fakespotter: a simple yet robust baseline for spotting ai-synthesized fake faces. Cryptography cryptography and security. arXiv:1909.06122

Wang S, Wang O, Zhang R, Owens A, Efros AA (2019b) Cnn-generated images are surprisingly easy to spot... for now. Computer vision and pattern recognition. arXiv:1912.11035

Xuan X, Peng B, Wang W, Dong J (2019) On the generalization of gan image forensics. In: Chinese conference on biometric recognition. Springer, pp 134–141

Yang X, Li Y, Lyu S (2019) Exposing deep fakes using inconsistent head poses. In: ICASSP

Chapter 13
Video Frame Deletion and Duplication

Chengjiang Long, Arslan Basharat, and Anthony Hoogs

Videos can be manipulated in a number of different ways, including object addition or removal, deep fake videos, temporal removal or duplication of parts of the video, etc. In this chapter, we provide an overview of the previous work related to video frame deletion and duplication and dive into the details of two deep-learning-based approaches for detecting and localizing frame deletion (Chengjiang et al. 2017) and duplication (Chengjiang et al. 2019) manipulations. This should provide the reader a brief overview of the related research and details of a couple of deep-learning-based forensics methods to defend against temporal video manipulations.

13.1 Introduction

Digital video forgery (Sowmya and Chennamma 2015) is referred to as intentional modification of the digital video for fabrication. A common digital video forgery technique is temporal manipulation, which includes frame sequence manipulations such as dropping, insertion, reordering and looping. By altering only the temporal aspect of the video the manipulation is not detectable by single-image forensic techniques; therefore, there is need for digital forensics methods that perform temporal analysis of videos to detect such manipulations.

In this chapter, we will first focus on the problem of video frame deletion detection in a given, possibly manipulated, video without the original video. As illustrated in Fig. 13.1, we define a frame drop to be a removal of any number of consecutive

C. Long
Meta Reality Labs, Burlingame, CA, USA

A. Basharat (✉) · A. Hoogs
Kitware, Inc., Clifton Park, NY, USA
e-mail: arslan.basharat@kitware.com

© The Author(s) 2022
H. T. Sencar et al. (eds.), *Multimedia Forensics*, Advances in Computer Vision and Pattern Recognition, https://doi.org/10.1007/978-981-16-7621-5_13

Original video

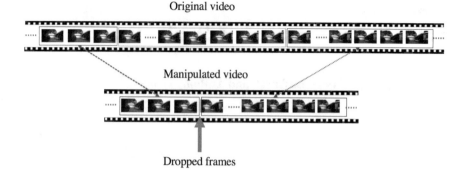

Manipulated video

Dropped frames

Fig. 13.1 The illustration of frame dropping detection challenge. Assuming that there are three consecutive frame sequences (marked in red, green and blue, respectively) in an original video, the manipulated video is obtained after removing the green frame sequence. Our goal is to identify the location of the frame drop at the end of the red frame sequence and the beginning of the blue frame sequence

frames within a video shot.[1] In our work (Chengjiang et al. 2017) to address this problem, we only consider videos with a single shot to avoid the confusion between frame drops and shot breaks. Single-shot videos are prevalent from various sources, like mobile phones, car dashboard cameras or body-worn cameras.

To the best of our knowledge, only a small amount of recent work (Thakur et al. 2016) has explored automatically detecting dropped frames without a reference video. In digital forgery detection, we cannot assume a reference video, unlike related techniques that detect frame drops for quality assurance. Wolf (2009) proposed a frame-by-frame motion energy cue defined based on the temporal information difference sequence for finding dropped/repeated frames, among which the changes are slight. Unlike Wolf's work, we detect the locations where frames are dropped in a manipulated video without being compared with the original video. Recently, Thakur et al. (2016) proposed an SVM-based method to classify tampered or non-tampered videos. In this work, we explore the authentication (Valentina et al. 2012; Wang and Farid 2007) of the scene or camera to determine if a video has one or more frame drops without a reference or original video. We expect such authentication is able to explore underlying spatio-temporal relationships across the video so that it is robust to digital-level attacks and conveys a consistency indicator across the frame sequences.

We believe that we can still use similar assumption that consecutive frames are consistent with each other and the consistency will be destroyed if there exists temporal manipulation. To authenticate a video, two-frame techniques such as color histogram, motion energy (Stephen 2009) and optical flow (Chao et al. 2012; Wang et al. 2014) have been used. By only using two frames these techniques cannot generalize to work on both videos with rapid scene changes (often from fast camera

---

[1] A shot is a consecutive sequence of frames captured between the start and stop operations of a single video camera.

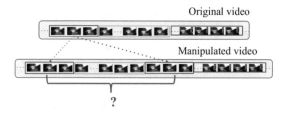

Fig. 13.2 An illustration of frame duplication manipulation in a video. Assume an original video has three sets of frames indicated here by red, green and blue rectangles. A manipulated video can be generated by inserting a second copy of the red set in the middle of the green and the blue sets. Our goal is to detect both instances of the red set as duplicated and also determine that the second instance is the one that's forged

motion) and videos with subtle scene changes such as static camera surveillance videos.

In the past few years, deep learning algorithms have made significant breakthroughs, especially in the image domain (Krizhevsky et al. 2012). The features computed by these algorithms have been used for image matching/classification (Zhang et al. 2014; Zhou et al. 2014). In this chapter, we evaluate approaches using these features for dropped frame detection using two to three frames. However, these image-based deep features still lack modeling the motion effectively.

Inspired by Tran et al.'s C3D network (Tran et al. 2015), which is able to extract powerful spatio-temporal features for action recognition, we propose a C3D-based network for detecting frame drops, as illustrated in Fig. 13.3. As we can observe, there are three aspects to distinguish our C3D-based network approach (Chengjiang et al. 2017) from Tran et al.'s work. (1) Our task is to check whether there exist frames dropped between the 8th and the 9th frame, which makes the center part more informative than the two ends of the 16-frame video clips; (2) the output of the network has two branches, which correspond to "frame drop" and "no frame drop", between the 8th and the 9th frame; (3) unlike most approaches, we use the output scores from the network as confidence score directly and define confidence score with a peak detection step and a scale term based on the output score curves; and (4) such a network is able to not only predict whether the video has frame dropping but also detect the exact location where the frame dropping occurs.

To summarize, the contributions of our work (Chengjiang et al. 2017) are:

- Proposed a 3D convolutional network for frame dropping detection, and the confidence score is defined with a peak detection step and a scale term based on the output score curves. It is able to identify whether there exists frame dropping and even determine the exact location of frame dropping without any information of the reference/original video.
- For performance comparison, we also compared a series of baselines, including cue-based algorithms (color histogram, motion energy and optical flow) and learning-based algorithms (an SVM algorithm and convolutional neural networks (CNNs) using two or three frames as input).

- The experimental results on both the Yahoo Flickr Creative Commons 100 Million (YFCC100m) dataset and the Nimble Challenge 2017 dataset clearly demonstrate the efficacy of the proposed C3D-based network.

An increasingly large volume of digital video content is becoming available in our daily lives through the internet due to the rapid growth of increasingly sophisticated, mobile and low-cost video recorders. These videos are often edited and altered for various purposes using image and video editing tools that have become more readily available. Manipulations or forgeries can be done for nefarious purposes to either hide or duplicate an event or content in the original video. Frame duplication refers to a video manipulation where a copy of a sequence of frames is inserted into the same video either replacing previous frames or as additional frames. Figure 13.2 provides an example of frame duplication where in the manipulated video the red frame sequence from the original video is inserted between the green and the blue frame sequences. As a real-world example, frame duplication forgery could be done to hide an individual leaving a building in a surveillance video. If such a manipulated video was part of a criminal investigation, without effective forensics tools the investigators could be misled.

Videos can also be manipulated by duplicating a sequence of consecutive frames with the goal of concealing or imitating specific content in the same video. In this chapter, we also describe a coarse-to-fine framework based on deep convolutional neural networks to automatically detect and localize such frame duplication (Chengjiang et al. 2019). First, an I3D network finds coarse-level matches between candidate duplicated frame sequences and the corresponding selected original frame sequences. Then a Siamese network based on ResNet architecture identifies fine-level correspondences between an individual duplicated frame and the corresponding selected frame. We also propose a robust statistical approach to compute a video-level score indicating the likelihood of manipulation or forgery. Additionally, for providing manipulation localization information we develop an inconsistency detector based on the I3D network to distinguish the duplicated frames from the selected original frames. Quantified evaluation on two challenging video forgery datasets clearly demonstrates that this approach performs significantly better than four state-of-the-art methods.

It is very important to develop robust video forensic techniques, to catch videos with increasing sophisticated forgeries. Video forensics techniques (Milani et al. 2012; Wang and Farid 2007) aim to extract and exploit features from videos that can distinguish forgeries from original, authentic videos. Like other areas in information security, the sophistication of attacks and forgeries continue to increase for images and videos, requiring a continued improvement in forensic techniques. Robust detection and localization of duplicated parts of a video can be a very useful forensic tool for those tasked with authenticating large volumes of video content.

In recent years, multiple digital video forgery detection approaches have been employed to solve this challenging problem. Wang and Farid (2007) proposed a frame duplication detection algorithm which takes the correlation coefficient as a measure of similarity. However, such an algorithm results in a heavy computational load due to a large number of correlation calculations. Lin et al. (2012) proposed to use histogram difference (HD) instead of correlation coefficients as the detection features. The drawback is that the HD features do not show strong robustness against common video operations or attacks. Hu et al. (2012) propose to detect duplicated frames using video sub-sequence fingerprints extracted from the DCT coefficients. Yang et al. (2016) propose an effective similarity-analysis-based method that is implemented in two stages, where the features are obtained via SVD. Ulutas et al. propose to use a BoW model (Ulutas et al. 2018) and binary features (Ulutas et al. 2017) for frame duplication detection. Although deep learning solutions, especially those based on convolution neural networks, have demonstrated promising performance in solving many challenging vision problems such as large-scale image recognition (Kaiming et al. 2016; Stock and Cisse 2018), object detection (Shaoqing et al. 2015; Yuhua et al. 2018; Tang et al. 2018) and visual captioning (Venugopalan et al. 2015; Aneja et al. 2018; Huanyu et al. 2018), no deep learning solutions were developed for this specific task at the time, which motivated us to fill this gap.

In Chengjiang et al. (2019) we describe a coarse-to-fine deep learning framework, called C2F-DCNN, for frame duplication detection and localization in forged videos. As illustrated in Fig. 13.4, we first utilize an I3D network (Carreira and Zisserman 2017) to obtain the candidate duplicate sequences at a coarse level; this helps narrow the search faster through longer videos. Next, at a finer level, we apply a Siamese network composed of two ResNet networks (Kaiming et al. 2016) to further confirm duplication at the frame level to obtain accurate corresponding pairs of duplicated and selected original frames. Finally, the duplicated frame range can be distinguished from the corresponding selected original frame range by our inconsistency detector that is designed as an I3D network with 16-frames as an input video clip.

Unlike other methods, we consider the consistency between two consecutive frames from a 16-frame video clip in which these two consecutive frames are at the center, i.e., 8th and 9th frame. This is aimed at capturing the temporal context for matching a range of frames for duplication. Inspired by Long et al. (2017), we design an inconsistency detector based on the I3D network to cover three categories, i.e., "none", "frame drop" and "shot break", which represent that between the 8th and 9th frame there are no manipulations, frames removal within one shot, and a shot boundary transition, respectively. Therefore, we are able to use output scores from the learned I3D network to formulate a confidence score of inconsistency between any two consecutive frames to distinguish the duplicated frame range from the selected original frame range, even in videos with multiple shots.

We also proposed a heuristic strategy to produce a video-level frame duplication likelihood score. This is built upon the measures like the number of possible frames duplicated, the minimum distance between duplicated frames and selected frames, and the temporal gap between the duplicated frames and the selected original frames.

To summarize, the contributions of this approach (Chengjiang et al. 2019) are as follows:

- A novel coarse-to-fine deep learning framework for frame duplication detection and localization in forged videos. This framework features fine-tuned I3D networks and the ResNet Siamese network, providing a robust yet efficient approach to process large volumes of video data.

- Designed an inconsistency detector based on a fine-tuned I3D network that covers three categories to distinguish duplicated frame range from the selected original frame range.
- A heuristic formulation for video-level detection score, which leads to significant improvement in detection benchmark performance.
- Evaluated performance on two video forgery datasets and the experimental results strongly demonstrate the effectiveness of the proposed method.

13.2 Related Work

13.2.1 Frame Deletion Detection

The most related prior work can be roughly split into two categories: *video inter-frame forgery identification* and *shot boundary detection*.

Video inter-frame forgery identification. Video inter-frame forgery involves frame insertion and frame deletion. Wang et al. proposed an SVM method (Wang et al. 2014) based on the assumption that the optical flows are consistent in an original video, while in forgeries the consistency will be destroyed. Chao's optical flow method (Chao et al. 2012) provides different detection schemes for inter-frame forgery based on the observation that the subtle difference between frame insertion and deletion. Besides optic flow, Wang et al. (2014) also extracted the consistency of correlation coefficients of gray values as distinguishing features to classify original videos and forgeries. Zheng et al. (2014) proposed a novel feature called block-wise brightness variance descriptor (BBVD) for fast detection of video inter-frame forgery. Different from this inter-frame forgery identification, our proposed C3D-based network (Chengjiang et al. 2017) is able to explore the powerful spatio-temporal relationships as the authentication of the scene or camera in a video for frame dropping detection.

Shot Boundary Detection. There is a large amount of work to solve the shot boundary detection problem (Smeaton et al. 2010). The task of shot boundary detection (Smeaton et al. 2010) is to detect the boundaries to separate multiple shots within a video. The TREC video retrieval evaluation (TRECVID) is an important benchmark dataset for automatic shot boundary detection challenge. And different research groups from across the world have worked to determine the best approaches to shot boundary detection using a common dataset and common scoring metrics. Instead of detecting where two shots are concatenated, we are focused on detecting a frame drop within a single shot.

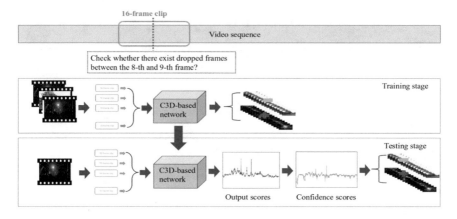

Fig. 13.3 The pipeline of the C3D-based method. At the training stage, the C3D-based network takes 16-frame video clips extracted from the video dataset as input, and produces two outputs, i.e., "frame drop" (indicated with "+") or "no frame drop" (indicated with "−"). At the testing stage, we decompose a testing video into a sequence of continuous 16-frame clips and then fit them into the learned C3D-based network to obtain the output scores. Based on the score curves, we use a peak detection step and introduce a scale term to define the confidence scores to detect/identify whether there exist dropped frames for per frame clip or per video. The network model consisted of 66 million parameters with $3 \times 3 \times 3$ filter size at all convolutional layers

13.2.2   Frame Duplication Detection

The research related to frame duplication can be broadly divided into *inter-frame forgery*, *copy-move forgery* and *convolutional neural networks*.

Inter-frame forgery refers to frame deletion and frame duplication. For features used for inter-frame forgery, either spatially or temporally, keypoints are extracted from nearby patches recognized over distinctive scales. Keypoint-based methodologies can be further subdivided into direction-based (Douze et al. 2008; Le et al. 2010), keyframe-based coordinating (Law-To et al. 2006) and visual-words-based (Sowmya and Chennamma 2015). In particular, keyframe-based feature has been shown to perform well for close video picture/feature identification (Law-To et al. 2006).

In addition to keypoint-based features, Wu et al. (2014) propose a velocity field consistency-based approach to detect inter-frame forgery. This method is able to distinguish the forgery types, identify the tampered video and locate the manipulated positions in forged videos as well. Wang et al. (2014) propose to make full use of the consistency of the correlation coefficients of gray values to classify original videos and inter-frame forgeries. They also propose an optical flow method (Wang et al. 2014) based on the assumption that the optical flows are consistent in an original video, while in forgeries the consistency will be destroyed. The optical flow is extracted as a distinguishing feature to identify inter-frame forgeries through a support vector machine (SVM) classifier to recognize frame insertion and frame deletion forgeries.

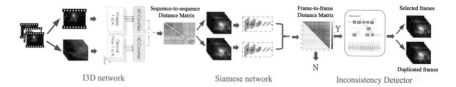

I3D network   Siamese network   Inconsistency Detector

Fig. 13.4 The C2F-DCNN framework for frame duplication detection and localization. Given a testing video, we first run the I3D network (Carreira and Zisserman 2017) to extract deep spatial-temporal features and build the coarse sequence-to-sequence distance to determine the possible frame sequences that are likely to have frame duplication. For the likely duplicated sequences, a ResNet-based Siamese network further confirms a frame duplication at the frame level. For the videos with duplication detected, temporal localization is determined with an I3D-based inconsistency detector to distinguish the duplicated frames from the selected frames

Huang et al. (2018) proposed a fusion of audio forensics detection methods for video inter-frame forgery. Zhao et al. (2018) developed a similarity analysis-based method to detect inter-frame forgery in a video shot. In this method, the HSV color histogram is calculated to detect and locate tampered frames in the shot, and then the SURF feature extraction and FLANN (Fast Library for Approximate Nearest Neighbors) matching are used for further confirmation.

Copy-move forgery is created by copying and pasting content within the same frame, and potentially post-processing it (Christlein et al. 2012; D'Amiano et al. 2019). Wang et al. (2009) propose a dimensionality reduction approach through principal component analysis (PCA) on the different pieces. Mohamadian et al. (2013) develop a singular value decomposition (SVD) based method in which the image is isolated into numerous little covering squares and after that SVD is requested to remove the copied frames. Yang et al. (2018) proposed a copy-move forgery detection based on a modified SIFT-based detector. Wang et al. (2018) presented a novel block-based robust copy-move forgery detection approach using invariant quaternion exponent moments. D'Amiano et al. (2019) proposed a dense-field method with a video-oriented version of PatchMatch for the detection and localization of copy-move video forgeries.

Convolutional neural networks (CNNs) have been demonstrated to learn rich, robust and powerful features for large-scale video classification (Karpathy et al. 2014). Various 3D CNN architectures (Tran et al. 2015; Carreira and Zisserman 2017; Hara et al. 2018; Xie et al. 2018) have been proposed to explore spatio-temporal contextual relations between consecutive frames for representation learning. Unlike the existing methods for inter-frame forgery and copy-move forgery which mainly use hand-crafted features or bag-of-words, we take advantage of convolutional neural networks to extract spatial and temporal features for frame duplication detection and localization.

13.3 Frame Deletion Detection

There is limited work exploring frame deletion or dropping detection problem without reference or original video. Therefore, we first introduce a series of baselines, including cue-based and learning-based methods, and then introduce our proposed C3D-based CNN.

13.3.1 Baseline Approaches

We studied three different cue-based baseline algorithms from the literature, i.e., (1) color histogram, (2) optical flow (Wang et al. 2014; Chao et al. 2012) and (3) motion energy (Stephen 2009) as follows:

- **Color histogram**. We calculate the histograms on all R, G and B three channels. Whether there are frames dropped between the two consecutive frames is detected by thresholding the score calculated by the L2 distances based on the color histograms of the two adjacent frames.
- **Optical flow**. We calculate the optical flow (Wang et al. 2014; Chao et al. 2012) from the two adjacent frames by the Lucas-Kanade method. Whether there exist frames dropped between the current frame and the next frame is detected by thresholding the L2 distance between the average moving direction between the previous frame and the current frame, and the average moving direction between the current frame and the next frame.
- **Motion energy**. Motion energy is the temporal information (TI) difference sequence (Stephen 2009), i.e., the difference of Y channel in the YCrCb color space. Whether there exist frames dropped between the current frame and the next frame is detected by thresholding the motion energy between the current frame and the next frame.

Note that each algorithm mentioned above compares two consecutive frames and estimates whether there are missing frames between them. We also developed four learning-based baseline algorithms as follows:

- **SVM**. We train an SVM model to predict whether there are frames dropped between two adjacent frames. The feature vector is the concatenation of the absolute difference of color histograms and the two-dimensional absolute difference of the optical flow directions. The optical flow dimensionality is much smaller than the color histogram, and therefore we give it a higher weight.
- **Pairwise Siamese Network**. We train a Siamese CNN that determines if the two input frames are consecutive or if there is frame dropping between them. Each CNN consists of two convolutional layers and three fully connected layers. The loss used is contrastive loss.
- **Triplet Siamese Network**. We extend the pairwise Siamese network to use three consecutive frames. Unlike the pairwise Siamese network, the triplet Siamese

Table 13.1 A list of related algorithms for temporal video manipulation detection. The first three algorithms are cue-based without any training work. The rest are learned-based algorithms, including the traditional SVM, the popular CNNs and the method we proposed in Chengjiang et al. (2019)

Method	Brief description	Learning?
Color histogram	RGB 3 channel histograms + L2 distance	No
Optical flow	The optic flow (Wang et al. 2014; Chao et al. 2012) with Lucas-Kanade method + L2 distance	No
Motion energy	Based on temporal information difference (Stephen 2009) sequence	No
SVM	770-D feature vector (3×256-D RGB histogram + 2-D optic flow)	Yes
Pairwise Siamese Network	Siamese network architecture (2 conv layers + 3 fc layers + contrastive loss)	Yes
Triplet Siamese Network	Siamese network architecture (Alexnet-variant + Euclidean&contrastive loss)	Yes
Alexnet (Krizhevsky et al. 2012) Network	Alexnet-variant network architecture	Yes
C3D-based Network (Chengjiang et al. 2019)	C3D-variant network architecture + confidence score	Yes

network consisted of three Alexnets (Krizhevsky et al. 2012) merging their output with Euclidean loss between the previous frame and the current frame, and with contrastive loss between the current frame and the next frame.

- **Alexnet-variant Network**. The input frames are converted to gray-scale and put into the RGB channels.

To facilitate the comparison of the competing algorithms, we summarize the above descriptions in Table 13.1.

13.3.2 C3D Network for Frame Deletion Detection

The baseline CNN algorithms we investigated lacked a strong temporal feature suitable to capture the signature of frame drops. These algorithms only used features from two to three frames that were computed independently. C3D network was originally designed for action recognition, however, we found that spatio-temporal signature produced by the 3D convolution is also very effective in capturing the frame drop signatures.

The pipeline of our proposed method is as shown in Fig. 13.3. As we can observe, there are three modifications from the original C3D network. First, the C3D network takes clips of 16 frames, therefore we check the center of the clip (between frames 8 and 9) for frame drops to give equal context on both sides of the drop. This is done by formulating our training data so that frame drops only occur in the center. Secondly, we have a binary output associated with "frames dropped" and "no frames dropped" between the 8th and 9th frame. Lastly, we further refine the per-frame network output scores into a confidence score using peak detection and temporal scaling to further suppress the noisy detections. With the refined confidence scores we are not only able to identify whether the video has frame drops but also localize them by applying the network to the video in a sliding window fashion.

13.3.2.1 Data Preparation

To obtain the training data, we used 2,394 iPhone 4 consumer videos from the World Dataset made available on the DARPA Media Forensics (MediFor) program for research. We pruned the videos such that all videos were of length 1–3 min. We get ended up with 314 videos, of which we randomly selected 264 videos for training, and the rest 50 videos for validation. We developed a tool that randomly drops fixed-length frame sequences from videos. It picks a random number of frame drops and random frame offsets in the video for each removal. The frame drops do not overlap, and it forces 20 frames to be kept around each drop. In our experiments, we manipulate each video many different times to create more data. We vary the fixed frame drop length to see how it affects detection we used 0.5 s, 1 s, 2 s, 5 s and 10 s as five different frame drop durations. We used the videos with these drop durations to train a general C3D-based network for frame drop detection.

13.3.2.2 Training

We use momentum $\mu = 0.9$, $\gamma = 0.0001$ and set power to be 0.075. We start training at a base learning rate of $\alpha = 0.001$ and the "inv" as the learning rate policy. We set the batch size to be 15 and use the 206000th iteration as the learned model for testing, which achieves about 98.2% validation accuracy.

13.3.2.3 Testing

The proposed C3D-based network is able to identify the temporal removal manipulation due to dropped frames in a video and also localize one or more frame drops within the video. We observe that some videos captured by moving digital cameras may have multiple changes due to quickly camera motion, zooming in/out, etc., which can be deceiving to the C3D-based network and can result in false frame dropping detections. In order to reduce such false alarms and increase the generalization ability of our proposed network, we propose an approach to refine the raw network output scores to the confidence scores using peak detection and introduction of a scale term based on the output score variation, i.e.,

1. We first detect the peaks on the output score curve obtained from the proposed C3D-based network per video. Among all the peaks, we only pick the top 2% peaks and ignore the rest of the peaks. Then we shift the time window to check the number of peaks (denoted as n_p) appearing in the time window with ith frame as the center (denoted as $W(i)$). If the number is more than one, i.e., other peaks in the neighborhood, the output score $f(i)$ will be penalized. The value will be penalized more if there are a lot of high peaks detected. The intuition behind is that we want to reduce the false alarms when there are multiple peaks occurring close just because the camera is moving or even zooming in/out.
2. We also introduce a scale term $\Delta(i)$ defined as the difference of the median score and the minimum score within the time window $W(i)$ to control the influence of the camera motion.

Based on the above statement, we can obtain the confidence score for the ith frame as

$$f_{conf}(i) = \begin{cases} f(i) - \lambda\Delta(i) & \text{when } n_p < 2 \\ \frac{f(i)}{n_p} - \lambda\Delta(i) & \text{otherwise} \end{cases}, \quad (13.1)$$

where

$$W(i) = \{i - \frac{w}{2}, \ldots, i + \frac{w}{2}\}. \quad (13.2)$$

Note that λ in Eq. 13.1 is a parameter to control how much the scale term affects the confidence score, and w in Eq. 13.2 indicates the width of the time window.

For testing per frame, say ith frame, we first form a 16-frame video clip and set the ith frame to be the 8th frame in the video clip, and then we can get the output score $f_{conf}(i)$. If $f_{conf}(i) > Threshold$, then we predict there are dropped frames between the ith frame and the $(i + 1)$th frame. For testing per video, we take it as a binary classification and confidence measure per video. To simplify, we use a simple confidence measure, i.e., $\max_i f_{conf}(i)$ across all frames. If $\max_i f_{conf}(i) > Threshold$, then there are temporal removal within the video. Otherwise, the video is predicted without any temporal removal. The results reported in this work are without any $Threshold$ as we are reporting the ROC curves.

13.3.3 Experimental Result

We conducted the experiments on a Linux machine with Intel(R) Xeon(R) CPU E5-2687 0 @ 3.10 GHz, 32 GB system memory and graphical card NVIDIA GTX 1080

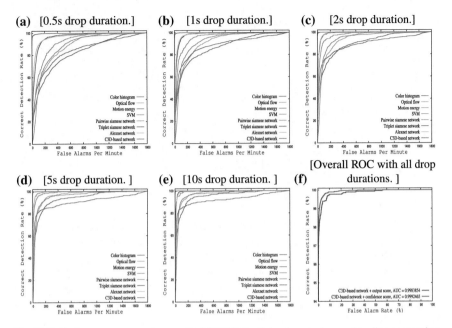

Fig. 13.5 Performance comparison on the YFCC100m dataset against seven baseline approaches, using per-frame ROCs for five different drop durations (**a–e**), and (**f**) is frame-level ROC for all the five drop durations combined

(Pascal). We report our results as the ROC curves based on the output score $f_{conf}(i)$ and accuracy as metrics. We present the ROC curves with false positive rate as well as false alarm rate per minute to provide and demonstrate the level of usefulness for a user that might have to adjudicate each detection reported by the algorithm. We present the ROC curves for both per-frame analysis where the ground truth data is available and per-video analysis otherwise.

To demonstrate the effectiveness of the proposed approach, we ran experiments on the YFCC100m dataset[2] and the Nimble Challenge 2017 (Development 2 Beta 1) dataset.[3]

13.3.3.1 YFCC100m Dataset

We download 53 videos tagged with iPhone from Yahoo Flickr Creative Commons 100 Million (YFCC100m) dataset and manually verified that they are single-shot videos. To create ground truth we used our automatic randomized frame dropping tool to generate the manipulated videos. For each video we generated manipulated

---

[2] YFCC100m dataset: http://www.yfcc100m.org.

[3] Nimble Challenge 2017 dataset: https://www.nist.gov/itl/iad/mig/nimble-challenge-2017-evaluation.

Table 13.2 The detailed results of our proposed C3D-based network. $\#_{pos}$ and $\#_{neg}$ are the number instances for the positive and the negative testing 16-frame video clips, respectively. The Acc_{pos} and Acc_{neg} are the corresponding accuracy. Acc is the total accuracy. All the accuracies use the unit %

Duration (s)	$\#_{pos} : \#_{neg}$	Acc_{pos}	Acc_{neg}	Acc
0.5	2816:416633	98.40	98.15	98.16
1	2333:390019	99.49	98.18	98.18
2	2816:416633	99.57	98.11	98.12
5	1225:282355	99.70	98.17	98.18
10	770:239210	100.00	98.12	98.13

videos with frame drops of 0.5, 1, 2, 5 or 10 s intervals at random locations. For each video and each drop duration, we randomly generate 10 manipulated videos. In this way we collect $53 \times 5 \times 10 = 2650$ manipulated videos as testing dataset.

For each drop duration, we run all the competing algorithms in Table 13.1 on the 530 videos with the parameter setting $w = 16$, $\lambda = 0.22$. The experimental results are summarized in the ROC curves for all these five different drop durations in Fig. 13.5.

One can note that (1) the traditional SVM outperforms the three simple cue-based algorithms; (2) the four convolution neural networks algorithms perform much better than the traditional SVM and all the cue-based algorithms; (3) among all the CNN-based networks, both the triplet Siamese network and the Alexnet-variant network perform similar and better than the pairwise Siamese network; and (4) our proposed C3D-based network performs the best. This provides some empirical support to the hypothesis that the proposed C3D-based method is able to take advantage of the temporal and spatial correlations, while the other CNN-based networks only explore the spatial information in the individual frames.

To better understand the C3D-based network, we provide more experimental details in Table 13.2. With the drop duration increase, both the number of positive and negative testing instances decrease and the positive accuracy keeps increasing. As one might expect, the shorter the frame drop duration, the more difficult it is to detect.

We also merge the results of the C3D-based network with five different drop durations in Fig. 13.5 together to plot a unified ROC curve. For comparison, we also plot another ROC curve that uses the output scores to detect whether there exist frame drops within a testing video. As we can see in Fig. 13.5(f), using output score from the C3D-based network, we can still achieve very good performance to 0.9983854 AUC. This observation can be explained by the fact that the raw phone videos from the YFCC100m dataset have less quick motion, no zooming in/out occurring and even no video manipulations. Also, the manipulated videos are generated in the same way as the generation of training manipulated videos with the same five drop durations. Since there are no overlaps on the video contents between training videos and testing videos, such a good performance demonstrates the power and the generalization

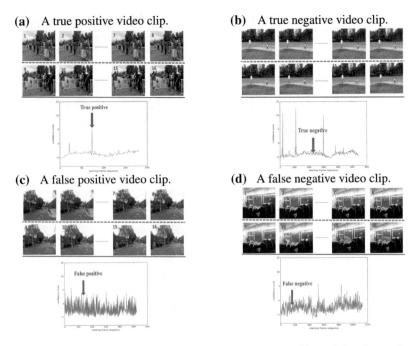

Fig. 13.6 The visualization of two successful examples (one true positive and the other one is true negative) and two failure examples (one false positive and the other one is false negative) from the YFCC100m dataset. The red dashed line indicates the location between the 8th frame and the 9th frame where we test for a frame drop. The red arrows point to the frame on the confidence score plots

ability of the trained network. Although using output score directly achieves a very good AUC, using the confidence score defined in Eq. 13.1 can still improve the AUC from 0.9983854 to 0.9992465. This demonstrates the effectiveness of our confidence score defined with such a peak detection step and a scale term.

We visualize both success and failure cases in our proposed C3D-based network, as shown in Fig. 13.6. Looking at the successful cases in Fig. 13.6(a), "frame drops" is identified correctly in the 16-frame video clip because a man stands at one side in the 8th frame and move to another side suddenly in the 9th frame, and the video clip in Fig. 13.6(b) is predicted as "no frame drops" correctly since a child follows his father in all 16 frames and the 8th frame and the 9th frame are consistent with each other.

Regarding the failures cases, as shown in Fig. 13.6(c), there is no frame drop but it is still identified as "frame drop" between the 8th frame and the 9th frame due to the camera shakes during the video capture of such a street scene. Also, "frame drop" in the top clip cannot be detected correctly between the 8th frame and the 9th frame in the video clip, as shown in Fig. 13.6(d), since the scene inside the bus has almost no visible changes between these two frames.

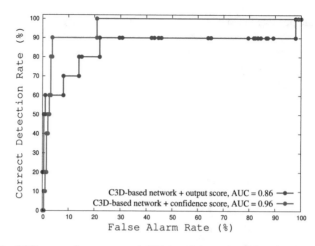

Fig. 13.7 The ROC curve of our proposed C3D-based network on the Nimble Challenge 2017 dataset

Note that our training stage is carried out off-line. Here we only offer the runtime for the testing stage under our experimental environment. For each testing video clip with a 16-frame length, it takes about 2 s. For a one-minute short video with 30 FPS, it requires about 50 min to complete the testing throughout all the frame sequence.

13.3.3.2   Nimble Challenge 2017 Dataset

In order to check whether our proposed C3D-based network is able to identify a testing video with unknown arbitrary drop duration, we also conducted experiments on the Nimble Challenge 2017 dataset, specifically the NC2017-Dev2Beta1 version, in which there are 209 probe videos with various video manipulations. Among these videos, there are six videos manipulated with "TemporalRemove", which is regarded as "frame dropping". Therefore, we run our proposed C3D-based network as a binary classifier to classify all these 209 videos into two groups, i.e., "frame dropping" and "no frame dropping", at the video level. In this experiment, the parameters are set as $w = 500, \lambda = 1.25$.

We first plot the output scores from the C3D-based network and the confidence score of each of the six videos is labeled with "TemporalRemove" in Fig. 13.9. It is clear that the video named "d3c6bf5f224070f1df74a63c232e360b.mp4" has the lowest confidence score smaller than zero.

To explain such a case, we further check the content of the video, as shown in Fig. 13.8. As we can observe, this video is even really hard for us to identify it as "TemporalRemoval" since it is taken by a static camera and only the lady's mouth and head are taking very slight changes across the whole video from the beginning to the end. As we trained purely on iPhone videos, our training network was biased

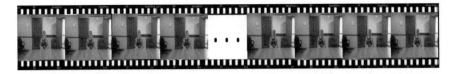

Fig.   13.8 The   entire   frame   sequence   of   the   34-second   video "d3c6bf5f224070f1df74a63c232e360b.mp4", which has 1047 frames and was captured by a static camera. We observe that only the lady's mouth and head are taking very slight change across the video from the beginning to the end

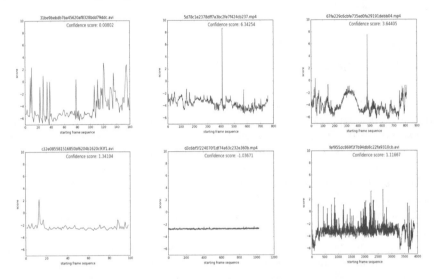

Fig. 13.9   The illustration of output scores from the C3D-based network and their confidence scores for six videos labeled with "TemporalRemove" from the Nimble Challenge 2017 dataset. The blue curve is the output score, the red "+" marks the detected peaks, and the red confidence score is used to determine whether the video can be predicted as a video with "frame drops"

toward videos with camera motion. With a larger dataset of static camera videos, we can train different networks for static and dynamic cameras to address this problem.

We plot the ROC curve in Fig. 13.7. As we can see, the AUC of the C3D-based network with confidence scores is high to 0.96, while the AUC of the C3D-based network with the output scores directly is only 0.86. The insight behind such a significant improvement is that there are testing videos with camera quick-moving, zooming in and out, as well as other types of video manipulations, and our confidence scores defined with the peak detection step and the scale term to penalize multiple peaks occurring too close and large scales is able to significantly reduces the false alarms. Such a significant improvement by 0.11 AUC strongly demonstrates the effectiveness of our proposed method.

13.4   Frame Duplication Detection

As shown in Fig. 13.4, given a probe video, our proposed C2F-DCNN framework is designed to detect and localize frame duplication manipulation. An I3D network is used to produce a sequence-to-sequence matrix and determine the candidate frame sequences at the coarse-search stage. A Siamese network is then applied for a fine-level search to verify whether frame duplications exist. After this, an inconsistency detector is applied to further distinguish duplicated frames from selected frames. All of these steps are described below in detail.

13.4.1   Coarse-Level Search for Duplicated Frame Sequences

In order to efficiently narrow the search space, we start by finding possible duplicate sets of frames throughout the video using a robust CNN representation. We split a video into overlapping frame sequences, where each sequence has 64 frames and the number of overlapped frames is 16. We choose I3D Network (Carreira and Zisserman 2017), instead of using C3D network (Tran et al. 2015) due to these reasons: (1) It inflates 2D ConvNets into 3D and makes filters from typically $N \times N$ square to $N \times N \times N$ cubic; (2) it bootstraps 3D filters from 2D filters to bootstrap parameters from the pre-trained ImageNet models; and (3) it paces receptive field growth in space, time and network depth.

In this work, we apply the pre-trained off-the-shell I3D network to extract the 1024-dimensional feature vector for $k = 64$ frame sequences since the input for the standard I3D network is 64 rgb-data and 64 flow-data. We observed that a lot of time was being spent on the pre-processing. To reduce the testing runtime, we only compute the first k rgb-data and k flow-data items. For the subsequent frame sequence, we can copy $(k - 1)$ rgb-data and $(k - 1)$ flow-data from the previous video clip, and only calculate the last rgb-data and flow-data. This significantly improved the testing efficiency.

Based on the sequence features, we calculate the sequence-to-sequence distance matrix over the whole video using L2 distance. If the distance is smaller than the threshold T_1, then this indicates that these two frame sequences are likely duplicated and we take them as two candidate frame sequences for further confirmation during the next fine-level search.

13.4.2   Fine-Level Search for Duplicated Frames

For the candidate frame sequences, detected by the previous stage described in Sect. 13.4.1, we evaluate the distance between all pairs of frames across the two sequences, i.e., a duplicated frame and the corresponding selected original frame.

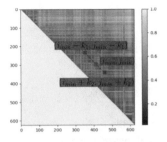

Fig. 13.10 A sample distance matrix based on the frame-to-frame distances computed by the Siamese network between a pair of frame sequences. The symbols shown on the line segment with low distance are used to compute the video-level confidence score for frame duplication detection

For this purpose we propose a Siamese neural network architecture, which learns to differentiate between two frames in the provided pair. It consists of two identical networks by sharing exactly the same parameters, each taking one of the two input frames. A contrastive loss function is applied to the last layers to calculate the distance between the pair. In principle, we can choose any neural network to extract features for each frame.

In this work, we choose the ResNet network (Kaiming et al. 2016) with 152 layers given its demonstrated robustness. We connect two ResNets in the Siamese architecture with a contrastive loss function, and each loss value associated with the distance between a pair of frames is formulated into the frame-to-frame distance matrix, in which the distance is normalized to the range $[0, 1]$. A distance smaller than the threshold T_2 indicates that these two frames are likely duplicated. For videos that have multiple consecutive frames duplicated we expect to see a line with low values parallel to the diagonal in the visualization of the distance matrix, as plotted in Fig. 13.10.

It is worth mentioning that we provide both frame-level and video-level scores to evaluate the likelihood of frame duplication. For the frame-level score, we can use the value in the frame-to-frame distance directly. For the video-level score, we propose a heuristic strategy to formulate the confidence value. We first find the minimal value of distance $d_{\min} = d(i_{\min}, j_{\min})$ where $i_{\min}, j_{\min} = \text{argmin}_{0 \le i < j \le n} d(i, j)$ is the frame-to-frame distance matrix. Then a search is performed in two directions to find the number of consecutive duplicated frames:

$$k_1 = \underset{k:k \le i_{\min}}{\text{argmax}} |d(i_{\min} - k, j_{\min} - k) - d_{\min}| \le \epsilon \tag{13.3}$$

and

$$k_2 = \underset{k:k \le n - j_{\min}}{\text{argmax}} |d(i_{\min} + k, j_{\min} + k) - d_{\min}| \le \epsilon \tag{13.4}$$

where $\epsilon = 0.01$ and the length of the interval with duplicated frames can be defined as

$$l = k_1 + k_2 + 1. \tag{13.5}$$

Finally, we can formulate the video-level confidence score as follows:

$$F_{video} = -\frac{d_{\min}}{l \times (j_{\min} - i_{\min})}. \tag{13.6}$$

The intuition here is that a more likely frame duplication is indicated by a smaller value of $d_{\min}$, a longer interval of duplicated frames and a larger temporal gap between the selected original frames and the duplicated frames.

13.4.3 Inconsistency Detector for Duplication Localization

We observe that the duplicated frames inserted into the source video usually yield artifacts due to temporal inconsistency at both the beginning frames and the end frames in a manipulated video. To automatically distinguish the duplicated frames from selected frames, we make use of both spatial and temporal information by training an inconsistency detector to locate this temporal discrepancy. For this purpose, we build upon our work discussed above, Long et al. (2017), which proposed a C3D-based network for frame-drop detection and only works for single-shot videos. Instead of using only one RGB stream data as input, we replace the C3D network with an I3D network to also incorporate the optical flow data stream. It is also worth mentioning that unlike the I3D network used in Sect. 13.4.1, input to the I3D network here is a 16-frame temporal interval, every frame in a sliding window, with RGB and optical flow data. The temporal classification provides insight into the temporal consistency between the 8th and the 9th frame within the 16-frame interval. In order to handle multiple shots in a video with hard cuts, we extend the binary classifier to three classes: "none"—no temporal inconsistency indicating manipulation; "frame drop"—there are frames removed within one-shot video; and "shot break" or "break"—there is a temporal boundary or transition between two video shots. Note that the training data with shot-break videos are obtained from TRECVID 2007 dataset (Kawai et al. 2007), and we only use the hard-cut shot-breaks since soft-cut changes gradually and has strong consistency between any two consecutive frames. The confusion matrix in Fig. 13.11 illustrates the high effectiveness of the proposed I3D network-based inconsistency detector.

Based on the output scores for the three categories from the I3D network, i.e., $S_{I3D}^{none}(i)$, $S_{I3D}^{drop}(i)$ and $S_{I3D}^{break}(i)$, we formulate the confidence score of inconsistency as the following function:

$$S(i) = S_{I3D}^{drop}(i) + S_{I3D}^{break}(i) - \lambda S_{I3D}^{none}(i), \tag{13.7}$$

where λ is the weight parameter, and for the results presented here, we use $\lambda = 0.1$. We assume the selected original frames have a higher temporal consistency with

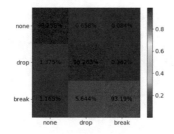

Fig. 13.11 The confusion matrix for three classes of temporal inconsistency within a video, used with the I3D-based inconsistency. We expect a high likelihood of "drop" class at the two ends of the duplicated frame sequence and a high "none" likelihood at the ends of the selected original frame sequence

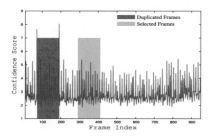

Fig. 13.12 Illustration of distinguishing duplicated frames from the selected frames. The index ranges for the red frame sequence and the green sequence are [72, 191] and [290, 409], respectively. s_1 and s_2 are the corresponding inconsistency scores for the red sequence and green sequence, respectively. Obviously, $s_1 > s_2$, which indicates that the red sequence is duplicated frames as expected

frames before and after such frames than the duplicated frames because the insertion of duplicated frames usually causes a sharp inconsistency at the beginning and the end of the duplicated interval, as illustrated in Fig. 13.12. Given a pair of frame sequences that are potentially duplicated, $[i, i + l]$ and $[j, j + l]$, we compare two scores,

$$s_1 = \sum_{k=-wind}^{wind} S(i - 1 + k) + S(i + l + k) \qquad (13.8)$$

and

$$s_2 = \sum_{k=-wind}^{wind} S(j - 1 + k) + S(j + l + k), \qquad (13.9)$$

where $wind$ is the window size. We check the inconsistency at both the beginning and the end of the sequence. In this work, we set $wind = 3$ to avoid the failure cases where a few start or end frames were detected incorrectly. If $s_1 > s_2$, then the duplicated frame segment is $[i, i + l]$. Otherwise, the duplicated frame segmentation

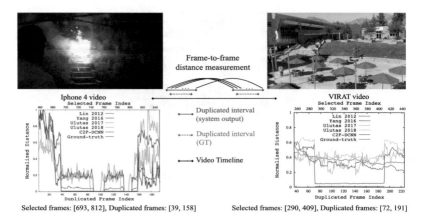

Selected frames: [693, 812], Duplicated frames: [39, 158]   Selected frames: [290, 409], Duplicated frames: [72, 191]

Fig. 13.13 Illustration of frame-to-frame distance between duplicated frames and the selected frames

is $[j, j + l]$. As shown in Fig. 13.12, our modified I3D network is able to measure the consistency between consecutive frames.

13.4.4 Experimental Results

We evaluate our proposed C2F-DCNN method on a self-collected video dataset and the Media Forensics Challenge 2018 (MFC18)[4] dataset (Guan et al. 2019).

Our self-collected video dataset is obtained through automatically adding frame duplication manipulation on the 12 raw static camera videos from VIRAT dataset (Oh et al. 2011) and 17 dynamic iPhone 4 videos. The duration of each video is in the range from 47 seconds to 3 minutes. In order to generate test videos with frame duplication, we randomly select frame sequences with the duration 0.5, 1, 2, 5 and 10 s, and then re-insert them into the same source videos. We use the X264 video codec and a frame rate of 30 fps to generate these manipulated videos. Note that we avoid any temporal overlap between the selected original frames and the duplicated frames in all generated videos. Since we have the frame-level ground truth, we can use it for frame-level performance evaluation.

The MFC18 dataset consists of two subsets, Dev dataset and Eval dataset, which we denote as the MFC18-Dev dataset and the MFC18-Eval dataset, respectively. There are 231 videos in the MFC18-Dev dataset and 1036 videos in the MFC18-Eval dataset. The duration of each video is in the range from 2 seconds to 3 minutes. The frame rate for most of the videos is 29–30 fps, while a smaller number of videos are 10 or 60 fps and only five videos in the MFC18-Eval dataset are larger than 240 fps. We opt out these five videos and another two videos which have less than 17

---

[4] https://www.nist.gov/itl/iad/mig/media-forensics-challenge-2018.

frames from the MFC18-Eval dataset because the input for the I3D network should have at least 17 frames. We use the remaining 1029 videos in the MFC18-Eval dataset to conduct the video-level performance evaluation.

The detection task is to detect whether or not a video has been manipulated with frame duplication manipulation, while the localization task to localize the duplicated frames index. For the measurement metrics, we use the performance measures of area under the ROC curve (AUC) for the detection task, and use the Matthews correlation coefficient

$$MCC = \frac{TP \times TN - FP \times FN}{\sqrt{(TP + FP)(TP + FN)(TN + FP)(TN + FN)}}$$

for localization evaluation, where TP, FP, TN and FN refer to frames which represent true positive, false positive, true negative and false negative, respectively. See Guan et al. (2019) for further details on the metrics.

13.4.4.1 Frame-Level Analysis on Self-collected Dataset

To better verify the effectiveness of deep learning solution in frame-duplication detection on the self-collected dataset, we consider four baselines: Lin et al.'s method (Guo-Shiang and Jie-Fan 2012) that uses histogram difference as the detection features, Yang et al.'s method (Yang et al. 2016) that is an effective similarity-analysis-based method with SVD features, Ulutas et al.'s method (Ulutas et al. 2017) based on binary features and another method by them (Ulutas et al. 2018) that uses bag-of-words with 130-dimensional SIFT descriptors. Different from our proposed C2F-DCNN method, all of these methods use traditional feature extraction without deep learning.

Note that the manipulated videos are generated by us, hence both selected original frames and duplicated frames are accessible to us. We treat these experiments as a white-box attack and evaluate the performance of frame-to-frame distance measurements.

We run the proposed C2F-DCNN approach and the above-mentioned four state-of-the-art approaches on our self-collected dataset and the results are summarized in Table 13.3. As we can see, due to the X264 codec, the contents of the duplicated frames have been affected so that the detection of a duplicated frame and its corresponding selected frame is very challenging. In this case, our C2F-DCNN method still outperforms the four previous methods.

To help the reader better understand the comparison, we provide a visualization of the normalized distances between the selected frames and the duplicated frames in Fig. 13.13. We can see our C2F-DCNN performs the best for both sample videos, especially with respect to the ability to distinguish the temporal boundary between duplicated frames and non-duplicated frames. All these observations strongly demonstrate the effectiveness of this deep learning approach for frame duplication detection.

Table 13.3 The AUC performance of frame-to-frame distance measurements for frame duplication detection on our self-collected video dataset.(unit: %)

Method	iPhone 4 videos	VIRAT videos
Lin 2012 (Guo-Shiang and Jie-Fan 2012)	80.81	80.75
Yang 2016 (Yang et al. 2016)	73.79	82.13
Ulutas 2017 (Ulutas et al. 2017)	70.46	81.32
Ulutas 2018 (Ulutas et al. 2018)	73.25	69.10
C2F-DCNN	**81.46**	**84.05**

Fig. 13.14 The ROC curves for video-level frame duplication detection on the MFC18-Dev dataset

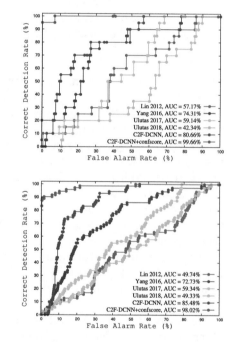

Fig. 13.15 The ROC curves for video-level frame duplication detection on the MFC18-Eval dataset

13.4.4.2   Video-Level Analysis on the MFC18 Dataset

It is worth mentioning that the duplicated videos in the MFC18 dataset usually include multiple manipulations, and this makes the content between the selected original frames and duplicated frames different at times. Therefore, the testing video in both the MFC18-Dev and the MFC18-Eval datasets are very challenging. Since we are not aware of the details about the generation of all the testing videos, we take this dataset as a black-box attack and evaluate its video-level detection and localization performance.

Table 13.4 The MCC metric in [−1.0, 1.0] range for video temporal localization on the MFC18 dataset. Our approach generates the best MCC score, where 1.0 is perfect

Method	MFC18-Dev	MFC18-Eval
Lin 2012 (Guo-Shiang and Jie-Fan 2012)	0.2277	0.1681
Yang 2016 (Yang et al. 2016)	0.1449	0.1548
Ulutas 2017 (Ulutas et al. 2017)	0.2810	0.3147
Ulutas 2018 (Ulutas et al. 2018)	0.0115	0.0391
C2F-DCNN w/ ResNet	0.4618	0.3234
C2F-DCNN w/ C3D	0.6028	0.3488
C2F-DCNN w/ I3D	**0.6612**	**0.3606**

Table 13.5 The video temporal localization performance on the MFC18 dataset. Note $\checkmark$, $\times$ and $\otimes$ indicate correct cases, incorrect cases and ambiguously incorrect cases, respectively. And #(.) indicates the number of a kind of specific cases

Dataset	#($\checkmark$)	#($\times$)	#($\otimes$)
MFC18-Dev	14	6	1
MFC18-Eval	33	38	15

We compare the proposed C2F-DCNN method and the above-mentioned four state-of-the-art methods, i.e., Lin 2012 (Guo-Shiang and Jie-Fan 2012), Yang 2016 (Yang et al. 2016), Ulutas 2017 (Ulutas et al. 2017) and Ulutas 2018 (Ulutas et al. 2018) on these two datasets. We use the negative minimum distance (i.e., $−d_{min}$) as a default video-level scoring method to generate a video-level score for each competing method, including ours. "C2F-DCNN+confscore" denotes our best configuration with C2F-DCNN along with the proposed video-level confidence score defined in Eq. 13.6. In contrast, "C2F-DCNNa" uses only $−d_{min}$ as the confidence score. The comparative manipulated video detection results are summarized in Figs. 13.14 and 13.15.

A few observations that we would like to point out: (1) C2F-DCNN always outperforms the four previous methods for the video-level frame duplication, with the video-level score as negative minimum distance; (2) with "+conf score", our "C2F-DCNN+confscore" method generates a significant boost in AUC as compared to the baseline score of $−d_{min}$ and achieves a high correct detection rate at a low false alarm rate; and (3) the proposed "C2F-DCNN+confscore" method achieves very high AUC scores on the two benchmark datasets: 99.66% on MFC18-Dev, and 98.02% on MFC18-Eval.

We also performed a quantified analysis of the temporal localization within a manipulated video with frame duplication. For comparison with the four previous methods, we use the feature distance between any two consecutive frames.

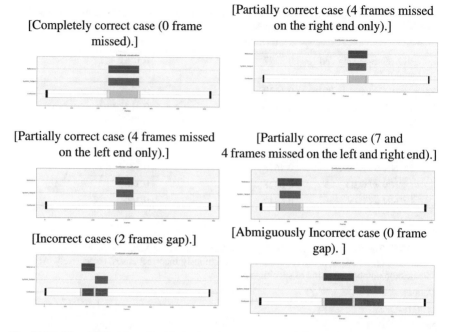

Fig. 13.16 The visualization of confusion bars in video temporal localization. For each subfigure, the top (purple) bar is ground truth indicating duplication, the middle bar (pink) is the system output from the proposed method and the bottom bar is the confusion calculated based on the above the truth and the system output. Note TN, FN, FP, TP and "OptOut" in the confusion are marked in white, blue, red, green and yellow/black, respectively. **a** and **b–d** are correct results, which include completely correct cases and partially correct cases. **e** and **f** show the failure cases

For the proposed C2F-DCNN approach, the best configuration "C2F-DCNN w/ I3D" includes the I3D network as the inconsistency detector. We also provide two baseline variants by replacing the I3D inconsistency detector with a ResNet network feature distance $S_{Res}(i)$ only ("C2F-DCNN w/ ResNet") or the C3D network's scores $S_{C3D}^{drop}(i) - \lambda S_{C3D}^{none}(i)$ from (Chengjiang et al. 2017) ("C2F-DCNN w/ C3D"). The temporal localization results are summarized in Table 13.4, from which we can observe that (1) our deep learning solutions, "C2F-DCNN w/ ResNet", "C2F-DCNN w/ C3D" or "C2F-DCNN w/ I3D" work better than the four previous methods and "C2F-DCNN w/ I3D" performs the best. These observations suggest that 3D convolutional kernel is able to measure the inconsistency between the consecutive frames, and both RGB data stream and optical flow data stream are complementary to further improve the performance.

To better understand the video temporal localization measurement, we plot the confusion bars on the video timeline based on the truth and the corresponding system output under different scenarios, as shown in Fig. 13.16. We would like to emphasize that no algorithm is able to distinguish duplicated frames from selected frames for the ambiguously incorrect cases indicated as $\otimes$ in Table 13.5, because such videos often break the assumption of temporal consistency and in many cases the duplicated frames are difficult to identify by the naked eye.

13.5   Conclusions and Discussion

We presented a C3D-based network with a confidence score defined with a peak detection step and a scale term for frame dropping detection. The method we proposed in Chengjiang et al. (2017) flexibly explores the underlying spatio-temporal relationship across the one-shot videos. Empirically, it is not only able to identify manipulation of temporal removal type robustly but also to detect the exact location where the frame dropping occurred.

Our future work includes revising frame dropping strategy to be more realistic for training video collection, evaluating an LSTM-based network for quicker runtime, and working on other types of video manipulation detection such as addressing shot boundaries and duplication in looping cases.

Multiple factors cause frame duplication detection and localization becoming more and more challenging in video forgeries. These factors include high frame rates, multiple manipulations (e.g., "SelectCutFrames", "TimeAlterationWarp", "AntiForensicCopyExif", "RemoveCamFingerprintPRNU"[5]) involved before and after, and gaps between the selected frames and the duplicated frames. In particular, zero gap between the selected frames and the duplicated frames renders the manipulation undetectable because the inconsistency which should exist at the end of the duplicated frames does not appear in the video temporal context.

Regarding the runtime, the I3D network for inconsistency detection is the most expensive component in our framework but we only apply it on the candidate frames that are likely to have frame duplication manipulations detected in the coarse-search stage. For each testing video clip with a 16-frame length, it takes about 2 s with our learned I3D network. For a one-minute short video with 30 FPS, it requires less than 5 min to complete the testing throughout all the frame sequences.

The coarse-to-fine deep learning approach is designed for frame duplication detection at both frame-level and video-level, as well as for video temporal localization. This work also included a heuristic strategy to formulate the video-level confidence score, as well as an I3D network-based inconsistency detector to distinguish the duplicated frames from the selected frames. The experimental results have demonstrated the robustness and effectiveness of the method.

Our future work includes continuing to extend multi-stream 3D neural networks for both frame drop, frame duplication and other video manipulation tasks like looping detection, working on frame-rate variations and train on multiple manipulations, investigating the effects of various video codecs on algorithm accuracy.

---

[5] These manipulation operation are defined in the MFC18 dataset.

Acknowledgements This work was supported by DARPA under Contract No. FA875016-C-0166. Any findings and conclusions or recommendations expressed in this material are solely the responsibility of the authors and does not necessarily represent the official views of DARPA of the U.S. Government. Approved for Public Release, Distribution Unlimited.

References

Aneja J, Deshpande A, Schwing AG (2018) Convolutional image captioning. In: The IEEE conference on computer vision and pattern recognition (CVPR)

Awad G, Le DD, Ngo CW, Nguyen VT, Quénot G, Snoek C, Satoh SI (2010) National institute of informatics, Japan at trecvid 2010. In: TRECVID

Carreira J, Zisserman A (2017) Quo vadis, action recognition? a new model and the kinetics dataset. In: 2017 IEEE conference on computer vision and pattern recognition (CVPR). IEEE, pp 4724–4733

Chao J, Jiang X, Sun T (2012) A novel video inter-frame forgery model detection scheme based on optical flow consistency. In: International workshop on digital watermarking, pp 267–281

Chen Y, Li W, Sakaridis C, Dai D, Van Gool L (2018) Domain adaptive faster r-cnn for object detection in the wild. In: The IEEE conference on computer vision and pattern recognition (CVPR)

Christlein V, Riess C, Jordan J, Riess C, Angelopoulou E (2012) An evaluation of popular copy-move forgery detection approaches. arXiv:1208.3665

Conotter V, O'Brien JF, Farid H (2013) Exposing digital forgeries in ballistic motion. IEEE Trans Inf Forensics Secur 7(1):283–296

D'Amiano L, Cozzolino D, Poggi G, Verdoliva L (2019) A patchmatch-based dense-field algorithm for video copy-move detection and localization. IEEE Trans Circuits Syst Video Technol 29(3):669–682

Douze M, Gaidon A, Jegou H, Marszalek M, Schmid C (2008) Inria-lear's video copy detection system. In: TRECVID 2008 workshop participants notebook papers. MD, USA, Gaithersburg

Guan H, Kozak M, Robertson E, Lee Y, Yates AN, Delgado A, Zhou D, Kheyrkhah T, Smith J, Fiscus J (2019) MFC datasets: large-scale benchmark datasets for media forensic challenge evaluation. In: 2019 IEEE winter applications of computer vision workshops (WACVW). IEEE, pp 63–72

Hara K, Kataoka H, Satoh Y (2018) Can spatiotemporal 3D CNNs retrace the history of 2D CNNs and imagenet? In: The IEEE conference on computer vision and pattern recognition (CVPR)

He K, Zhang X, Ren S, Sun J (2016) Deep residual learning for image recognition. In: The IEEE conference on computer vision and pattern recognition (CVPR)

Huang T, Zhang X, Huang W, Lin L, Weifeng S (2018) A multi-channel approach through fusion of audio for detecting video inter-frame forgery. Comput Secur 77:412–426

Karpathy A, Toderici G, Shetty S, Leung T, Sukthankar R, Fei-Fei L (2014) Large-scale video classification with convolutional neural networks. In: Proceedings of the IEEE conference on computer vision and pattern recognition, pp 1725–1732

Kawai Y, Sumiyoshi H, Yagi N (2007) Shot boundary detection at TRECVID 2007. In: TRECVID 2007 workshop participants notebook papers. MD, USA, Gaithersburg

Krizhevsky A, Sutskever I, Hinton GE (2012) Imagenet classification with deep convolutional neural networks. In: Advances in neural information processing systems, vol 25, pp 1097–1105

Law-To J, Buisson O, Gouet-Brunet V, Boujemaa N (2006) Robust voting algorithm based on labels of behavior for video copy detection. In: Proceedings of the 14th ACM international conference on multimedia. ACM, pp 835–844

Lin G-S, Chang J-F (2012) Detection of frame duplication forgery in videos based on spatial and temporal analysis. Int J Pattern Recognit Artif Intell 26(07):1250017

Long C, Basharat A, Hoogs A (2019) A coarse-to-fine deep convolutional neural network framework for frame duplication detection and localization in forged videos. In: IEEE international conference on computer vision and pattern recognition workshop (CVPR-W) on media forensics

Long C, Smith E, Basharat A, Hoogs A (2017) A C3D-based convolutional neural network for frame dropping detection in a single video shot. In: IEEE international conference on computer vision and pattern recognition workshop (CVPR-W) on media forensics

Milani S, Fontani M, Bestagini P, Barni M, Piva A, Tagliasacchi M, Tubaro S (2012) An overview on video forensics. APSIPA Trans Signal Inf Process 1

Mohamadian Z, Pouyan AA (2013) Detection of duplication forgery in digital images in uniform and non-uniform regions. In: 2013 UKSim 15th international conference on computer modelling and simulation (UKSim). IEEE, pp 455–460

Oh S, Hoogs A, Perera A, Cuntoor N, Chen CC, Lee JT, Mukherjee S, Aggarwal JK, Lee H, Davis L, et al (2011) A large-scale benchmark dataset for event recognition in surveillance video. In: 2011 IEEE conference on computer vision and pattern recognition (CVPR). IEEE, pp 3153–3160

Ren S, He K, Girshick R, Sun J (2015) Faster r-cnn: towards real-time object detection with region proposal networks. In Cortes C, Lawrence ND, Lee DD, Sugiyama M, Garnett R (eds) Advances in neural information processing systems (NIPS), pp 91–99

Smeaton AF, Over P, Doherty AR (2010) Video shot boundary detection: seven years of trecvid activity. CVIU 114(4):411–418

Sowmya KN, Chennamma HR (2015) A survey on video forgery detection. Int J Comput Eng Appl 9(2):17–27

Stock P, Cisse M (2018) Convnets and imagenet beyond accuracy: understanding mistakes and uncovering biases. In: The European conference on computer vision (ECCV)

Tang P, Wang X, Wang A, Yan Y, Liu W, Huang J, Yuille A (2018) Weakly supervised region proposal network and object detection. In: The European conference on computer vision (ECCV)

Thakur MK, Saxena V, Gupta JP (2016) Learning based no reference algorithm for dropped frame identification in uncompressed video, pp 451–459

Thakur MK, Saxena V, Gupta JP (2016) Learning based no reference algorithm for dropped frame identification in uncompressed video. In: Information systems design and intelligent applications: proceedings of third international conference INDIA 2016, vol 3, pp 451–459

Tran D, Bourdev L, Fergus R, Torresani L, Paluri M (2015) Learning spatiotemporal features with 3d convolutional networks. In: 2015 IEEE international conference on computer vision (ICCV). IEEE, pp 4489–4497

Ulutas G, Ustubioglu B, Ulutas M, Nabiyev V (2017) Frame duplication/mirroring detection method with binary features. IET Image Process 11(5):333–342

Ulutas G, Ustubioglu B, Ulutas M, Nabiyev VV (2018) Frame duplication detection based on bow model. Multimed Syst 1–19

Venugopalan S, Xu H, Donahue J, Rohrbach M, Mooney R, Saenko K (2015) Translating videos to natural language using deep recurrent neural networks. In: North American chapter of the association for computational linguistics—human language technologies (NAACL-HLT)

Wang J, Liu G, Zhang Z, Dai Y, Wang Z (2009) Fast and robust forensics for image region-duplication forgery. Acta Autom Sin 35(12):1488–1495

Wang Q, Li Z, Zhang Z, Ma Q (2014) Video inter-frame forgery identification based on optical flow consistency. Sens Transducers 166(3):229

Wang Q, Li Z, Zhang Z, Ma Q (2014) Video inter-frame forgery identification based on consistency of correlation coefficients of gray values. J Comput Commun 2(04):51

Wang X, Liu Y, Huan X, Wang P, Yang H (2018) Robust copy-move forgery detection using quaternion exponent moments. Pattern Anal Appl 21(2):451–467

Wang W, Farid H (2007) Exposing digital forgeries in video by detecting duplication. In: Proceedings of the 9th workshop on multimedia & security. ACM, pp 35–42

Wolf S (2009) A no reference (NR) and reduced reference (RR) metric for detecting dropped video frames. In: National telecommunications and information administration (NTIA)

Wu Y, Jiang X, Sun T, Wang W (2014) Exposing video inter-frame forgery based on velocity field consistency. In: 2014 IEEE International Conference on acoustics, speech and signal processing (ICASSP). IEEE, pp 2674–2678

Xie S, Sun C, Huang J, Tu Z, Murphy K (2018) Rethinking spatiotemporal feature learning: speed-accuracy trade-offs in video classification. In: The European conference on computer vision (ECCV)

Yang J, Huang T, Lichao S (2016) Using similarity analysis to detect frame duplication forgery in videos. Multimed Tools Appl 75(4):1793–1811

Yang B, Sun X, Guo H, Xia Z, Chen X (2018) A copy-move forgery detection method based on CMFD-SIFT. Multimed Tools Appl 77(1):837–855

Yongjian H, Li C-T, Wang Y, Liu B (2012) An improved fingerprinting algorithm for detection of video frame duplication forgery. Int J Digit Crime Forensics (IJDCF) 4(3):20–32

Yu H, Cheng S, Ni B, Wang M, Zhang J, Yang X (2018) Fine-grained video captioning for sports narrative. In: The IEEE conference on computer vision and pattern recognition (CVPR)

Zhang N, Paluri M, Ranzato MA, Darrell T, Bourdev L (2014) Pose aligned networks for deep attribute modeling. In: CVPR. Panda

Zhao DN, Wang RK, Lu ZM (2018) Inter-frame passive-blind forgery detection for video shot based on similarity analysis. Multimed Tools Appl 1–20

Zheng L, Sun T, Shi YQ (2014) Inter-frame video forgery detection based on block-wise brightness variance descriptor. In: International workshop on digital watermarking, pp 18–30

Zhou B, Lapedriza A, Xiao J, Torralba A, Oliva A (2014) Learning deep features for scene recognition using places database. In: NIPS

Chapter 14
Integrity Verification Through File Container Analysis

Alessandro Piva and Massimo Iuliani

In the previous chapters, multimedia forensics techniques based on the analysis of the data stream, i.e., the audio-visual signal, aimed at detecting artifacts and inconsistencies in the (statistics of the) content were presented. Recent research highlighted that useful forensic traces are also left in the file structure, thus offering the opportunity to understand a file's life-cycle without looking at the content itself. This Chapter is then devoted to the description of the main forensic methods for the analysis of image and video file formats.

14.1 Introduction

Most forensic techniques look for traces within the content itself that are, however, mostly ineffective in some scenarios, for example, when dealing with strongly compressed or low resolution images and videos. Recently, it has been shown that also the image or video file container maintain some traces of the content history. The analysis of the file format and metadata allows to determine their compatibility, completeness, and consistency based on the expected media's history. This analysis proved to be strongly promising since the image and video compression standards leave some freedom in the file container's generation. This fact allows forensic practitioners to identify media's container signatures related to a specific brand, model, social media platform, etc. Furthermore, the cost for analyzing a file header is usually negligible. This is even more relevant when dealing with digital videos where the stream analy-

A. Piva (✉) · M. Iuliani
Department of Information Engineering, University of Florence, Via di S. Marta, 3, 50139 Florence, Italy
e-mail: alessandro.piva@unifi.it

© The Author(s) 2022
H. T. Sencar et al. (eds.), *Multimedia Forensics*, Advances in Computer Vision and Pattern Recognition, https://doi.org/10.1007/978-981-16-7621-5_14

sis can quickly become unfeasible. Indeed, recent methods highlight that video can be analyzed in seconds, independently of their length and resolution.

It is worth also mentioning that most of the achieved results highlighted that most interesting containers' signatures usually have a low intra-variability, thus suggesting that this analysis can be promising even when only a few training data are available. On the other side, these techniques' main drawback is that they only allow forgery detection while localization is usually beyond its capabilities.

Furthermore, the design of practical media containers' parsers is not to be underestimated since it requires a deep comprehension of image and video formats, and the parser should be updated when new models are introduced in the market or unknown elements are added to the media header. Finally, we are currently not aware of any publicly available software that would allow users to consistently forge such information without advanced programming skills. This makes the creation of plausible forgeries undoubtedly a highly non-trivial task and thereby reemphasizes that file characteristics and metadata must not be dismissed as unreliable source of evidence for the purpose of file authentication per se. The Chapter is organised as follows: in Sects. 14.1.1 and 14.1.2 the main image and video file format specifications are summarized. Then, in Sect. 14.2 and 14.3 the main methods for the analysis of image and video file formats are described. Finally, in Sect. 14.4 the main findings of this research area are provided, along with some possible future works.

14.1.1 Main Image File Format Specifications

Multimedia technologies allow digital image generation based on several image formats. DSLR cameras and proficient acquisition devices can store images in uncompressed formats (e.g., DNG, CR2, BMP). When necessary, lossless compression schemes (e.g., PNG) can be applied to reduce the image size without impacting its quality. Furthermore, lossy compression schemes (e.g., JPEG, HEIF) are available to strongly reduce image size with minimal impact on visual quality.

Nowadays, most of the devices on the market and social media platforms generate and store JPEG images. For this reason, most of the forensic literature focuses on this image type. The JPEG standard itself JPEG (1992) defines the pixel data encoding and decoding, while the full-featured file format implementation is defined in the *JPEG Interchange Format* (JIF) ISO/IEC (1992). This format is quite complicated, and some simpler alternatives are generally preferred for handling and encapsulating JPEG-encoded images: the *JPEG File Interchange Format* (JFIF) Eric Hamilton (2004) and the *Exchangeable Image File Format* (Exif) JEITA (2002), both built on JIF.

Each of the formats defines the basic structure of JPEG files as a set of marker segments, either mandatory or optional. An identifier of the marker id indicates the beginning of each marker segment. Abbreviations for marker ids are denoted by 16 bit short values starting with 0xFF. Marker segments can encapsulate either data compression parameters (e.g., in DQT, DHT, or SOF) or application-specific metadata, like thumbnails or EXIF metadata (e.g., in APP0(JFIF) or APP1(EXIF)). Additionally, some markers consist of the marker id only and indicate state transitions necessary to parse the file format. For example, SOI and EOI indicate the start and end of a JPEG file, requiring all other markers to be placed between these two mandatory markers.

The JIF standard also allows application markers (APPn), populated with entries in the form of key-value pairs. The values can vary from human-readable strings to complex structures like binary data. The APP0 segment defines pixel densities and pixel ratios, and an optional thumbnail of the actual image can be placed here. In APP1, the EXIF standard enables cameras to save a vast range of optional information (EXIF-Tags) related to the camera's photographic settings when and where the image was taken. The information are split into five main image file directories (IFDs): (i) Primary; (ii) Exif; (iii) Interoperability; (iv) Thumbnail; and (v) GPS. However, the content of each IFD is customizable by camera manufacturers, and the EXIF standard also allows for the creation of additional IFDs. Other metadata are customizable within the file header, such as XMP and IPTC that provide additional information like copyright or the image's editing history. Furthermore, image processing software can introduce a wide variety of marker segment sequences.

These differences in the file formats and the not strictly standardization of several sequences may expose forensic traces within the image file container Thomas Gloe (2012). Indeed, not all segments are mandatory and different combinations thereof can occur. Some segments can appear multiple times (e.g., quantization tables can be either encapsulated in one single or multiple separate DQT segments). Furthermore, the sequence of segments is generally not fixed (with some exceptions) and customizable. For example, JPEG thumbnails exist in different segments and employ their own complete sequence of marker segments. Eventually, arbitrary data can exist after EOI.

In the next sections, we report the main technologies and the results achieved based on the technical analysis of these formats.

14.1.2 Main Video File Format Specifications

When a camera acquires a digital video, image sequence processing and audio sequence processing are performed in parallel. After compression and synchronization, the streams are encapsulated in a multimedia container, simply called a video container from now on. There are several compression standards; however, H264/AVC or MPEG-4 Part 10 International Telecommunication Union (2016), and H265/HEVC or MPEG-H Part 2 International Telecommunication Union (2018) are the most relevant since most mobile devices implement them. Video and audio tracks, plus additional information (such as metadata and sync information), are then encapsulated in the video container based on specific format standards. In the following, we describe the two most adopted formats.

MP4-like videos

Most smartphones and compact cameras output videos in mp4, mov, or 3gp format.
This video packaging refers to the same standard, ISO/IEC 14496 Part 12 ISO/IEC
(2008), that defines the main features of *MP4 File Format* ISO/IEC (2003) and *MOV
File Format* Apple Computer, Inc (2001). The *ISO Base format* is characterized by
a sequence of atoms or boxes. A unique 4-byte code identifies each node (atom)
and each atom consists of a header and a data box, and possibly nested atoms. As
an example, we report in Fig. 14.1 a typical structure of an MP4-like container. The
file type box, ftyp, is a four-letter code that is used to identify the type of encoding, the compatibility, or the intended usage of a media file. According to the latest
ISO standards, it is considered a semi-mandatory atom, i.e., the ISO expects it to be
present and explicit as soon as possible in the file container. In the example given in
Fig. 14.1 (a), the fields of the ftyp descriptor explain that the video file is MP4-like
and it is compliant to the MP4 Base Media v1 [IS0 14496-12:2003] (here *isom*)
and the 3GPP Media (.3GP) Release 4 (here *3gp4*) specifications.[1] The movie box,
moov, is a nested atom containing the metadata needed to decode the data stream,
which is embedded in the following mdat atom. It is important to note that moov

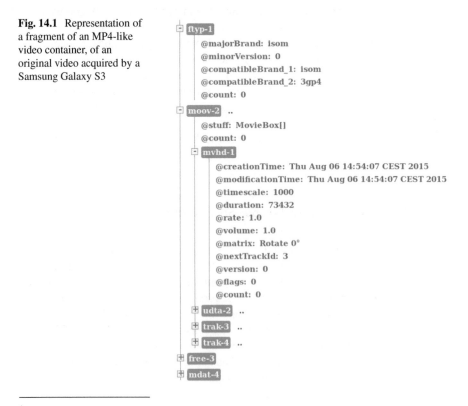

Fig. 14.1 Representation of a fragment of an MP4-like video container, of an original video acquired by a Samsung Galaxy S3

---

[1] The reader can refer to http://www.ftyps.com/ for further details.

may contain multiple `trak` box instances, as shown in Fig. 14.1. The `trak` atom is mandatory, and its numerosity depends on the number of streams included in the file; for example, if the video contains a visual-stream and an audio-stream, there will be two independent `trak` atoms. A more detailed description of these structures can be found in Carlos Quinto Huamán et al. (2020).

AVI videos

Audio video interleave (AVI) is a container format developed by Microsoft in 1992 for Windows software Microsoft. Avi riff file (1992). AVI files are based on the RIFF (resource interchange file format) document format consisting of a RIFF header followed by zero or more lists and chunks.

A chunk is defined as *ckID*, *ckSize*, and *ckData*, where *ckID* identifies the data contained in the chunk, *ckSize* is a 4-byte value giving the size of the data in *ckData*, and *ckData* is zero or more bytes of data.

A list consists of *LIST*, *listSize*, *listType*, and *listData*, where *LIST* is the literal code LIST, *listSize* is a 4-byte value giving the size of the list, *listType* is a 32-bit unsigned integer, and *listData* consists of chunks or lists, in any order.

An AVI file might also include an index chunk, which gives the location of the data chunks within the file. All AVI files must keep these three components in the proper sequence: the *hdrl* list defining the format of the data and is the first required LIST chunk; the *movi* list containing the data for the AVI sequence; the *idx1* list containing the index.

Depending on the specific camera or phone model, additional lists and JUNK chunks may exist between the *hdrl* and *movi* lists. These optional data segments are either used for padding or to store metadata. The *idx1* chunk indexes the data chunks and their location in the file.

14.2 Analysis of Image File Formats

The first studies in the image domain considered JPEG quantization tables and image resolution values Hany Farid (2006); Jesse D. Kornblum (2008); H. Farid (2008). Just with these few features, the authors demonstrated that it is possible to link a probe image to a set of devices or editing software. Then, in Kee Eric (2011), the authors increased the set of features by also considering JPEG coding data and Exif metadata. These studies provided better results in terms of integrity verification and device identification. The main drawback of such approaches is that a user can easily edit the metadata's information, limiting these methods' reliability. Following studies demonstrated that the file structure also contains a lot of traces of the content's history. These data are also more reliable, being much more challenging to extract and modify for a user than metadata. Furthermore, available editing software and metadata editors do not allow the modification of such low-level information Thomas Gloe (2012). However, like the previous ones, this method is based on a manual analysis of the extracted features, to be compared with a set of collected signatures. In Mullan Patrick et al. (2019), the authors present the first automatic method for characterizing the source device, based on the analysis of features extracted from the JPEG header information. Finally, another branch of research demonstrated that the JPEG file format can be exploited to understand whether a probe image has been exchanged through a social network service. In the following, we describe the primary studies of this domain.

14.2.1   Analysis of JPEG Tables and Image Resolution

In Hany Farid (2006), the author shows, for the first time, that manufacturers typically configure the compression of devices differently according to their own needs and preferences. This difference, primarily manifested in the JPEG quantization table, can be used to identify the device that acquired a probe image.

To carry out this analysis, the authors collect a single image, at the highest quality, from each of 204 digital cameras. Then, they extract the JPEG quantization table, noticing that 62 cameras show a unique table.

The remaining cameras fall into equivalence classes of sizes between 2 and 28 (i.e., between 2 and 28 devices share the same quantization tables). Usually, cameras that share the same tables belong to the same manufacturer. Conversely, different makes and models usually share the same table. These results highlight that JPEG quantization tables can partially characterize the source device. Thus, they effectively narrow the source of an image, if not to a single camera make and model, at least to a small set of possible cameras (on average, this set size is 1.43).

This study was performed by also saving the images with five Photoshop versions (at that time, version CS2, CS, 7, 4, and 3) at each of the 13 available levels of compression available. As expected, quantization tables at each compression level were different from one another. Moreover, at each compression level, the tables were the same for all five versions of Photoshop. More importantly, no one of these tables can be found in the images belonging to the 204 digital cameras. These findings allow arguing the possibility to link an image to specific editing software, or at least to a set of possible editing tools.

Similar results have been presented in Jesse D. Kornblum (2008). The authors examined images from devices (a Motorola KRZR K1m, a Canon PowerShot 540, a FujiFilm Finepix A200, a Konica Minolta Dimage Xg, and a Nikon Coolpix 7900) and images edited by several software programs such as libjpeg, Microsoft Paint, the Gimp, Adobe Photoshop, and Irfanview. The analysis carried out on this dataset shows that, although some cameras always use the same quantization tables, most of them use a different set of quantization tables in each image. Moreover, these tables usually differ according to the source or editing software.

In H. Farid (2008), the author expands the earlier work by considering a much larger dataset of images and adding the image resolution to the quantization table as discriminating features for source identification.

The author analyzes over 330 thousand Flickr images from 859 different camera models from 48 manufacturers. Quantization tables and resolutions allowed to obtain 10153 different image classes.

The resolution and quantization table were then combined to narrow the search criteria to identify the source camera. The analysis revealed that 2704 out of 10153 entries (26.6%) have a unique paired resolution and quantization table. In these cases, the features can correctly discriminate the source device. Moreover, 37.2% of the classes have at most two matches, and 44.1% have at most three matches. These results show that the combination of JPEG quantization table and image resolution allows matching the source of an image to a single camera make and model or at least to a small set of possible devices.

14.2.2 Analysis of Exif Metadata Parameters

In Kee Eric (2011), the authors expand the previous analysis by creating a larger image signature for identifying the source camera with greater accuracy than only using quantization tables and resolution. This approach considers a set of features of the JPEG format, namely properties of the run-length encoding employed by the JPEG standard and some Exif header format characteristics. In particular, the signature is composed of three kinds of features: image parameters, thumbnail parameters, and Exif metadata parameters. The image parameters consist of the image size, quantization tables, and the Huffman codes; the Y, Cb, and Cr 8×8 quantization tables are represented as a vector of 192 values; the Huffman codes are specified as six vectors (one for the dc DCT coefficients and one for the ac DCT coefficients of each of the three channels Y, Cb, and Cr) storing the number of codes of length from 1 to 15. This part of the signature is then composed of 284 values: 2 image dimensions, 192 quantization values, and 90 Huffman codes.

The same components are extracted from the image thumbnail, usually embedded in the JPEG header. Thus, the thumbnail parameters are other 284 values: 2 thumbnail dimensions, 192 quantization values, and 90 Huffman codes.

A compact representation of EXIF Metadata is finally obtained by counting the number of entries in each of the five image file directories (IFDs) that compose the EXIF metadata, as well as the total number of any additional IFDs, and the total number of entries in each of these. It worth noticing that some manufacturers adopt metadata representation that does not conform to the EXIF standard, yielding errors when parsing the metadata. The authors also consider these errors as discriminating features, so the total number of parser errors, as specified by the EXIF standard, is also stored. Thus, they consider 8 values from the metadata: 5 entry values from the standard IFDs, 1 for the number of additional IFDs, 1 for the number of entries in these additional IFDs, 1 for the number of parser errors.

The image signature is then composed of 284 header features extracted from the full resolution image, 284 header features from the thumbnail image, and another 8 from the EXIF metadata, for a total of 576 features. These signatures are extracted

Table 14.1 The percentage of camera configurations with an equivalence class size from 1 to 5. Each row corresponds to different subsets of the complete signature

	Equivalence class size				
	1	2	3	4	5
Image	12.9%	7.9%	6.2%	6.6%	3.4%
Thumbnail	1.1%	1.1%	1.0%	1.1%	0.7%
EXIF	8.8%	5.4%	4.2%	3.2%	2.6%
Image+thumbnail	24.9%	15.3%	11.3%	7.9%	3.7%
All	69.1%	12.8%	5.7%	4.0%	2.9%

from the image files and analyzed to verify if they can assign the image to a given camera make and model.

Tests are carried out on over 1.3 million Flickr images, acquired by 773 camera and cell phone models, belonging to 33 different manufacturers.

These images span 9, 163 different distinct pairings of the camera make, model, and signature, referred to as a camera configuration. It is worth noting that each camera model can produce different signatures based on the acquisition resolutions and quality settings. Indeed, in the collected dataset, we find an average of 12 different signatures for each camera make and model.

The camera signature analysis on the above collection shows that the image, thumbnail, and EXIF parameters are not distinctive individually. However, when combined, they provide a highly discriminative signature. This result indicates that the three classes of parameters are not highly correlated, and hence their combination improves overall distinctiveness, as show in Table 14.1.

Eventually, the signature was tested on images edited by Adobe Photoshop (versions 3, 4, 7, CS, CS2, CS3, CS4, CS5 at all qualities), by exploiting, in this case, image and thumbnail quantization tables and Huffman codes only. No overlaps were found between any Photoshop version/quality and camera manufacturer. The Photoshop signatures, each residing in an equivalence class of size 1, are unique. This means that any photo-editing with Photoshop can be easily and unambiguously detected.

14.2.3 Analysis of the JPEG File Format

In Thomas Gloe (2012), the author makes a step forward with respect to the previous studies. They analyze the JPEG file format structure to identify new characteristics that are more discriminating than JPEG metadata only. In this study, a subset of images extracted from the Dresden Dataset (Thomas Gloe and Rainer Bähme 2014) was used. In particular, 4, 666 images belonging to the JPEG scenes (a part of the dataset built to specifically study model-specific JPEG compression algorithms) and 16, 956 images of natural scenes were considered. Images were captured with one device of each available camera model while iterating over all combinations of compression, image size, and flash settings) Overall, they considered images acquired by 73 devices from 26 camera models of 14 different brands. A part of these authentic images was re-saved with each of the eight investigated image processing software (ExifTool, Gimp, IrfanView, Jhead, cjpeg, Paint.NET, PaintShop, and Photoshop), with all available combinations of JPEG compression settings. Finally, they obtained and analyzed more than 32, 000 images.

The study focuses firstly on analyzing the sequence and occurrence of marker segments in the JPEG data structures. The main findings of the author are that:

1. some optional segments can occur with different combinations;
2. segments can appear multiple times;
3. the sequence of segments is generally not fixed, with the exception of some required combinations;
4. JPEG thumbnails exist in different segments and employ their complete sequence of marker segments;
5. arbitrary data can exist after the marker end of image (EOI) of the main image.

The markers' analysis allows to distinguish between pristine and edited images correctly. Furthermore, none of the investigated software allows to recreate the sequence of markers consistently.

The author also considered the sequence of EXIF data structures that store camera properties and image acquisition settings. Interestingly, they found differences between digital cameras, image processing software, and metadata editors. In particular, the following four characteristics appear to be relevant:

1. the byte order is different between camera models (and the default setting employed by ExifTool);
2. sequences of image file directories (IFDs) and corresponding entries (including tag, data type and often also offsets) appear constant in images acquired with the same model and differ between different sources;
3. some manufacturers use different data types for the same entry;
4. raw values of rational types differ between different models while resulting in the same interpreted values (e.g., 200/10 or 20/1).

These findings allow to conclude that the analysis of EXIF data structures can be useful to discriminate between authentic and edited images.

14.2.4   Automatic Analysis of JPEG Header Information

All previous works analyzed images acquired by digital cameras, whereas today, most of the visual content is generated by smartphones. These two classes of devices are somewhat different. While traditional digital cameras typically have a fixed firmware, modern smartphones keep updating their acquisition software and operating system, making it harder to achieve a unique characterization of device models based on the image file format.

In Mullan Patrick et al. (2019), the authors performed a large-scale study on Apple smartphones to characterize the source device based on JPEG header information. The results confirm that identifying the hardware is much harder for smartphones than for traditional cameras, while we can identify the operating system version and selected apps.

The authors exploit Flickr images following the rules adopted in Kee Eric (2011). Starting from a crawling of one million images, they filtered the downloaded media to remove edited images, obtaining at the end a dataset of 432, 305 images. All images belong to Apple devices, including all models comprised between iPhone 4 and iPhone X, between iPad Air and iPad Air 2, with operating system versions from iOS 5 to iOS 12.

They consider two families of features: the number of entries in the image file directories (IFDs) and the quantization tables. The selected directories include the standard IFDs "ExifIFD", "IFD0", "IFD1", "GPS", and the special directories "ICC_Profile" and "MakerNotes". The Y and Cb quantization tables are concatenated, and their hash is computed, obtaining one alphanumeric number, called QC-Table. Differently from Kee Eric (2011), the Huffman tables were not considered since, in the collected database, they are identical in the vast majority of cases. For similar reasons, they also discarded the sequence of JPEG syntax elements studied by Thomas Gloe (2012).

First, the authors verified the modification of the header information for Apple smartphones over time. They showed that image metadata change more often than compression matrices, and that different hardware platforms with same operating system exhibit very similar metadata information. These findings highlight that it is more complicated to analyse smartphones than digital cameras.

After that, the authors design an algorithm for determining the smartphone hardware model from the header information in an automatic way. This is the first algorithm designed to perform an automatic source identification on images through file format information. The classification is performed with a random forest, that uses numbers of entries per directory and quantization tables, as described before, as features.

Firstly, a random forest classifier was trained to identify seven different iPhone models marketed in 2017. Secondly, the same algorithm was trained to verify its ability to identify the operating system from header information, considering all versions from iOS 4 to iOS 11. In all the two cases, experiments were performed by considering as input features: (i) the Exif directories only, (ii) the quantization tables only, (iii) metadata and quantization features together. We report the achieved results in Table 14.2. It worth noticing that using only the quantization tables gives the lowest results, whereas the highest accuracy is reached with all the considered features. Moreover, it is easier to discriminate among operating systems than among hardware models.

Table 14.2   Overall accuracy of models and iOS version classification when using EXIF directories (EXIF-D), quantization matrices (QT), or both

	EXIF-D	QT	Both
iPhone Models classification	61.9%	35.4%	65.6%
iOS version Classification	80.4%	33.7%	81.7%

14.2.5   Methods for the Identification of Social Networks

Several works investigated how JPEG file format modifications can be exploited to understand whether a probe image has been exchanged through a social network service. Indeed, a social network usually recompresses, resizes, and alters image metadata.

Castiglione et al. Castiglione et al. (2011) made the first study on image resizing, compression, renaming, and metadata modifications introduced by Facebook, Badoo, and Google+.

Giudice et al. Oliver Giudice et al. (2017) built a dataset of images from different devices, including digital cameras and smartphones with Android and iOS operating systems. Then, they exchanged all the images through ten different social networks.

The considered features are a 44-dimensional vector composed of pixel width and height, an array containing the Exif metadata, the number of JPEG markers, the first 32 coefficients of the luminance quantization table, and the first 8 coefficients of the chrominance quantization table. These features are fed to an automatic detector based on two K-NN classifiers and a decision tree fitted on the built dataset. The method can predict the social network the image belongs to and the client application with an accuracy of 96% and 97.69% respectively.

Summarizing, the authors noticed that modifications left by the process in the JPEG file depend on both the platform (server-side) and the application used for the upload (client-side).

In Phan et al. (2019), the previous method was extended by merging JPEG metadata information (including quantization tables, Huffman coding tables, image size) with content-related features, namely the histograms of DCT coefficients. A CNN is trained on these features to detect an image's multiple exchanges through Facebook, Flickr, and Twitter.

The adopted dataset consists of two parts: R-SMUD (RAISE Social Multiple Up-Download), generated by the RAISE dataset Duc-Tien Dang-Nguyen et al. (2015), containing images shared over the three Social Networks; V-SMUD (VISION Social Multiple Up-Download), obtained from a part of the VISION dataset Dasara Shullani et al. (2017), shared three times via the three social networks, following different testing configurations.

The experiments demonstrated that JPEG metadata's information improves the performance with respect to using content-based information only. When considering single-shared contents, the accuracy improves from 85.63% to 99.87% on R-SMUD and is in both cases 100.00% on V-SMUD. When considering images shared both once and twice, the accuracy improves from 43.24% to 65.91% on R-SMUD and from 58.82% to 77.12% on V-SMUD. When dealing with three times shared images also, they achieve an accuracy of 36.18% on R-SMUD and 49.72% on V-SMUD.

14.3 Analysis of Video File Formats

Gloe et al. proposed the first more in-depth analysis of video file containers Gloe Thomas et al. (2014). The authors observe that videos from different cameras and phone models expose different container formats and compression. This allows a forensic practitioner to extract this information manually and, based on a comparison with a reference, expose forensic findings. In Jieun Song et al. (2016) the authors studied 296 video files in AVI format acquired with 43 video event data recorders (VEDRs), and their manipulated version with 5 different video editing software tools. All videos were classified according to a sequence of field data structure types. This study showed that the field data structure represents a valid feature set to determine whether video editing software was used to process a probe video. However, these methods require the manual analysis of the video container information, thus needing high effort and high programming skills. Some attempts were proposed to identify the containers' signature of the most relevant brands and social media platforms Carlos Quinto Huamán et al. (2020). Another path tried to automatize the forensic analysis of video file structures by verifying their multimedia stream descriptors with simple binary classifiers through a supervised approach David Güera et al. (2019). However, in Iuliani Massimo (2019), the authors propose a way to formalize the MP4-like (.mp4, .mov, .3gp) video file structure. A probe video is converted in an XML tree-based structure; then, the comparison between the parsed data and a reference dataset addresses both integrity verification and brand classification. A similar approach has been also proposed in Erik Gelbing (2021), but the experiments have been applied only to the scenario of brand classification.

Both Iuliani Massimo (2019) and David Güera et al. (2019) allow the integrity assessment of a digital video with negligible computational costs. A further improvement was finally proposed in Yang Pengpeng et al. (2020), where decision trees were employed to reduce further the computational effort required to a fixed amount of time, independently of the reference dataset's size. These latest works also introduced an open-world scenario where the tested device was not available in the training set. The issue is also addressed in Raquel Ramos López et al. (2020), where information extracted from the video container was used to cluster sets of videos by data source, consisting in brand, model, or social media. In the next sections, we briefly summarize the main works: (i) the first relevant paper on video container Gloe Thomas et al. (2014), (ii) the first video container formalization and usage Iuliani Massimo (2019), (iii) the most recent work for efficient video file format analysis Yang Pengpeng et al. (2020).

14.3.1 Analysis of the Video File Structure

In Gloe Thomas et al. (2014), the authors identify the manufacturer and model-specific video file format characteristics and show how processing software further modifies those features. In particular, they consider 19 digital cameras and 14 mobile phones belonging to different models, and they consider from 3 to 14 videos per device, with all available video quality settings (e.g., frame size and frame rate). In Table 14.3, we report the list of these devices. Most of these digital cameras

Table 14.3 The list of devices analyzed in the study

Make	Model	Container
Digital camera models		
Agfa	DC-504, DC-733s, DC-830i, Sensor530s	AVI
Agfa	Sensor505-X	AVI
Canon	S45,S70, A640, Ixus IIs	AVI
Canon	EOS-7D	MOV
Casio	EX-M2	AVI
Kodak	M1063	MOV
Minolta	DiMAGE Z1	MOV
Nikon	CoolPix S3300	AVI
Pentax	Optio A40	AVI
Pentax	Optio W60	AVI
Praktica	DC2070	MOV
Ricoh	GX100	AVI
Samsung	NV15	AVI
Mobile phone models		
Apple	IPhone 4	MOV
Benq Siemens	S88	3GP
BlackBerry	8310	3GP
Google	Nexus 7	3GP
LG	KU990	3GP, AVI
Motorola	MileStone	3GP
Nokia	6710	3GP, MP4
Nokia	E61i	3GP, MP4
Nokia	E65	3GP, MP4
Nokia	X3-00	3GP
Palm	Pre	MP4
Samsung	GT-5500i	MP4
Samsung	SGH-D600	MP4
Sony Ericsson	K800i	3GP

store videos in AVI format, and some use Apple Quicktime MOV containers and compress video data using motion JPEG (MJPEG), and only three use more efficient codecs (DivX, Xvid, or H.264). All the mobile phones store video data in MOV-based container formats (MOV, 3GP, MP4), except for the LG KU990 camera phone, which supports AVI. On the contrary, the mobile phones adopt H.263, H.264, MPEG-4 or DivX, and not MJPEG compression.

A subset of the camera models has also been used to generate cut short videos, 10 seconds long, through video editing tools (FFmpeg, Avidemux, FreeVideoDub, VirtualDub, Yamb, and Adobe Premiere. Except for Adobe Premiere, all of them have allowed lossless video editing, i.e., the edited files have been saved without re-compressing the original stream. After that, they have analyzed the extracted data. This analysis reveals considerable differences between device models and software vendors, and some general findings are summarized here.

The authors carry out the first analysis by grouping the devices according to the file structure, i.e., AVI or MP4. Original AVI videos do not share the same components. Both the content and the format of specific chunks may vary, depending on the particular Camera model. Their edited versions, with all software tools, even when lossless processing is applied, leave distinct traces in the file structure, which do not match any of the camera's features in the dataset. It is then possible to use these differences to spot if a video file was edited with these tools by comparing a questioned file's file structure with a device's reference container.

Original MP4 videos, due to the by far more complex file format, exhibit an even more considerable degree of source-dependent internal variations than AVI files. However, none of the cut videos produced through the editing tools is compatible with any source device considered in the study. Dealing with original MJPEG-compressed video frames, different camera models adopt different JPEG marker segment sequences when building MJPEG-compressed video frames. Moreover, content-adaptive quantization tables are generally used in a video sequence, such that 2914 unique JPEG luminance quantization tables in this small dataset have been found. On these contents, lossless video editing leaves compression settings unaltered, but they introduce very distinctive artifacts in container files' structure. These findings stated that videos from different digital cameras and mobile phones appear to employ different container formats and compression. It worth noticing that this study is not straightforward since the authors had to write customized file parsers to extract the file format information from the videos. On the other side, this difficulty is also advantageous since we are currently unaware of any publicly available software that consistently allows users to forge such information without advanced programming skills. This fact makes the creation of realistic forgeries undoubtedly a highly non-trivial undertaking and thereby reemphasizes that file characteristics and metadata must not be dismissed as unreliable sources of evidence to address file authentication.

14.3.2 Automated Analysis of mp4-like Videos

The previous method is subjected to two main limitations: Firstly, being manual is time demanding. Secondly, it is prone to error since the forensic practitioner have to subjectively assess the finding's value. In Iuliani Massimo (2019), an automated method for unsupervised analysis of video file containers to assess video integrity and classify the source device brand. The authors developed an MP4 Parser library Apache (2021) that automatically extracted a video container and store it in an XML file. The video container is represented as a labeled tree where internal nodes are labeled by atoms names (e.g.,*moov-2*), and leaves are labeled by field-value attributes (e.g.,@**stuff:MovieBox[]**). To take into account the order of the atoms, each XML-node is identified by a 4-byte code of the corresponding atom along with an index that represents the relative position with respect to the other siblings at a certain level.

Technical Description

A video file container, extracted from a video X, can be represented as an ordered collection of atoms $a_1, \ldots, a_n$, possibly nested.

We can describe each atom in terms of a set of field-value attributes, as $a_i = \left(\omega_1(a_i), \ldots, \omega_{m_i}(a_i)\right)$.

By combining the two previous descriptions, the video container can be characterized by the set of field-value attributes $X = \left(\omega_1, \ldots, \omega_m\right)$, each with its associated path $p_X(\omega)$, that is the ordered list of atoms to be crossed to reach ω in X starting from the root. As an example, consider the video X, whose fragments are reported in Fig. 14.1. The path to reach the field-value $\omega = @timescale: 1000$ in X is the sequence of atoms (*ftyp-1, moov-2, mvhd-1*). In this sense, we state that $p_X(\omega) = p_{X'}(\omega')$ if the same ordered list of atoms is crossed to reach the field-values in the two trees, respectively, which is not the case in the previous example.

In summary, the video container structure is completely described by a list of m field-value attributes $X = \left(\omega_1, \ldots, \omega_m\right)$, and their corresponding paths $p_X(\omega_1), \ldots, p_X(\omega_m)$.

From now on, we will denote with $\mathcal{X} = \{X_1, \ldots, X_N\}$ the world set of digital videos, and with $\mathcal{C} = \{C_1, \ldots, C_s\}$ the set of disjoint possible origins, e.g., device *Huawei* P9, iPhone 6s, and so on.

When a video is processed in any way (with respect to its native form), its integrity is compromised SWGDE-SWGIT (2017).

More generally, given a query video X and a native reference video X' coming from the same supposed device model, their container structure dissimilarities can be exploited to expose integrity violation evidence, as follows.

Given two containers $X = (\omega_1, \ldots, \omega_m)$, $X' = (\omega'_1, \ldots, \omega'_m)$ with the same cardinality[2] m, we define a similarity core function between two field-values as

---

[2] If the two containers have a different cardinality, we pad the smaller one with empty field-values to obtain the same value m.

$$S(\omega_i, \omega_j^{'}) = \begin{cases} 1 & \text{if } \omega_i = \omega_j^{'} \text{ and } p_X(\omega_i) = p_{X^{'}}(\omega_j^{'}) \\ 0 & \text{otherwise} \end{cases} \tag{14.1}$$

We can easily extend the measure to the comparison between a single field-value $\omega_i \in X$ and the whole $X^{'}$ as

$$\mathbf{1}_{X^{'}}(\omega_i) = \begin{cases} 1 & \text{if } \exists\, \omega_j^{'} \in X^{'} : S(\omega_i, \omega_j^{'}) = 1 \\ 0 & \text{otherwise} \end{cases} \tag{14.2}$$

Then, the dissimilarity between X and $X^{'}$ can be computed as the mismatching percentage of all field-values, i.e.,

$$mm(X, X^{'}) = 1 - \frac{\sum_{i=1}^{m} \mathbf{1}_{X^{'}}(\omega_i)}{m} \tag{14.3}$$

and, to preserve symmetry, the degree of dissimilarity between X and $X^{'}$ can be computed as

$$D(X, X^{'}) = \frac{mm(X, X^{'}) + mm(X^{'}, X)}{2}. \tag{14.4}$$

Based on its definition, $D(X, X^{'}) \in [0, 1]$, $D(X, X^{'}) = 1$ when $X = X^{'}$, and $D(X, X^{'}) = 0$, when they have no elements in common.

Experiments

The method is tested on 31 portable devices from VISION dataset Dasara Shullani et al. (2017) that leads to an available collection of 578 videos in the native format plus their corresponding social versions (YouTube and WhatsApp are considered).

The metric in Eq. (14.4) is applied to determine whether a video's integrity is preserved or violated. The authors consider four different scenarios of integrity violation[3]:

- exchange through *WhatsApp*;
- exchange through *YouTube*;
- cut without re-encoding through *FFmpeg*;
- date editing through *ExifTool*.

For each of the four cases, we consider the set of videos $X_1, \ldots, X_{N_{C_i}}$, available for each device C_i, and we compute the intra-class dissimilarities between two native videos X_i, X_j belonging to the same model $D_{i,j} = D(X_i, X_j), \forall i \neq j$ based on Eq. (14.4). For simplicity, we denote with D_{oo} this set of dissimilarities. Then, we consider the corresponding inter-class dissimilarities $D_{i,j}^t = D(X_i, X_j^t), \forall i \neq j$,

---

[3] See the paper for the implementation details.

Table 14.4 AUC values computed for all devices for each of the four attacks, and when the native video is compared to videos from other devices. In the case of ExifTool and Other Devices, the achieved min and max AUC values are represented. For all other cases, an AUC equal to 1 is always obtained

Case	WhatsApp	YouTube	FFmpeg	ExifTool
AUC	1.00	1.00	1.00	[0.64 1.00]

between a native video X_i and its corrupted version X_j^t obtained with the tool-t (WhatsApp, YouTube, *ffmpeg*, or *Exiftool*). We denote with D_{oa}^t this set of dissimilarities. By applying this procedure to all the considered devices, we collected 2890 samples for both D_{oo} and any of the four D_{oa}^t.

In most of the performed tests the maximum value achieved for D_{oo} is lower than the minimum value of D_{oa}^t, thus indicating that the two classes can be separated. In Table 14.4 we report the AUCs for each of the considered cases. Noticeably, the integrity violation through *ffmpeg*, YouTube, and WhatsApp, the AUC is 1 for all devices. These values clearly show that the exchange through WhatsApp and YouTube strongly compromises the container structure, and thus it is easily possible to detect the corresponding integrity violation. Interestingly, cutting the video using *ffmpeg*, without any re-encoding, also results in a substantial container modification that we can detect. On the contrary, the date change with *Exiftool* induced on some containers a modification that is comparable with the observed intra-variability of native videos; in these cases, the method can not detect it. Fourteen devices still achieve unitary AUC, eight more devices yield an AUC greater than 0.8, and the remaining nine devices yield AUC below 0.8, with the lowest value of 0.64 obtained for D02 and D20 (*Apple* iPhone 4s and iPad mini).

To summarize, for the video integrity verification tests, the method obtains perfect discrimination for videos altered by social networks or *ffmpeg*, while for *Exiftool* it achieves an AUC greater than 0.82 on 70% of the considered devices. It worth also noticing that analyzing a video query requires less than a second.

More detailed results are reported in the corresponding paper. The authors also show that this container description can be immersed in a likelihood ratio framework to determine which atoms and field-values are highly discriminative for specific classes. In this way, they also show how to classify the brand of the originating device Iuliani Massimo (2019).

14.3.3 Efficient Video Analysis

The method described in Iuliani Massimo (2019), although effective, has a linear computational cost since it requires checking the dissimilarity of the probe video with all available reference containers. As a consequence, increasing the reference dataset size leads to a higher computational effort.

In this section, we summarize the method proposed in Yang Pengpeng et al. (2020) that provides an efficient way to analyze video file containers independently on the reference dataset's size.

The method, called *EVA* from now on, is based on Decision Trees Ross Quinlan (1986), a non-parametric learning method used for classification problems in many signal processing fields.

Their key feature is breaking down a complex decision-making process into a collection of more straightforward decisions.

The process is extremely efficient since a decision can be taken by checking a small number of features.

Furthermore, *EVA* allows both to characterize the identified manipulations and to provide an explanation for the outcome.

The method is also enriched with a likelihood ratio framework designed to automatically clean up the container elements that only contribute to source intra-variability (for the sake of brevity, we do not report these details here).

In short, *EVA* can identify the manipulating software (e.g., *Adobe Premiere*, *ffmpeg*, ...) and provide additional information related to the original content history such as the source device operating system.

Technical Description

Similarly to the previous case, we consider a video container X as a labeled tree where internal nodes and leaves correspond to, respectively, atoms and field-value attributes. It can be characterised by the set of symbols $\{s_1, \ldots s_m\}$, where s_i can be: (i) the path from the root to any field (value excluded), also called *field-symbols*; (ii) the path from the root to any field-value (value included), also called *value-symbols*. An example of this representation can be[4]:

$s_1 = [\texttt{ftyp/@majorBrand}]$
$s_2 = [\texttt{ftyp/@majorBrand/isom}]$
...
$s_i = [\texttt{moov/mvhd/@duration}]$
$s_{i+1} = [\texttt{moov/mvhd/@duration/73432}]$
...

Overall, we denote with Ω the set of all unique symbols $s_1, \ldots, s_M$ available in the world set of digital video containers $\mathcal{X} = \{X_1, \ldots, X_N\}$. Similarly, $\mathcal{C} = \{C_1, \ldots, C_s\}$ denotes a set of possible origins (e.g., *Huawei* P9, *Apple* iPhone 6s). Let a container X, the different structure of its symbols $\{s_1, \ldots, s_m\}$ can be exploited to assign the video to a specific class C_u.

For this purpose, binary decision trees Rasoul Safavian and Landgrebe (1991) are employed to build a set of hierarchical decisions. In each internal tree node the input data is tested against a specific condition; the test outcome is used to select a child as the next step in the decision process. Leaf nodes represent decisions taken by the

---

[4] Note that @ is used to identify atom parameters, and `root` is used for visualization purposes. Still, it is not part of the container data.

algorithm. More specifically, *EVA* adopts the growing-pruning-based Classification And Regression Trees (CART) Leo Breiman (2017).

Given the size of unique symbols $|\Omega| = M$, a video container X is converted into a vector of integers $X \mapsto (x_1 \ldots x_M)$ where x_i is the number of times that s_i occurs into X. This approach is inspired by the bag-of-words representation Hinrich Schütze et al. (2008) which is used to reduce variable-length documents to a fixed-length vectorial representation.

For the sake of example, we consider two classes:

C_u: iOS devices, native videos;
C_v: iOS devices, dates modified through *Exiftool* (see Sect. 14.3.3 for details).

With these settings, the decision tree highlights that the usage of *Exiftool* can be simply decided by looking, for instance, at the presence of `moov/udta/XMP_/ @stuff`.

Experiments

Tests were performed on 34 smartphones of 10 different brands. Over a thousand tampered videos were tested, considering both automated (*ffmpeg*, *Exiftool*) and manual editing operations (*Kdenlive*, *Avidemux* and *Adobe Premiere*). The manipulations mainly include cut with and without re-encoding, speed up, and slow motion.

Achieved videos (both native and manipulated) were also exchanged through YouTube, Facebook, Weibo, and TikTok.

All tests were performed by adopting an exhaustive leave-one-out cross-validation strategy: the dataset is partitioned in 34 subsets, each one of them containing pristine, manipulated, and social-exchanged videos belonging to a specific device. In this way, test accuracy collected after each iteration is computed on videos belonging to an unseen device. When compared to state of the art, *EVA* exposes competitive effectiveness at the lowest computational effort (see 14.5).

In comparison with David Güera et al. (2019), it achieves higher accuracy. We can reasonably attribute this performance to their use of a smaller feature space; indeed, only a subset of the available pieces of information are extracted without considering their position within the video container. On the contrary, *EVA* features also include the path from the root to the value, thus providing a higher discriminating power. Indeed, this approach distinguishes between two videos where the same information

Table 14.5 Comparison of our method with the state of the art. Values of accuracy and time are averaged over the 34 folds

	Balanced accuracy	Training time	Test time
Guera et al. David Güera et al. (2019)	0.67	347 s	< 1 s
Iuliani et al. Iuliani Massimo (2019)	0.85	N/A	8 s
EVA	0.98	31 s	< 1 s

is stored in different atoms. When compared with Iuliani Massimo (2019), *EVA* is capable of obtaining better classification performance with a lower computational cost. In Iuliani Massimo (2019), $O(N)$ comparisons are required since all the N reference-set examples must be compared with a tested video; on the contrary, the cost for a decision tree analysis is $O(1)$ since the output is reached in a constant number of steps. Furthermore, *EVA* allows a simple explanation for the outcome. For the sake of example, we report in Fig. 14.2a a sample tree from the integrity verification experiments: the decision is taken by up to four checks, just based on the presence of the symbols `ftyp/@minorVersion = 0`, `uuid/@userType`, `moov/udta/XMP_` and `moov/udta/auth`. Then, a single decision tree can handle both easy- and hard-to-classify cases at the same time.

Manipulation Characterization

EVA is also capable of identifying the manipulating software and the operating system of the originating device. More specifically, three main questions were posed:

A **Software identification:** Is the proposed method capable of identifying the software used to manipulate a video? If yes, is it possible to identify the operating system of the original video?

B **Integrity Verification on Social Media:** Given a video from a social media platform (YouTube, Facebook, TikTok or WeiBo), can we determine whether the original video was pristine or tampered?

C **Blind scenario:** Given a video that may or may not has been exchanged through a social media platform, is it possible to retrieve some information on the video origin?

In the **Software Identification**, the experiment comprises the following classes: "native" (136 videos), "*Avidemux*" (136 videos), "*Exiftool*" (136 videos), "*ffmpeg*" (680 videos), "*Kdenlive*" (136 videos), and "*Premiere*" (136 videos). As shown in Table 14.6, *EVA* obtains a balanced global accuracy of 97.6% with slightly lower accuracy in identifying *ffmpeg* with respect to the other tools. These wrong classifications are reasonable because *ffmpeg* library is used by other software and, internally, by Android devices.

The algorithm maintains a high discriminative power even when trained to identify the originating operating system (Android vs. iOS).

In a few cases, *EVA* wrongly classifies only the source's operating system. This mistake happens explicitly in the case of *Kdenlive* and *Adobe Premiere*. At the same time, these programs are always correctly identified. This result indicates that the container's structure of videos saved by *Kdenlive* and *Adobe Premiere* is probably reconstructed in a software-specific way. More detailed performance is reported in the related paper.

With regard to **Integrity Verification on Social Media**, *EVA* was also tested on YouTube, Facebook, TikTok, and Weibo videos to determine whether they were pristine or manipulated before the upload (Table 14.7).

Results highlight poor performance due to the social media transcoding process that flattens the containers almost independently on the video origin. For example,

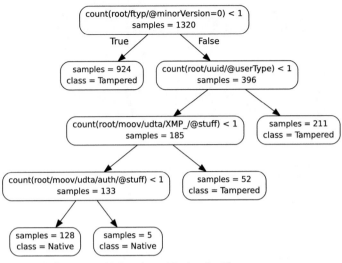

(a) Integrity verification classifier.

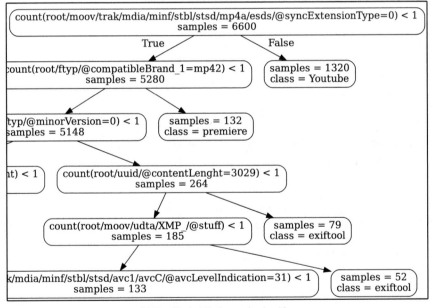

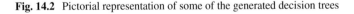

(b) Detail of a blind scenario classifier.

Fig. 14.2 Pictorial representation of some of the generated decision trees

Table 14.6 Confusion matrix under the software identification scenario for native contents (Na), *Avidemux* (Av), *Exiftool* (Ex), *ffmpeg* (Ff), *Kdenlive* (Kd), *Premiere* (Pr)

	Na	Av	Ex	Ff	Kd	Pr
Na	0.97	–	0.03	–	–	–
Av	–	1.00	–	–	–	–
Ex	0.01	–	0.99	–	–	–
Ff	–	0.01	–	0.90	0.09	–
Kd	-	–	–	–	1.00	–
Pr	–	–	–	–	–	1.00

Table 14.7 Performance achieved for integrity verification on social media contents. We report for each social network the obtained accuracy, true positive rate (TPR), and true negative rate (TNR). All these performance measures are balanced

	Accuracy	TNR	TPR
Facebook	0.76	0.40	0.86
TikTok	0.80	0.51	0.75
Weibo	0.79	0.45	0.82
YouTube	0.60	0.36	0.74

after YouTube transcoding, videos produced by *Avidemux* and by *Exiftool* are shown to have exactly the same container representation. We do not know how the considered platforms process the videos due to the lack of public documentation, but we can assume that uploaded videos undergo custom/multiple processing. Indeed, social media videos need to be viewable on a great range of platforms and thus need to be transcoded to multiple video codecs and adapted for various resolutions and bitrates. Thus, it seems plausible that those operations could discard most of the original container structure.

Eventually, in the *Blind Scenario*, the authors show that we can remove the prior information (whether the video belongs to social media or not) without degrading detection performances.

In summary, *EVA* can be considered an efficient forensic method for checking video integrity. If manipulation is detected, *EVA* can also identify the editing software and, in most cases, the video source device's operating system.

14.4 Concluding Remarks

In this chapter, we described how features extracted from image and video file containers can be exploited for integrity verification. In some cases, these features also provide insights into the probe's previous history. In several cases, they allow the characterization of the source device's brand or model. Similarly, they allow deriving that the content was exchanged through a particular social network.

This kind of information shows several advantages with respect to content-based traces. Firstly, it is usually much faster to be extracted and analyzed than content-based ones. It also requires an almost fixed computational effort, independently on the media size and resolution (this characteristic is particularly relevant for videos). Secondly, it is also much more challenging to perform anti-forensics operations on these features since the structure of an image, or video file is overly complicated. Thus, it is not straightforward to mimic a native file container. There are, however, several drawbacks to be taken into account. First of all, current editing tools and metadata editors cannot access such low-level information. It is required to design a proper parser to extract the information stored in the file container. Moreover, these features' effectiveness strongly depends on the availability of updated reference datasets of native and manipulated contents. Eventually, these features cannot discriminate the full history of a media since some processing operations, like uploading to a social network, tend to remove the traces left by previous steps. It is also worth mentioning that the forensic analysis of social media content is fragile since the providers' settings keep changing. This fact makes collected data and achieved results outdated in a short time. Furthermore, collecting large datasets can be challenging depending on the social media provider's upload and download protocol. Future work should be devoted to the fusion of content-based and format-based features since it is expected that the exploitation of these two domains can provide a better performance, as witnessed in Phan et al. (2019), where images exchanged on multiple social networks were studied.

References

Apache. Java mp4 parser. *http://www.github.com/sannies/mp4parser*

Apple Computer, Inc. Quicktime file format. 2001

Carlos Quinto Huamán, Ana Lucila Sandoval Orozco, and Luis Javier García Villalba. Authentication and integrity of smartphone videos through multimedia container structure analysis. *Future Generation Computer Systems*, 108:15–33, 2020

A. Castiglione, G. Cattaneo, and A. De Santis. A forensic analysis of images on online social networks. In *2011 Third International Conference on Intelligent Networking and Collaborative Systems*, pages 679–684, 2011

Dasara Shullani, Marco Fontani, Massimo Iuliani, Omar Al Shaya, and Alessandro Piva. VISION: a video and image dataset for source identification. *EURASIP Journal on Information Security*, 2017(1):15, 2017

David Güera, Sriram Baireddy, Paolo Bestagini, Stefano Tubaro, and Edward J Delp. We need no pixels: Video manipulation detection using stream descriptors. In *International Conference on Machine Learning (ICML), Synthetic Realities: Deep Learning for Detecting AudioVisual Fakes Workshop*, 2019

Duc-Tien Dang-Nguyen, Cecilia Pasquini, Valentina Conotter, and Giulia Boato. Raise: A raw images dataset for digital image forensics. In *Proceedings of the 6th ACM multimedia systems conference*, pages 219–224, 2015

Eric Hamilton. Jpeg file interchange format. 2004

Eric Kee, Johnson Micah K, Hany Farid (2011) Digital image authentication from JPEG headers. IEEE Trans. Inf. Forensics Secur. 6(3–2):1066–1075

Erik Gelbing, Leon Würsching, Sascha Zmudzinski, and Martin Steinebach. Video source identification from mp4 data based on field values in atom/box attributes. In *Proceedings of Media Watermarking, Security, and Forensics 2021*, 2021

H. Farid. Digital image ballistics from jpeg quantization: a followup study. *[Technical Report TR2008-638] Department of Computer Science, Dartmouth College, Hanover, NH, USA*, 2008

Hany Farid. Digital image ballistics from jpeg quantization. *[Technical Report TR9-1-2006] Department of Computer Science, Dartmouth College, Hanover, NH, USA*, 2006

Hinrich Schütze, Christopher D Manning, and Prabhakar Raghavan. *Introduction to information retrieval*, volume 39. Cambridge University Press Cambridge, 2008

International Telecommunication Union. Advanced video coding for generic audiovisual services h.264. https://www.itu.int/rec/T-REC-H.264/, 2016

International Telecommunication Union. High efficiency video coding. https://www.itu.int/rec/T-REC-H.265, 2018

ISO/IEC 10918-1. Information technology. digital compression and coding of continuous-tone still images. *ITU-T Recommendation T.81*, 1992

ISO/IEC 14496. Information technology. coding of audio-visual objects, part 12: Iso base media file format, 3rd ed. 2008

ISO/IEC 14496. Information technology. coding of audio-visual objects, part 14: Mp4 file format. 2003

JEITA CP-3451. Exchangeable image file format for digital still cameras: Exif version 2.2. 2002

Jesse D. Kornblum. Using jpeg quantization tables to identify imagery processed by software. *Digital Investigation*, 5:S21–S25, 2008. The Proceedings of the Eighth Annual DFRWS Conference

Jieun Song, Kiryong Lee, Wan Yeon Lee, and Heejo Lee. Integrity verification of the ordered data structures in manipulated video content. *Digital Investigation*, 18:1–7, 2016

JPEG: Still image data compression standard. Springer Science & Business Media, 1992

Leo Breiman. *Classification and regression trees*. Routledge, 2017

Massimo Iuliani, Dasara Shullani, Marco Fontani, Saverio Meucci, Alessandro Piva (2019) A video forensic framework for the unsupervised analysis of mp4-like file container. IEEE Trans. Inf. Forensics Secur. 14(3):635–645

Microsoft. Avi riff file. https://docs.microsoft.com/en-us/windows/win32/directshow/avi-riff-file-reference, 1992

Oliver Giudice, Antonino Paratore, Marco Moltisanti, and Sebastiano Battiato. A classification engine for image ballistics of social data. In Sebastiano Battiato, Giovanni Gallo, Raimondo Schettini, and Filippo Stanco, editors, *Image Analysis and Processing - ICIAP 2017*, pages 625–636, Cham, 2017. Springer International Publishing

Patrick Mullan, Christian Riess, Felix Freiling (2019) Forensic source identification using jpeg image headers: The case of smartphones. Digital Investigation 28:S68–S76

Q. Phan, G. Boato, R. Caldelli, and I. Amerini, Tracking multiple image sharing on social networks. In: 2019 IEEE International Conference on Acoustics, Speech and Signal Processing (ICASSP 2019), pages 8266–8270, 2019

Pengpeng Yang, Daniele Baracchi, Massimo Iuliani, Dasara Shullani, Rongrong Ni, Yao Zhao, Alessandro Piva (2020) Efficient video integrity analysis through container characterization. IEEE J. Sel. Top. Signal Process. 14(5):947–954

Raquel Ramos López, Elena Almaraz Luengo, Ana Lucila Sandoval Orozco, and Luis Javier García-Villalba. Digital video source identification based on container's structure analysis. *IEEE Access*, 8:36363–36375, 2020

J. Ross Quinlan. Induction of decision trees. *Machine learning*, 1(1):81–106, 1986

S Rasoul Safavian and David Landgrebe. A survey of decision tree classifier methodology. *IEEE transactions on systems, man, and cybernetics*, 21(3):660–674, 1991

SWGDE-SWGIT. SWGDE best practices for maintaining the integrity of imagery, Version: 1.0, July 2017

Thomas Gloe. Forensic analysis of ordered data structures on the example of JPEG files. In *2012 IEEE International Workshop on Information Forensics and Security, WIFS 2012, Costa Adeje, Tenerife, Spain, December 2-5, 2012*, pages 139–144. IEEE, 2012

Thomas Gloe, and Rainer Bähme (2010) The Dresden image database for benchmarking digital image forensics. J Digit Forensic Pract 3(2–4):150–159

Thomas Gloe, André Fischer, Matthias Kirchner (2014) Forensic analysis of video file formats. Digit. Investig. 11(1):S68–S76

Chapter 15
Image Provenance Analysis

**Daniel Moreira, William Theisen, Walter Scheirer, Aparna Bharati,
Joel Brogan, and Anderson Rocha**

The literature of multimedia forensics is mainly dedicated to the analysis of single assets (such as sole image or video files), aiming at individually assessing their authenticity. Different from this, *image provenance analysis* is devoted to the joint examination of multiple assets, intending to ascertain their history of edits, by evaluating pairwise relationships. Each relationship, thus, expresses the probability of one asset giving rise to the other, through either global or local operations, such as data compression, resizing, color-space modifications, content blurring, and content splicing. The principled combination of these relationships unveils the provenance of the assets, also constituting an important forensic tool for authenticity verification. This chapter introduces the problem of provenance analysis, discussing its importance and delving into the state-of-the-art techniques to solve it.

15.1 The Problem

Consider a questioned media asset, namely a *query* (such as a digital image whose authenticity is suspect), and a large corpus of media assets (such as the Internet). Provenance analysis comprises the problem of (i) finding, within the available corpus,

D. Moreira · W. Theisen · W. Scheirer (✉)
University of Notre Dame, Notre Dame, IN, USA
e-mail: walter.scheirer@nd.edu

A. Bharati
Lehigh University, Bethlehem, PA, USA

J. Brogan
Oak Ridge National Laboratory, Oak Ridge, TN, USA

A. Rocha
University of Campinas, Campinas, Brazil

© The Author(s) 2022
H. T. Sencar et al. (eds.), *Multimedia Forensics*, Advances in Computer Vision and Pattern
Recognition, https://doi.org/10.1007/978-981-16-7621-5_15

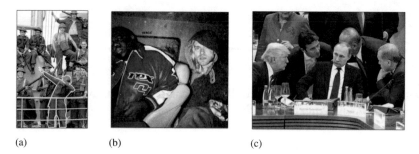

(a)   (b)   (c)

Fig. 15.1 Images that became viral on the web in the last decade, all with unknown sources. In **a**, the claimed world's first dab, supposedly captured during WWII. The dabbing soldier was highlighted, to make him more noticeable. In **b**, the claimed proof of an unlikely friendship between the Notorious B.I.G., on the left, and Kurt Cobain, on the right. In **c**, a photo of a supposed NATO meeting aimed at supporting a particular political narrative

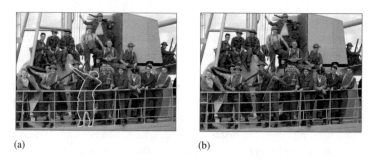

(a)   (b)

Fig. 15.2 A reverse image search of Fig. 15.1a leads to the retrieval of these two images, among others. Figure 15.1a is probably the result of cropping (**a**), which in turn is a color transformation of (**b**). The dabbing soldier was highlighted in **a**, to make him more noticeable. Image (b) is, thus, the source, comprising a behind-the-scenes picture from Dunkirk (2017)

the assets that directly and transitively share content with the query, as well as of (ii) establishing the derivation and content-donation processes that explain the existence of the query. Take, for example, the three queries depicted in Fig. 15.1, which became viral images in the last decade. Reasons for their virality range from the popularity of harmless pop-culture jokes and historical oddities (such as in the case of Fig. 15.1a, b) to interest in more critical political narratives and agendas (such as in Fig. 15.1c). Provenance analysis offers a principled and automated framework to debunk such media types by retrieving and associating other assets that help to elucidate their authenticity.

To get a glimpse of the expected outcome of performing provenance analysis, one could manually submit each one of the three queries depicted in Fig. 15.1 to a reverse image search engine, such as TinEye (2021), and try to select and associate the retrieved images to the queries by hand. This process is illustrated through Figs. 15.2, 15.3, and 15.4. Based on Fig. 15.2, for instance, one can figure out that the claimed world's first dab, supposedly captured during WWII, is a crop of Fig. 15.2a, which

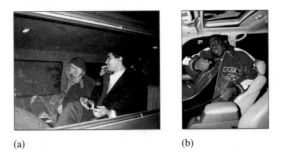

(a)   (b)

Fig. 15.3 A reverse image search of Fig. 15.1b leads to the retrieval of these two images, among others. Figure 15.1b is probably a composition of (**a**) and (**b**), where **a** donates the background, while **b** donates the Notorious B.I.G. on his car's seat. Cropping and color corrections are also performed to complete the forgery

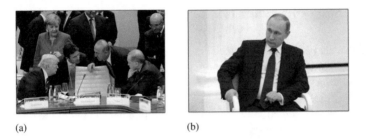

(a)   (b)

Fig. 15.4 A typical reverse image search of Fig. 15.1c leads to the retrieval of **a** but not **b**. Image **b** was obtained through one of the content retrieval techniques presented in Sect. 15.2 (context incorporation). Figure 15.1c is, thus, a composition of **a** and **b**, where **a** donates the background and **b** donates Vladimir Putin. Cropping and color corrections are also performed to complete the forgery

in turn is a color modified version of Fig. 15.2b, a well-known picture of the cast of the Hollywood movie Dunkirk (2017).

Figure 15.3, in turn, helps to reveal that Fig. 15.1b is actually a forgery (composition), where Fig. 15.3a probably serves as the background for the splicing of the Notorious B.I.G. on his car's seat, taken from Fig. 15.3a. In a similar fashion, Fig. 15.4 points out that Fig. 15.1c is also a composition, this time using Fig. 15.4a as background, and Fig. 15.4b as the donor of the portrayed individual. In this particular case, Fig. 15.4b is not easily found by a typical reverse image search. To do so, we had to perform *context incorporation*, a content retrieval strategy adapted to provenance analysis that is explained in Sect. 15.2.

In the era of misinformation and "fake news", there is a symptomatic crisis of trust in the media assets shared online. People are aware of editing software, with which even unskilled users can quickly fabricate and manipulate content. Although many of these manipulations have benign purposes (no, there is nothing wrong with the silly memes you share), some content is generated with malicious intent (general public deception and propaganda), and some modifications may undermine the ownership of the media assets. Under this scenario, provenance analysis reveals itself as a convenient tool to expose the provenance of the assets, aiding in the verification of their authenticity, protecting their ownership, and restoring credibility.

In the face of the massive amount of data produced and shared online, though, there is no space for performing provenance analysis manually, such as formerly described. Besides the need for particular adaptations such as context incorporation (see Sect. 15.2), provenance analysis must also be performed at scale automatically and efficiently. By combining ideas from image processing, computer vision, graph theory, and multimedia forensics, provenance analysis constitutes an interesting interdisciplinary endeavor, into which we delve into in detail in the following sections.

15.1.1 The Provenance Framework

Provenance analysis can be executed at scale and fully automated by following a basic framework that involves two stages. Such a framework is depicted in Fig. 15.5. As one might observe, the first stage is always related to the activity of content retrieval, which incorporates a questioned media asset (*a.k.a.* the query) and a corpus of media assets to retrieve a selection of assets of interest that are related to the query.

Figure 15.6 depicts the expected outcome of content retrieval for Fig. 15.1b as the query. In the case of provenance analysis, the content retrieval activity must retrieve not only the objects that directly share content with the query but also transitively. Take, for example, images 1, 2, and 3 within Fig. 15.6, which all share some visual

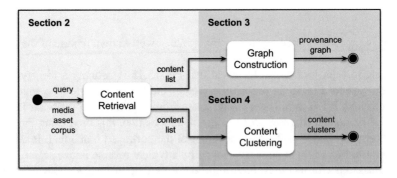

Fig. 15.5 The provenance framework. Provenance analysis is usually executed in two stages. Starting from a given query and a corpus of media assets, the first stage is always related to the content retrieval activity, which is herein explained in Sect. 15.2. The first stage's output is a list of assets of interest (content list), which is fed to the second stage. The second stage, in turn, may either comprise the activity of graph construction (discussed within Sect. 15.3) or the activity of content clustering (discussed within Sect. 15.4). While the former activity aims at organizing the retrieved assets in a provenance graph, the latter focuses on establishing meaningful asset clusters

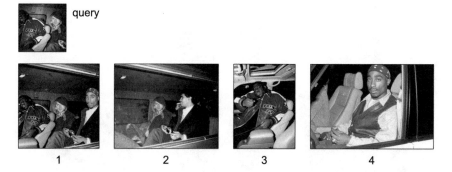

query

1   2   3   4

Fig. 15.6 Content retrieval example for a given query. The desired output of content retrieval for provenance analysis is a list of media assets that share content with the query, either directly (such as objects 1, 2, and 3) or transitively (object 4, through object 1). Methods to perform content retrieval are discussed in Sect. 15.2

elements with the query. Image 4, however, has nothing in common with the query. But its retrieval is still desirable because it shares visual content with image 1 (the head of Tupac Shakur, on the right), hence it is related to the query transitively. Techniques to perform content retrieval for the provenance analysis of images are presented and discussed in Sect. 15.2.

Once a list of related media assets is available, the provenance framework's typical execution moves forward to the second stage, which has two alternate modes. The first, provenance graph construction, aims at computing the directed acyclic graph whose nodes individually represent the query and the related media assets and whose edges express the edit and content-donation story (e.g., cropping, blurring, splicing, etc.) between pairs of assets, linking seminal to derived elements. It, thus, embodies the provenance of the objects it contains. Figure 15.7 provides an example of provenance graph, constructed for the media assets depicted in Fig. 15.6. As shown, the query is probably a crop of image 1, which is a composition. It uses image 2 as background and splicing objects from images 3 and 4 to complete the forgery. Methods to construct provenance graphs from sets of images are presented in Sect. 15.3.

The provenance framework may be changed by replacing the second stage of graph construction with a content clustering approach. This setup is sound in the study of contemporary communication on the Internet, such as exchanging memes and the reproduction of viral movements. Dabbing (see Fig. 15.1a), for instance, is an example of this phenomenon. In these cases, the users' intent is not limited to retrieving near-duplicate variants or compositions that make use of a given query. They are also interested in the retrieval and grouping of semantically similar objects, which may greatly vary in appearance to elucidate the provenance of a trend. This situation is represented in Fig. 15.8. Since graph construction may not be the best approach to organize semantically similar media assets obtained during content retrieval, content clustering reveals itself as an interesting option, as we discuss in Sect. 15.4.

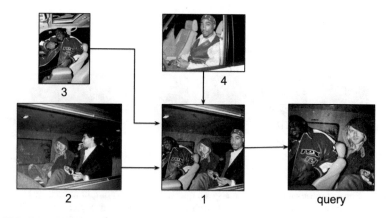

Fig. 15.7 Provenance graph example. This graph is constructed using the media assets retrieved in the example given through Fig. 15.6. In a nutshell, it expresses that the query is probably a crop of image 1, which in turn is a composition forged by combining the contents of images 2, 3, and 4. Methods to perform provenance graph construction are presented in Sect. 15.3

Fig. 15.8 Content retrieval of semantically similar assets. Some applications might aim at identifying particular behaviors or gestures that are depicted in a meme-style viral query. That might be the case for someone trying to understand the trend of dabbing, which is reproduced in all of the images above. In such a case, provenance content retrieval might still be helpful to fetch related objects. Graph construction, however, may be rendered useless due to the dominant presence of semantically similar objects rather than near-duplicates or compositions. In such a situation, we propose using content clustering as the second stage of the provenance framework, which we discuss in Sect. 15.4

15.1.2 Previous Work

Early Work: De Rosa et al. (2010) have mentioned the importance of considering groups of images instead of single images while performing media forensics. Starting with a set of images of interest, they have proposed to express pairwise image dependencies through the analysis of the mutual information between every pair of images. By combining these dependencies into a single correlation adjacency matrix for the entire image set, they have suggested the generation of a *dependency graph*, whose edges should be individually evaluated as being in accordance with a particular set of image transformation assumptions. These assumptions should presume a known set of operations, such as rotation, scaling, and JPEG compression. Edges unfit to these assumptions should be removed.

Similarly, Kennedy and Chang (2008) have explored solutions to uncover the processes through which near-duplicate images have been copied or manipulated. By relying on the detection of a closed set of image manipulation operations (such as cropping, text overlaying, and color changing), they have proposed a system to construct *visual migration maps*: graph data structures devised to express the parent-child derivation operations between pairs of images, being equivalent to the dependency graphs proposed in De Rosa et al. (2010).

Image Phylogeny Trees: Rather than modeling an exhaustive set of possible operations between near-duplicate images, Dias et al. (2012) designed and adopted a robust image similarity function to compute a pairwise image similarity matrix **M**. For that, they have introduced an image similarity calculation protocol to generate **M**, which is widely used across the literature, including provenance analysis. This method is detailed in Sect. 15.3. To obtain a meaningful *image phylogeny tree* from **M**, to represent the evolution of the near-duplicate images of interest, they have introduced *oriented Kruskal*, a variation of Kruskal's algorithm that extracts an oriented optimum spanning tree from **M**. As expected, phylogeny trees are analogous to the aforementioned dependency graphs and visual migration maps.

In subsequent work, Dias et al. (2013) have reported a large set of experiments with their methodology in the face of a family of six possible image operations, namely scaling, warping, cropping, brightness and contrast changes, and lossy compression. Moreover, Dias et al. (2013) have also explored the replacement of oriented Kruskal with other phylogeny tree-building methods, such as *oriented Prim* and *Edmond's optimum branching*.

Melloni et al. (2014) have contributed to the topic of image phylogeny tree reconstruction by investigating ways to combine different image similarity metrics. Bestagini et al. (2016) have focused on the clues left by local image operations (such as object splicing, object removal, and logo insertion) to reconstruct the phylogeny trees of near-duplicates. More recently, Zhu and Shen (2019) have proposed heuristics to improve phylogeny trees by correcting local image inheritance relationship edges. Castelletto et al. (2020), in turn, have advanced the state of the art by training a denoising convolutional autoencoder that takes an image similarity adjacency matrix as input and returns an optimum spanning tree as the desired output.

Image Phylogeny Forests: All the techniques mentioned so far were conceived to handle near-duplicates. Aiming at providing a solution to deal with semantically similar images, Dias et al. (2013) have extended the oriented Kruskal method proposed in Dias et al. (2012) to what they named the *automatic oriented Kruskal*. This technique is an algorithm to compute a family of disjoint phylogeny trees (hence a phylogeny forest) from a given set of near-duplicate and semantically similar images, such that each disjoint tree describes the relationships of a particular group of near-duplicates.

In the same direction, Costa et al. (2014) have provided two extensions to the optimum branching algorithm proposed in Dias et al. (2013), namely *automatic optimum branching* and *extended automatic optimum branching*. Both solutions are based on the automatic calculation of branching cut-off execution points. Alternatively, Oikawa et al. (2015) have proposed the use of clustering techniques for finding the various disjoint phylogeny trees. Images coming from the same source (near-duplicates) should be placed in the same cluster, while semantically similar images should be placed in different clusters.

Milani et al. (2016), in turn, have suggested relying on the estimation of the geometric localization of captured viewpoints within the images as a manner to distinguish between near-duplicates (which should share viewpoints) from semantically similar objects (which should present different viewpoints). Lastly, Costa et al. (2017) have introduced solutions to improve the creation of the pairwise image similarity matrices, even in the presence of semantically similar images and regardless of the graph algorithm used to construct the phylogeny trees.

Multiple Parenting: Previously mentioned phylogeny work did not address the critical scenario of image compositions, in which objects from one image are spliced into another. Aiming at dealing with these cases, de Oliveira et al. (2015) have modeled every composition as the outcome of two parents (one *donor*, which provides the spliced object, and one *host*, which provides the composition background). Extended automatic optimum branching, proposed in Costa et al. (2014), should then be applied for the reconstruction of ideally three phylogeny trees: one for the near-duplicates of the donor, one for the near-duplicates of the host, and one for the near-duplicates of the composite.

Other Types of Media: Besides processing still images, some works in the literature have addressed the phylogeny reconstruction of assets belonging to other types of media, such as video (see Dias et al. 2011; Lameri et al. 2014; Costa et al. 2015, 2016; Milani et al. 2017) and even audio (see Verde et al. 2017). In particular, Oikawa et al. (2016) have investigated the role of similarity computation between digital objects (such as images, video, and audio) in multimedia phylogeny.

Provenance Analysis: The herein-mentioned literature of media phylogeny has made use of diverse metrics, individually focused on retrieving either the root, the leaves, or the ancestors of a node within a reconstructed phylogeny tree, evaluating the tree as a whole. Moreover, the datasets used in the experiments presented different types of limitations, such as either containing only images in JPEG format or lacking compositions with more than two sources (cf. object 1 depicted in Fig. 15.7).

Aware of these limitations and aiming to foster more research in the topic of multi-asset forensic analysis, the American Defense Advanced Research Projects Agency (DARPA) and the National Institute of Standards and Technology (NIST) have joined forces to introduce new terminology, metrics, and datasets (all herein presented, in the following sections), within the context of the Media Forensics (MediFor) project (see Turek 2021). Therefore, they coined the term *Provenance Analysis* to express a broader notion of phylogeny reconstruction, in the sense of including not only the task of reconstructing the derivation stories of the assets of interest but also the fundamental step of retrieving these assets (as we discuss in Sect. 15.2). Furthermore, DARPA and NIST have suggested the use of directed acyclic graphs (*a.k.a. provenance graphs*), instead of groups of trees, to represent the derivation story of the assets better.

15.2 Content Retrieval

With the amount of content on the Internet being so vast, performing almost any type of computationally expensive analysis across the web's entirety or even smaller subsets for that matter is simply intractable. Therefore, when setting out to tackle the task of provenance analysis, a solution must start with an algorithm for retrieving a reasonably-sized subset of relevant data from a larger corpus of media. With this in mind, effective strategies for content retrieval become an integral module within the provenance analysis framework.

This section will focus specifically on *image* retrieval algorithms that provide results contingent on one or multiple images as queries into the system. This image retrieval is commonly known as reverse image search or more technically *Content-Based Image Retrieval* (CBIR). For provenance analysis, an appropriate CBIR algorithm should produce a corpus subset that contains a rich collection of images with relationships relevant to the provenance of the query image.

15.2.1 Approaches

Typical CBIR: Typical CBIR solutions employ multi-level representations of the processed images to reduce the semantic gap between the image pixel values (in the low level) and the system user's retrieval intent (in the highlevel, see Liu et al. 2007). Having provenance analysis in mind, the primary intent is to trace connections between images that mutually share visual content. For instance, when performing the query from Fig. 15.6, one would want a system to retrieve images that contain identical or near-identical structural elements (such as the corresponding images, respectively, depicting the Notorious B.I.G. and Kurt Cobain, both used to generate the composite query, but not other images of unrelated people sitting in cars). Considering that, we can describe an initial concrete example of a CBIR system that is not entirely suited to provenance analysis and further gradually provide methods to make it more suitable to the problem at hand.

A typical and moderately useful CBIR system for provenance analysis may rely on local features (*a.k.a.* keypoints) to obtain a low-level representation of the image content. Hence, it may consist of four steps, namely: (1) feature extraction, (2) feature compression, (3) feature retrieval, and (4) result ranking. Feature extraction comprises the task of computing thousands of n-dimensional representations for

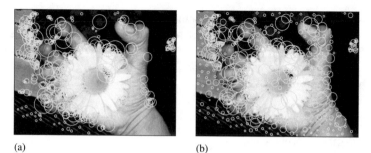

(a)   (b)

Fig. 15.9 SURF keypoints extracted from a given image. Each yellow circle represents a keypoint location, whose pixel values generate an n-dimensional feature vector. In **a**, a regular extraction of SURF keypoints, as defined by Bay et al. (2008). In **b**, a modified keypoint extraction dubbed distributed keypoints, proposed by Moreira et al. (2018), whose intention is also to describe homogeneous regions, such as the wrist skin and clapboard in the background. Images adapted from Daniel Moreira et al. (2018)

each image, ranging from classical handcrafted technologies, such as Scale Invariant Feature Transform (SIFT, see Lowe 2004) and Speeded-up Robust Features (SURF, see Bay et al. 2008), to neural network learned methods, such as Learned Invariant Feature Transform (LIFT, see Yi et al. 2016) and Deep Local Features (DELF, see Noh et al. 2017). Figure 15.9a depicts an example of SURF keypoints extracted from a target image.

Although the feature-based representation of images drastically reduces the amount of space needed to store their content, there is still a necessity for reducing their size, a task performed during the second CBIR step of feature compression. Take, for example, a CBIR system that extracts 1,000 64-dimensional floating-point SURF features from each image. Using 4 bytes for each dimension, each image occupies a total of $4 \times 64 \times 1000 = 256,000$ bytes (or 256 kB) of memory space. While that may seem relatively low from a single-image standpoint, consider an image database containing ten million images (which is far from being an unrealistic number, considering the scale of the Internet). This would mean the need for an image index on the order of 25 terabytes. Instead, we recommend utilizing Optimized Product Quantization (OPQ, see Ge et al. 2013) to reduce the size of each feature vector. In summary, OPQ learns grids of bins arranged along different axes within the feature vector space. These bin grids are rotated and scaled within the feature vector space to distribute optimally feature vectors extracted from an example training set. This provides a significant reduction in the number of bits required to describe a vector while keeping relatively high fidelity. For the sake of illustration, a simplified two-dimensional example of OPQ bins is provided in Fig. 15.10.

The third CBIR step (feature retrieval) aims at using the local features to index and compare, within the optimized n-dimensional space they constitute, and through Euclidean distance or similar method, pairs of image localities. Inverted File Indices (IVF, see Baeza-Yates and Ribeiro-Neto 1999) are the index structure commonly used in the feature retrieval step. Utilizing a feature vector binning scheme, such

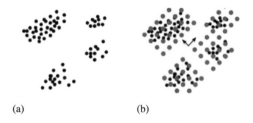

(a)   (b)

Fig. 15.10 A simplified 2D representation of OPQ. A set of points in the feature space **a** is re-described by rotating and translating grids of m bins to high-density areas (**b**), in red. The grid intervals and axis are learned separately for each dimension. Each dimension of a feature point can then be described using only m bits

as that of OPQ, an index is generated such that it stores a list of features contained within each bin. Each stored feature, in turn, points back to the image it was extracted from (hence the *inverted* terminology). The only calculation required is to determine the query feature's respective bin within the IVF to retrieve the nearest neighbors to a given query feature. Once this bin is known, the IVF can return the feature vectors' list in that bin and the nearest surrounding neighbor bins. Given that each feature vector points back to its source image, one can trace back the database images that are similar to the query. This simple yet powerful method provides easy scalability and distributability to index search. This is the main storage and retrieval structure used within powerful state-of-the-art search frameworks, such as the open-source Facebook Artificial Intelligence Similarity Search (FAISS) library introduced by Jeff Johnson et al. (2019).

Lastly, the result ranking step takes care of polling the feature-wise most similar database images to the query. The simplest metric with which one can rank the relatedness of the query image with the other images is feature voting (see Pinto et al. 2017). To perform it, one must iterate through each query feature and its retrieved nearest neighbor features. Then, by checking which database image each returned feature belongs to, one must accumulate a tally of how many features are matched to the query for each image. This final tally is then utilized as the votes for each database image, and these images are ranked (or ordered) accordingly. An example of this method can be seen in Fig. 15.11.

With these four steps implemented, one already has access to a rudimentary image search system capable of retrieving both near-duplicates and semantically similar images. While operational, this system is not particularly powerful when tasked with finding images with nuanced inter-image relationships relevant to provenance analysis. In particular, images that share only small amounts of content with the query image will not receive many votes in the matching process and may not be retrieved as a relevant result. Instead, this system can be utilized as a foundation for a retrieval system more fine-tuned to the task of image provenance analysis. In this regard, four methods to improve a typical CBIR solution are discussed in the following.

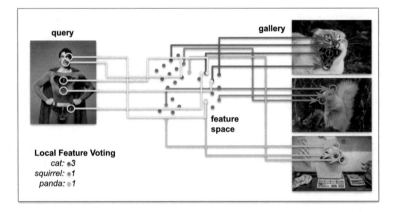

Fig. 15.11 A 2D simplified representation of IVF and feature voting. Local features are extracted from three images, representing a small gallery. The feature vector space is partitioned into a grid, and gallery features are indexed within this grid. Local features are then extracted from the query image of *Reddit Superman* (whose chest symbol and underwear are modified versions of the yawning cat's mouth depicted within the first item of the gallery). Each local query feature is fed to the index, and nearest neighbors are retrieved. Colored lines represent how each local feature matches the IVF features and subsequently map back to individual gallery features. We see how these feature matches can be used in a voting scheme to rank image similarities in the bottom left

Distributed Keypoints: To take the best advantage of these concepts, we must ensure that the extracted keypoints for the local features used within the indexing and retrieval pipeline do a good job describing the entire image. Most keypoint detectors return points that lie on corners, edges, or other high-entropy patches containing visual information. This, unfortunately, means that some areas within an image may be given too many vital points and may be over-described, while other areas may be entirely left out, receiving no keypoints at all. An area with no keypoints and, thus, no representation within the database image index has no chance of being correctly retrieved if a query with similar features comes along.

To mitigate over-description and under-description of image areas, Moreira et al. (2018) proposed an additional step in the keypoint detection process, which is the avoidance and removal of keypoints that present too much overlap with others. Such elements are then replaced with keypoints coming from weaker-entropy image regions, allowing for a more distributed content description. Figure 15.9b depicts an example of applying this strategy, in comparison with the regular SURF extraction approach.

Context Incorporation: One of the main reasons a typical CBIR solution performs poorly in an image provenance scenario is the nature behind many image manipulations within provenance cases. These manipulations often consist of composite images with small objects coming from donor images. An example of these types of relationships is shown in Fig. 15.12.

Fig. 15.12 An example of composite from the *r/photoshopbattles* subreddit. The relations of the composite with the images donating small objects, such as the hummingbird and the squirrel, challenge the retrieval capabilities of a typical CBIR solution. Context incorporation comes in handy in these situations. Image adapted from Joel Brogan et al. (2021)

These types of image relationships do not lend themselves to a naive feature voting strategy, as the size of the donated objects is often too small to garner enough local feature matches to impart a high vote score with the query image. We can augment the typical CBIR pipeline with an additional step, namely context incorporation, to solve this problem. Context incorporation takes into consideration the top N retrieved images for a given query, to accurately localize areas within the query and the retrieved images that differ from each other (most likely due to a composite or manipulation), to generate attention masks, and to re-extract features over only the distinct regions of the query, for a second retrieval execution. By using only these features, the idea is that the additional retrieval and voting steps will be driven towards finding the images that have donated small regions to the query due to the absence of distractors.

Figure 15.13 depicts an example where context incorporation is crucial to obtain the image that has donated Vladimir Putin (Fig. 15.13d) to a questioned query (Fig. 15.13a), in the first positions of the result rank. Different approaches for performing context incorporation, including attention mask generation, were proposed and benchmarked by Brogan et al. (2017), while an end-to-end CBIR pipeline employing such a strategy was discussed by Pinto et al. (2017).

Iterative Filtering: Another requirement of content retrieval within provenance analysis is the recovery of images directly related to the query and transitively related to it. That is the case of image 4 depicted within Fig. 15.6, which does not share content

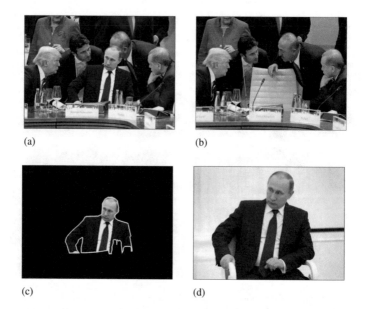

Fig. 15.13 Context incorporation example. In **a**, the query image. In **b**, the most similar retrieved near-duplicate (top 1 image) through a typical CBIR solution. In **c**, the attention mask highlighting the different areas between **a** and **b**, after proper content registration. In **d**, the top 1 retrieved image after using **c** as a new query (a.k.a., context incorporation). The final result rank may be a combination of the two ranks after using **a** and **c** as a query, respectively. Image **d** is only properly retrieved, from the standpoint of provenance analysis, thanks to the execution of context incorporation

directly with the query, but is related to it through image 1. Indeed, any other near-duplicates of image 4 should ideally be retrieved by a flawless provenance-aware CBIR solution. Aiming at also retrieving transitively related content to the query, Moreira et al. (2018) introduced iterative filtering. After retrieving the first rank of images, the results are iteratively refined by suppressing near-duplicates of the query and promoting non-near-duplicates as new queries to the next retrieval iteration. This process is executed a number of times, leading to a set of image ranks for each iteration, which are then combined into a single one, at the end of the process.

Object-Level Retrieval: Despite the modifications above to improve typical CBIR solutions towards provenance analysis, local-feature-based retrieval systems do not inherently incorporate structural aspects into the description and matching process. For instance, any retrieved image with a high match vote score could still, in fact, be completely dissimilar to the query image. That happens because the matching process does not take into account the position of local features with respect to each other within an image. As a consequence, unwanted database images that contain features individually similar to the query's features, but in a different pattern, may still be ranked highly.

(a)   (b)

(c)   (d)

Fig. 15.14 An example of how OS2OS matching is accomplished. In **a**, a query image has the local feature keypoints mapped relative to a computed centroid location. In **b**, matching features from a database image are projected to an estimated centroid location using the vectors shown in **a**. In **c**, the subsequent vote accumulation matrix is created. In **d**, the accumulation matrix is overlayed on the database image. The red area containing many votes shows that the query and database images most likely share an object in that location

To avoid this behavior, Brogan et al. (2021) modified the retrieval pipeline to account for the structural layout of local features. To do so, they proposed a method to leverage the scale and orientation components calculated as part of the SURF keypoint extraction mechanism and to perform a transform estimation similar to the *generalized Hough voting* (see Ballard 1981), relative to a computed centroid. Because each feature's scale and orientation are known, for both the query's and database images' features, database features can be projected to the query keypoint layout space. Areas that contain similar structures accumulate votes in a given location on a Hough accumulator matrix. By clustering highly active accumulation areas, one is able to quickly determine local areas shared between images that have structural consistency with each other. A novel ranking algorithm then leverages these areas within the accumulator to subsequently score the relationship between images. This entire process is called "objects in scenes to objects in scenes" (OS2OS) matching. An example of how the voting accumulation works is depicted in Fig. 15.14.

15.2.2 Datasets and Evaluation

The literature for provenance analysis has reported results of content retrieval over two major datasets, namely the Nimble Challenge (NC17, 2017) and the Media Forensics Challenge (MFC18, 2018) datasets.

NC17 (2017): On the occasion of promoting the Nimble Challenge 2017, NIST released an image dataset containing a development partition (Dev1-Beta4) specifically curated to support both research tasks for provenance analysis. Namely, provenance content retrieval (named provenance filtering by the agency) and provenance graph construction. This partition contains 65 image queries and 11,040 images either related to the queries through provenance graphs, or completely unrelated material (named distractors). The provenance graphs were manually created by image edition experts and include operations such as splicing, removal, cropping, scaling, blurring, and color transformations. As a consequence, the partition offers content retrieval ground truth composed of 65 expected image ranks. Aiming to increase the challenge offered by this partition, Moreira et al. (2018) extended the set of distractors by adding one million unrelated images, which were randomly sampled from Eval-Ver1, another partition released by NIST as part of the 2017 challenge. We rely on this configuration to provide some results of content retrieval and explain how the different provenance CBIR add-ons explained in Sect. 15.2.1 contribute to solve the problem at hand.

MFC18 (2018): Similar to the 2017 challenge, NIST released another image dataset in 2018, with a partition (Eval-Ver1-Part1) also useful for provenance content retrieval. Used to officially evaluate the participants of the MediFor program (see Turek 2021), this set contains 3,300 query images and over one million images, including content related to the queries and distractors. Many of these queries are composites, with the expected content retrieval image ranks provided as ground truth. Moreover, this dataset also provides ground-truth annotations as to whether a related image contributes only a particular small object to the query (such as in the case of image 4 donating Tupac Shakur's head to image 1, within Fig. 15.7), instead of an entire large background. These cases are particularly helpful to assess the advantages of using the object-level retrieval approach presented in Sect. 15.2.1 in comparison to the other methods.

As suggested in the protocol introduced by NIST (2017), the metric used to evaluate the performance of a provenance content retrieval solution is the CBIR recall of the images belonging to the ground truth rank, at three specific cut-off points. Namely, (i) $R@50$ (i.e., the percentage of ground-truth expected images that are retrieved among the top 50 assets returned by the content retrieval solution), (ii) $R@100$ (i.e., the percentage of ground truth images retrieved among the top 100 assets returned by the solution), and (iii) $R@200$ (i.e., the percentage of ground truth images retrieved among the top 200 assets returned by the solution). Since the recall expresses the percentage of relevant images being effectively retrieved, the method delivering higher recall is considered better.

In the following section, results in terms of a recall are reported for the different solutions presented in Sect. 15.2.1, over the aforementioned datasets.

15.2.3 Results

Table 15.1 summarizes the results of provenance content retrieval reported by Moreira et al. (2018) over the NC17 dataset. It helps to put into perspective some of the techniques detailed in Sect. 15.2.1. As one might observe, by comparing rows 1 and 2 of Table 15.1, the simple modification of using more keypoints (from 2,000 to 5,000 features) to describe the images within the CBIR base module already provides a significant improvement on the system recall. Distributed Keypoints, in turn, improve the recall of larger ranks (for $R@100$ and $R@200$), while Iterative Filtering alone allows the system to reach an impressive recall of 90% of the expected images among the top 50 retrieved ones. At the time of their publication, Moreira et al. (2018) found that a combination of Distributed Keypoints and Iterative Filtering led to the best content retrieval solution, represented by the last row of Table 15.1.

More recently, Brogan et al. (2021) performed new experiments on the MFC18 dataset, this time aiming at evaluating the performance of their proposed OS2OS approach. Table 15.2 compares the results of the best solution previously identified by Moreira et al. (2018), in row 1, with the addition of OS2OS, in row 2, and

Table 15.1 Results of provenance content retrieval over the NC17 dataset. Reported here are the average recall values of 65 queries at the top 50 ($R@50$), top 100 ($R@100$), and top 200 ($R@200$) retrieved images. Provenance add-ons on top of the CBIR base module were presented in Sect. 15.2.1

CBIR Base	Provenance Add-ons	$R@50$	$R@100$	$R@200$	Source
2,000 SURF features, OPQ	Context Incorporation	71%	72%	74%	Daniel Moreira et al. (2018)
5,000 SURF features, OPQ	Context Incorporation	88%	88%	88%	Daniel Moreira et al. (2018)
5,000 SURF features, OPQ	Distributed Keypoints	88%	90%	90%	Daniel Moreira et al. (2018)
5,000 SURF features, OPQ	Iterative Filtering	90%	90%	92%	Daniel Moreira et al. (2018)
5,000 SURF features, OPQ	Distrib. Keypoints, Iterative Filtering	91%	91%	92%	Daniel Moreira et al. (2018)

Table 15.2 Results of provenance content retrieval over the MFC18 dataset. Reported here are the average recall values of 3,300 queries at the top 50 ($R@50$), top 100 ($R@100$), and top 200 ($R@200$) retrieved images. Provenance add-ons on top of the CBIR base module were presented in Sect. 15.2.1. OS2OS stands for "objects in scene to objects in scene", previously presented as object-level retrieval

CBIR Base		Provenance Add-ons	$R@50$	$R@100$	$R@200$	Source
5,000 SURF features, OPQ		Distrib. Keypoints, Iterative Filtering	77%	81%	82%	Joel Brogan et al. (2021)
5,000 SURF features, OPQ		Distrib. Keypoints, OS2OS	83%	83%	84%	Joel Brogan et al. (2021)
1,000 DELF features, OPQ		Iterative Filtering	87%	90%	91%	Joel Brogan et al. (2021)
1,000 DELF features, OPQ		OS2OS	91%	93%	95%	Joel Brogan et al. (2021)

replacement of the SURF features with DELF (see Noh et al. 2017) features, in rows 3 and 4, over the MFC18 dataset. OS2OS alone (compare rows 1 and 2) improves the system recall for all the three rank cut-off points. Also, the usage of only 1,000 DELF features (a data-driven image description approach that relies on learned attention models, in contrast to the 5,000 handcrafted SURF features) significantly improves the system recall (compare rows 1 and 3). In the end, the combination of a DELF-based CBIR system and OS2OS leads to the best content retrieval approach, whose recall values are shown in the last row of Table 15.2.

Lastly, as explained before, the MFC18 dataset offers a unique opportunity to understand how the content retrieval solutions work for the recovery of dataset images that eventually donated small objects to the queries at hand since it contains specific annotations in this regard. Table 15.3 summarizes the results obtained by Brogan et al. (2021) when evaluating this aspect. As expected, the usage of OS2OS greatly improves the recall of small-content donors, as it can be seen through rows 2 and 4 of Table 15.3. Overall, the best content retrieval approach (a combination of DELF features and OS2OS) is able to retrieve only 55% of the expected small-content donors among the top 200 retrieved assets (see the last row of Table 15.3). This result indicates that more work still needs to be done in this specific aspect: improving the provenance content retrieval of small-content donor images.

Table 15.3 Results of provenance content retrieval over the MFC18 dataset, with focus on the retrieval of images that donated small objects to the queries. Reported here are the average recall values of 3,300 queries at the top 50 ($R@50$), top 100 ($R@100$), and top 200 ($R@200$) retrieved images. Provenance add-ons on top of the CBIR base module were presented in Sect. 15.2.1. OS2OS stands for "objects in scene to objects in scene", previously presented as object-level retrieval

CBIR Base			Provenance Add-ons	$R@50$	$R@100$	$R@200$	Source
5,000	SURF	features, OPQ	Distrib. Keypoints, Iterative Filtering	28%	34%	42%	Joel Brogan et al. (2021)
5,000	SURF	features, OPQ	Distrib. Keypoints, OS2OS	45%	48%	52%	Joel Brogan et al. (2021)
1,000	DELF	features, OPQ	Iterative Filtering	41%	45%	49%	Joel Brogan et al. (2021)
1,000	DELF	features, OPQ	OS2OS	51%	54%	55%	Joel Brogan et al. (2021)

15.3 Graph Construction

A provenance graph depicts the story of edits and manipulations underwent by a media asset. This section focuses on the provenance graph of images, whose vertices individually represent the image variants and whose edges represent the direct pairwise image relationships. Depending on the transformations applied to one image to obtain another, the two connected images can share partial to full visual content. In the case of partial content sharing, the source images of the shared content are called the *donor images* (or simply donors), while the resultant manipulated image is called the *composite* image. In full-content sharing, we have near-duplicate variants when one image is created from another through a series of transformations such as cropping, blurring, and color changes. Once a set of related images is collected from the first stage of content retrieval (see Sect. 15.2), a fine-grained analysis of pairwise relationships is required to obtain the full provenance graph. This analysis involves two major steps, namely (1) image similarity computation and (2) graph building.

Similarity computation involves understanding the degree of similarity between two images. It is a fundamental task for any visual recognition problem. Image matching methods are at the core of vision-based applications, ranging from handcrafted approaches to modern deep-learning-based solutions. A matching method is a similarity (or dissimilarity) score that can be used for further decision-making and classification. For provenance analysis, computing pairwise image similarity helps distinguish between direct versus indirect relationships. A selection of a feasible set of pairwise relationships creates a provenance graph. To analyze the closest provenance match to an image in the provenance graph, pairwise matching is performed for all possible image pairs in the set of k retrieved images. The similarity scores are then recorded in a matrix $\mathbf{M}$ of size $k \times k$ where each cell indexed $\mathbf{M}(i, j)$ represents the similarity between image I_i and image I_j.

Graph building, in turn, comprises the task of constructing the provenance graph after similarity computation. The matrix containing the similarity scores for all pairs of images involved in the provenance analysis for each case can be interpreted as an adjacency matrix. This implies that each similarity score in this matrix is the weight of an edge in a complete graph of k vertices. Extracting a provenance graph requires selecting a minimal set of edges that span the entire set of relevant images or vertices (this can be different from k). If the similarity measure used for the previous stage is symmetric, the final graph will be undirected, whereas an asymmetric measure of the similarity will lead to a directed provenance graph. The provenance cases considered in the literature, so far, are spanning trees. This implies that there are no cycles within graphs, and there is at most one path to get from one vertex to another.

15.3.1 Approaches

There are multiple aspects of a provenance graph. Vertices represent the different variants of an image or visual subject, pairwise relationships between images (i.e., undirected edges) represent atomic manipulations that led to the evolution of the manipulated image, and directions for these relationships provide more precise information about the change. Finally, the last details are the specific operations performed on one image to create the other. The fundamental task for an image-based provenance analysis is, thus, performing image comparison. This stage requires describing an image using a global or a set of local descriptors. Depending on the methods used for image description and matching, the similarity computation stage can create different types of adjacency weight matrices. The edge selection algorithm then depends on the nature of the computed image similarity. In the rest of this section, we present a series of six graph construction techniques that have been proposed in the literature and represent the current state of the art in image provenance analysis.

Undirected Graphs: A simple and yet-effective graph construction solution was proposed by Bharati et al. (2017). It takes the top k retrieved images for the given query and computes the similarity between the two elements of every image pair, including the query, through keypoint extraction, description, and matching. Keypoint extractors and descriptors, such as SIFT (see Lowe 2004) or SURF (see Bay et al. 2008), offer a manner to highlight the important regions within the images (such as corners and edges), and to describe their content in a way that is robust to several of the transformations manipulated images might have been through (such as scaling, rotating, and blurring.). The quantity of keypoint matches that are geometrically consistent with the others in the match set can act as an image similarity score for each image pair.

As depicted in Fig. 15.15, two images that share visual content will present more keypoint matches (see Fig. 15.15a) than the ones that have nothing in common (see

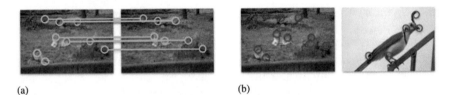

(a)   (b)

Fig. 15.15 Examples of geometrically consistent keypoint matching. In **a**, the matching of keypoints over two images that share visual content. In **b**, the absence of matches between images that do not look alike. The number of matching keypoint pairs can be used to express the similarity between two images

Fig. 15.15b). Consequently, a symmetric pairwise image adjacency matrix can be built by simply using the number of keypoint matches between every image pair. Ultimately, a maximum spanning tree algorithm, such as Kruskal's (1956) or Prim's (1957), can be used to generate the final undirected image provenance graph.

Directed Graphs: The previously described method has the limitation of generating only symmetric adjacency matrices, therefore, not providing enough information to compute the direction of the provenance graphs' edges. As explained in Sect. 15.1, within the problem of provenance analysis, the direction of an edge within a provenance graph expresses the important information of which asset gives rise to the other.

Aiming to mitigate this limitation and inspired by the early work of Dias et al. (2012), Moreira et al. (2018) proposed an extension to the keypoint-based image similarity computation alternative. After finding the geometrically consistent keypoint matches for each pair of images (I_i, I_j), the obtained keypoints can be used for estimating the homography H_{ij} that guides the registration of image I_i onto image I_j, as well as the homography H_{ji} that analogously guides the registration of image I_j onto image I_i.

In the particular case of H_{ij}, after obtaining the transformation $T_j(I_i)$ of image I_i towards I_j, $T_j(I_i)$ and I_j are properly registered, with $T_j(I_i)$ presenting the same size of I_j and the matched keypoints relying on the same position. One can, thus, compute the bounding boxes that enclose all the matched keypoints within each image, obtaining two correspondent patches R_1, within $T_j(I_i)$, and R_2, within I_j. With the two aligned patches at hand, the distribution of the pixel values of R_1 can be matched to the distribution of R_2, before calculating the similarity (or dissimilarity) between them.

Considering that patches R_1 and R_2 have the same width W and height H after content registration, one possible method of patch dissimilarity computation is the pixel-wise mean squared error (MSE):

$$MSE(R_1, R_2) = \frac{\sum_w^W \sum_h^H (R_1(w, h) - R_2(w, h))^2}{H \times W}, \qquad (15.1)$$

where $R_1(w, h) \in [0, 255]$ and $R_2(w, h) \in [0, 255]$ are the pixel values of R_1 and R_2 at position (w, h), respectively.

Alternatively to MSE, one can express the similarity between $R1$ and $R2$ as the mutual information (MI) between them. From the perspective of information theory, MI is the amount of information that one random variable contains about another. From the point of view of probability theory, it measures the statistical dependence of two random variables. In practical terms, assuming each random variable as, respectively, the aligned and color-corrected patches R_1 and R_2, the value of MI can be given by the entropy of discrete random variables:

$$MI(R_1, R_2) = \sum_{x \in R_1} \sum_{y \in R_2} p(x, y) \log \left(\frac{p(x, y)}{\sum_x p(x, y) \sum_y p(x, y)} \right), \quad (15.2)$$

where $x \in [0, 255]$ refers to the pixel values of R_1, and $y \in [0, 255]$ refers to the pixel values of R_2. The $p(x, y)$ value regards the joint probability distribution function of R_1 and R_2. As explained by Costa et al. (2017), it can be satisfactorily approximated by

$$p(x, y) = \frac{h(x, y)}{\sum_{x, y} h(x, y)}, \quad (15.3)$$

where $h(x, y)$ is the joint histogram that counts the number of occurrences for each possible value of the pair (x, y), evaluated on the corresponding pixels for both patches R_1 and R_2.

As a consequence of their respective natures, while MSE is inversely proportional to the two patches' similarity, MI is directly proportional. Aware of this, one can either use (i) the inverse of the MSE scores or (ii) the MI scores directly as the similarity elements s_{ij} within the pairwise image adjacency matrix $\mathbf{M}$, to represent the similarity between image I_j and the transformed version of image I_i towards I_j, namely $T_j(I_i)$.

The homography H_{ji} is calculated in an analogous way to H_{ij} with the difference that $T_i(I_j)$ is manipulated by transforming I_j towards I_i. Due to this, the size of the registered images, the format of the matched patches, and the matched color distributions will be different, leading to unique MSE (or MI) values for setting s_{ji}. Since $s_{ij} \neq s_{ji}$, the resulting similarity matrix $\mathbf{M}$ will be asymmetric. Figure 15.16 depicts this process.

Upon computing the full matrix, the assumption introduced by Dias et al. (2012) is that, in the case of $s_{ij} > s_{ji}$, it would be easier to transform image I_i towards image I_j, than the contrary (i.e., I_j towards I_i). Analogously, $s_{ij} < s_{ji}$ would mean the opposite. This information can, thus, be used for edge selection. The oriented Kruskal (2012) solution (with a preference for higher adjacency weights) would help construct the final provenance graph.

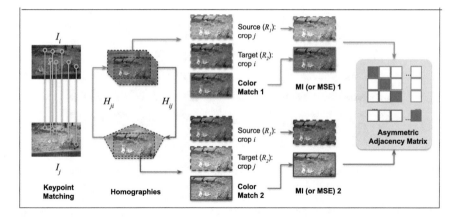

Fig. 15.16 Generation of an asymmetric pairwise image adjacency matrix. Based on the comparison of two distinct content transformations for each image pair (I_i, I_j), namely I_i towards I_j and vice-versa, this method allows the generation of an asymmetric pairwise image similarity matrix, which is useful for computing provenance graphs with directed edges

Clustered Graph Construction: As an alternative to the oriented Kruskal, Moreira et al. (2018) introduced a method of directed provenance graph building, which leverages both symmetric keypoint-based and asymmetric mutual-information-based image similarity matrices.

Inspired by Oikawa et al. (2015) and dubbed *clustered graph construction*, the idea behind such a solution is to group the available retrieved images in a way that only near-duplicates of a common image are added to the same cluster. Starting from the image query I_q as the initial expansion point, the remaining images are sorted according to the number of geometrically consistent matches shared with I_q, from the largest to the smallest. The solution then clusters probable near-duplicates around I_q, as long as they share *enough content*, which is decided based upon the number of keypoint matches (see Daniel Moreira et al. 2018). Once the query's cluster is finished (i.e., the remaining images do not share enough keypoint matches with the query), a new cluster is computed over the remaining unclustered images, taking another image of the query's cluster as the new expansion point. This process is repeated iteratively by trying different images as the expansion point until every image belongs to a near-duplicate cluster.

Once all images are clustered, it is time to establish the graph edges. Images belonging to the same cluster are sequentially connected into a single path without branches. This makes sense in scenarios containing sequential image edits where one near-duplicate is obtained on top of the other. As a consequence of the iterative execution and selection of different images as expansion points, the successful ones (i.e., the images that were helpful in the generation of new image clusters) fatally belong to more than one cluster, hence serving as graph bifurcation points. Orthogonal edges are established in such cases, allowing every near-duplicate image branch to be connected to the final provenance graph through an expansion point image as a

Fig. 15.17 Usage of metadata information to refine the direction of pairwise image provenance relationships. For the presented images, the executed manipulation operation could be either the splicing or the removal of the male lion. According to the image generation date metadata, the operation is revealed to be a splice since the image on the left is older. Adapted from Aparna Bharati et al. (2019)

joint. To determine the direction of every single edge, Moreira et al. (2018) suggested using the mutual information similarity asymmetry in the same way as depicted in Fig. 15.16.

Leveraging Metadata: Image comparison techniques may be limited depending upon the transformations involved in any given image's provenance analysis. In cases where the transformations are reversible or collapsible, the visual content analysis may not suffice for edge selection during graph building. Specifically, the homography estimation and color mapping steps involved in asymmetric matrix computation for edge direction inference could be noisy. To make this process more robust, it is pertinent to utilize other evidence sources to determine connections. As can be seen from the example in Fig. 15.17, it is difficult to point out the plausible direction of manipulation with visual correspondence, but auxiliary information related to the image, mostly accessible within the image files (*a.k.a.* image metadata), can increase confidence in predicting the directions.

Image metadata, when available, can provide additional evidence for directed edge inference. Bharati et al. (2019) identify highly relevant tags for the task. Specific tags that provide the time of image acquisition and editing, location, editing operation, etc. can be used for metadata analysis that corroborates visual evidence for provenance analysis. An asymmetric heuristic-based metadata comparison parallel to a symmetric visual comparison is proposed. The metadata comparison similarity scores are higher for image pairs (ordered) with consistency from more sets of metadata tags. The resulting visual adjacency matrix is used for edge selection, while the metadata-based comparison scores are used for edge direction inference. As explained in clustered graph construction, there are three parts of the graph building method, namely node cluster expansion, edge selection, and assigning directions to edges. The metadata information can supplement the last two depending on the specific stage at which it is incorporated. As metadata tags can be volatile in the world of intelligent forgeries, a conservative approach is to use them to improve the confidence of the provenance graph obtained through visual analysis. The proposed design enables the usage of metadata when available and consistent.

Transformation-Aware Embeddings: While metadata analysis can improve the fidelity of edge directions in provenance graphs when available and not tampered with, local keypoint matching for visual correspondence faces challenges in image ordering for provenance analysis. Local matching is efficient and robust to finding shared regions between related images. This works well for connecting donors with composite images but can be insufficient in capturing subtle differences between near-duplicate images, which affect the ordering of long chains of operations. Establishing sequences of images that vary slightly based on the transformations requires differentiating between slightly modified versions of the same content.

Towards improving the reconstructions of globally-related image chains in provenance graphs, Bharati et al. (2021) proposed encoding awareness of the transformation sequence in the image comparison stage. Specifically, the devised method learns transformation-aware embeddings to better order related images in an edit sequence or provenance chain. The framework uses a patch-based siamese structure trained with an Edit Sequence Loss (ESL) using sets of four image patches. Each set is expressed as *quadruplets* or *edit sequences*, namely (i) the *anchor* patch, which represents the original content, (ii) the *positive* patch, a near-duplicate of the anchor after M image processing transformations, (iii) the *weak positive* patch, the positive patch after N transformations, and (iv) the *negative* patch, a patch that is unrelated to the others. The quadruplets of patches are obtained for training using a specific set of image transformations that are of interest to image forensics, particularly image phylogeny and provenance analysis, as suggested in Dias et al. (2012). For each anchor patch, random unit transformations are sequentially applied, one on top of the other's result, allowing to generate positive and weak positive patches from the anchor, after M and $M + N$ transformations, respectively. The framework aims at providing distance scores to pairs of patches, where the output score between the anchor and the positive patch is smaller than the one between the anchor and the weak positive, which, in turn, is smaller than the score between the anchor and the negative patch (as shown in Fig. 15.18).

Given a feature vector for an anchor image patch a, two transformed derivatives of the anchor patch p (positive) and p' (weak positive) where $p = T_M(a)$ and $p' = T_N(T_M(a))$, and an unrelated image patch from a different image n, ESL is a pairwise margin ranking loss computed as follows:

$$SL(a, p, p', n) = \max(0, -y \times (d(a, p') - d(a, n)) + \mu_1) + \\ \max(0, -y \times (d(p, p') - d(p, n)) + \mu_2) + \quad (15.4) \\ \max(0, -y \times (d(a, p) - d(a, p')) + \mu_3)$$

Here, y is the truth function which determines the rank order (see Rudin and Schapire 2009) and μ_1, μ_2, and μ_3 are margins corresponding to each pairwise distance term and are treated as hyperparameters. Both terms having the same sign

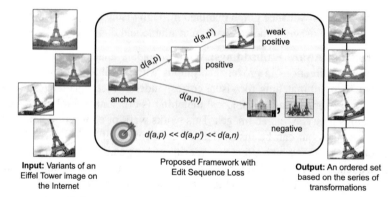

Fig. 15.18 Framework to learn transformation-aware embeddings in the context of ordering image sequences for provenance. Specifically, to satisfy an edit-based similarity precision constraint, *i.e.*, $d(a, p) < d(a, p') < d(a, n)$. Adapted from Aparna Bharati et al. (2021)

implies ordering is correct, and the loss is zero. A positive loss is accumulated when the ordering is wrong, and they are of opposite signs.

The above loss is optimized, and the model corresponding to the best measure for validation is used for feature extraction from patches of test images. Features learned with the proposed technique are used to provide pairwise image similarity scores. The value d_{ij} between images I_i and I_j is computed by matching the set of features (extracted from patches) from one image to the other using an iterative greedy brute-force matching strategy. At each iteration, the best match is selected as the pair of patches between image I_i and image I_j whose $l2$-distance is the smallest and whose patches did not participate in a match on previous iterations. This guarantees a deterministic behavior regardless of the order of the images, meaning that either comparing the patches of I_i against I_j or vice-versa will lead to the same consistent set of patch pairs. Once all patch pairs are selected, the average $l2$-distance is calculated and finally set as d_{ij}. The inverse of d_{ij} is then used to set both s_{ij} and s_{ji} within the pairwise image similarity matrix $\mathbf{M}$, which in this case is a symmetric one. Upon computing all values within $\mathbf{M}$ for all possible image pairs, a greedy algorithm (such as Kruskal's 1956) is employed to order these pairwise values and create an optimally connected undirected graph of images.

Leveraging Manipulation Detectors: A challenging aspect of image provenance analysis is establishing high-confidence direct relationships between images that share a small portion of content. Keypoint-based approaches may not suffice as there may not be enough keypoints in the shared regions, and global matching approaches may not appropriately capture the matching region's importance. To improve analysis of composite images where source images have only contributed a small region and determine the source image among a group of image variants, Zhang et al. (2020) proposed to combine a pairwise ancestor-offspring classifier with manipulation detection approaches. They build the graph by combining edges based on both local feature matching and pixel similarity.

Their proposed algorithm attempts to balance global and local features and matching scores to boost performance. They start by using a weighted combination of the matched SIFT keypoints and the matched pixel values for image pairs that can be aligned, and null for the ones that cannot be aligned. A hierarchical clustering approach is used to group images coming from the same source together. For graph building within each determined cluster, the authors combine the likelihood of images being manipulated or extracted from a holistic image manipulation detector (see Zhang et al. 2020) and the pairwise ancestor score extracted by an L2-Net (see Tian et al. 2017). The image manipulation detector uses a patch-based convolutional neural network (CNN) to predict manipulations from a median-filtered residual image. For ambiguous cases where the integrity score may not be assigned accurately, a lightweight CNN-based ancestor-offspring network takes patch pairs as input and predicts one's scores to be derived from the other. The similarity scores used as edge weights are the average of the integrity and the ancestor scores from the two used networks. The image with the highest score among the smaller set of images is considered as the source. All incoming links to this vertex are removed to reduce confusion in directions. This one is then treated as the root of the arborescence built by applying Chu-Liu/Edmonds' algorithm (see Chu 1965; Edmonds 1967) on pairwise image similarities.

The different arborescences are connected by finding the best-matched image pair among the image clusters. If the matched keypoints are above a threshold, these images are connected, indicating a splicing or composition possibility. As reported in the following section, this method obtains state-of-the-art results on the NIST challenges (MFC18 2018 and MFC19 2019), and it significantly improves the computation of the edges of the provenance graphs over the Reddit Photoshop Battles dataset (see Brogan 2021).

15.3.2 Datasets and Evaluation

With respect to the step of provenance graph construction, four datasets stand out as publicly available and helpful benchmarks, namely NC17 (2017), MFC18 (2018) (both discussed in Sect. 15.2.2), MFC19 (2019), and the Reddit Photoshop Battles dataset (2021).

NC17 (2017): As mentioned in Sect. 15.2.2, this dataset contains an interesting development partition (Dev1-Beta4), which presents 65 image queries, each one belonging to a particular manually curated provenance graph. As expected, these provenance graphs are provided within the partition as ground truth. The number of images per provenance graph ranges from two to 81, with the average graph order being equal to 13.6 images.

MFC18 (2018): Besides providing images and ground truth for content retrieval (as explained in Sect. 15.2.2), the Eval-Ver1-Part1 partition of this dataset also pro-

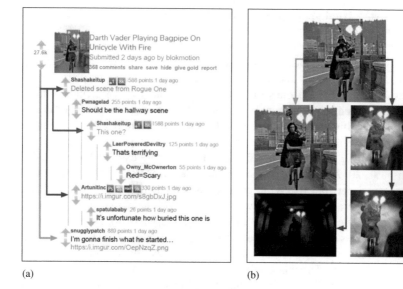

(a)   (b)

Fig. 15.19 A representation of how provenance graphs were obtained from the users' interactions within the Reddit r/photoshopbattles subreddit. In **a**, the parent-child posting structure of comments, which are used to generate the provenance graph depicted in **b**. Equivalent edges across the two images have the same color (either green, purple, blue, or red). Adapted from Daniel Moreira et al. (2018)

vides provenance graphs and 897 queries aiming at evaluating graph construction. For this dataset, the average graph order is 14.3 images, and the resolution of its images is larger, on an average, when compared to NC17. Moreover, its provenance cases encompass a larger set of applied image manipulations.

MFC19 (2019): A more recent edition of the NIST challenge released a larger set of provenance graphs. The Eval-Part1 partition within the MFC19 (2019) dataset has 1,027 image queries, and the average order of the provided ground truth provenance graphs is equal to 12.7 image vertices. In this group, the number of types of image manipulations used to generate the edges of the graphs was almost twice the number of MFC18.

Reddit Photoshop Battles (2021): Aiming at testing the image provenance analysis solutions over more realistic scenarios, Moreira et al. (2018) introduced the Reddit Photoshop Battles dataset. This dataset was collected from images posted to the Reddit community known as r/photoshopbattles (2012), where professional and amateur image manipulators share doctored images. Each "battle" starts with a teaser image posted by a user. Subsequent users post modifications of either the teaser or previously submitted manipulations in comments to the related posts. By using the underlying tree comment structure, Moreira et al. (2018) were able to infer and collect 184 provenance graphs, which together contain 10,421 original and composite images. Figure 15.19 illustrates this provenance graph inference process.

To evaluate the available graph construction solutions, two configurations are proposed by the NIST challenge (2017). In the first one, named the "oracle" scenario, there is a strong focus on the graph construction task. It assumes that a flawless content retrieval solution is available, thus, starting from the ground-truth content retrieval image ranks to build the provenance graphs, with neither missing images nor distractors. In the second one, named "end-to-end" scenario, content retrieval must be performed before graph construction, thus, delivering imperfect image ranks (with missing images or distractors) to the step of graph construction. We rely on both configurations and on the aforementioned datasets to report results of provenance graph construction, in the following section.

Metrics: As suggested by NIST (2017), given a provenance graph $G(V, E)$ generated by a solution whose performance we want to assess, we compute the $F1$-measure (i.e., the harmonic mean of precision and recall) of the (i) retrieved image vertices V and of the (ii) established edges E, when compared to the ground truth graph $G'(V', E')$, with its V' and E' homologous components. The first metric is named vertex overlap (VO) and the second one is named edge overlap (EO), respectively:

$$VO(G', G) = 2 \times \frac{|V' \cap V|}{|V'| + |V|}, \tag{15.5}$$

$$EO(G', G) = 2 \times \frac{|E' \cap E|}{|E'| + |E|}. \tag{15.6}$$

Moreover, we compute the vertex and edge overlap (VEO), which is the $F1$-measure of retrieving both vertices and edges simultaneously:

$$VEO(G', G) = 2 \times \frac{|V' \cap V| + |E' \cap E|}{|V'| + |V| + |E'| + |E|}. \tag{15.7}$$

In a nutshell, these metrics aim at assessing the overlap between G and G'. The higher the values of VO, EO, and VEO, the better the performance of the solution. Finally, in the particular case of EO and VEO, when they are both assessed for an approach that does not generate directed graphs (such as Undirected Graphs and Transformation-Aware Embeddings, presented in Sect. 15.3.1), an edge within E is considered a hit (i.e., a correct edge) when there is a homologous edge within E' that connects equivalent image vertices, regardless of the edges' directions.

15.3.3 Results

Table 15.4 puts in perspective the different provenance graph construction approaches explained in Sect. 15.3.1, when executed over the NC17 dataset. The provided results were all collected in oracle mode, hence the high values of VO (above 0.9), since there are neither distractors nor missing images in the rank lists used to build the provenance graphs. A comparison between rows 1 and 2 within this table shows the efficacy of leveraging image metadata as additional information to compute the edges of the provenance graphs. The values of EO (and VEO, consequently) have a significant increase (from 0.12 to 0.45, and from 0.55 to 0.70, respectively), when metadata is available. In addition, by comparing rows 3 and 4, one can observe the contribution of the data-driven Transformation-Aware Embeddings approach, in the scenario where only undirected graphs are being generated. In both cases, the generated edges have no direction by design, making their edge overlap conditions easier to be achieved (since the order of the vertices within the edges become irrelevant for the computation of EO and VEO, justifying their higher values when compared to rows 1 and 2). Nevertheless, contrary to the first two approaches, these solutions are not able to define which image gives rise to the other within the established provenance edges.

Table 15.5 compares the current state-of-the-art solution (Leveraging Manipulation Detectors by Xu Zhang et al. 2020) with the official NIST challenge participation results of the Purdue-Notre Dame team (2018), for both MFC18 and MFC19 datasets. In both cases, the reported results refer to the more realistic end-to-end scenario, where performers must execute content retrieval prior to building the provenance graphs. As a consequence, the image ranks fed to the graph construction step are noisy, since they contain both missing images and distractors. For all the reported cases, the image ranks had 50 images and presented an average $R@50$ of around 90% (i.e., nearly 10% of the needed images are missing). Moreover, nearly 35% of the images within the 50 available ones in a rank are distractors, on average. The

Table 15.4 Results of provenance graph construction over the NC17 dataset. Reported here are the average vertex overlap (VO), edge overlap (EO), and vertex and edge overlap (VEO) values of 65 queries. These experiments were executed in the "oracle" scenario, where the image ranks fed to the graph construction step are perfect (i.e., with neither distractors nor missing images)

Graph construction approach	VO	EO	VEO	Source
Clustered graph construction	0.93	0.12	0.55	Daniel Moreira et al. (2018)
Clust. Graph Const., Leveraging Metadata	0.93	0.45	0.70	Aparna Bharati et al. (2019)
Undirected Graphs	0.90	†0.65	†0.78	Aparna Bharati et al. (2021)
Transformation-Aware Embeddings	1.00	†0.68	†0.85	Aparna Bharati et al. (2021)

†Values collected over undirected edges

Table 15.5 Results of provenance graph construction over the MFC18 and MFC19 datasets. Reported here are the average vertex overlap (VO), edge overlap (EO), and vertex and edge overlap (VEO) values of 897 queries, in the case of MFC18, and of 1,027 queries, in the case of MFC19. These experiments were executed in the "end-to-end" scenario, thus, building graphs upon imperfect image ranks (i.e., with distractors or missing images)

Dataset	Graph Const. approach	VO	EO	VEO	Source
MFC18	Clustered Graph Construction	0.80	0.27	0.54	MFC19 (2019)
	Leveraging Manipulation Detectors	0.82	0.40	0.61	Xu Zhang et al. (2020)
MFC19	Clustered Graph Construction	0.70	0.30	0.52	MFC19 (2019)
	Leveraging Manipulation Detectors	0.85	0.42	0.65	Xu Zhang et al. (2020)

Table 15.6 Results of provenance graph construction over the Reddit Photoshop Battles dataset. Reported here are the average vertex overlap (VO), edge overlap (EO), and vertex and edge overlap (VEO) values of 184 queries. These experiments were executed in the "oracle" scenario, where the image ranks fed to the graph construction step are perfect. "N.R." stands for not-reported values

Graph construction approach	VO	EO	VEO	Source
Clustered Graph Construction	0.76	0.04	0.40	Daniel Moreira et al. (2018)
Clust. Graph Const., Leveraging Metadata	0.76	0.09	0.42	Aparna Bharati et al. (2019)
Leveraging Manipulation Detectors	N.R.	0.21	N.R.	Xu Zhang et al. (2020)

best solution (contained in rows 2 and 4 within Table 15.5) still delivers low values of EO when compared to VO, revealing an important limitation of the available approaches.

Table 15.6, in turn, reports results on the Reddit Photoshop Battles dataset. As one might observe, especially in terms of EO, this set is a more challenging one for the graph construction approaches, except for the state-of-the-art solution (Leveraging Manipulation Detectors by Xu Zhang et al. (2020). While methods that have worked fairly well on the NC17, MFC18, and MFC19 datasets drastically fail in the case of the Reddit dataset (see EO values below 0.10 in the case of rows 1 and 2), the state-of-the-art approach (in the last row of the table) more than doubles the results of EO. Again, even with this improvement, increasing the values of EO within graph construction solutions is still an open problem that deserves attention from researchers.

15.4 Content Clustering

In the study of human communication on the Internet and the understanding of the provenance of trending assets, such as memes and other forms of viral content, the users' intent is mainly focused on the retrieval, selection, and organization of semantically similar objects, rather than the gathering of near-duplicate variants or compositions that are related to a query. Under this scenario, although the step of content retrieval may be useful, the step of graph construction loses its purpose, since the available content is preferably related through semantics (e.g., different people on diverse scenes doing the same action, such as the "dabbing" trend depicted on Fig. 15.8), greatly varying in appearance, making the techniques presented on Sect. 15.3 less suitable.

Humans can only process so much data at once—if too much is present, they begin to be overwhelmed. In order to help facilitate the human processing and perception of the retrieved content, Theisen et al. (2020) proposed a new stage to the provenance pipeline focused on image clustering. Clustering the images based on shared elements helps triage the massive amounts of data that the pipeline has grown to accommodate. Nobody can reasonably review several million images to find emerging trends and similarities, especially without ordering in the image collection. However, when these images are grouped based on shared elements using the provenance pipeline, the number of items a reviewer would have to look at can decrease by several magnitude orders.

15.4.1 Approach

Object-Level Content Indexing: From the content retrieval techniques discussed in Sect. 15.2, Theisen et al. (2020) recommended the use of OS2OS matching (see Brogan et al. 2021) to index the millions of images eventually available, due to two major reasons. Firstly, to obtain a fast and scalable content retrieval engine. Secondly, to benefit from the OS2OS capability of comparing images through either large and global content matching or through many small object-wise local matches.

Affinity Matrix Creation: In the task of analyzing a large corpus of assets shared on the web to understand the provenance of a trend, the definition of a query (i.e., a questioned asset) is not always as straightforward as it is in the image provenance analysis case. For example, the memes shared during the 2019 Indonesian elections and discussed in William Theisen et al. (2020). In such cases, a natural question to ask would be which memes in the dataset one should use as the queries, for performing the first step of content retrieval.

Inspired by a simplification of iterative filtering (see Sect. 15.2.1), Theisen et al. (2020) identified the cheapest option as being randomly sampling images from the dataset and iteratively using them as queries, for executing content retrieval until all the dataset images (or a sufficient number of them) are "touched" (i.e., they are retrieved by the content retrieval engine). There are several advantages to this, other than being easy to implement. Randomly sampling means that end-users would need to have no prior knowledge of the dataset and potential trends they are looking for. The cluster created at the end of the process would show "emergent" trends, which could even surprise the reviewer.

On the other hand, from a more forensics-driven perspective, it is straightforward to imagine a system in which "informed queries" are used. If the users already suspect that several specific trends may exist in the data, cropped images of the objects pertaining to a trend may be used as a query, thus, prioritizing the content that the reviewers are looking for. This might be a demanding process because the user must already have some sort of idea of the landscape of the data and must produce the query images themselves.

Following the suit in the solution proposed in William Theisen et al. (2020), prior to performing clustering, a particular type of image pairwise adjacency matrix (dubbed *affinity matrix*) must first be generated. By leveraging the provenance pipeline's pre-existing steps, this matrix can be constructed based on the retrieval of the many selected queries. To prepare for the clustering step, a number of queries need to be run through the content retrieval system, the number of queries, and the recall of them depending on what type of graph the user wants to model for the analyzed dataset. Using the retrieval step's output, a number of query-wise "hub" nodes are naturally generated, each of them having many connections. Consider, for example, that the user has decided to have a recall of the top 100 best matches for any selected query. This means that for every query submitted, there are 100 connections to other assets. These connections link the query image to each of the 100 matches, thus, imposing a "hubness" onto the query image. For the sake of illustration, Fig. 15.20 shows an example of this process, for the case of memes shared during the 2019 Indonesian elections.

By varying the recall and number of queries, the affinity matrix can be generated, and an order can be imposed. Lowering the recall will result in a decrease in hub-like nodes in the structure but will require more queries to be run to connect all the available images. The converse is also true. Once enough queries have been run, the final step of clustering can be executed.

Spectral Clustering: After the affinity matrix has been generated, Theisen et al. (2020) suggested the use of multiclass spectral clustering (see Stella and Shi 2003) to organize the available images. In the case of memes, this step has the effect of assigning each image to a hypothesized *meme genre*, similar to the definition presented by Shifman (2014). The most important and perhaps trickiest part is deciding on a number of clusters to create. While more clusters may allow for a more targeted

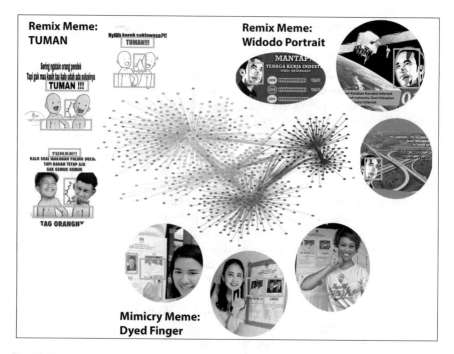

Fig. 15.20 A demonstration of the type of structure that may be inferred from the output of content retrieval, with many selected image queries. According to global and object-level local similarity, images with similar sub-components should be clustered together, presenting connections to their respective queries. Image adapted from William Theisen et al. (2020)

look into the dataset, it increases processing time for both the computer generating the clusters and the human reviewing them at the end. In their studies, Theisen et al. (2020) pointed out that 150 clusters appeared sufficient when the number of total dataset available images was on the order of millions of samples. This allowed for a maximum amount of images per cluster that a human could briefly review, but not so few images that trends could not be noticed.

15.4.2 Datasets and Evaluation

Indonesian Elections Dataset: The alternative content clustering stage inside the framework of provenance analysis aims to unveil emerging trends in large image datasets collected around a topic or an event of interest. To evaluate the capability of the proposed solutions, Theisen et al. (2020) collected over two million images from social media, concentrated around the 2019 Indonesian presidential elections. Harvested from Twitter (2021) and Instagram (2021) over 13 months (from March 31, 2018, to April 30, 2019), this dataset spans an extensive range of emotion and

Fig. 15.21 Imposter-host test to measure the quality of the computed image clusters. Five images are presented at a time to a human subject, four of which belong to the same host cluster. The remaining one belongs to an unrelated cluster, therefore, being named imposter. The human subject is always asked to identify the imposter, which in this example is the last image. A high number of correct answers given by humans indicate that the clustering solution was successful. Image adapted from William Theisen et al. (2020)

thought surrounding the elections. The images are publicly available at https://bit. ly/2Rj0odI.

Human Understanding Assessments: Keeping the objective of aiding the human understanding of large image datasets in mind, a metric is needed to measure how meaningful an image cluster is for a human. Inspired by the work of Weninger et al. (2012), Theisen et al. (2020) proposed an imposter-host test, which is performed by showing a person a collection of N images, all but one of which are from a common "host" cluster, which was previously computed by the proposed solution, whose quality needs to be measured. The other item, the "imposter", is randomly selected from one of the other established clusters. The idea is that the more related the images in a single cluster are, the easier it should be for a human to pick out the imposter image. Figure 15.21 depicts an example of this configuration. To test the results of their studies, Theisen et al. (2020) hired Amazon Mechanical Turk workers (2021) to perform 25 of these imposter-host tasks each subject, with N equal to 5 images, in order to measure the ease with which a human can pick out an imposter from the clusters that were generated.

Fig. 15.22 Accuracy of the imposter-host tasks given to Amazon Mechanical Turk workers, relative to cluster size. The larger the point on the accuracy axis, the more images were in the cluster. The largest cluster is indicated as having 15.89% of all the dataset images. Image adapted from William Theisen et al. (2020)

15.4.3 Results

The Amazon Mechanical Turk experiments reported by Theisen et al. (2020) demonstrated that the provenance pipeline's clustering step can produce human interpretable clusters while minimizing the average cluster size. The average accuracy for the imposter-host test was 62.42%. If the worker is shown five images, the chance of correctly guessing would be $1/5$ (20%). Therefore, the reported average is far above the baseline, thus, demonstrating the salience of the trends discovered in the clusters to human reviewers. The median cluster size was only 132 images per cluster. A spread of the cluster sizes as related to the worker accuracy for a cluster can be seen in Fig. 15.22. Surprisingly even the largest cluster, containing 15.89% of all the images, has an accuracy still higher than random chance. Three examples of what an individual cluster could look like can be seen in Fig. 15.23.

15.5 Open Issues and Research Directions

State of the Art: Based on the results presented and discussed in the previous sections within this chapter, one can safely conclude that, in the current state of the art of provenance analysis, the available solutions are indeed proven to be effective for at least two tasks, namely (1) authenticity verification of images and (2) the understanding of image-sharing trends online.

As for the verification of the authenticity of a questioned image (i.e. the query), this is done through the principled inspection of a given corpus of potentially related images to the query. In this case, whenever a non-empty set of related images is retrieved and a non-empty provenance graph (either directed or undirected) is generated and presented, one might infer the edit history of the query and take it as evidence of potential conflicts that might attest against its authenticity, such as inadvertent manipulations or source misattributions.

Regarding the understanding of trending content online, this can be done during the unfolding of a target event, such as national elections or international sports competitions. In this case, the quick retrieval of related images and subsequent content-based clustering have the potential to unveil emergent trends and surface

(a)   (b)

(c)

Fig. 15.23 Three examples of what the obtained clusters may look like. In **a** and **b**, it is very easy to see the similarity shared among all the images. In **c**, it is not quite as clear, but if one were to look a little more closely, they might notice the stylized "01" logo in each of the images. This smaller, shared component is what makes object-level matching and local feature analysis a compelling tool. Image adapted from William Theisen et al. (2020)

their provenance, allowing for a better comprehension of the event itself, as well as its relevance. A summary of what the public is thinking about an event and who the actors are trying to steer public opinion is only a couple of the possibilities that provenance content clustering may allow one to do.

While results have been very promising, no solution is without flaws, and much work is left to be done. In particular, image provenance analysis still has many caveats, which are briefly described below. Similarly, multimodal provenance analysis is an unexplored field, which deserves a great deal of attention from researchers shortly.

Image Provenance Analysis Caveats: There are open problems in image provenance analysis that still need attention from the scientific community. For instance, the solutions proposed to compute the directions of the edges of the provenance graphs still deliver values of edge overlap that are inferior to the results for vertex recall and vertex overlap. This aspect indicates that the proposed solutions are better at determining which images must be added to the provenance graphs as vertices, but there is still plenty of room to improve the computation of how these vertices must be connected. In this regard, novel techniques of learning which image might have been acquired or generated first and leveraging the output of single-image manipulation detectors (tackled throughout this book) are desired.

Moreover, an unexplored aspect of image provenance analysis is understanding, representing, and detecting the space of image transformations used to generate one image from the other. By doing so, one will determine what transformations might have been performed during the establishment of an edge, a problem that currently still lacks solutions. Thankfully, through the recent joint efforts of DARPA and NIST within the MediFor program (2021), a viable regime for data generation and annotation at the level of registering the precisely applied image transformations that have been performed from one asset to the other has emerged. This regime's outcome is available to the scientific community as a useful benchmark (see Guan et al. 2019) for further research.

Multimodal Provenance Analysis: As explained in Sect. 15.3, we have already witnessed that additional information such as image metadata helps to improve provenance analysis. Moving forward, another important research direction that requires progress is the identification and principled usage of information coming from other asset modalities (such as the text of image captions, in the case of questionable assets that are rich documents), as well as the development of provenance analysis for media types other than images (*e.g.*, the provenance of text documents, videos, audios, etc.).

When one considers the multimodal aspect, many questions appear and are open for research. One of them is how to analyze complex assets, such as the texts coming from different suspect documents (looking for cases of plagiarism and attribution to the original author), or the images extracted from scientific papers (which may be inadvertently reused to fabricate scientific findings, such as what was recently described in PubPeer Foundation (2020)). Another question is how to leverage images and captions in a document, or the video frames and their respective audio subtitles, within a movie.

Video provenance analysis, in particular, is a topic that deserves a great deal of attention. While one might assume image-based methods could be extended to video by being run over multiple frames, such a solution would fail to glean videos' inherent temporal dimension. Sometimes videos shared on social media such as Instagram (2021) or TikTok (2021) are composites of one or more pieces of viral footage. Tracking the provenance of videos is as important as tracking the provenance of still images.

Document provenance analysis, in turn, has also gained attention lately due to the recent *pandemic of bad science* (see Scheirer 2020). The COVID-19 crisis has caused a subsequent explosion in scientific publications surrounding the pandemic, not all of which might be considered highly rigorous. Using a provenance pipeline to aid in document analysis could allow reviewers to find repeated figures across many publications. It would be the reviewers' decision, though, to determine if the repetition is something as simple as citing the previous work or something more nefarious like trying to claim pre-existing work as one's own.

Towards an End-user Provenance Analysis System: Finally, for the solutions discussed in this chapter to be truly useful, they require front-end development and back-end integration, which are currently missing. Such a system would allow users to perform tasks similar to reverse image search, with the explicit intent of finding the origins of all related content within the query asset in question. If a front-end requiring minimal user skill to use could be designed, provenance analysis could be consolidated as a powerful tool for fighting fake news and misinformation.

Table 15.7 List of datasets useful for provenance analysis

Dataset	Year	Description	Link	QR Code
Reddit Photoshop Battles	2018	184 provenance graphs involving 10,421 images collected from the Reddit PhotoshopBattles community.	`https://bit.ly/3jfrLSJ`	
Open Media Forensics Challenge	2020	Reunion of NC17 MFC18, MFC19 and more recent sets, with useful partitions for provenance analysis.	`https://bit.ly/3svh81R`	
Indonesian Election Memes	2021	Over two million images collected from Twitter and Instagram about the 2019 Indonesian elections.	`https://bit.ly/2Rj0odI`	

15.6 Summary

The current state of the art of provenance analysis is mature enough for aiding image authenticity verification by processing a corpus of images rather than the processing of a single and isolated asset. However, more work needs to be done to improve the quality of the generated provenance graphs concerning the direction and identification of image transformations associated with the graphs' edges. Besides, provenance analysis still needs to be extended to media types other than images, such as video, audio, and text. In this regard, both the development of algorithms and the collection of datasets are yet to be done, revealing a unique research opportunity.

Provenance Analysis Datasets: Table 15.7 enlists the currently available and useful datasets for image provenance analysis.

Table 15.8 List of implementations of provenance analysis solutions

Solution	Year	Description	Link	QR Code
Context Incorporation	2017	Implementation of content retrieval based on context incorporation.	https://bit.ly/3daU2Ju	
Undirected Graph Construction	2017	Implementation of graph construction that generates undirected graphs.	https://bit.ly/3wRj2gM	
Provenance Analysis at Scale	2018	Implementation of content retrieval based on distributed keypoints and iterative filtering, and of graph construction that generates directed graphs, including the clustered approach.	https://bit.ly/3mKlVev	
NIST MedisScore	2018	Implementation of provenance analysis metrics and evaluation toolkit provided by NIST.	https://bit.ly/3anUMsN	
Leveraging Manipulation Detectors	2020	Implementation of content retrieval and of graph construction supported by image manipulation detectors.	https://bit.ly/3g44nZv	
Object-Level Content Retrieval	2021	Implementation of the OS2OS object-level content matching and retrieval strategy.	https://bit.ly/2Pv8OBM	
Transform-Aware Embeddings	2021	Implementation of graph construction based on transformation-aware embeddings.	https://bit.ly/3sgfrFo	

Provenance Analysis Source Code: Table 15.8 summarizes the currently available source code for image provenance analysis.

Acknowledgements Anderson Rocha thanks the financial support of the São Paulo Research Foundation (FAPESP, Grant #2017/12646-3).

References

Advance Publications, Inc (2012) Reddit PhotoshopBattles. https://bit.ly/3ty7pJr. Accessed on 19 Apr 2021

Amazon Mechanical Turk, Inc (2021) Amazon Mechanical Turk. https://www.mturk.com/. Accessed 11 Apr 2021

Baeza-Yates R, Ribeiro-Neto B (1999) Modern information retrieval, vol 463, Chap 8. ACM Press, New York, pp 191–227

Ballard D (1981) Generalizing the Hough transform to detect arbitrary shapes. Elsevier Pattern Recognit 13(2):111–122

Bay H, Ess A, Tuytelaars T, Van Gool L (2008) Speeded-up robust features (SURF). Elsevier Comput Vis Image Understand 110(3):346–359

Bestagini P, Tagliasacchi M, Tubaro S (2016) Image phylogeny tree reconstruction based on region selection. In: IEEE international conference on acoustics, speech and signal processing, pp 2059–2063

Bharati A, Moreira D, Flynn P, Rocha A, Bowyer K, Scheirer W (2021) Transformation-aware embeddings for image provenance. IEEE Trans Inf Forensics Secur 16:2493–2507

Bharati A, Moreira D, Brogan J, Hale P, Bowyer K, Flynn P, Rocha A, Scheirer W (2019) Beyond pixels: Image provenance analysis leveraging metadata. In: IEEE winter conference on applications of computer vision, pp 1692–1702

Bharati A, Moreira D, Pinto A, Brogan J, Bowyer K, Flynn P, Scheirer W, Rocha A (2017) U-phylogeny: Undirected provenance graph construction in the wild. In: IEEE international conference on image processing, pp 1517–1521

Brogan J (2019) Reddit Photoshop Battles Image Provenance. https://bit.ly/3jfrLSJ. Accessed on 16 Apr 2021

Brogan J, Bestagini P, Bharati A, Pinto A, Moreira D, Bowyer K, Flynn P, Rocha A, Scheirer W (2017) Spotting the difference: Context retrieval and analysis for improved forgery detection and localization. In: IEEE international conference on image processing, pp 4078–4082

Brogan J, Bharati A, Moreira D, Rocha A, Bowyer K, Flynn P, Scheirer W (2021) Fast local spatial verification for feature-agnostic large-scale image retrieval. IEEE Trans Image Process 30:6892–6905

ByteDance Ltd (2021) About TikTok. https://www.tiktok.com/about?lang=en. Accessed 11 Apr 2021

Castelletto R, Milani S, Bestagini P (2020) Phylogenetic minimum spanning tree reconstruction using autoencoders. In: IEEE international conference on acoustics, speech and signal processing, pp 2817–2821

Chu Y-J (1965) On the shortest arborescence of a directed graph. Sci Sini 14:1396–1400

Costa F, Oikawa M, Dias Z, Goldenstein S, Rocha A (2014) Image phylogeny forests reconstruction. IEEE Trans Inf Forensics Secur 9(10):1533–1546

Costa F, De Oliveira A, Ferrara P, Dias Z, Goldenstein S, Rocha A (2017) New dissimilarity measures for image phylogeny reconstruction. Springer Pattern Anal Appl 20(4):1289–1305

Costa F, Lameri S, Bestagini P, Dias Z, Rocha A, Tagliasacchi M, Tubaro S (2015) Phylogeny reconstruction for misaligned and compressed video sequences. In: IEEE international conference on image processing, pp 301–305

Costa F, Lameri S, Bestagini P, Dias Z, Tubaro S, Rocha A (2016) Hash-based frame selection for video phylogeny. In: IEEE international workshop on information forensics and security, pp 1–6

De Oliveira A, Ferrara P, De Rosa A, Piva A, Barni M, Goldenstein S, Dias Z, Rocha A (2015) Multiple parenting phylogeny relationships in digital images. IEEE Trans Inf Forensics Secur 11(2):328–343

De Rosa A, Uccheddu F, Costanzo A, Piva A, Barni M (2010) Exploring image dependencies: a new challenge in image forensics. In: IS&T/SPIE electronic imaging, SPIE vol 7541, Media forensics and security II, pp 1–12

Dias Z, Rocha A, Goldenstein S (2012) Image phylogeny by minimal spanning trees. IEEE Trans Inf Forensics Secur 7(2):774–788

Dias Z, Goldenstein S, Rocha A (2013) Toward image phylogeny forests: automatically recovering semantically similar image relationships. Elsevier Forensic Sci Int 231(1–3):178–189

Dias Z, Goldenstein S, Rocha A (2013) Large-scale image phylogeny: tracing image ancestral relationships. IEEE Multimed 20(3):58–70

Dias Z, Goldenstein S, Rocha A (2013) Exploring heuristic and optimum branching algorithms for image phylogeny. Elsevier J Vis Commun Image Represent 24(7):1124–1134

Dias Z, Rocha A, Goldenstein S (2011) Video phylogeny: Recovering near-duplicate video relationships. In: IEEE international workshop on information forensics and security, pp 1–6

Edmonds J (1967) Optimum branchings. J Res Natl Bure Stand B 71(4):233–240

Facebook, Inc (2021) About Instagram. https://about.instagram.com/about-us. Accessed 11 Apr 2021

Ge T, He K, Ke Q, Sun J (2013) Optimized product quantization for approximate nearest neighbor search. In: IEEE conference on computer vision and pattern recognition, pp 2946–2953

Guan H, Kozak M, Robertson E, Lee Y, Yates AN, Delgado A, Zhou D, Kheyrkhah T, Smith J, Fiscus J (2019) Mfc datasets: Large-scale benchmark datasets for media forensic challenge evaluation. In: 2019 IEEE winter applications of computer vision workshops (WACVW). IEEE, pp 63–72

Idée, Inc. TinEye (2020) Reverse image search. https://tineye.com/. Accessed on 17 Jan 2021

Johnson J, Douze M, Jégou H (2019) Billion-scale similarity search with gpus. IEEE Trans Big Data 1–12

Kennedy L, Chang S-F (2008) Internet image archaeology: Automatically tracing the manipulation history of photographs on the web. In: ACM international conference on multimedia, pp 349–358

Kruskal J (1956) On the shortest spanning subtree of a graph and the traveling salesman problem. Proc Am Math Soc 7(1):48–50

Lameri S, Bestagini P, Melloni A, Milani S, Rocha A, Tagliasacchi M, Tubaro S (2014) Who is my parent? reconstructing video sequences from partially matching shots. In: IEEE international conference on image processing, pp 5342–5346

Liu Y, Zhang D, Guojun L, Ma W-Y (2007) A survey of content-based image retrieval with high-level semantics. Elsevier Pattern Recognit 40(1):262–282

Lowe D (2004) Distinctive image features from scale-invariant keypoints. Springer Int J Comput Vis 60(2):91–110

Melloni A, Bestagini P, Milani S, Tagliasacchi M, Rocha A, Tubaro S (2014) Image phylogeny through dissimilarity metrics fusion. In: IEEE European workshop on visual information processing, pp 1–6

Milani S, Bestagini P, Tubaro S (2016) Phylogenetic analysis of near-duplicate and semantically-similar images using viewpoint localization. In *IEEE international workshop on information forensics and security*, pp 1–6

Milani S, Bestagini P, Tubaro S (2017) Video phylogeny tree reconstruction using aging measures. In: IEEE European signal processing conference, pp 2181–2185

Moreira D, Bharati A, Brogan J, Pinto A, Parowski M, Bowyer K, Flynn P, Rocha A, Scheirer W (2018) Image provenance analysis at scale. IEEE Trans Image Process 27(12):6109–6123

National Institute of Standards and Technology (2017) Nimble Challenge 2017 Evaluation. https://bit.ly/3e8GeOP. Accessed on 16 Apr 2021

National Institute of Standards and Technology (2018) Nimble Challenge 2018 Evaluation. https://bit.ly/3mY2sHA. Accessed on 16 Apr 2021

National Institute of Standards and Technology (2019) Media Forensics Challenge 2019. https://bit.ly/3susnrw. Accessed on 19 Apr 2021

Noh H, Araujo A, Sim J, Weyand T, Han B (2017) Large-scale image retrieval with attentive deep local features. In: IEEE international conference on computer vision, pp 3456–3465

Oikawa M, Dias Z, Rocha A, Goldenstein S (2015) Manifold learning and spectral clustering for image phylogeny forests. IEEE Trans Inf Forensics Secur 11(1):5–18

Oikawa M, Dias Z, Rocha A, Goldenstein S (2016) Distances in multimedia phylogeny. Int Trans Oper Res 23(5):921–946

Pinto A, Moreira D, Bharati A, Brogan J, Bowyer K, Flynn P, Scheirer W, Rocha A (2017) Provenance filtering for multimedia phylogeny. In: IEEE international conference on image processing, pp 1502–1506

Prim R (1957) Shortest connection networks and some generalizations. Bell Syst Tech J 36(6):1389–1401

PubPeer Foundation (2020) Traditional Chinese medicine for COVID-19 treatment. https://bit.ly/2Su3g8U. Accessed on 11 Apr 2021

Rudin C, Schapire R (2009) Margin-based ranking and an equivalence between adaboost and rankboost. J Mach Learn Res 10(10):2193–2232

Scheirer W (2020) A pandemic of bad science. Bull Atom Sci 76(4):175–184

Shifman L (2014) Memes in digital culture. MIT Press

Stella Y, Shi J (2003) Multiclass spectral clustering. In: IEEE international conference on computer vision, pp 313–319

Theisen W, Brogan J, Thomas PB, Moreira D, Phoa P, Weninger T, Scheirer W (2021) Automatic discovery of political meme genres with diverse appearances. In: AAAI International Conference on Web and Social Media, pp 714–726

Tian Y, Fan B, Wu F (2017) L2-net: Deep learning of discriminative patch descriptor in Euclidean space. In: IEEE conference on computer vision and pattern recognition, pp 661–669

Turek M (2020) Media Forensics (MediFor). https://www.darpa.mil/program/media-forensics. Accessed on 24 Feb 2021

Twitter, Inc (2021) About Twitter. https://about.twitter.com/. Accessed 11 Apr 2021

University of Notre Dame, Computer Vision Research Laboratory (2018) MediFor (Media Forensics). https://bit.ly/3txaZU7. Accessed on 16 Apr 2021

Verde S, Milani S, Bestagini P, Tubaro S (2017) Audio phylogenetic analysis using geometric transforms. In: IEEE workshop on information forensics and security, pp 1–6

Weninger T, Bisk Y, Han J (2012) Document-topic hierarchies from document graphs. In: ACM international conference on information and knowledge management, pp 635–644

Yi KM, Trulls E, Lepetit V, Fua P (2016) LIFT: learned invariant feature transform. In: Springer European conference on computer vision, pp 467–483

Zhang X, Sun ZH, Karaman S, Chang S-F (2020) Discovering image manipulation history by pairwise relation and forensics tools. IEEE J Sel Top Signal Process 1012–1023

Zhu N, Shen J (2019) Image phylogeny tree construction based on local inheritance relationship correction. Springer Multimed Tools Appl 78(5):6119–6138

Part IV
Counter-Forensics

Chapter 16
Adversarial Examples in Image Forensics

Mauro Barni, Wenjie Li, Benedetta Tondi, and Bowen Zhang

Abstract We describe the threats posed by adversarial examples in an image forensic context, highlighting the differences and similarities with respect to other application domains. Particular attention is paid to study the transferability of adversarial examples from a source to a target network and to the creation of attacks suitable to be applied in the physical domain. We also describe some possible countermeasures against adversarial examples and discuss their effectiveness. All the concepts described in the chapter are exemplified with results obtained in some selected image forensics scenarios.

16.1 Introduction

Deep neural networks, most noticeably Convolutional Neural Networks (CNNs), are increasingly used in image forensic applications due to their superior accuracy in detecting a wide number of image manipulations, including double and multiple JPEG compression (Wang and Zhang 2016; Barni et al. 2017), median filtering (Chen et al. 2015), resizing (Bayar and Stamm 2018), contrast manipulation (Barni et al. 2018), splicing (Ahmed and Gulliver 2020). Good performance of CNNs have also been reported for image source attribution, i.e., to identify the model of the camera which acquired a certain image (Bondi et al. 2017; Freire-Obregon et al. 2018; Bayar and Stamm 2018), and deepfake detection (Rossler et al. 2019). Despite the good performance they achieve, the use of CNNs in security-oriented applications, like image forensics, is put at risk by the easiness with which adversarial examples can be built (Szegedy et al. 2014; Carlini and Wagner 2017; Papernot et al. 2016). As a

M. Barni (✉) · B. Tondi
Department of Information Engineering and Mathematics, University of Siena, Siena, Italy
e-mail: barni@dii.unisi.it

W. Li
Institute of Information Science, Beijing Jiaotong University, Beijing, China

B. Zhang
School of Cyber Engineering, Xidian University, Xi'an, China

© The Author(s) 2022
H. T. Sencar et al. (eds.), *Multimedia Forensics*, Advances in Computer Vision and Pattern Recognition, https://doi.org/10.1007/978-981-16-7621-5_16

matter of fact, an attacker who has access to the internal details of the CNN used for a certain image recognition task can easily build an attacked image which is visually indistinguishable from the original one but is misclassified by the CNN. Such a problem is currently the subject of an intense research activity, yet no satisfactory solution has been found yet (see Akhtar and Ajmal (2018) for a survey on this topic). The problem is worsened by the observation that adversarial attacks are often transferable from the target network to other networks designed for the same task (Papernot et al. 2016). This means that even in a Limited Knowledge (LK) scenario, wherein the attacker has only partial information about the to-be-attacked network, he can attack a surrogate network mimicking the target one and the attack will be effective also on the target network with good probability. Such a property opens the way toward very powerful attacks that can be used in real applications wherein the attacker does not have full access to the attacked system (Papernot et al. 2016).

While some recent works have shown that CNN-based image forensics tools are also susceptible to adversarial examples (Guera et al. 2017; Marra et al. 2018), some differences exist between the generation of adversarial examples targeting computer vision networks and networks trained to solve image forensic problems. For instance, in Barni et al. (2019), it is shown that adversarial examples aiming at deceiving image forensic networks are less transferable than those addressing computer vision tasks. A possible explanation for such a different behavior could be the different kinds of features image forensic networks rely on with respect to those used in computer vision applications, however a definitive answer to this question is not known yet.

With the above ideas in mind, the goal of this chapter is to give a brief introduction to adversarial examples and illustrate their applicability to image forensics networks. Particular attention is given to the transferability of the examples, since it is mainly thanks to this property that adversarial examples can be exploited to develop practical attacks against image forensic detectors. We also describe some possible defenses against adversarial examples, even if, up to date, a definitive strategy to make the creation of adversarial examples impossible, or at least impractical, is not available.

16.2 Adversarial Examples in a Nutshell

In this section we briefly review the various approaches proposed so far to generate adversarial examples, without referring specifically to a multimedia forensics scenario. After a rigorous formalization of the problem, we review the main approaches proposed so far, then we consider the generation of adversarial examples in the physical domain. Eventually, we discuss the problems associated with the generation of adversarial attacks in a black-box setting. Throughout the rest of this chapter we will always refer to the case of deep neural networks aiming at classifying an input image into a number of possible classes, since this is by far the most relevant setting in a multimedia forensic context.

16.2.1 Problem Definition and Review of the Most Popular Attacks

Let $x \in [0, 1]^n$ denote a clean image[1] whose true label is y and let ϕ be a CNN-based image classifier providing a correct classification of x, namely, $\phi(x) = y$. An adversarial example is a perturbed image $x' = x + \delta$ for which $\phi(x') \neq y$. More specifically, an untargeted attack aims at generating an adversarial example for which $\phi(x') = y'$ for any $y' \neq y$, while for a targeted attack the wrong label y' must correspond to a specific target class y_t.

In practice, we require that δ is a very small quantity, so that the perturbed image x' is perceptually identical to x. Such a goal is usually achieved by constraining the L_p norm of δ to be lower than a small positive quantity, that is $\|\delta\|_p \leq \varepsilon$, where ε is a very small value.

Starting from the seminal work by Szegedy, many approaches to generate the adversarial perturbation δ have been proposed, leading to a wide variety of different algorithms an attacker can rely on. The proposed approaches include gradient-based attacks (Szegedy et al. 2014; Goodfellow et al. 2015; Kurakin et al. 2017; Madry et al. 2018; Carlini and Wagner 2017), transfer-based attacks (Dong et al. 2018, 2019), decision-based attacks (Chen et al. 2017; Ilyas et al. 2018; Li et al. 2019), and so on. Most methods assume that the attacker has a full knowledge of the to-be-attacked model ϕ a situation usually referred to as a white-box attack scenario. It goes without saying that this may not be the case in practical applications for which more flexible black-box or gray-box attacks are needed. We will discuss the main difficulties associated with black-box attacks in Sect. 16.2.2.

In the following we review some of the most popular gradient-based white-box attacks.

- **L-BFGS Attack**. In Szegedy et al. (2014), Szegedy et al. first proposed to generate adversarial examples aiming at fooling a CNN-based classifier and minimizing the perturbation of the image at the same time. The problem is formulated as

$$\begin{aligned} \text{minimize } & \|\delta\|_2^2 \\ \text{subject to } & \phi(x + \delta) = y_t \\ & x + \delta \in [0, 1]^n \end{aligned} \qquad (16.1)$$

Here, the perturbation introduced by the attack is limited by using the squared L_2 norm. Solving the above minimization is not an easy task, so the adversarial examples are looked for by solving more manageable problem:

$$\begin{aligned} \underset{\delta}{\text{minimize }} & \lambda \cdot \|\delta\|_2^2 + \mathcal{L}(x + \delta, y_t) \\ \text{subject to } & x + \delta \in [0, 1]^n \end{aligned} \qquad (16.2)$$

---

[1] Hereafter, we assume that image pixels assume values in [0, 1]. For gray-level images n is equal to the number of pixels, while for color images n accounts for all the color channels.

where $\mathcal{L}(x + \delta, y_t)$ is a loss term forcing the CNN to assign the target label y_t to $x + \delta$ and λ is a parameter balancing the importance of two terms. In Szegedy et al. (2014), the usual cross-entropy loss is adopted as $\mathcal{L}(x + \delta, y_t)$. The latter optimization problem is solved by the box-constraint L-BFGS method. Moreover, the optimal value of λ is determined by conducting a binary search within a pre-defined range of values.

- **Fast Gradient Sign Method (FGSM).** A drawback of the L-BFGS method is its computational complexity. For this reason, Goodfellow et al. (2015) proposed a faster attack. The new method starts from the observation that, for small values of the perturbation, the output of the CNN can be computed by using a linear approximation, and that, due to the large dimensionality of x, a small perturbation of the input in the direction of gradient can result in a large modification of the output. Accordingly, the adversarial examples are generated by adding a small perturbation to the clean image as follows:

$$x' = x + \varepsilon \cdot \text{sign}(\nabla_x \mathcal{L}(x, y)) \tag{16.3}$$

where $\nabla_x \mathcal{L}$ is the gradient of $\mathcal{L}$ with respect to x, and the adversarial perturbation is constrained by $\|\delta\|_\infty \leq \varepsilon$. Using the sign of the gradient, ensures that the perturbation is within the allowed limits. Since the adversarial examples are generated based on the gradient of the loss function with respect to the input image x, they can be efficiently calculated by using the back-propagation algorithm, thus resulting in a very efficient attack.

- **Iterative FGSM (I-FGSM).** The FGSM attack is a one-step attack whose strength can be modulated only by increasing the value of ε, with the risk that the linearity approximation the method relies on is no more valid. In Kurakin et al. (2017), the one-step FGSM is extended to an iterative, multi-step, method by applying FGSM multiple times with a small step size, each time by recomputing the gradient. The perturbed pixels are then clipped in an $\varepsilon-$neighborhood of the original input, to ensure that the distortion constraint is satisfied. The N-th step of the resulting method, referred to as I-FGSM (or sometimes BIM—Basic Iterative Method), is defined by

$$x_0' = x, \quad x_N' = x_{N-1}' + \text{clip}_\varepsilon \{\alpha \cdot \text{sign}(\nabla_x \mathcal{L}(x_{N-1}', y))\} \tag{16.4}$$

where α is a small step size used to update the attack at each iteration. Similarly to I-FGSM, a universal first-order adversary, called projected gradient descent (PGD), has been proposed in Madry et al. (2018). Compared with I-FGSM, in which the starting point of the iterations is exactly the input image, the starting point of PGD is obtained by randomly projecting the input image within an allowed perturbation ball, which can be done by adding small random perturbations to the input image. Then, the following iterations are conducted in the same way as I-FGSM.

- **Jacobian-based Saliency Map Attack (JSMA).** According to the JSMA algorithm proposed by Papernot et al. in (2016), the adversarial examples are generated on the basis of an adversarial saliency map, which indicates the contribution of the image pixels to the classification result. In particular, the saliency map is computed based on the forward propagation of the pixel values to the output of the to-be-attacked network. According to the saliency map, at each iteration, a pair of pixels whose modification leads to a significant increase of the output values assigned to the target class and/or a decrease of the values assigned to the other classes are perturbed by an amount θ. The iterations stop when the perturbed image is classified as belonging to the target class or a maximum distortion level is reached. To keep the perturbation imperceptible, each pixel can be modified up to a maximum of T times.

- **Carlini & Wagner Attack (C&W).** In Carlini and Wagner (2017), Carlini and Wagner proposed to solve the optimization problem in (16.2) in a more efficient way. Specifically, two modifications of the basic optimization algorithm are proposed. On one hand, the loss term is replaced by a properly designed function directly related to the maximum difference between the logit of the target class and those of the other classes. On the other hand, the box-constraint used to keep the perturbed pixels in a valid range is automatically satisfied by letting $\delta = \frac{1}{2}(\tanh(\omega) + 1) - x$, where ω is a new variable used for the transformation of δ. In this way, the box-constraint optimization problem in (16.2) is transformed into the following unconstrained problem:

$$\underset{\omega}{\text{minimize}} \left\| \tfrac{1}{2}(\tanh(\omega) + 1) - x \right\|_2^2 + \lambda \cdot g(\tfrac{1}{2}(\tanh(\omega) + 1)) \tag{16.5}$$

where the function g is defined as

$$g(x') = \max(\max_{i \neq t} z_i(x') - z_t(x'), -\kappa), \tag{16.6}$$

and where z_i is the logit corresponding to the i-th class, and the parameter $\kappa > 0$ is used to encourage the attack to generate adversarial examples classified with high confidence. Finally, the adversarial examples are generated by solving the modified optimization problem using the Adam optimizer.

16.2.2 Adversarial Examples in the Physical Domain

The algorithms described in the previous sections work in the digital domain, since they directly modify the digital images fed to the CNN. In many applications, however, it is required that the adversarial examples are generated in the physical domain, producing real-world objects that, when sensed by the sensors of the attacked system and fed to the CNN, cause a misclassification error. In this setting, and by focusing on the case of still images, which is the most relevant case for multimedia forensics applications, the images with the adversarial examples must first be printed or displayed on a screen and then shown to a camera, which will digitize them again before feeding them to the CNN classifier. This process, sometimes referred to as

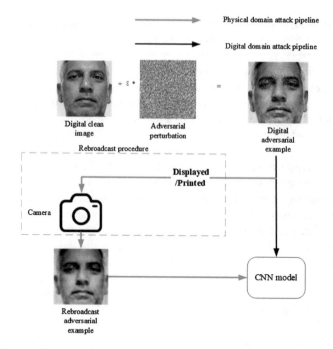

Fig. 16.1 Physical domain and digital domain attacks

image *rebroadcast*, involves a digital-to-analog and an analog-to-digital conversions that may degrade the adversarial perturbation thus making the attack ineffective. The rebroadcast process involved by a physical domain attack is exemplified in Fig. 16.1.

As shown in the figure, during the rebroadcast process, the image is affected by several forms of degradation including noise addition, light changes and geometric transformations. If no countermeasure is taken, it is very likely that the invisible perturbations introduce to generate the adversarial examples are damaged up to a point to make the attack ineffective.

Recognizing the above difficulties, several methods have been proposed to generate physical domain adversarial examples. Kurakin et al. first brought up this concept in Kurakin et al. (2017). In this paper, the new images taken after rebroadcasting are geometrically calibrated before being fed to the classifier. In this way, the adversarial perturbation is not affected by any geometric distortion, hence easing the attack. More realistic scenarios are considered in later works. In the seminal paper by Athalye et al. (2018) a method is proposed to generate physical adversarial examples that are robust to various distortions, including changes of the viewing angle and distance. Specifically, a procedure, named Expectation Over Transformation (EOT), is applied to generate the adversarial examples by ensuring that they are effective across a number of possible transformations.

Let x be the to-be-attacked image, with ground truth label y, and let y_t be target class of the attack. Without loss of generality, let the output of the classifier be obtained by computing the probability $P(y_i|x)$ that the input image x belongs to class y_i and then choosing for the class resulting in the maximum probability,[2] namely,

$$\phi(x) = \arg\max_i P(y_i|x). \tag{16.7}$$

The generation of adversarial examples is formalized as an optimization problem that maximizes the likelihood of the target class y_t over an ϵ-radius ball around the original image x, i.e.,

$$\arg\max_\delta \quad \log P(y_t|x + \delta) \tag{16.8}$$

$$\text{subject to} \quad ||\delta||_p \leq \epsilon. \tag{16.9}$$

To ensure robustness of the attack, EOT considers a set T of transformations (usually consisting of geometric transformations) and a probability distribution P_T over T, and maximizes the likelihood of the target class averaged over a transformed version of the input images, generated according to P_T. A similar strategy is applied to the constraint, i.e., EOT constrains the expected distance between the transformed versions of the adversarial example and the original image. As a result, with EOT the optimization problem is reformulated as

$$\arg\max_\delta \quad E_{P_T}[\log P(y_t|t(x + \delta))] \tag{16.10}$$

$$\text{subject to} \quad E_{P_T}[||t(x + \delta) - t(x)||_p] \tag{16.11}$$

where t is a transformation in T and E_{P_T} denotes expectation over T. The basic EOT approach has been extended in many ways, including a wider variety of transformations and by optimizing the perturbation over more than a single image. For instance, Sharif et al. (2016) have proposed a way to generate adversarial examples in the physical domain for face recognition systems. They restrict the perturbation within a small area around the eye-glass frame and increase the robustness of the perturbations by optimizing them over a set of properly chosen images. Eykholt et al. (2018) have proposed an attack capable of generating adversarial examples under even more realistic conditions. The attack targets a road sign classification system possibly deployed on board of autonomous vehicles. In addition, to sample the images from synthetic transformations as done by the basic EOT attack, they also generated experimental data containing actual physical condition variability. Their experiments show that the generated adversarial examples are so effective that can be printed on stickers and applied to road signs, and fool a vehicular traffic sign recognition system in driving-by tests.

Zhang et al. (2020) further enlarged the set of transformations to implement a physical domain attack against a face authentication system equipped with a spoofing-

---

[2] This is the common case of CNN classifiers adopting one-hot-encoding to represent the result of the classification.

detection module, whose aim is to distinguish real faces and rebroadcast face photos. This situation presents an additional challenge, as explained in the following. To attack the face authentication system in the physical world, the attacker must present the perturbed images in front of the system by printing them or showing them on a screen. The rebroadcasting procedure, then, introduces new physical traces into the image fed to the classifier, which, having been trained to recognize rebroadcast traces, can recognize them and classify the images correctly despite the presence of the adversarial perturbation. To exit this apparent deadlock, the authors propose to utilize the EOT framework to design an attack that pre-emptively takes into account the new traces introduced by the rebroadcast procedure. The experiments shown in Zhang et al. (2020) demonstrate that EOT effectively allows to carry out an adversarial attack even in such difficult conditions.

16.2.3 White Versus Black-Box Attacks

An adversary may have different levels of knowledge about the algorithm used by the forensic analyst. From this point of view, adversarial examples in DL are commonly classified as *white-box* and *black-box attacks* (Yuan et al. 2019). In a white-box scenario, the attacker knows everything about the target forensic classifier, that is, he knows all the details of $\phi(\cdot)$. In a black-box scenario, instead, the attacker knows nothing about it, or, has only a very limited knowledge. In some works, following the taxonomy in Zheng and Hong (2018), the class of gray-box attacks is also considered, referring to a situation wherein the attacker has a partial knowledge of the classifier, for instance, he knows the architecture of the classifier, but has no knowledge of its internal parameters (e.g., he does not know some of the hyper-parameters or, more in general, he does not know the training data $\mathcal{D}_{tr}$ used to train the classifier).

In the white-box scenario, the attacker is also assumed to know the defence strategy adopted by the forensic analyst $\phi(\cdot)$, hence he can take into account the presence of such a defence during the attack. Having full access to $\phi(\cdot)$, the most common white-box attacks rely on some form of gradient-descent (Goodfellow et al. 2015; Kurakin et al. 2017; Dong et al. 2018), or directly solve the optimization problem underlying the attack as in Carlini and Wagner (2017); Chen et al. (2018).

When the attacker has a partial knowledge of the algorithm (gray-box setting), he can build its own version of the classifier $\phi'(\cdot)$, usually referred to as substitute or surrogate classifier, providing a hopefully good approximation of $\phi(\cdot)$, and then use $\phi'(\cdot)$ to carry out the attack. Hence, the attacker implements a white-box attack against $\phi'(\cdot)$, hoping that the attack will also work against $\phi(\cdot)$ (attack transferability, see Sect. (16.3.1)). Mathematically speaking, the difference between white-box and black-box attacks is the loss function that the adversary seeks to maximize, namely $\mathcal{L}(\phi(\cdot), \cdot)$ in the white-box case and $\mathcal{L}(\phi'(\cdot), \cdot)$ in the black-box case. As long as an attack thought to work against $\phi(\cdot)$ can be transferred to $\phi'(\cdot)$, adversarial samples crafted for $\phi'(\cdot)$ will also be misclassified by $\phi(\cdot)$ (Papernot et al. 2016). The substitute classifier $\phi'(\cdot)$ is built by exploiting the available information and making an educated guess about the unknown parameters.

Black-box attacks often assume that the target network can be queried as an oracle, for a limited number of times, to get useful information to build the substitute classifier (Liu et al. 2016). When the score values or the logit values z_i can be obtained by querying the target model (score-based black-box attacks), the gradient can be estimated by drawing some random samples and acquiring the corresponding loss values (Ilyas et al. 2018). A more challenging scenario is when only hard-label predictions, i.e., the predicted classes, are returned by the queried model (Brendel et al. 2017).

In the DL literature, several approaches have been proposed to craft adversarial examples against substitute models, in such a way to maximize the probability that they can be transferred to the original model. One approach is the translation-invariant (Dong et al. 2019) attack method, that, instead of optimizing the objective function directly on the perturbed image, optimizes it over a set of shifted versions of the image. Another closely related approach improves the transferability of adversarial examples by creating diverse input patterns (Xie et al. 2019). Instead of only using the original images to generate adversarial examples, the method applies image some transformations to the inputs with a given probability at each iteration of the gradient-based attack. The transformations include resizing and padding.

A general and simple method to increase the transferability of adversarial examples is described in Sect. 16.3.2.

16.3 Adversarial Examples in Multimedia Forensics

Given the recent trend toward the use of deep learning architectures in Multimedia Forensics, several Counter-Forensic (CF) attacks against deep learning models have been developed in the last years. In this case, one of the main advantages of CNN-based classifiers and detectors, namely, their ability to learn forensic features directly from the input data (be their images, videos, or audio tracks), turns into a weakness. An informed attacker can in fact generate his attacks directly in the sample domain without the need to map them from the feature domain to the sample domain, as it happens with conventional methods based on hand-crafted features.

The vulnerability of multimedia forensics tools based on deep learning has been addressed by many works as summarized in Marra et al. (2018). The underlying motivations for the vulnerability of CNNs used in multimedia forensics are basically the same characterizing image recognition applications, with a prominent role played by the huge dimension of the space of possible inputs, which is substantially larger than the set of images used to train the model. As a result, in a white-box scenario, the attacker can easily create slightly modified images that fall into an 'unseen' region of the image space thus forcing a misclassification error. Such a weakness of deep learning-based approaches represents a serious threat in multimedia forensics where the tools are designed to work under intrinsically adversarial conditions.

Adversarial examples that have been developed against CNNs in multimedia forensics are reviewed in the following.

The first targeted attack based on adversarial examples was proposed in Guera et al. (2017) to fool a CNN-based camera model identification system. By relying on adversarial examples, the authors propose a counter-forensic method for slightly altering an image in such a way to change the estimated camera model. A state-of-the-art CNN-based camera model detector based on the DenseNet architecture (Huang et al. 2017) is considered in the analysis. The counter-forensic method uses both the FGSM attack and the JSMA attack to craft adversarial images obtaining good performance in both cases.

Adversarial attacks against CNN-based image forensics methods have been derived in Gragnaniello et al. (2018) for several common manipulation detection tasks. The following common processing has been considered: image blurring, applied with different variance of the Gaussian filter; image resizing, both downscaling and upscaling; JPEG compression, with different qualities; and median filtering detection, with window sizes 3×3 and 7×7. Several CNN models have been successfully attacked by means of adversarial perturbations obtained via the FGSM algorithm applied with different strengths ε.

Adversarial perturbations have been shown to offer poor robustness to image processing operations, e.g., image compression (Marra et al. 2018). In particular, rounding to integers is sometimes sufficient to wash out the perturbation and make an adversarial example ineffective. A gradient-inspired pixel domain attack to generate adversarial examples against CNN-based forensic detectors in the integer domain has been proposed in Tondi (2018). In contrast to standard gradient-based attacks, which perturb the attacked image based on the gradient of the loss function w.r.t. the input, in the attack proposed in Tondi (2018), the gradient of the output score function is approximated with respect to the pixel values, incorporating the constraint that the input is an integer in the optimization. The performance of the integer-based attack has been assessed against three image manipulation detectors, for median filtering, resizing, and contrast adjustment. Moreover, while common gradient-based adversarial attacks, like FGSM, I-FGSM, C&W, etc. ..., work in a white-box setting, the attack proposed in Tondi (2018) is a black-box one, since it does not need any knowledge of the internal parameters of the model, requiring only that the network can be queried as an oracle and the output observed.

All the above attacks are targeted to a specific CNN classifier. Carrying out an effective attack in a completely black-box or no-knowledge scenario, that is, generating an attack that can be effectively transferred to the (unknown) model used by the analyst, turns out to be a difficult task, that is exacerbated by the complexity of the decision functions learnt by neural networks. For this reason, and given that white-box attacks can be hardly applied in practical applications, studying the transferability of adversarial examples plays a major role to assess the security of DL-based multimedia forensics tools.

16.3.1 Transferability of Adversarial Examples in Multimedia Forensics

It has been demonstrated in several studies concerning computer vision applications (Tramèr et al. 2017; Papernot et al. 2016) that, in a gray-box or black-box scenario, adversarial examples crafted to attack a given CNN model are also effective against other models designed for the same task. In this section, we focus on the transferability of adversarial examples in an image forensics context and review the main results reported so far regarding the transferability of adversarial examples in image forensics.

Several works have investigated the transferability of adversarial examples in a network mismatch scenario (Marra et al. 2018; Gragnaniello et al. 2018). Focusing on camera model identification, in Marra et al. (2018), Marra et al. analyzed the transferability of adversarial examples generated by means of FGSM among different networks. Four different network architectures for camera model identification trained on the VISION dataset (Shullani et al. 2017) are utilized for experiments, i.e., Shallow CNN (Bondi et al. 2017), DenseNet-40 (Huang et al. 2017), DenseNet-121 (Huang et al. 2017), and XceptionNet (Chollet 2017). The classification accuracy achieved by these models is 80.77% for Shallow CNN, 87.96% for DenseNet-40, 93.88% for DenseNet-121, and 95.15% for XceptionNet. The transferability performance for the case of cross-network mismatch is shown in Table 16.1. As it can be seen, the classification accuracy on attacked images remains pretty high when the network used to build the adversarial examples is not the same targeted by the attack, thus proving a poor transferability of the attacks.

A more comprehensive investigation of the transferability of adversarial examples in different mismatch scenarios has been carried out in Barni et al. (2019). As in Barni et al. (2019), in the following we report the transferability of adversarial examples in various settings, according to the kind of attack used to create the examples and the mismatch existing between the source and the target models.

Table 16.1 Classification accuracy (%) for cross-network mismatch in the context of camera model identification. The FGSM attack is used

		Source network			
		Shallow CNN	DenseNet-40	DenseNet-121	XceptionNet
TN	Shallow CNN	2.19	40.87	42.13	42.51
	DenseNet-40	37.23	3.88	37.13	53.42
	DenseNet-121	62.81	60.60	4.67	44.35
	XceptionNet	68.19	79.80	57.21	2.63

Type of attacks

We consider the transferability of attacks generated according to two of the most popular general attacks proposed so far, namely I-FGSM and JSMA (see Sect. 16.2.1 for a detailed description of such attacks). The attacks were implemented by relying on the Foolbox package (Rauber et al. 2017).

Source/target model mismatch

To carry out a comprehensive evaluation of the transferability of adversarial examples, three different mismatch situations between the network used to create the adversarial examples (hereafter referred to as source network—SN) and the network the adversarial examples should be transferred to (hereafter named target network—TN) are considered. Specifically, we consider the following situations: (i) cross-network mismatch, wherein different network architectures are trained on the same dataset; (ii) cross-training-set mismatch, according to which the same network architecture is trained on different datasets; and (iii) cross-network-and-training-set mismatch, wherein different architectures are trained on different datasets.

Network architectures

For cross-network mismatch, we consider BSnet (Bayar and Stamm 2016), designed for the detection of a number of widely used image manipulations, and BC+net (Barni et al. 2018) proposed to detect generic contrast adjustment operations. In particular, BSnet consists of 3 convolutional layers, 3 max-pooling layers, and 3 fully-connected layers, and the filters used in the first convolutional layer are constrained to extract residual-based features from images. As for BC+net, it has 9 convolutional layers, and no constraint is applied to the filters used in the network. The results reported in the following refer to two common image manipulations, namely, median filtering (with window size 5×5) and image resizing (by a factor of 0.8).

Datasets

For cross-training-set mismatch, we consider two datasets: the RAISE (R) dataset (Dang-Nguyen et al. 2021) and the VISION (V) dataset (Shullani et al. 2017). More in details, RAISE consists of 8156 high-resolution uncompressed images with size 4288×2848 in RAISE, 2000 of which were used for the experiments. With regard to the VISION dataset, it contains 11,732 JPEG images, with 2000 images used for experiments, with a minimum resolution equal to 2336×4160 and a maximum one of 3480×4640. Results refer to JPEG images compressed with a quality factor larger than 97. The images were split into training, validation, and test sets without overlap. All color images were first converted to gray-scale.

Table 16.2 Detection accuracy (%) of the models in the absence of attacks

Model	Median filtering			Image resizing		
	ϕ_{BSnet}^{R}	ϕ_{BSnet}^{V}	ϕ_{BC+net}^{R}	ϕ_{BSnet}^{R}	ϕ_{BSnet}^{V}	ϕ_{BC+net}^{R}
Accuracy	98.1	99.5	98.4	97.5	96.6	98.5

Experimental methodology

Before evaluating the transferability of adversarial examples, the detection models were first trained. The input patch size was set to 128×128. To train the BSnet models, 200,000 patches per class were considered while 10,000 patches were used for testing. As to BC+net models, 10^{6} patches were used for training, 10^{5} for validation, and 5×10^{4} for testing. The Adam optimizer was used with a learning rate of 10^{-4} for both networks, and the training batch size was set to 32 patches. BSnet was trained for 30 epochs and BC+net for 3 epochs. The detection accuracies of the models are given in Table 16.2. For sake of clarity, the symbol ϕ_{net}^{DB} is used to denote the trained models, where net $\in$ {BSnet, BC+net} corresponds to the network architecture, and DB $\in$ {R, V} is the dataset used for training.

In counter-forensic applications, it is reasonable to assume that the attacker is interested only in passing off the manipulated images as original ones to hide the traces of manipulations. Therefore, for all the experiments, 500 manipulated images were attacked to generate the adversarial examples.

For I-FGSM, the number of iterations was set to 10, and the optimal attack strength used in each iteration was determined by spanning the range $[0 : \varepsilon_s : 0.1]$, where ε_s was set to 0.01 and 0.001 in different cases. As for JSMA, the perturbation strength θ was set to 0.01 and 0.1. For each attack, two different attack strengths were considered. The average PSNR of the adversarial examples is always larger than 40 dB.

Transferability results

For each setting, we show the average PSNR of the adversarial examples, and the attack success rate on the target network, ASR_{TN}, which corresponds to the transferability degree (the attack success rate on the source network—ASR_{SN}—is always close to 100%, hence it is not reported in the tables).

Table 16.3 reports the results regarding cross-network transferability. All models were trained on the RAISE dataset. In most of the cases, the adversarial examples are not transferable, indicating a likely failure of the attacks in a black-box scenario. The only exception is median filtering when the attack is carried out by I-FGSM, for which we have $ASR_{TN} = 82.5\%$ with an average PSNR of 40 dB.

With regard to cross-training-set transferability, the BSnet network was trained on the RAISE and the VISION datasets, obtaining the results shown in Table 16.4. For the cases of median filtering, the values of ASR_{TN} are relatively high for I-FGSM with $\varepsilon_s = 0.01$, while the transferability is poor with the other attacks. With regard

Table 16.3 Attack transferability for cross-network mismatch

$SN = \phi^R_{BSnet}$, $TN = \phi^R_{BC+net}$

Attack	Parameter	Median filtering		Image resizing	
		ASR$_{TN}$ %	PSNR (dB)	ASR$_{TN}$ %	PSNR (dB)
I-FGSM	$\varepsilon_s = 0.01$	82.5	40.0	0.2	40.0
	$\varepsilon_s = 0.001$	18.1	59.7	0.2	58.5
JSMA	$\theta = 0.1$	1.0	49.6	1.6	46.1
	$\theta = 0.01$	1.6	58.5	0.6	55.0

Table 16.4 Transferability of adversarial examples for cross-training-set mismatch

$SN = \phi^R_{BSnet}$, $TN = \phi^V_{BSnet}$

Attack	Parameter	Median filtering		Image resizing	
		ASR$_{TN}$ %	PSNR (dB)	ASR$_{TN}$ %	PSNR (dB)
I-FGSM	$\varepsilon_s = 0.01$	84.5	40.0	69.2	40.0
	$\varepsilon_s = 0.001$	4.5	59.7	4.9	58.5
JSMA	$\theta = 0.1$	1.2	49.6	78.2	46.0
	$\theta = 0.01$	0.2	58.5	11.5	55.0

$SN = \phi^V_{BSnet}$, $TN = \phi^R_{BSnet}$

Attack	Parameter	Median filtering		Image resizing	
		ASR$_{TN}$ %	PSNR (dB)	ASR$_{TN}$ %	PSNR (dB)
I-FGSM	$\varepsilon_s = 0.01$	94.1	40.0	0.2	40.0
	$\varepsilon_s = 0.001$	7.7	59.9	0.0	59.6
JSMA	$\theta = 0.1$	1.0	49.6	0.0	50.6
	$\theta = 0.01$	0.8	58.1	0.0	57.8

to image resizing, when the source network is trained on RAISE, the ASR$_{TN}$ reaches a high level when the stronger attacks are employed (IFGSM with $\varepsilon_s = 0.01$, and JSMA with $\theta = 0.1$). However, the adversarial examples are never transferable when the SN is trained on VISION, showing that the transferability between models trained on different datasets is not symmetric.

The results for the case of strongest mismatch (cross-network-and-training-set) are illustrated in Table 16.5. Only the case of median filtering attacked by I-FGSM with $\varepsilon_s = 0.01$ achieves a large transferability, while for all the other cases the transferability is nearly null.

In summary, in contrast to computer vision applications, adversarial examples targeting image forensics networks are generally non-transferable. Accordingly, from the attacker's side, it is important to investigate if the attack transferability can be improved by increasing the attack strength used to generate adversarial examples lying deeper inside the target region of the attack.

Table 16.5 Attacks transferability for cross-network-and-training-set mismatch

$SN = \phi_{BSnet}^{V}$, $TN = \phi_{BC+net}^{R}$

Attack	Parameter	Median filtering		Image resizing	
		ASR_{TN} %	PSNR (dB)	ASR_{TN} %	PSNR (dB)
I-FGSM	$\varepsilon_s = 0.01$	79.6	40.0	0.4	40.0
	$\varepsilon_s = 0.001$	0.8	59.9	0.2	59.6
JSMA	$\theta = 0.1$	0.8	49.6	0.0	50.2
	$\theta = 0.01$	1.2	58.1	0.0	57.4

16.3.2 Increased-Confidence Adversarial Examples with Improved Transferability

The lack of transferability of adversarial examples in image forensic applications is partly due to the fact that, in most implementations, the attacks aim at generating adversarial examples with minimum distortion. Consequently, the resulting adversarial examples are close to the decision boundary and a small difference between the decision boundaries of the source and target networks can lead to the failure of the attack on the TN. To overcome this problem, the attacker may want to generate adversarial examples lying deeper into the target region of the attack. However, given the complexity of the decision boundary learned by CNNs, it is hard to control the distance of the adversarial examples from the boundary. In addition, increasing the attack strength by simply going on with the attack iterations until a limit value of PSNR is reached may not be effective, since a lower PSNR does not necessarily result in a stronger attack with higher transferability (see Table 16.3).

In order to generate adversarial examples with higher transferability that can be used for counter-forensics applications, a general strategy consists in increasing the confidence of the misclassification, where the confidence is defined as the maximum difference between the logit of the target class and those of the classes (Li et al. 2020). Specifically, by focusing on the binary case, for a clean image x with label $y = i$ ($i = 0, 1$), a perturbed image x' is looked for which $z_{1-i} - z_i > c$, where $c > 0$ is the desired confidence value. By implementing the above stop condition, the most popular attacks (such as I-FGSM and PGD) can be modified to generate adversarial examples with increased confidence.

In the following, we evaluate the transferability of increased-confidence adversarial examples by adopting a methodology similar to that used in Sect. 16.3.1.

Attacks

We report the results obtained by applying the increased-confidence strategy to four popular attacks, namely, I-FGSM, Momentum-based I-FGSM (MI-FGSM) (Dong et al. 2018), PGD, and C&W attack. In particular, the MI-FGSM is a method proposed explicitly to improve the transferability of the attacks by mitigating the momentum term in each iteration of the I-FGSM with a decay factor μ.

Networks

Three network architectures are considered in the cross-network experiments, that is: BSnet (Bayar and Stamm 2016), BC+net (Barni et al. 2018), and VGG-16 network (VGGnet) commonly used in computer vision applications (VGGnet consists of 13 convolutional layers and 3 fully connected layers, more information can be found in Simonyan and Zisserman (2015)). With regard to the image manipulation tasks, we report results referring to median filtering, image resizing, and addition of white Gaussian noise (AWGN) with zero mean and unitary standard deviation.

Datasets

The experiments on cross-training-set mismatch are based on the RAISE and VISION datasets.

Experimental setup

BSnet and BC+net were trained by using the same setting described in Sect. 16.3.1. With regard to VGGnet, 10^5 patches were used for training and validation, and 10^4 patches were used for testing. The models were trained for 50 epochs by using the Adam optimizer with a learning rate of 10^{-5}. The range of detection accuracies in the absence of attacks are [98.1, 99.5%] for median filtering detection, [96.6, 99.0%] for image resizing, and [98.3, 99.9%] for AWGN detection.

The attacks were applied to 500 manipulated images for each task. The Foolbox toolbox (Rauber et al. 2017) was employed to generate increased-confidence adversarial examples with the new stop condition. Specifically, for C&W attack, all the parameters were set to their default values. For PGD, the binary search was conducted with initial values of $\varepsilon = 0.3$ and $\alpha = 0.005$, for 100 iterations. For I-FGSM, we also applied 100 steps, with the optimal stepsize determined in the range [0.001 : 0.001 : 0.1]. Eventually, for MI-FGSM, all the parameters were set to the same values of I-FGSM except the decay factor, for which we let $\mu = 0.2$.

Transferability results

In the following we show some results demonstrating the improved transferability of adversarial examples with increased confidence. For each attack, we report the ASR_{TN} and the average PSNR. The ASR_{SN} is always close to 100% and hence it is omitted in the tables. For the confidence value c that depends on the logits of the SN, different values were chosen for different networks.

Table 16.6 reports the results for cross-network mismatch, showing the transferability when (SN,TN) = $(\phi_{\text{VGGnet}}^{\text{R}}, \phi_{\text{BSnet}}^{\text{R}})$ for three different image manipulations. As expected, by using an increased confidence, the transferability of the adversarial examples improves significantly in all these cases. Specifically, the ASR_{TN} reaches a

Table 16.6 Attacks transferability for cross-network mismatch (SN = ϕ_{VGGnet}^{R}, TN = ϕ_{BSnet}^{R}). The ASR_{TN} (%) and average PSNR (dB) of adversarial examples is reported

Median filtering

c	I-FGSM		MI-FGSM		PGD		C&W	
	ASR_{TN}	PSNR	ASR_{TN}	PSNR	ASR_{TN}	PSNR	ASR_{TN}	PSNR
0	27.2	58.1	27.2	58.1	5.4	67.5	8.4	69.1
12	71.0	47.9	71.6	47.8	55.0	52.0	50.6	54.6
12.5	88.2	45.3	88.6	45.3	79.0	48.9	70.1	51.5
13	96.4	42.5	96.6	42.7	94.6	45.4	91.2	48.1

Image resizing

c	I-FGSM		MI-FGSM		PGD		C&W	
	ASR_{TN}	PSNR	ASR_{TN}	PSNR	ASR_{TN}	PSNR	ASR_{TN}	PSNR
0	0.4	59.3	0.4	59.2	1.8	75.4	1.2	71.5
17	25.0	33.3	24.8	33.3	22.0	33.4	40.4	36.8
18	39.6	30.9	37.4	30.9	39.8	30.9	53.6	34.3
19	51.6	28.9	52.6	28.9	52.0	28.9	64.4	32.2

AWGN

c	I-FGSM		MI-FGSM		PGD		C&W	
	ASR_{TN}	PSNR	ASR_{TN}	PSNR	ASR_{TN}	PSNR	ASR_{TN}	PSNR
0	10.6	56.2	10.8	56.0	5.8	60.6	1.2	64.2
10	40.4	52.1	41.4	51.9	20.6	54.8	19.2	57.8
15	79.4	49.4	77.8	49.2	50.8	51.4	52.0	53.8
20	91.0	45.4	92.2	45.4	82.8	46.8	79.9	49.5

maximum of 96.6, 64.4, and 92.2% for median filtering, image resizing, and AWGN, respectively. Moreover, a larger confidence always results in adversarial examples with higher transferability and larger image distortion, and different degrees of transferability are obtained on different attacks. Among the four attacks, the C&W attack always achieves the highest PSNR for similar values of ASR_{TN}. A noticeable observation regards the MI-FGSM attack. While (Dong et al. 2018) reports that MI-FGSM attack improves the transferability of attacks against computer vision networks, a similar improvement of ASR_{TN} is not observed in Table 16.6 (w.r.t. I-FGSM). The reason could be that the gradients between subsequent iterations are highly correlated in image forensics models, and thus the advantage of gradient stabilization sought for by MI-FGSM is reduced.

Considering cross-training-set transferability, the results for the cases of median filtering, image resizing, and AWGN addition are shown in Table 16.7. The transferability of adversarial examples from BSnet trained on RAISE to the same architecture trained on VISION is reported. According to the table, increasing the confidence always helps to improve the transferability of the adversarial examples. Moreover, although the transferability is improved in all the cases, transferring adversarial examples for the case of image resizing detection still tends to be more difficult, as a larger

Table 16.7 Attacks transferability for cross-training-set mismatch (SN: ϕ^R_{BSnet}, TN: ϕ^V_{BSnet}). The ASR$_{TN}$ (%) and the average PSNR (dB) of adversarial examples is reported

Median filtering

c	I-FGSM		MI-FGSM		PGD		C&W	
	ASR$_{TN}$	PSNR	ASR$_{TN}$	PSNR	ASR$_{TN}$	PSNR	ASR$_{TN}$	PSNR
0	4.8	59.7	4.8	59.7	0.2	74.5	0.2	72.0
50	65.0	48.9	65.4	48.8	61.2	50.3	60.0	52.2
80	88.0	45.1	88.0	44.8	84.4	45.8	82.0	47.8
100	97.4	42.9	97.4	42.7	96.6	43.5	95.0	45.2

Image resizing

c	I-FGSM		MI-FGSM		PGD		C&W	
	ASR$_{TN}$	PSNR	ASR$_{TN}$	PSNR	ASR$_{TN}$	PSNR	ASR$_{TN}$	PSNR
0	12.8	58.7	12.8	58.6	9.8	66.9	9.8	68.3
30	53.0	48.0	48.6	47.8	38.6	49.6	23.8	52.5
40	64.0	44.7	60.2	44.6	54.2	45.7	32.8	48.9
50	67.2	41.7	64.2	41.7	59.8	42.3	39.2	45.9

AWGN

c	I-FGSM		MI-FGSM		PGD		C&W	
	ASR$_{TN}$	PSNR	ASR$_{TN}$	PSNR	ASR$_{TN}$	PSNR	ASR$_{TN}$	PSNR
0	0.6	57.7	0.6	57.6	0.2	62.4	0.2	65.0
20	13.8	49.7	15.0	49.6	11.0	52.2	12.2	54.4
30	54.4	44.8	72.8	45.1	77.2	47.3	78.0	49.4
40	88.2	41.0	93.0	41.3	94.0	43.0	95.4	45.5

Table 16.8 Attacks transferability for cross-network-and-training-set mismatch (SN = ϕ^V_{BSnet}, TN = ϕ^R_{BC+net}), including ASR$_{TN}$ (%) and average PSNR (dB) of adversarial samples

Median filtering

c	I-FGSM		MI-FGSM		PGD		C&W	
	ASR$_{TN}$	PSNR	ASR$_{TN}$	PSNR	ASR$_{TN}$	PSNR	ASR$_{TN}$	PSNR
0	2.6	60.0	2.6	60.0	0.0	71.9	0.0	70.5
100	84.6	42.4	85.6	42.4	83.0	43.1	78.0	45.1

Image resizing

c	I-FGSM		MI-FGSM		PGD		C&W	
	ASR$_{TN}$	PSNR	ASR$_{TN}$	PSNR	ASR$_{TN}$	PSNR	ASR$_{TN}$	PSNR
0	2.0	59.8	2.0	59.7	0.8	74.7	0.8	73.3
400	37.4	31.2	33.4	31.1	40.0	31.2	17.0	33.6

image distortion is needed to achieve a transferability comparable to that obtained for median filtering and AWGN addition.

Finally we consider the strongest mismatch case, i.e., cross-network-and-training-set mismatch, in the case of median filtering and image resizing detection. Only the attack transferability with $c = 0$ and that with largest confidences are reported in Table 16.8. A higher transferability is achieved (with a good image quality) for the case of median filtering, while adversarial examples targeting image resizing detection are less transferable.

To summarize, in contrast to computer vision applications, using adversarial examples for counter-forensics in a black-box or gray-box scenario, requires that proper measures are taken to ensure the transferability of the attacks (Li et al. 2020). As we have shown by referring to the method proposed in Li et al. (2020), however, attack transferability can be ensured in a wide variety of settings, the price to pay being a slight increase of the distortion introduced by the attack, thus calling for the development of suitable defences.

16.4 Defenses

As a response to the threats posed by the existence of adversarial examples and by the ease with which they can be crafted, many defence mechanisms have been proposed to develop CNN-based forensic methods that can work in adversarial conditions. Generally speaking, defences can work in a *reactive* or *proactive* manner. In the first case, the defence mechanisms are designed to work in conjunction with the original CNN algorithm ϕ, e.g., by revealing whether an adversarial example is being fed to the CNN by means of a dedicated detector, or by mitigating the (possibly present) adversarial perturbation applying some input transformations before feeding the CNN (this latter approach has been tested in the general DL literature—e.g., Xu et al. (2017)—but has not been adopted as a defence strategy in forensic applications). The second branch of defences aims at building more secure CNN models from scratch or by properly modifying the original models. The large majority of the methods developed in the ML and DL-based forensic literature belongs to this category. Among the most popular approaches, we mention adversarial training, multiple classification, and detector randomization. With regard to detector randomization, we refer to methods wherein the detector is built by including within it some random elements, thus qualifying this approach as a proactive one. This contrasts with randomization techniques commonly adopted in DL literature, like, for instance, stochastic activation pruning (Dhillon et al. 2018), that randomly drops some neurons of the CNN during the forward pass.

In the following, we review some of the methods developed in the forensic literature to defend against adversarial examples.

16.4.1 Detect Then Defend

The most common defence approach in early anti-counter-forensics works consisted in performing adversarial examples detection to rule out adversarial images, followed by the application of standard (unaware) detectors for the analysis of the samples deemed to be benign.

The analyst, aware of the counter-forensic method the system may be subject to, develops a new detector ϕ_A capable to expose the attacked images by looking for the traces left by the counter-forensic algorithm. This goal is achieved by resorting to new, tailored, features. The new algorithm ϕ_A is explicitly designed to reveal whether the image underwent a certain attack or not. If this is the case, the analyst may refuse the image or try to clean it to remove the effect of the attack. In this kind of approaches, the algorithm ϕ_A is used in conjunction with the original, unaware, algorithm ϕ. Such a view is adopted in Valenzise et al. (2013); Zeng et al. (2014), to address the adversarial detection of JPEG compression and median filtering, when the attacker tries to hinder the detection by removing the traces of JPEG compression and median filtering. Among the examples for the case of model-based analysis, we mention the algorithm in Costanzo et al. (2014) against the keypoint removal and injection attack against copy-move detectors. The "detect then defend" approach has also been adopted recently in the case of CNN-based forensic detectors, to defend against adversarial examples. For instance, Carrara et al. (2018) proposes to tell apart adversarial examples from benign images by looking at their behavior in the feature space. The method focuses on the analysis of the trajectory of the internal representations of the network (i.e., the activation of the neurons of the hidden layers), arguing that, for adversarial inputs, the representations follow a different evolution with respect to the case of genuine (non-adversarial) inputs. Detection is achieved by defining a feature distance to encode these trajectories into a fixed length sequence, used to train an Long Short Term Memory (LSTM) neural network to discern adversarial inputs from genuine ones.

Modeling normal data distributions for the activations has also been considered in the general DL literature to reveal abnormal behavior of adversarial examples (Tao et al. 2018).

It goes without saying that, this kind of defences can be easily circumvented if the attacker has some information about the method used by the analyst to expose the attack, thus entering a never-ending loop where attacks and forensic algorithms are iteratively developed.

16.4.2 Adversarial Training

A simple, yet effective, approach to improve the robustness of machine learning classifiers against adversarial attacks is *adversarial training*. Adversarial training consists in augmenting the training dataset with examples of adversarial images.

Such an approach implicitly assumes that the attack algorithm is known to the analyst and that the attack is not carried out on the retrained version of the detector. This identifies adversarial training as a white-box defense carried out against a gray-box attack.

Let $\mathcal{D}_A$ be the set of attacked images used for training. The adversarial trained classifier is $\phi(\cdot, \mathcal{T}; \mathcal{D}_{tr} \cup \mathcal{D}_A)$, where $\mathcal{T} = \{t_1, t_2, \ldots\}$ denotes the set of all the hyperparameters (i.e., the non-trainable parameters) of the network, including, for instance, the type of algorithm, its structure, the loss function, the internal parameters, the training procedure, etc. …Adversarial training has been widely adopted in the general DL literature to improve the robustness of DL classifiers against adversarial examples (Goodfellow et al. 2015; Madry et al. 2018). DL-based forensics is no exception, with the proposal of several approaches that resort to adversarial training to defend against adversarial examples (see, for instance, (Schöttle et al. 2018; Zhao et al. 2018)). In machine-learning-based forensic literature, JPEG-aware training is often considered to achieve robustness against JPEG compression. The interest in JPEG compression is motivated by the fact that JPEG compression is a common post-processing operation applied to images, either innocently (e.g., when uploading images on a social network), or maliciously (in some cases, in fact, JPEG compression can be considered as a simple and effective laundering attack (Barni et al. 2019)). The algorithms in Barni et al. (2017, 2016), for double JPEG detection, and in Boroumand and Fridrich (2017) for a variety of manipulation detection problems, based on Support Vector Machines (SVM), are examples of this approach. Several examples have also been proposed more recently in CNN-based forensics, e.g., for camera model identification (Marra et al. 2018), contrast adjustment detection (Barni et al. 2018), and GAN detection (Barni et al. 2020; Nataraj et al. 2019).

Resizing is another common processing operator applied to images, either innocently or maliciously. In a splicing scenario, for instance, image resizing can be applied to delete the traces of the splicing operation. Accordingly, a resizing-aware detector can be trained to detect splicing in the presence of resizing. From a different perspective, resizing can also be used as a defence against adversarial perturbations. Like other geometric transformations, resizing can be applied to the input images before testing, so to disable the effectiveness of the adversarial perturbations (He et al. 2017).

16.4.3 Detector Randomization

Detector randomization is another defense strategy that has been proposed to make life difficult for the attacker. With regard to computer vision applications, many randomization-based approaches have been proposed to hinder the transferability of adversarial examples in gray-box or black-box scenarios, and hence improve the security of the applications (Xie et al. 2018; Dhillon et al. 2018; Taran et al. 2019; Liu et al. 2018).

A similar approach can also be employed for securing image forensics detectors. The method proposed in Chen et al. (2019) applies randomization in the feature space by randomly selecting a subset of features to be used by the subsequent forensic detector. Randomization is based on a secret key, unknown to the attacker, who, then, cannot gain full knowledge of the to-be-attacked system (thus being forced to operate in a gray-box attack scenario). In Chen et al. (2019), the effectiveness of random feature selection (RFS) is proven theoretically, under some simplifying assumptions, and verified experimentally for two detectors based on support vector machines.

The approach proposed in Chen et al. (2019) has been extended to detectors based on deep learning by means of a techniques called Random Deep Feature Selection (RDFS) (Barni et al. 2020). In detail, a CNN-based forensic detector is first divided into two parts, i.e., the convolutional layers, playing the role of feature extractor, and the fully connected layers, to be regarded as the final detector. Let N denote the number of features extracted by the convolutional layers. To train a secure detector, a subset of K features is randomly selected according to a secret key. The fully connected layers are then retrained based on the selected K features. To maintain a good classification accuracy, the same model with the same secret key is applied during the training and the testing phases. In this case, the convolutional layers are only used for feature extraction and are frozen in the retraining phase to keep the feature space unchanged.

Assuming that the attacker has no knowledge of the RDFS strategy, he will target the original CNN-based model during his attack and the effectiveness of the randomized network will ultimately depend on the transferability of the attack.

The effectiveness of RDFS has been evaluated in Barni et al. (2020). Specifically, the experiments were conducted based on the RAISE dataset (Dang-Nguyen et al. 2021). Two different network architectures (BSnet (Bayar and Stamm 2016) and BC+net (Barni et al. 2018)) are utilized to detect three different image manipulations, namely, median filtering, image resizing, and adaptive histogram equalization (CL-AHE).

To build the original models based on BSnet, for each class, a number of 100,000 patches was considered for training, and 3000 and 10,000 patches were considered for validation and testing, respectively. As to the models based on BC+net, 500,000 patches per class were used for training, 5000 for validation, and 10,000 for testing. For both networks, the input patch size was set to 64×64, and the batch size was set to 32. The Adam optimizer with learning rate of 10^{-4} was used for training. The BSnet models were trained for 40 epochs and the BC+net models for 4 epochs. For BSnet, the classification accuracy of the models are 98.83, 91.30, and 90.45% for median filtering, image resizing, and CL-AHE, respectively. The corresponding accuracy values are 99.73, 95.05, and 98.30% for BC+net.

To build the RDFS-based detectors, first, a subset of the original training set with 20,000 patches (per class) was randomly selected. Then, for each model, the FC layers were retrained based on K features randomly selected from the full feature set. The Adam optimizer with learning rate of 10^{-5} is utilized. An early stop condition was adopted with a maximum number of 50 epochs, and the training process will stop when the validation loss changes less than 10^{-3} within 5 epochs. The number of K considered in the experiments are $K = \{5, 10, 30, 50, 200, 400\}$ as well as the full feature with $K = N$. For each K, 50 models trained on randomly selected K features were utilized in the experiments.

Three popular attacks were applied to generate adversarial examples, namely L-BFGS, I-FGSM and PGD. All the attacks were conducted based on Foolbox (Rauber et al. 2017). The detailed parameters of the attacks are given below. For L-BFGS, the parameters were set as default. I-FGSM was conducted with 10 iterations, and the best strength was found by spanning the range $[0 : 0001 : 0.1]$. For PGD, a binary search was conducted with initial values of $\varepsilon = 0.3$ and $\alpha = 0.05$, to find the optimal parameters. An exception for PGD is that the parameters used for attacking the models for the detection of CL-AHE were set as $\varepsilon = 0.01$ and $\alpha = 0.025$ without using binary search.

In the absence of attacks, the classification performance of the models trained on K features was evaluated on 4000 patches per class, while the defense performance was evaluated based on 500 manipulated images.

Tables 16.9 and 16.10 show the classification accuracy for the cases of BSnet and BC+net on three different manipulations. The RDFS strategy is helpful to hinder the transferability of adversarial examples on both networks at the expense of a slight decrease of the classification accuracy on clean images (in absence of attacks). Considering the BSnet, with the decrease of K, the detection accuracy of adversarial examples increases. For instance, the gain of the accuracy reaches 20–30% for $K = 30$ and 30–50% for $K = 10$, while the classification accuracy on clean images decreases by only 2–4% in these two cases. A similar behavior can be observed in the cases of BC+net in Table 16.10. In some cases, the detection accuracy for $K = N$ is already pretty high, which means that the adversarial examples built on the original CNN cannot transfer to the new detector trained on the full feature set. This phenomenon confirms the conclusion that adversarial examples have a poor transferability in the context of image forensics.

Further research has been carried out to evaluate the effectiveness of RDFS against increased-confidence adversarial examples (Li et al. 2020). Both the original CNNs and the RDFS detectors utilized in Barni et al. (2020) were tested. Two different attacks were conducted, namely, C&W attack and PGD, to generate adversarial examples with confidence value c. Similarly to Li et al. (2020), different confidence values are chosen for different SN. The detection accuracies of the RDFS detectors for the median filtering and image resizing task in the presence of increased-confidence adversarial examples are shown in Tables 16.11 and 16.12. It can be observed that decreasing K also helps improving the defense performance of the RDFS detectors against increased-confidence adversarial examples. For example, considering the case of BSnet for median filtering detection, for adversarial examples with $c = 10$ generated based on C&W attack, by decreasing K from N to 10, the detection accuracy is improved by 34.5%. However, with the increase of the confidence, the attack tends to be stronger, and the detection accuracy tends to be very low, even for small value of K.

Table 16.9 Accuracy (%) of the RDFS detector based on BSnet. 'No' corresponds to the absence of attacks

K	CL-AHE				Median filtering				Image resizing			
	No	L-BFGS	I-FGSM	PGD	No	L-BFGS	I-FGSM	PGD	No	L-BFGS	I-FGSM	PGD
N	80.5	35.1	93.9	91.8	99.0	39.7	4.3	81.9	98.0	20.3	0.6	31.5
400	81.3	41.0	94.5	91.8	98.8	42.6	7.5	76.6	97.7	9.1	7.2	20.7
200	81.5	42.8	94.0	91.6	98.7	44.8	10.8	77.6	97.8	17.4	13.7	31.0
50	80.7	56.3	91.3	90.2	97.7	53.5	24.6	80.0	97.4	40.1	35.9	52.0
30	80.1	64.7	90.7	89.5	96.8	56.1	30.8	79.7	97.0	48.8	43.4	58.5
10	78.0	78.6	89.1	88.0	93.2	67.1	44.5	80.6	95.0	62.0	55.7	68.0
5	73.0	88.0	89.2	87.4	88.7	73.0	51.0	79.8	91.0	65.6	61.6	69.9

Table 16.10 Accuracy (%) of the RDFS detector based on BC+net. 'No' corresponds to the absence of attacks

K	CL-AHE				Median filtering				Image resizing			
	No	L-BFGS	I-FGSM	PGD	No	L-BFGS	I-FGSM	PGD	No	L-BFGS	I-FGSM	PGD
N	98.2	26.2	34.0	33.5	99.7	71.3	13.7	85.2	100	81.2	75.2	89.8
400	97.1	21.0	83.6	30.1	99.6	75.6	15.6	88.1	99.8	80.0	71.8	89.3
200	96.9	26.0	83.0	48.5	99.6	76.2	17.0	88.6	99.7	77.9	69.6	88.0
50	95.1	35.3	80.0	50.6	99.6	76.6	21.9	87.4	96.8	73.0	66.8	85.2
30	94.3	39.8	76.3	56.7	99.4	79.6	30.0	88.5	92.7	70.7	65.5	81.8
10	91.1	48.3	68.8	55.6	98.8	79.2	44.3	86.1	78.6	63.0	59.9	71.9
5	87.4	47.0	63.7	47.2	97.2	77.1	48.3	83.3	74.4	60.7	58.7	67.7

Table 16.11 Detection accuracy (%) of the RDFS detector (based on BSnet) on increased-confidence adversarial examples

	Median filtering						Image resizing					
	C&W			PGD			C&W			PGD		
c	0	10	20	0	10	20	0	5	10	0	5	10
$K = N$	62.9	0.2	0.2	84.7	0.2	0.2	32.1	1.0	0.0	30.1	0.7	0.0
$K = 400$	53.6	0.2	0.2	75.2	0.2	0.2	18.4	0.7	0.0	20.5	0.7	0.0
$K = 200$	56.7	0.4	0.2	74.7	0.4	0.2	26.7	3.9	0.1	29.0	4.0	0.1
$K = 50$	62.1	6.7	0.5	80.1	9.2	0.5	48.0	17.5	2.9	52.0	20.2	3.4
$K = 30$	64.5	14.9	2.2	79.1	19.4	2.7	57.5	29.9	9.1	59.8	30.6	9.5
$K = 10$	72.0	34.7	15.5	81.1	38.2	16.4	66.9	47.1	29.2	69.0	48.8	30.0
$K = 5$	75.5	46.2	29.2	80.0	53.7	33.0	64.0	51.6	39.9	65.1	52.7	40.6

Table 16.12 Detection accuracy (%) of the RDFS detector (based on the BC+net) on increased-confidence adversarial examples

	Median filtering						Image resizing					
	C&W			PGD			C&W			PGD		
c	0	30	60	0	30	60	0	10	20	0	10	20
$K = N$	46.1	1.4	0.0	84.3	0.7	0.0	89.3	33.1	5.4	89.8	23.7	3.3
$K = 400$	52.6	3.1	0.0	87.3	2.7	0.0	88.3	35.4	5.4	89.4	28.1	3.2
$K = 200$	56.0	4.9	0.4	87.7	4.9	0.0	86.2	35.2	6.4	88.4	31.0	4.5
$K = 50$	61.9	10.5	0.7	86.8	13.8	0.5	80.1	37.1	11.0	85.0	37.3	8.7
$K = 30$	68.2	18.1	2.5	87.9	24.8	2.5	76.7	39.4	14.4	80.4	39.7	13.5
$K = 10$	71.7	36.0	14.8	85.8	42.5	14.2	72.0	46.3	26.2	75.4	47.3	25.2
$K = 5$	71.1	39.8	20.9	82.8	46.9	20.1	61.1	38.8	23.4	63.1	39.9	23.0

16.4.4 Multiple-Classifier Architectures

Another possibility to combat counter-forensics attacks is to resort to intrinsically more secure architectures. This approach has been widely considered in the general literature of ML security and in ML-based forensics, however there are only few methods resorting to such an approach in the case of CNNs.

Among the approaches developed for general ML applications we mention (Biggio et al. 2015), where a multiple classifier architecture is proposed, referred to as a one-and-a-half classifier. The one-and-a-half class architecture consists of two-class classifiers and two one-class classifiers run in parallel followed by a final one-class classifiers. It has been shown that, when properly trained, this architecture can effectively limit the damage of an attacker with perfect knowledge. The one-and-a-half architecture has also been considered in forensic applications, to improve the security of SVM-based detectors (Barni et al. 2020). In particular, considering several manipulation detection tasks, the authors of Barni et al. (2020) showed that the one-and-a-half class architecture can outperform two-class architectures in terms of security against white-box attacks, for a fixed level of robustness. The effectiveness of such an approach for CNN-based classifiers has not been investigated yet.

Another simple possibility to design a more secure classifier consists in building an ensemble of individual algorithms using some convenient ensemble strategies, as proposed in Strauss et al. (2017) in the general DL literature. A similar approach has been consider in Fontani et al. (2014) for ML-based forensics, where the authors propose to improve the robustness of the decision by fusing the outputs of several forensic algorithms looking for different traces. The method has been assessed for the representative forensic task of splicing detection in JPEG images. Even in this case, the effectiveness of such an approach for CNN-based forensic applications has still to be investigated.

Other approaches that have been proposed in security-oriented applications for general ML-based classification consist in using multiple classifiers in conjunction with randomization. Randomness can pertain to the selection of the training samples of the individual classifiers, like in Breiman (1996), or it can be associated with the selection of the features used by the classifiers, as in Ho (1998). Another strategy resorting to multiple classification and randomization has been proposed in Biggio et al. (2008) for spam-filtering applications, where the source of randomness is in the choice of the weights assigned to the filtering modules of the individual classifiers.

Regarding the DL literature, an approach that goes in this direction is model switching (Wang et al. 2020), according to which a certain number of trained sub-models are randomly selected at test time.

All considered, combining multiple classification with randomization in the attempt to improve the security of DL-based forensic classifiers is something that has not been studied much and is worth of further investigation.

16.5 Final Remarks

Deep learning architectures provide new powerful tools enriching the toolbox available to image forensics designers. At the same time, they introduce new vulnerabilities due to some inherent security weaknesses of DL techniques. In this chapter, we have reviewed the threats posed by adversarial examples and their possible use for counter-forensics purposes. Even if adversarial examples targeting image forensics networks tend to be less transferable than those created for computer vision applications, we have shown that, by properly increasing the strength of the attacks, a transferability level which is sufficient for practical applications can be reached. We have also presented some possible remedies against attacks based on adversarial examples, even if a definitive solution capable to prevent such attacks in the most challenging white-box scenario has not been found yet. We are convinced that, together with the necessity of developing image forensics techniques suitable to be applied outside controlled laboratory settings, robustness against intentional attacks is one of the most pressing needs, if we want that image forensics is finally used in real-life applications contributing to restore the credibility of digital media.

References

Ahmed B, Gulliver TA (2020) Image splicing detection using mask-RCNN. Signal, Image Video Process

Akhtar N, Ajmal M (2018) Threat of adversarial attacks on deep learning in computer vision: a survey. IEEE Access 2018(6):14410–14430

Athalye A, Engstrom L, Ilyas A, Kwok K (2018) Synthesizing robust adversarial examples. In: International conference on machine learning, pp 284–293

Barni M, Bondi L, Bonettini N, Bestagini P, Costanzo A, Maggini M, Tondi B, Tubaro S (2017) Aligned and non-aligned double JPEG detection using convolutional neural networks. J Vis Commun Image Represent 49:153–163

Barni M, Nowroozi E, Tondi B (2020) Improving the security of image manipulation detection through one-and-a-half-class multiple classification. Multimed Tools Appl 79(3):2383–2408

Barni M, Chen Z, Tondi B (2016) Adversary-aware, data-driven detection of double jpeg compression: how to make counter-forensics harder. In: 2016 IEEE international workshop on information forensics and security (WIFS). IEEE, pp 1–6

Barni M, Costanzo A, Nowroozi E, Tondi B (2018) CNN-based detection of generic contrast adjustment with jpeg post-processing. In: 2018 IEEE international conference on image processing, ICIP 2018, Athens, Greece, Oct 7–10, 2018. IEEE, pp 3803–3807

Barni M, Huang D, Li B, Tondi B (2019) Adversarial CNN training under jpeg laundering attacks: a game-theoretic approach. In: 2019 IEEE international workshop on information forensics and security (WIFS), pp 1–6

Barni M, Kallas K, Nowroozi E, Tondi B (2019) On the transferability of adversarial examples against CNN-based image forensics. In: IEEE international conference on acoustics, speech and signal processing, ICASSP 2019, Brighton, United Kingdom, 12–17 May 2019. IEEE, pp 8286–8290

Barni M, Kallas K, Nowroozi E, Tondi B (2020) CNN detection of gan-generated face images based on cross-band co-occurrences analysis. In: IEEE international workshop on information forensics and security

Barni M, Nowroozi E, Tondi B (2017) Higher-order, adversary-aware, double jpeg-detection via selected training on attacked samples. In: 2017 25th European signal processing conference (EUSIPCO), pp 281–285

Barni M, Nowroozi E, Tondi B, Zhang B (2020) Effectiveness of random deep feature selection for securing image manipulation detectors against adversarial examples. In: 2020 IEEE international conference on acoustics, speech and signal processing, ICASSP 2020, Barcelona, Spain, 4–8 May 2020. IEEE, pp 2977–2981

Bayar B, Stamm M (2018) Constrained convolutional neural networks: a new approach towards general purpose image manipulation detection. IEEE Trans Inf Forensics Secur 13(11):2691–2706

Bayar B, Stamm MC (2016) A deep learning approach to universal image manipulation detection using a new convolutional layer. In: Pérez-González F, Bas P, Ignatenko T, Cayre F (eds) Proceedings of the 4th ACM workshop on information hiding and multimedia security, IH&MMSec 2016, Vigo, Galicia, Spain, 20–22 June 2016. ACM, pp 5–10

Bayar B, Stamm MC (2018) Towards open set camera model identification using a deep learning framework. In: 2018 IEEE international conference on acoustics, speech and signal processing (ICASSP), pp 2007–2011

Biggio B, Corona I, He ZM, Chan PP, Giacinto G, Yeung DS, Roli F (2015) One-and-a-half-class multiple classifier systems for secure learning against evasion attacks at test time. In: International workshop on multiple classifier systems. Springer, pp 168–180

Biggio B, Fumera G, Roli F (2008) Adversarial pattern classification using multiple classifiers and randomisation. In: Joint IAPR international workshops on statistical techniques in pattern recognition (SPR) and structural and syntactic pattern recognition (SSPR). Springer, pp 500–509

Bondi L, Baroffio L, Guera D, Bestagini P, Delp EJ, Tubaro S (2017) First steps toward camera identification with convolutional neural networks. IEEE Signal Process Lett 24(3):259–263

Boroumand M, Fridrich J (2017) Scalable processing history detector for jpeg images. Electron Imaging 2017(7):128–137

Breiman L (1996) Bagging predictors. Mach Learn 24(2):123–140

Brendel W, Rauber J, Bethge M (2017) Decision-based adversarial attacks: reliable attacks against black-box machine learning models. arXiv:abs/1712.04248

Carlini N, Wagner DA (2017) Towards evaluating the robustness of neural networks. In: 2017 IEEE symposium on security and privacy, SP 2017, San Jose, CA, USA, 22–26 May 2017. IEEE, Computer Society, pp 39–57

Carrara F, Becarelli R, Caldelli R, Falchi F, Amato G (2018) Adversarial examples detection in features distance spaces. In: Proceedings of the European conference on computer vision (ECCV) workshops

Chen PY, Sharma Y, Zhang H, Yi J, Hsieh CJ (2018) Ead: elastic-net attacks to deep neural networks via adversarial examples. In: Proceedings of the AAAI conference on artificial intelligence, vol 32

Chen PY, Zhang H, Sharma Y, Yi J, Hsieh CJ (2017) ZOO: zeroth order optimization based black-box attacks to deep neural networks without training substitute models. In: Thuraisingham BM, Biggio B, Freeman DM, Miller B, Sinha A (eds) Proceedings of the 10th ACM workshop on artificial intelligence and security, AISec@CCS 2017, Dallas, TX, USA, 3 Nov 2017. ACM, pp 15–26

Chen J, Kang X, Liu Y, Wang ZJ (2015) Median filtering forensics based on convolutional neural networks. IEEE Signal Process Lett 22(11):1849–1853 Nov

Chen Z, Tondi B, Li X, Ni R, Zhao Y, Barni M (2019) Secure detection of image manipulation by means of random feature selection. IEEE Trans Inf Forensics Secur 14(9):2454–2469

Chollet F (2017) Xception: deep learning with depthwise separable convolutions. In: 2017 IEEE conference on computer vision and pattern recognition, CVPR 2017, Honolulu, HI, USA, 21–26 July 2017. IEEE Computer Society, pp 1800–1807

Costanzo A, Amerini I, Caldelli R, Barni M (2014) Forensic analysis of sift keypoint removal and injection. IEEE Trans Inf Forensics Secur 9(9):1450–1464

Dang-Nguyen DT, Pasquini C, Conotter V, Boato G (2021) RAISE: a raw images dataset for digital image forensics. In: Proceedings of the 6th ACM multimedia systems conference, MMSys '15, New York, NY, USA, 2015. ACM, pp 219–224

Dhillon GS, Azizzadenesheli K, Lipton ZC, Bernstein J, Kossaifi J, Khanna A, Anandkumar A (2018) Stochastic activation pruning for robust adversarial defense. In: Conference track proceedings 6th international conference on learning representations, ICLR 2018, Vancouver, BC, Canada, April 30–May 3, 2018. OpenReview.net

Dong Y, Liao F, Pang T, Su H, Zhu J, Hu X, Li J (2018) Boosting adversarial attacks with momentum. In: 2018 IEEE conference on computer vision and pattern recognition, CVPR 2018, Salt Lake City, UT, USA, 18–22 June 2018. IEEE Computer Society, pp 9185–9193

Dong Y, Pang T, Su H, Zhu J (2019) Evading defenses to transferable adversarial examples by translation-invariant attacks. In: IEEE conference on computer vision and pattern recognition, CVPR 2019, Long Beach, CA, USA, 16–20 June 2019. Computer Vision Foundation/IEEE, pp 4312–4321

Fontani M, Bonchi A, Piva A, Barni M (2014) Countering anti-forensics by means of data fusion. In: Media watermarking, security, and forensics 2014, vol 9028. International Society for Optics and Photonics, p 90280Z

Freire-Obregon D, Narducci F, Barra S, Castrillon-Santana M (2018) Deep learning for source camera identification on mobile devices. Pattern Recognit Lett

Goodfellow IJ, Shlens J, Szegedy C (2015) Explaining and harnessing adversarial examples. In: Bengio Y, LeCun Y (eds) Conference track proceedings 3rd international conference on learning representations, ICLR 2015, San Diego, CA, USA, 7–9 May 2015

Gragnaniello D, Marra F, Poggi G, Verdoliva L (2018) Analysis of adversarial attacks against CNN-based image forgery detectors. In: 26th European signal processing conference, EUSIPCO 2018, Roma, Italy, 3–7 Sept 2018. IEEE, pp 967–971

Guera D, Wang Y, Bondi L, Bestagini P, Tubaro S, Delp EJ (2017) A counter-forensic method for CNN-based camera model identification. In: IEEE computer vision and pattern recognition workshops, pp 1840–1847

He W, Wei J, Chen X, Carlini N, Song D (2017) Adversarial example defense: ensembles of weak defenses are not strong. In: 11th {USENIX} workshop on offensive technologies ({WOOT} 17)

Ho TK (1998) The random subspace method for constructing decision forests. IEEE Trans Pattern Anal Mach Intell 20(8):832–844

Huang G, Liu Z, Van Der Maaten L, Weinberger KQ (2017) Densely connected convolutional networks. In 2017 IEEE conference on computer vision and pattern recognition, CVPR 2017, Honolulu, HI, USA, 21–26 July 2017. IEEE Computer Society, pp 2261–2269

Ilyas A, Engstrom L, Athalye A, Lin J (2018) Black-box adversarial attacks with limited queries and information. In: Dy JG, Krause A (eds) Proceedings of the 35th international conference on machine learning, ICML 2018, Stockholm, Stockholmsmässan, Stockholm, Sweden, 10–15 July 2018, vol 80 of Proceedings of machine learning research. PMLR, pp 2142–2151

Kevin Eykholt K, Evtimov I, Fernandes E, Li B, Rahmati A, Xiao C, Prakash A, Kohno T, Song D (2018) Robust physical-world attacks on deep learning visual classification. In: Proceedings of the IEEE conference on computer vision and pattern recognition, pp 1625–1634

Kurakin A, Goodfellow I, Bengio S (2017) Adversarial examples in the physical world. In: Workshop track proceedings 5th international conference on learning representations, ICLR 2017, Toulon, France, 24–26 April 2017. OpenReview.net

Li Y, Li L, Wang L, Zhang T, Gong B (2019) NATTACK: learning the distributions of adversarial examples for an improved black-box attack on deep neural networks. In: Chaudhuri K, Salakhutdinov R (eds) Proceedings of the 36th international conference on machine learning, ICML 2019, 9–15 June 2019, Long beach, California, USA, vol 97 of Proceedings of machine learning research. PMLR, pp 3866–3876

Li W, Tondi B, Ni R, Barni M (2020) Increased-confidence adversarial examples for deep learning counter-forensics. In: Del Bimbo A, Cucchiara R, Sclaroff S, Farinella GM, Mei T, Bertini M, Escalante HJ, Vezzani R (eds) Pattern recognition. ICPR international workshops and challenges—virtual event, 10–15 Jan 2021, Proceedings, Part VI, vol 12666 of Lecture notes in computer science. Springer, pp 411–424

Liu X, Cheng M, Zhang H, Hsieh CJ (2018) Towards robust neural networks via random self-ensemble. In: Ferrari V, Hebert M, Sminchisescu C, Weiss Y (eds) Computer vision—ECCV 2018—15th European conference, Munich, Germany, 8–14 Sept 2018, Proceedings, Part VII, vol 11211 of Lecture notes in computer science. Springer, pp 381–397

Liu Y, Chen X, Liu C, Song D (2016) Delving into transferable adversarial examples and black-box attacks. arXiv:1611.02770

Madry A, Makelov A, Schmidt L, Tsipras D, Vladu A (2018) Towards deep learning models resistant to adversarial attacks. In: Conference track proceedings 6th international conference on learning representations, ICLR 2018, Vancouver, BC, Canada, April 30–May 3, 2018. OpenReview.net

Marra F, Gragnaniello D, Verdoliva L (2018) On the vulnerability of deep learning to adversarial attacks for camera model identification. Signal Process Image Commun 65:240–248

Nataraj L, Mohammed TM, Manjunath BS, Chandrasekaran S, Flenner A, Bappy JH, Roy-Chowdhury AK (2019) Detecting gan generated fake images using co-occurrence matrices. Electron Imaging 2019(5):532–1

Papernot N, McDaniel P, Goodfellow IJ (2016) Transferability in machine learning: from phenomena to black-box attacks using adversarial samples. CoRR. arXiv:abs/1605.07277

Papernot N, McDaniel P, Jha S, Fredrikson M, Celik ZB, Swami A (2016) The limitations of deep learning in adversarial settings. In: IEEE European symposium on security and privacy, EuroS&P 2016, Saarbrücken, Germany, 21–24 Mar 2016. IEEE, pp 372–387

Rauber J, Brendel W, Bethge M (2017) Foolbox v0. 8.0: a python toolbox to benchmark the robustness of machine learning models. arXiv:1707.04131

Rossler A, Cozzolino D, Verdoliva L, Riess C, Thies J, Niessner M (2019) Faceforensics++: learning to detect manipulated facial images. In: IEEE/CVF international conference on computer vision

Schöttle P, Schlögl A, Pasquini C, Böhme R (2018) Detecting adversarial examples-a lesson from multimedia security. In: 2018 26th European signal processing conference (EUSIPCO). IEEE, pp 947–951

Sharif M, Bhagavatula S, Bauer L, Reiter MK (2016) Accessorize to a crime: real and stealthy attacks on state-of-the-art face recognition. In: ACM conference on computer and communications security, p 1528–1540

Shullani D, Fontani M, Iuliani M, Shaya OA, Piva A (2017) VISION: a video and image dataset for source identification. EURASIP J Inf Secur 1–16

Simonyan K, Zisserman A (2015) Very deep convolutional networks for large-scale image recognition. In: Bengio Y, LeCun Y (eds) Conference track proceedings 3rd international conference on learning representations, ICLR 2015, San Diego, CA, USA, 7–9 May 2015

Strauss T, Hanselmann M, Junginger A, Ulmer H (2017) Ensemble methods as a defense to adversarial perturbations against deep neural networks. arXiv:1709.03423

Szegedy C, Zaremba W, Sutskever I, Bruna J, Erhan D, Goodfellow I, Fergus R (2014) Intriguing properties of neural networks. In: Bengio Y, LeCun Y (eds) Conference track proceedings 2nd international conference on learning representations, ICLR 2014, Banff, AB, Canada, 14–16 April 14–16 2014

Tao G, Ma S, Liu Y, Zhang X (2018) Attacks meet interpretability: attribute-steered detection of adversarial samples. arXiv:1810.11580

Taran O, Rezaeifar S, Holotyak T, Voloshynovskiy S (2019) Defending against adversarial attacks by randomized diversification. In: Proceedings of the IEEE conference on computer vision and pattern recognition, pp 11226–11233

Tondi B (2018) Pixel-domain adversarial examples against CNN-based manipulation detectors. Electron Lett

Tramèr F, Papernot N, Goodfellow I, Boneh D, McDaniel P (2017) The space of transferable adversarial examples. CoRR. arXiv:abs/1704.03453

Valenzise G, Tagliasacchi M, Tubaro S (2013) Revealing the traces of JPEG compression anti-forensics. IEEE Trans Inf Forensics Secur 8(2):335–349

Wang Q, Zhang R (2016) Double JPEG compression forensics based on a convolutional neural network. EURASIP J Inf Secur 1:2016

Wang X, Wang S, Chen PY, Lin X, Chin P (2020) Block switching: a stochastic approach for deep learning security. arXiv:2002.07920

Xie C, Wang J, Zhang Z, Ren Z, Yuille A (2018) Mitigating adversarial effects through randomization. In: Conference track proceedings 6th international conference on learning representations, ICLR 2018, Vancouver, BC, Canada, April 30–May 3, 2018. OpenReview.net

Xie C, Zhang Z, Zhou Y, Bai S, Wang J, Ren Z, Yuille AL (2019) Improving transferability of adversarial examples with input diversity. In: Proceedings of the IEEE conference on computer vision and pattern recognition, pp 2730–2739

Xu W, Evans D, Qi Y (2017) Feature squeezing: detecting adversarial examples in deep neural networks. arXiv:1704.01155

Yuan X, He P, Zhu Q, Li X (2019) Adversarial examples: attacks and defenses for deep learning. IEEE Trans Neural Netw Learn Syst 30(9):2805–2824

Zeng H, Qin T, Kang X, Liu L (2014) Countering anti-forensics of median filtering. In: IEEE international conference acoustics, speech and signal processing. IEEE, pp 2704–2708

Zhang B, Tondi B, Barni M (2020) Adversarial examples for replay attacks against CNN-based face recognition with anti-spoofing capability. Comput Vis Image Underst 102988:197–198

Zhao W, Yang P, Ni R, Zhao Y, Wu H (2018) Security consideration for deep learning-based image forensics. CoRR arXiv:abs/1803.11157

Zheng Z, Hong P (2018) Robust detection of adversarial attacks by modeling the intrinsic properties of deep neural networks. In: Proceedings of the 32nd international conference on neural information processing systems, NIPS'18, USA, 2018. Curran Associates Inc., pp 7924–7933

Chapter 17
Anti-Forensic Attacks Using Generative Adversarial Networks

Matthew C. Stamm and Xinwei Zhao

17.1 Introduction

The rise of deep learning has led to rapid advances in multimedia forensics. Algorithms based on deep neural networks are able to automatically learn forensic traces, detect complex forgeries, and localize falsified content with increasingly greater accuracy. At the same time, deep learning has expanded the capabilities of anti-forensic attackers. New anti-forensic attacks have emerged, including those discussed in Chap. 14 based on adversarial examples, and those based on generative adversarial networks (GANs).

In this chapter, we discuss the emerging threat posed by GAN-based anti-forensic attacks. GANs are a powerful machine learning framework that can be used to create realistic, but completely synthetic data. Researchers have recently shown that anti-forensic attacks can be built by using GANs to create synthetic forensic traces. While only a small number of GAN-based anti-forensic attacks currently exist, results show these early attacks are both effective at fooling forensic algorithms and introduce very little distortion into attacked images. Furthermore, by using the GAN framework to create new anti-forensic attacks, attackers can dramatically reduce the time required to create a new attack.

For simplicity, we will assume in this chapter that GAN-based anti-forensic attacks will be launched against images. This also aligns with the current state of research, since all existing attacks are targeted at images. However, the ideas and methodologies presented in this chapter can be generalized to a variety of media modalities including video and audio. We expect new GAN-based anti-forensic attacks targeted on these modalities will likely arise in the near future.

This chapter begins with Sect. 17.2, which provides a brief background on GANs. Section 17.3 provides background on anti-forensic attacks, including how they are

M. C. Stamm (✉) · X. Zhao
Department of Electrical and Computer Engineering, Drexel University, Philadelphia, PA, USA
e-mail: mcs382@drexel.edu

© The Author(s) 2022
H. T. Sencar et al. (eds.), *Multimedia Forensics*, Advances in Computer Vision and Pattern Recognition, https://doi.org/10.1007/978-981-16-7621-5_17

traditionally designed as well as their shortcomings. In Sect. 17.4, we discuss how GANs are used to create anti-forensic attacks. This gives a high-level overview of the components of an anti-forensic GAN, differences between attacks based on GANs and adversarial examples, as well as an overview of existing GAN-based anti-forensic attacks. Section 17.5 provides more details about how these attacks are trained, including different techniques that can be employed based on the attacker's knowledge of and access to the forensic algorithm under attack. We discuss known shortcomings of anti-forensic GANs and future research directions in Sect. 17.6.

17.2 Background on GANs

Generative adversarial networks (GANs) are a machine learning framework that is used to train generative models (Goodfellow et al. 2014). In their standard form, GANs consist of two components: a generator G and a discriminator D. The generator and the discriminator can be viewed as competitors in a two-player game. The goal of the generator is to learn to produce deceptively synthetic data that can mimic the real data, while the goal of the discriminator aims to learn to distinguish the synthetic data from the real data. The two parties are trained alternatively and each party attempts to improve its performance to defeat the other until the synthetic data is indistinguishable from the real data.

The generator is a generative model that creates synthetic data $\mathbf{x}'$ by mapping some input $\mathbf{z}$ in a latent space to an output as a set of real data $\mathbf{x}$. Ideally, the generator produces synthetic data $\mathbf{x}'$ with the same distribution as real data $\mathbf{x}$ such that the two cannot be differentiated. The discriminator is a discriminative model that will assign a scalar to each input data. The discriminator can output any scalar between 0 and 1, with 1 representing the real data and 0 representing the synthetic data. During training phase, the discriminator aims to maximize the probability of assigning correct label to both real data $\mathbf{x}$ and synthetic data $\mathbf{x}'$, while the generator aims to reduce the difference between the discriminator's decision on the synthetic data and 1 (i.e., it must be real data). Training a GAN is equivalent to finding the solution of the min-max function

$$\min_{G} \max_{D} \mathbb{E}_{\mathbf{x} \sim f_X(\mathbf{x})}[\log D(\mathbf{x})] + \mathbb{E}_{\mathbf{z} \sim f_Z(\mathbf{z})}[\log(1 - D(G(\mathbf{z})))] \qquad (17.1)$$

where $f_X(x)$ represents the distribution of real data, $f_Z(z)$ represents the distribution of input noise, $\mathbb{E}$ represents the operation of calculating expected value.

17.2.1 GANs for Image Synthesis

While GANs can be used to synthesize many types of data, one of their most common uses is to synthesize images or modify images. The development of several specific GANs has provided important insights and guidelines into the design of GANs for images. For example, while the GAN framework does not specify the specific form of the generator or discriminator, the development of DCGAN showed that deep convolutional generators and discriminators yield substantial benefits when performing image synthesis (Radford et al. 2015). Additionally, this work suggested constraints on the architectures of the generator and discriminator that can result in stable training. The creation of Conditional GANs (CGANs) showed the benefit of using information such as labels as auxiliary inputs to both the generator and the discriminator (Mehdi Mirza and Simon Osindero 2014). This can help can improve the visual quality and control the appearance of synthetic images generated by a GAN. InfoGAN showed that the latent space can be structured to both select and control generation of images (Chen et al. 2016). Pix2Pix (Isola et al. 2018) showed that GANs can learn to translate from one image space to another (Isola et al. 2018), while StackGAN demonstrated that text can be used to guide the generation of images via a series of CGANs (Zhang et al. 2017).

Since their first appearance in literature, the development of GANS for image synthesis has proceeded at an increasingly rapid pace. Research has been performed to improve the architectures of the generator and discriminator (Zhang et al. 2019; Brock et al. 2018; Karras et al. 2017; Zhu et al. 2017; Karras et al. 2020; Choi et al. 2020), improve the training procedure (Arjovsky et al. 2017; Mao et al. 2017), and build publicly available datasets to aid the research community (Yu et al. 2016; Liu et al. 2015; Karras et al. 2019; Caesar et al. 2018). Because of these efforts, GANs have been used to achieve state-of-the-art performance on various computer vision tasks, such as super-resolution (Ledig et al. 2017), photo inpainting (Pathak et al. 2016), photo blending (Wu et al. 2019), image-to-image translation (Isola et al. 2018; Zhu et al. 2017; Jin et al. 2017), text-to image-translation (Zhang et al. 2017), semantic-image-to-photo translations (Park et al. 2019), face aging (Antipov et al. 2017), generating human faces (Karras et al. 2017; Choi et al. 2018, 2020; Karras et al. 2019, 2020), generating 2-D (Jin et al. 2017; Karras et al. 2017; Brock et al. 2018) and 3-D objects (Wu et al. 2016), video prediction (Vondrick et al. 2016), and many more applications.

17.3 Brief Overview of Relevant Anti-Forensic Attacks

In this section, we provide a brief discussion of what anti-forensic attacks are and how they are designed.

17.3.1 What Are Anti-Forensic Attacks

Anti-forensic attacks are countermeasures that a media forger can use to fool forensic algorithms. They operate by falsifying the traces that forensic algorithms rely upon to make classification or detection decisions. A forensic classifier that is targeted by an anti-forensic attack is referred to as a *victim classifier.*

Just as there are many different forensic algorithms, there are similarly a wide variety of anti-forensic attacks (Stamm et al. 2013; Böhme and Kirchner 2013). Many of them, however, can be broadly grouped into two categories:

- **Attacks designed to disguise evidence of manipulation and editing.** Attacks have been developed to fool forensic algorithms designed to detect a variety of manipulations such as resampling (Kirchner and Bohme 2008; Gloe et al. 2007), JPEG compression (Stamm and Liu 2011; Stamm et al. 2010; Fan et al. 2013, 2014; Comesana-Alfaro and Pérez-González 2013; Pasquini and Boato 2013; Chu et al. 2015), median filtering (Fontani and Barni 2012; Wu et al. 2013; Dang-Nguyen et al. 2013; Fan et al. 2015), contrast enhancement (Cao et al. 2010), and unsharp masking (Laijie et al. 2013), as well as algorithms detect forgeries by identifying inconsistencies in lateral chromatic aberration (Mayer and Stamm 2015), identify copy-move forgeries (Amerini et al. 2013; Costanzo et al. 2014), and detect video frame deletion (Stamm et al. 2012a, b; Kang et al. 2016).
- **Attacks designed to falsify the source of a multimedia file.** Attacks have been developed to falsify demosaicing traces associated with an image's source camera model (Kirchner and Böhme 2009; Chen et al. 2017)??, as well as PRNU traces that can be linked to a specific device (Lukas et al. 2005; Gloe et al. 2007; Goljan et al. 2010; Barni and Tondi 2013; Dirik et al. 2014; Karaküçük and Dirik 2015)

To understand how anti-forensic attacks operate, it is helpful to first consider a simple model of how forensic algorithms operate. Forensic algorithms extract forensic traces from an image, then associate those traces with a particular forensic class. That class can be associated with editing operation or manipulation that an image has undergone, the image's source (i.e., it's camera model, its source device, it's distribution channel, etc.,) or some other property that a forensic investigator wishes to ascertain. It is important to note that when performing manipulation detection, even unaltered images contain forensic traces. While they do not contain traces associated with editing, splicing, or falsification, they do contain traces associated with an image being generated by a digital camera and not undergoing subsequent processing.

Anti-forensic attacks create synthetic forensic traces within an image that are designed to fool a forensic classifier. This holds true even for attacks designed to remove evidence of editing. These attacks synthesize traces in manipulated images that are associated with unaltered images. Most anti-forensic attacks are targeted, i.e., they are designed to trick the forensic classifier into associating an image with a particular target forensic class. For example, attacks against manipulation detectors typically specify the class of 'unaltered' images as the target class. Attacks designed to fool source identification algorithms synthesize traces associated with a target source that the image did not truly originate from. Additionally, a small number of anti-forensic attacks are untargeted. This means that the attack does not care which class the forensic algorithm associates the image with, so long as it is not the true class. Untargeted attacks are more commonly used to fool source identification algorithms, since an untargeted attack still tricks the forensic algorithm into believing that an image can form an untrue source. They are typically not used against manipulation detectors, since a forensic investigator will still decide that image is inauthentic even if their algorithm mistakes manipulation A for manipulation B.

17.3.2 Anti-Forensic Attack Objectives and Requirements

When creating an anti-forensic attack, an attacker must consider several design objectives. For an attack to be successful, it must satisfy the following design requirements;

- **Requirement 1**: *Fool the victim forensic algorithm*
  An anti-forensic attack should cause the victim algorithm to classify an image as belonging to the attacker's target class. If the attack is designed to disguise manipulation or content falsification, then the target class is the class of unaltered images. If the attack is designed to falsify an image's source, then the target class is the desired fake source chosen by the attacker such as a particular source camera.
- **Requirement 2**: *Introduce no visually perceptible artifacts.*
  An attack should not introduce visually distinct telltale signs that a human can easily identify. If this occurs, a human will quickly disregard an attacked image as fake even if it fools a forensic algorithm. This is particularly important if the image is to be used as part of a misinformation campaign, since images that are not plausibly real will be quickly flagged.
- **Requirement 3**: *Maintain high visual quality of the attacked image.*
  Similar to Requirement 2, if an attack fools a detector but significantly distorts an image, it is not useful to an information attacker. Some distortion is allowable, since ideally no one other than the attacker will be able to compare the attacked image to its unattacked counterpart. Additionally, this requirement is put in place to ensure that an attack does not undo desired manipulations that an attacker has previously performed to an image. For example, localized color corrections and contrast adjustments may be required to make a falsified image appear visually plausible. If an anti-forensic attack reverses these intentional manipulations, then the attack is no longer successful.

In addition to these requirements, there are several other highly desirable properties for an attack to possess. These include

- **Desired Goal 1**: *Be rapidly deployable in practical scenarios.*
  Ideally, an attack should be able to be launched quickly and efficiently, thus
  enabling rapid attacks or attacks at scale. It is flexible enough to attack images of
  any arbitrary size, not only images of a fixed predetermined size. Additionally, it
  should not require prior knowledge of which region of an image will be analyzed,
  such as a specific image block or block grid (it is typically unrealistic to assume
  that an investigator will only examine certain fixed image locations).
- **Desired Goal 2**: *Achieve attack transferability.*
  An attack achieves transferability if it is able to fool other victim classifiers that
  it has not been explicitly designed or trained to fool. This is important because an
  investigator can typically utilize multiple different forensic algorithms to perform
  a particular task. For example, several different algorithms exist for identifying an
  image's source camera model. An attack designed to fool camera model identification algorithm A is maximally effective if it also transfers to fool camera model
  identification algorithms B and C.

  While it is not necessary that an attack satisfy these additional goals, attacks that
do are significantly more effective in realistic scenarios.

17.3.3   Traditional Anti-Forensic Attack Design Procedure and Shortcomings

While anti-forensic attacks target different forensic algorithms, at a high level the
majority of them operate in a similar manner. First, an attack builds or estimates a
model of the forensic trace that it wishes to falsify. This could be associated with
the class of unaltered images in the case that the attack is designed to disguise
evidence of manipulation or falsification, or it could be associated with a particular
image source. Next, this model is used to guide a technique that synthesizes forensic
traces associated with this target class. For example, anti-forensic attacks to falsify
an image's source camera model synthesize demosaicing traces or other forensic
traces associated with a target camera model. Alternatively, techniques designed to
remove evidence of editing such as multiple JPEG compression or median filtering
synthesize traces that manipulation detectors associate with unaltered images.

Traditionally, creating a new anti-forensic attack has required a human expert to
first design an explicit model of the target trace that the attack wishes to falsify. Next,
the expert must create an algorithm capable of synthesizing this target trace such that it
matches their model. While this approach has led to the creation of several successful
anti-forensic attacks, it is likely not scalable enough to keep pace with the rapid
development of new forensic algorithms. It is very difficult and time consuming for
humans to construct explicit models of forensic traces that are accurate enough to for
an attack to be successful. Furthermore, the forensic community has widely adopted
the use machine learning and deep learning approaches to learn sophisticated implicit
models directly from data. In this case, it may be intractable for human experts to

explicitly model a target trace. Even if a model can be successfully constructed, a new algorithm must be designed to synthesize each trace under attack. Again, this is also a challenging and time-consuming process.

In order to respond to the rapid advancement of forensic technologies, adversaries would like to develop some automated means of creating new anti-forensic attacks. Tools from deep learning such as convolutional neural networks have allowed forensics researchers to successfully automate how forensic traces are learned. Similarly, intelligent adversaries have begun to look to deep learning for approaches to automate the creation of new attacks. As a result, new attacks based upon GANs have begun to emerge.

17.3.4 Anti-Forensic Attacks on Parametric Forensic Models

WIth the advances of software and apps, it is very easy for people to manipulate images the way they want. To make an image deceptively convincing to human eyes, post-processing operations such as resampling, blending, denoising are often applied to the falsified images. Recent years, deep learning-based algorithms are also used to create innovation contents. In the hand of an attacker, these techniques can be used for malicious purposes. Previous research shows that manipulations and attacks leave traces that can be modeled or characterized by forensic algorithms. While these traces may be invisible to human eyes, through forensic algorithms forged images can be detected or distinguished from the real images.

Anti-forensic attacks are countermeasures attempting to fool target forensic algorithms. Some anti-forensic attacks aim to remove forensic traces left by manipulation operations or attacks that the forensic algorithms analysis. Some anti-forensic attacks synthesize fake traces to make the forensic algorithms make a wrong decision. For example, anti-forensic attacks can remove traces left by resampling and make the forensic algorithms designed for characterizing resampling traces to believe the anti-forensically attacked images as "unaltered". Anti-forensic attacks can also synthesize fake forensic traces associated with a particular target camera model B into an images, and make a camera model identification algorithm to believe an image was captured by the camera model B, while the image was actually captured by camera model A. Anti-forensic attacks can also hide forensic traces to obfuscate their true origins. Anti-forensic attacks are expected to leave invisible distortion, since an visible distortion often flag "fake" for investigators. However, anti-forensic attacks do not consider countermeasures of itself, and only examine fooling target forensic algorithms.

Additionally, anti-forensic attacks are often designed to falsify particular forensic traces. A common methodology is typically used to create anti-forensic attacks similar to those described in the previous section. First, a human expert designs a parametric model of the target forensic trace that they wish to falsify during the attack. Next, an algorithm is created to introduce a synthetic trace into the attacked image so that it matches this model. Often this is accomplished by creating a generative noise model which is used to introduce specially designed noise either directly into the image or into some transform of the image. Many anti-forensic attacks have been developed following this methodology, such as removing traces left by median filtering, JPEG compression, or resampling, or synthesizing demosaicing information of camera models.

This approach to creating attacks has several important drawbacks. First, it is both challenging and time consuming for human experts to create accurate parametric models of forensic traces. In some cases it is extremely difficult or even infeasible to create models accurate enough to fool state-of-the-art detectors. Furthermore, models of one forensic cannot typically be reused to attack a different forensic trace. This creates an important scalability challenge for the attacker.

17.3.5   Anti-Forensic Attacks on Deep Neural Networks

An even greater challenge arises from the widespread adoption of deep-learning-based forensic algorithms such as convolutional neural networks (CNNs). The approach described above cannot be used to attack forensic CNNs since they don't utilize an explicit feature set or model of a forensic trace. Anti-forensic attacks targeting on deep-learning-based forensic algorithms now draw an increasing amount of attentions both from academia and industries.

While deep learning has achieved many state-of-the-art performances on many machine learning tasks, including computer vision and multimedia forensics, researchers found that non-ideal properties of neural networks can be exploited and cause misclassifications of an input image. Several explanations have been posited for the non-ideal properties of neural networks that adversarial examples exploit, such as imperfections caused by the locally-linear nature of neural networks (Goodfellow et al. 2014) or that there is misalignment between the features used by humans and those learned by neural networks when performing the same classification task (Ilyas et al. 2019). These findings facilitate the development of adversarial example generation algorithms.

Adversarial example generation algorithms operate by adding adversarial perturbations to an input image. The adversarial perturbations are some noises produced by optimizing a certain distance metric, usually L_p norm (Goodfellow et al. 2014; Madry et al. 2017; Kurakin et al. 2016; Carlini and Wagner 2017). The adversarial perturbations aim to push the representation in latent feature space just across the decision boundary of the desired class (i.e., targeted attack) or push the data away from its true class (i.e., untargeted attack). Forensic researchers found that adversarial example generation algorithms can be used as anti-forensic attacks on forensic algorithms. In 2017, Güera et al. showed that camera model identification CNNs (Güera et al. 2017) can be fooled by Fast Gradient Sign method (FGSM) (Goodfellow et al. 2014) and Jacobian-based Saliency Map (JSMA) (Papernot et al. 2016).

There are several drawbacks for using adversarial example generation algorithms as anti-forensic attacks. To ensure the CNN can be successfully fooled, the adversarial perturbations added to the image may be strong and result in fuzzy or distorted output images. While using some tricks, researchers could control the visual quality of anti-forensic attacked images to a certain level, maintaining high visual quality is still a challenge for further study. Another drawback is that adversarial perturbations have to be optimized for particular given input image. As a result, it could be time-consuming for producing large volume of anti-forensically attacked images. Additionally, the adversarial perturbations are not invariant to detection block alignment.

17.4   Using GANs to Make Anti-Forensic Attacks

Recently, researchers have begun adapting the GAN framework to create a new class of anti-forensic attacks. Though research into these new GAN-based attacks is still in its early stages, recent results have shown that these attacks can overcome many of the problems associated with both traditional anti-forensic attacks and attacks based on adversarial examples. In this section, we will describe at a high level how GAN-based anti-forensic attacks are constructed. Additionally, we will discuss the advantages of utilizing these types of attacks.

17.4.1   How GANs Are Used to Construct Anti-Forensic Attacks

GAN-based anti-forensic attacks are designed to fool forensic algorithms built upon neural networks. At a high level, they operate by using an adversarially trained generator to synthesize a target set of forensic traces in an image. Training occurs first, before the attack is launched. Once the anti-forensic generator has been trained, an image can be attacked simply by passing it through the generator. The generator does not need to be re-trained for each image, however, it must be re-trained if a new target class is selected for the attack.

Researchers have shown that GANs can be adapted to both automate and generalize the anti-forensic attack design methodology described in Sect. 17.5 of this chapter. The first step in the traditional approach to creating an anti-forensic attack involves the creation of a model of the target forensic trace. While it is difficult or potentially impossible to build an explicit model of the traces learned by forensic neural networks, GANs can exploit the model that has already been learned by the forensic neural network. This is done by adversarially training the generator against a pre-trained version of the victim classifier C. The victim classifier can either take the place of the discriminator in the traditional GAN formulation, or it can be used in conjunction with a traditional discriminator. Currently, there is not a clear consensus on whether a discriminator is necessary or beneficial when creating GAN-based anti-forensic attacks. Additionally, in some scenarios the attacker may not have direct access to the victim classifier. If this occurs, the victim classifier cannot be directly integrated into the adversarial training process. Instead, alternate means of training the generator must be used. These are discussed in Sect. 17.5.

When the generator is adversarially trained against the victim classifier, it learns the model of the target forensic trace implicitly used by the victim classifier. Because of this, the attacker does not need to explicitly design a model of the target trace, as would typically be done in the traditional approach to creating an anti-forensic attack. This dramatically reduces the time required to construct an attack. Furthermore, it increases the accuracy of the trace model used by the attack, since it is directly matched to the victim classifier.

Additionally, GAN-based attacks significantly reduce the effort required for the second step of the traditional approach to creating an anti-forensic attack, i.e., creating a technique to synthesize the target trace. This is because the generator automatically learns to synthesize the target trace through adversarial training. While existing GAN-based anti-forensic attacks utilize different generator architectures, the generator still remains somewhat generic in that it is a deep convolutional neural network. CNNs can be quickly and easily implemented using deep learning frameworks such as TensorFlow or Pytorch, making it easy for an attacker to experimentally optimize the design of their generator.

Currently, there are no well-established guidelines for designing anti-forensic generators. However, research by Zhang et al. has shown that when an upsampling component is utilized in a GAN's generator, it leaves behind "checkerboard" artifacts in the synthesized image. Zhang et al. (2019). As a result, it is likely important to avoid the use of upsampling in the generator so that visual or statistically detectable artifacts that may act as giveaways are not introduced to an attacked image. This is reinforced by the fact that many existing attacks utilize fully convolutional generators that do not include any pooling or upsampling layers.

Convolutional generators also possess the useful property that once they are trained, they can be used to attack an image of any size. Fully convolutional CNNs are able to accept images of any size as an input and produce output images of the same size. As a result, these generators can be deployed in realistic scenarios in which the size of the image under attack may not be known a prior, or may vary in the case of attacking multiple images. They also provide an advantage in that synthesized forensic traces are distributed throughout the attacked image and do not depend on a blocking grid. As a result, the attacker does not require advanced knowledge of which image region or block that an investigator will analyze. This is a distinct advantage over anti-forensic attacks based upon adversarial examples.

Research also suggests that a well-constructed generator can be re-trained to synthesize different forensic traces than the ones which it was initially designed for. For example, the MISLGAN attack was initially designed to falsify an image's origin by synthesizing source camera model traces linked with a different camera model.

Recent work has also used this generator architecture to construct a GAN-based attack that removes evidence of multiple editing operations.

17.4.2 Overview of Existing GAN-Based Attacks

Recently, several GAN-based anti-forensic attacks have been developed to falsify information about an image's source and authenticity. Here we provide a brief overview of existing GAN-based attacks.

In 2018, Kim et al. proposed a GAN-based attack to remove median filtering traces from an image (Kim et al. 2017). This GAN was built using a generator and a discriminator. The generator was trained to remove the median filtering traces, and the discriminator was trained to distinguish between the restored images and the unaltered images. The author introduced loss computed from a pre-trained VGG network to improve the visual quality of attacked images. This attack was able to remove media filtering traces from gray-scale images, and fool many traditional forensic algorithms using hand-designed features, and CNN detectors.

Chen et al. proposed a GAN-based attack in 2018 named MISLGAN to falsify an image's source camera model (Chen et al. 2018). In this work, the authors assumed that the attacker has access to the forensic CNN that the investigator will use to identify an image's source camera model. Their attack attempts to modify the traces in images associated with a target camera model such that the forensic CNN mis-classifies an attacked image as having been captured by the target camera model. MISLGAN is consists of three major components: a generator, a discriminator, and the pre-trained forensic camera model CNN. The generator was adversarially trained against both the discriminator and the camera model CNN for each target camera model. The trained generator was shown to be able to falsify fake camera models for any color images with high success attack rate (i.e., the percentage of an image being classified as the target camera model after being attacked). This attack was demonstrated to be effective even when images under attack were captured by camera models not used during training. Additionally, the attacked images also maintain high visual qualities even comparing with the original unaltered image side by side.

In 2019, the authors of MISLGAN extended this work to create a black box attack against forensic source camera model identification CNNs (Chen et al. 2019). To accomplish this, the authors integrated a substitute network to approximate the camera model identification CNNs under attack. By training MISLGAN against a substitute network, this attack can successfully fool state-of-the-art camera model CNNs even in black-box scenarios, and outperformed adversarial example generation algorithms such as FGSM (Goodfellow et al. 2014). This GAN-based attack has been adapted to remove traces left by editing operations and to create transferable anti-forensic attacks (Zhao et al. 2021). Additionally, by improving the generator's architecture, Zhao et al. were able to falsify traces left in synthetic images by the GAN generation process (Xinwei Zhao and Matthew C. Stamm 2021). This attack was shown to fool existing synthetic image detectors targeted at detecting both GAN-generated faces and objects.

In 2021, Cozzolino et al. proposed a GAN-based attack, SpoC (Cozzolino et al. 2021), to falsify camera model information of a GAN-synthesized image. SpoC is consisted of a generator, a discriminator, and a pre-trained embedder. The embedder is used to learn the reference vectors of real images or the images of the target camera models in the latent spaces. This attack can successfully fool GAN-synthesized image detectors or the camera model identification CNNs, and outperformed adversarial example generation algorithm such as FGSM (Goodfellow et al. 2014) and PGD (Madry et al. 2017). The authors also showed that the attack is robust to JPEG compression of a certain level.

In 2021, Wu et al. proposed a GAN-based attack, JRA-GAN, to restore JPEG compressed images (Wu et al. 2021). The authors used a GAN to restore the high frequency information in the DCT domain, and remove the blocking effect caused by JPEG compression to fool JPEG compression detection algorithms. The authors showed that the attack can successfully restore gray-scale JPEG compressed images and fool traditional JPEG detection algorithms using hand-designed features, and CNN detectors.

17.4.3 Differences Between GAN-Based Anti-Forensic Attacks and Adversarial Examples

While attacks built from GANs and from adversarial examples both use deep learning to create anti-forensic countermeasures, it is important to note that the GAN-based attacks described in this chapter are fundamentally different from adversarial examples

As discussed in Chap. 16, adversarial examples operate by exploiting non-ideal properties of a victim classifier. There are several explanations that have been considered by the research community as to why adversarial examples exist and are successful. These include "blind spots" caused by non-ideal training or overfitting, a mismatch between features used by humans and those learned by neural networks when performing classification (Ilyas et al. 2019), and inadvertent effects caused by the locally-linear nature of neural networks (Goodfellow et al. 2014). In all of these explanations, adversarial examples cause a classifier to misclassify due to unintended behavior.

By contrast, GAN-based anti-forensic attacks do not try to trick the victim classifier into misclassifying an image by exploiting its non-ideal properties. Instead, they replace the forensic traces present in an image with a synthetic set of traces that accurately match the victim classifier's model of the target trace. In this sense, they attempt to synthesize traces that the victim classifier *correctly* uses to perform tasks such as manipulation detection and source identification. This is possible because forensic traces, such as those linked to editing operations or an image's source camera, are typically not visible to the human eye. As a result, GANs are able to create new forensic traces in an image without altering its content or visual appearance.

Additionally, adversarial example attacks are customized to each image under attack. Often, this is done through an iterative training or search process that must be repeated each time a new image is attacked. It is worth noting that GAN-based techniques for generating adversarial examples that have recently been proposed in the computer vision and machine learning literature (Poursaeed et al. 2018; Xiao et al. 2018). These attacks are distinct from the GAN-based attacks discussed in this chapter, specifically because they generate adversarial examples and not synthetic versions of features that should accurately be associated with the target class. For these attacks, the GAN framework is instead adapted to search for an adversarial example that can be made from a particular image in the same manner that iterative search algorithms are used.

17.4.4 Advantages of GAN-Based Anti-Forensic Attacks

There are several advantages to using GAN-based anti-forensic attacks over both traditional human-designed approaches and those based on adversarial examples. We briefly summarize these below. Open problems and weaknesses of GAN-based adversarial attacks are further discussed in Sect. 17.6.

- **Rapid and automatic attack creation**: As discussed earlier, GANs both automate and generalized the traditional approach to designing anti-forensic attacks. As a result, it is much easier and quicker to create new GAN-based attacks than through traditional human-designed approaches.
- **Low visual distortion**: In practice, existing GAN-based attacks tend to introduce little-to-no visually detectable distortion. While this condition is not guaranteed, visual distortion can be effectively controlled during training by carefully balancing the weight placed on the generator's loss term controlling visual quality. This is discussed in more detail in Sect. 17.5. By contrast, several adversarial example attacks can leave behind visually detectable noise-like artifacts (Güera et al. 2017).
- **Attack only requires training once**: After training, GAN-based attack can be launched on any image as long as the target class remains constant. This differs from attacks based on adversarial examples, which need to create a custom set of modifications for each image, often involving an iterative algorithm. Additionally, no image-specific parameters need to be learned, unlike for several traditional anti-forensic attacks.
- **Scalable deployment in realistic scenarios**: Attack requires only to pass the image through a pre-trained convolutional generator. It launches very rapidly and can be deployed at scale. Additionally, since the attack is launched
- **Ability to attack images of arbitrary size**: Because the attack is launched via a convolutional generator, the attack can be applied to images of any size. By contrast, attacks based on adversarial examples only produce attacked images of

the same size as the input to the victim CNN, which is typically a small patch and not a full sized image.

- **Does not require advanced knowledge of the analysis region**: One way to adapt adversarial attacks to larger images is to break the image into blocks, then attack each block separately. This however requires advanced knowledge of the particular blocking grid that will be used during forensic analysis. If the blocking grid used by the attacker and the detector do not align, or if the investigator analyzes an image multiple times with different blocking grids, adversarial example attacks are unlikely to be successful.

17.5 Training Anti-Forensic GANs

In this section, we give an overview of how to train an anti-forensic generator. We begin by discussing the different terms of the loss function used during training. We shown how each term of the loss function is formulated in the perfect knowledge scenario (i.e., when a white box attack is launched). After this, we show how to modify the loss function to train the attack to create black box attacks in limited knowledge scenarios, as well transferable attacks in the zero knowledge scenario.

17.5.1 Overview of Adversarial Training

During adversarial training, the generator G used in GAN-based attack should be incentivized to produce visually convincing anti-forensic attacked images that can fool the victim forensic classifier C, as well as a discriminator D if it is used. These goals are achieved by properly formulating the generator's loss function $\mathcal{L}_G$. A typical loss function for adversarially training an anti-forensic generator consists of three major loss terms: the classification loss $\mathcal{L}_C$, the adversarial loss $\mathcal{L}_A$ (if a discriminator is used), and perceptual loss $\mathcal{L}_P$,

$$\mathcal{L}_G = \alpha \mathcal{L}_P + \beta \mathcal{L}_C + \gamma \mathcal{L}_A, \qquad (17.2)$$

where α, β, γ are weights to balance the performance of the attack and the visual quality of the attacked images.

Like all anti-forensic attacks, the generator's primary goal is to produce output images that fool a particular forensic algorithm. This is done by introducing a term into the generator's loss function that we describe as the *forensic classification loss*, or classification loss for short. Classification loss is the key element to guide the generator to learn the forensic traces of the target class. Depending on the attacker's access to and knowledge level of the forensic classifier, particular strategies may need to be adopted during training We will elaborate on this later in this section. Typically, the classification loss is defined as 0 if the output of the generator is classified by the forensic classifier as belonging to the target class t and 1 if the output is classified as any other class.

If the generator is trained to fool a discriminator as well, then the adversarial loss provided by the feedback of the discriminator is consolidated into the total loss function. Typically, the adversarial loss is 0 if the generated anti-forensically attacked images fools the discriminator and 1 if not.

At the same time, the anti-forensic generator should introduce minimal distortion into the attacked image. Some amount of distortion is acceptable, since in an anti-forensic setting, the unattacked image will not be presented to the investigator. However these distortions should not undo or significantly alter any modifications that were introduced by the attacker before the image is passed through the anti-forensic generator. To control the visual quality of anti-forensically attack images, the perceptual loss measures the pixel-wise difference between the image under attack and the anti-forensically attacked image produced by the generator. Minimizing the perceptual loss during the adversarial training process helps to ensure that attacked images produced by the generator contain minimal, visually imperceptible distortions.

When constructing an anti-forensic attack, it is typically advantageous to exploit as much knowledge of the algorithm under attack as possible. This holds especially true with GAN-based attacks, which directly integrates a forensic classifier into training the attack. Since gradient information from a victim forensic classifier is needed to compute the classification loss, different strategies must be adopted to train the anti-forensic generator depending on the attacker's knowledge about the forensic classifier under attack. As a result, it is necessary to provide more detail about the different knowledge levels that an attacker may have of the victim classifier before we are able to completely formulate the loss function. We note that the knowledge level typically only has an influence on the classification loss. The perceptual loss and discriminator loss remain unchanged for all knowledge levels.

17.5.2 Knowledge Levels of the Victim Classifier

As mentioned previously, we refer to the forensic classifier under attack as the victim classifier C. Depending on the attacker's knowledge level of the victim classifier, it is common to categorize the knowledge scenarios into the perfect knowledge (i.e., white box) scenario, the limited knowledge (i.e., black box) scenario, and the zero knowledge scenario.

In the perfect knowledge scenario, the attacker has full access to the victim classifier or an exact identical copy of the victim classifier. This is an ideal situation for the attacker. The attacker can launch a white box attack in which they train directly against the victim classifier that they wish to fool.

In many other situations, however, the attacker has only partial knowledge or even zero knowledge of the victim classifier. Partial knowledge scenarios may include a variety of specific settings in real life. A common aspect is that the attacker does not have full access or control over the victim classifier. However, the attacker is allowed to probe the victim classifier as a black box. For example, they can query the victim classifier using an input, then observe the value of its output. Because of this, the partial knowledge scenario is also referred to as the black box scenario. This scenario is a more challenging, but potentially more realistic situation for the attacker. For example, a piece of forensic software may or online service may be encrypted to reduce the risk of attack. As a result, its users would only be able to provide input images and observe the results that the system output.

An even more challenging situation is when the attacker has zero knowledge of the victim classifier. Specifically, the attacker not only has no access to the victim classifier, but also the attacker is not allowed to observe the victim classifier by any means. This is also a realistic situation for the attacker. For example, an investigator may develop private forensic algorithms and keep this information classified for security purposes. In this case, we assume that the attacker knows what the victim classifier's goal is (i.e., manipulation detection, source camera model identification, etc.). A broader concept of zero knowledge may include the attacker not knowing if a victim classifier exists, or what the specific goals of the victim classifier are (i.e., the attacker will not know anything about an image will be analyzed). Currently, however, this is beyond the scope of existing antiforensics research.

In both the black box and the zero knowledge scenarios, the attacker cannot use the victim classifier C to directly train the attack. The attacker needs to modify the classification loss in (17.2) such that a new classification loss is provided by other classifiers that can accessed during attack training. We note that the perceptual loss and the adversarial loss usually remain the same for all knowledge scenarios, or may need trivial adjustment based on specific cases. The main change, however, is the formulation of the classification loss.

17.5.3 White Box Attacks

We start by discussing the formulation of each loss term in the perfect knowledge scenario, i.e., when creating a white box attack. Each term in the generator's loss is defined below. The perceptual loss and the adversarial loss will remain the same for all subsequent knowledge scenarios.

- **Perceptual Loss**: The perceptual loss L_P can be formulated in many different ways. The mean squared error (L_2 loss) or mean absolute difference (L_1 loss) between an image before and after attack are the most commonly used formulations, i.e.,

$$\mathcal{L}_P = \tfrac{1}{N}\|I - G(I)\|_p, \tag{17.3}$$

where N is the number of pixels in the reference image I and anti-forensically attacked image $G(I)$, $\|\cdot\|p$ denotes the p norm and p equals to 1 or 2. Empirically, the L_1 loss usually yields better visual quality for GAN-based anti-forensic attacks. This is potentially because the L_1 loss penalizes small pixel differences comparatively more than the L_2 loss, which puts greater emphasis on penalizing large pixel differences.

- **Adversarial Loss**: The adversarial loss $\mathcal{L}_A$ is used to ensure the anti-forensically attacked image can fool the discriminator D. Ideally, the discriminator's output of the anti-forensically attacked images is 1 when the attacked images fool the discriminator, 0 otherwise. Therefore, the adversarial loss is typically formulated as the sigmoid cross-entropy between the discriminator's output of the generated anti-forensically attacked image and 1.

$$\mathcal{L}_A = \log(1 - D(G(I))), \tag{17.4}$$

As mentioned previously, a separate discriminator is not always used when creating a GAN-based anti-forensic attack. If this is the case, then the adversarial loss term can be discarded.

- **Classification Loss**: In the perfect knowledge scenario, since the victim classifier C is fully accessible by the attacker, the victim classifier can be directly used to calculate the forensic classification loss $\mathcal{L}_C$. Typically this is done by computing the softmax cross-entropy between the classifier's output and the target class t, i.e.,

$$\mathcal{L}_C = -\log(C(G(I))_t) \tag{17.5}$$

where $C(G(I))_t$ is the softmax output of victim classifier at location t.

17.5.4 Black Box Scenario

The perfect knowledge scenario is often used to evaluate the baseline performance of an anti-forensic attack. However, it is frequently unrealistic to assume that an attacker would have full access to the victim classifier. More commonly, the victim classifier may be presented as a black box to the attacker. Specifically, the attacker does not have full control over the victim classifier, nor do they have enough knowledge of its architecture to construct a perfect replica. As a result, the victim classifier can not be used to produce the classification loss as formulated in the perfect knowledge scenario. The victim classifier can, however, be probed by the attacker as a black box, then the output can be observed. This is done by providing it input images, then recording how the victim classifier classifies those images. The attacker can use this information and modify the classification loss to exploit it.

One approach is to build a substitute network C_s to mimic the behavior of the victim classifier C. The substitute network is a forensic classifier built and fully controlled by the attacker. Ideally, if the substitute network is trained properly, it perfectly mimics the decisions made by the victim classifier. The substitute network can then be used to adversarially train the generator instead of the victim classifier by reformulating the classification loss as

$$\mathcal{L}_C = -\log(C_s(G(I))_t) \qquad (17.6)$$

where $C_s(G(I))_t$ the softmax output of the substitute network C_s at location t.

When training a substitute network, it is important that it is trained to reproduce the forensic decisions produced by the victim classifier even if these decisions are incorrect. By doing this, the substitute network can approximate the latent space in which the victim classifier encodes forensic traces. For an attack to be successful, it is important to match this space as accurately as possible. Additionally, there are no strict rules guiding the design of the substitute network's architecture. Research suggests that as long as the substitute network is deep and expressive enough to reproduce the victim classifier's output, a successful black box attack can be launched. For example, Chen et al. demonstrated that successful black box attacks against source camera model identification algorithms can be trained using different substitute networks can built primarily from residual blocks and dense blocks (Chen et al. 2019).

17.5.5 Zero Knowledge

In the zero knowledge scenario, the attacker has no access to the victim classifier C. Furthermore, the attacker cannot observe or probe the victim classifier like in black box scenario. As a result, the substitute network approach described in the black box scenario does not translate to the zero knowledge scenario. In this circumstance, the synthetic forensic traces created by the attack must be transferable enough to fool the victim classifier.

Creating transferable attacks is challenging and remains an open research area. One recently proposed method of achieving attack transferability is to adversarially train the anti-forensic generator against an ensemble of forensic classifiers. Each forensic classifier C_m in the ensemble is built and pre-trained for a forensic classification by the attacker and acts as a *surrogate* for the victim classifier. For example, if the attacker's goal is to fool a manipulation detection CNN, then each surrogate classifier in the ensemble should be trained to perform manipulation detection as well. These surrogate classifiers, however, should be as diverse as possible. Diversity can be achieved by varying the architecture of the surrogate classifiers, the classes that they distinguish between, or other factors.

Together, these surrogate classifiers can be used to replace the victim classifier and compute a single classification loss in the similar fashion as in the white box scenario. The final classification loss is formulated as a weighted sum of individual classification losses pertaining to each surrogate classifier in the ensemble, such that

$$\mathcal{L}_C = \sum_{m=1}^{M} \beta_m \mathcal{L}_{C_m} \qquad (17.7)$$

where M is the number of surrogate classifiers in the ensemble, $\mathcal{L}_{C_m}$ corresponds to individual classification loss of the mth surrogate classifier, β_m corresponds to the weight of mth individual classification loss.

While there exists no strict mathematical justification for why this approach can achieve robustness, the intuition is that each surrogate classifier in the ensemble learns to partition the forensic feature space into separate regions for the target and other classes. By defining the classification loss in this fashion, the generator is trained to synthesize forensic features that lie in the intersection of these regions. If a diverse set of classifiers are used to form the ensemble, this intersection will likely lie inside the decision region that other classifiers (such as the victim classifier) associate with the target class.

17.6   Known Problems with GAN-Based Attacks & Future Directions

While GAN-based anti-forensic attacks pose a serious threat to forensic algorithms, this research is still in its very early stages. There are several known shortcomings of these new attacks, as well as open research questions.

One important issue facing GAN-based anti-forensic attacks is transferability. While these attacks are strongest in the white box scenario where the attacker has full access to the victim classifier, this scenario is the least realistic. More likely, the attacker knows only the general goal of the victim classifier (i.e., manipulation detection, source identification, etc.) and possibly the set of classes this classifier is trained to differentiate between. This corresponds to the zero knowledge scenario described in Sect. 17.5.5. In this case, the attacker must rely entirely on attack transferability to launch a successful attack.

Recent research has shown, however, that achieving attack transferability against forensic classifiers is a difficult task (Barni et al. 2019; Zhao and Stamm 2020). For example, small implementation differences between two forensic CNNs *with the same architecture* can prevent a GAN-based white box attack trained against one CNN from transferring to the other nearly identical CNN (Zhao and Stamm 2020). These differences can include the data used to train each CNN or how each CNN's classes are defined, i.e., distinguish unmanipulated versus manipulated (binary) or unmanipulated versus each possible manipulation (multi-class). For GAN-based attacks to work in realistic scenarios, new techniques must be created for achieving transferrable attacks. While the ensemble-based strategy training discussed in Sect. 17.5.5 is a new way to create transferrable attacks, there is still much work to be done. Additionally, it is possible that there are theoretical limits on attack transferability. As of yet, these limits are unknown, but future research on this topic may help reveal them.

Another important topic that research has not yet addressed is the types of forensic algorithms that GAN-based attacks are able to attack. Currently, anti-forensic attacks based on both GANs and adversarial examples are targeted against forensic classifiers. However, many forensic problems are more sophisticated than simple classification. For example, state-of-the-art techniques to detect locally falsified or spliced content are use sophisticated approaches built upon Siamese networks, not simple classifiers (Huh et al. 2018; Mayer and Stamm 2020, ?). Siamese networks are able to compare forensic traces in two local regions of an image and produce a measure of how similar or different these traces are, rather than simply providing a class decision. Similarly, state-of-the-art algorithms built to localize fake or manipulated content also utilize Siamese networks during training or localization (Huh et al. 2018; Mayer and Stamm 2020, ?; Cozzolino and Verdoliva 2018). Attacking these forensic algorithms is much more challenging than attacking comparatively simple CNN-based classifiers. It remains to be seen if and how GAN-based anti-forensic attacks can be constructed to fool these algorithms.

At the other end of the spectrum, little-to-no work currently exists aimed specifically at detecting or defending against GAN-based anti-forensic attacks. Past research has shown that traditional anti-forensic attacks often leave behind their own forensically detectable traces (Stamm et al. 2013; Piva 2013; Milani et al. 2012; Barni et al. 2018). Currently, it is unknown if these new GAN-based attacks leave behind their own traces, or what these traces may be. Research into adversarial robustness may potentially provide some protection against GAN-based anti-forensic attacks (Wang et al. 2020; Hinton et al. 2015). However, because GAN-based attacks operate differently than adversarial examples, it is unclear if these techniques will be successful. New techniques may need to be constructed to allow forensic algorithms to correctly operate in the presence of a GAN-based anti-forensic attack. Clearly, much more research is needed to provide protection against these emerging threats.

Acknowledgements Research was sponsored by the Army Research Office and was accomplished under Cooperative Agreement Number W911NF-20-2-0111, the Defense Advanced Research Projects Agency (DARPA) and the Air Force Research Laboratory (AFRL) under agreement numbers and HR001120C0126, and by the National Science Foundation under Grant No. 1553610. The U.S. Government is authorized to reproduce and distribute reprints for Governmental purposes notwithstanding any copyright notation thereon. The views and conclusions contained herein are those of the authors and should not be interpreted as necessarily representing the official policies or endorsements, either expressed or implied, of DARPA and AFRL, the Army Research Office, the National Science Foundation, or the U.S. Government.

References

Amerini I, Barni M, Caldelli R, Costanzo A (2013) Counter-forensics of sift-based copy-move detection by means of keypoint classification. EURASIP J Image Video Process 1:1–17

Antipov G, Baccouche M, Dugelay JL (2017) Face aging with conditional generative adversarial networks. In: 2017 IEEE international conference on image processing (ICIP), pp 2089–2093

Arjovsky M, Chintala S, Bottou L (2017) Wasserstein GAN

Barni M, Kallas K, Nowroozi E, Tondi B (2019) On the transferability of adversarial examples against CNN-based image forensics. In: ICASSP 2019-2019 IEEE international conference on acoustics, speech and signal processing (ICASSP). IEEE, pp 8286–8290

Barni M, Stamm MC, Tondi B (2018) Adversarial multimedia forensics: Overview and challenges ahead. In: 2018 26th European signal processing conference (EUSIPCO). IEEE, pp 962–966

Barni M, Tondi B (2013) The source identification game: an information-theoretic perspective. IEEE Trans Inf Forensics Secur 8(3):450–463

Böhme R, Kirchner M (2013) Counter-forensics: attacking image forensics. In: Sencar HT, Memon N (eds) Digital image forensics: there is more to a picture than meets the eye. Springer, pp 327–366

Brock A, Donahue J, Simonyan K (2018) Large scale GAN training for high fidelity natural image synthesis. arXiv:1809.11096

Caesar H, Uijlings J, Ferrari V (2018) Coco-stuff: thing and stuff classes in context. In: 2018 IEEE conference on computer vision and pattern recognition (CVPR). IEEE

Cao G, Zhao Y, Ni R, Tian H (2010) Anti-forensics of contrast enhancement in digital images. In: Proceedings of the 12th ACM workshop on multimedia and security, pp 25–34

Carlini N, Wagner D (2017) Towards evaluating the robustness of neural networks. In: 2017 IEEE symposium on security and privacy, pp 39–57

Chen X, Duan Y, Houthooft R, Schulman J, Sutskever I, Abbeel P (2016) Interpretable representation learning by information maximizing generative adversarial nets, Infogan

Chen C, Zhao X, Stamm MC (2017) Detecting anti-forensic attacks on demosaicing-based camera model identification. In: 2017 IEEE international conference on image processing (ICIP), pp 1512–1516

Chen C, Zhao X, Stamm MC (2018) Mislgan: an anti-forensic camera model falsification framework using a generative adversarial network. In: 2018 25th IEEE international conference on image processing (ICIP). IEEE, pp 535–539

Chen C, Zhao X, Stamm MC (2019) Generative adversarial attacks against deep-learning-based camera model identification. IEEE Trans Inf Forensics Secur 1

Chen C, Zhao X, Stamm MC (2019) Generative adversarial attacks against deep-learning-based camera model identification. IEEE Trans Inf Forensics Secur 1

Choi Y, Choi M, Kim M, Ha JW, Kim S, Choo J (2018) Stargan: unified generative adversarial networks for multi-domain image-to-image translation. In: Proceedings of the IEEE conference on computer vision and pattern recognition

Choi Y, Uh Y, Yoo J, Ha JW (2020) Stargan v2: diverse image synthesis for multiple domains. In: Proceedings of the IEEE conference on computer vision and pattern recognition

Chu X, Stamm MC, Chen Y, Liu KR (2015) On antiforensic concealability with rate-distortion tradeoff. IEEE Trans Image Process 24(3):1087–1100

Comesana-Alfaro P, Pérez-González F (2013) Optimal counterforensics for histogram-based forensics. In: 2013 IEEE international conference on acoustics, speech and signal processing. IEEE, pp 3048–3052

Costanzo A, Amerini I, Caldelli R, Barni M (2014) Forensic analysis of sift keypoint removal and injection. IEEE Trans Inf Forensics Secur 9(9):1450–1464

Cozzolino D, Thies J, Rossler A, Nießner M, Verdoliva L (2021) Spoofing camera fingerprints, Spoc

Cozzolino D, Verdoliva L (2018) Noiseprint: a CNN-based camera model fingerprint. arXiv:1808.08396

Dang-Nguyen DT, Gebru ID, Conotter V, Boato G, De Natale FG (2013) Counter-forensics of median filtering. In: 2013 IEEE 15th international workshop on multimedia signal processing (MMSP). IEEE, pp 260–265

Dirik AE, Sencar HT, Memon N (2014) Analysis of seam-carving-based anonymization of images against PRNU noise pattern-based source attribution. IEEE Trans Inf Forensics Secur 9(12):2277–2290

Fan W, Wang K, Cayre F, Xiong Z (2013) A variational approach to jpeg anti-forensics. In: 2013 IEEE international conference on acoustics, speech and signal processing. IEEE, pp 3058–3062

Fan W, Wang K, Cayre F, Xiong Z (2014) JPEG anti-forensics with improved tradeoff between forensic undetectability and image quality. IEEE Trans Inf Forensics Secur 9(8):1211–1226

Fan W, Wang K, Cayre F, Xiong Z (2015) Median filtered image quality enhancement and anti-forensics via variational deconvolution. IEEE Trans Inf Forensics Secur 10(5):1076–1091

Fontani M, Barni M (2012) Hiding traces of median filtering in digital images. In: Proceedings of the 20th European signal processing conference (EUSIPCO). IEEE, pp 1239–1243

Gloe T, Kirchner M, Winkler A, Böhme R (2007) Can we trust digital image forensics? In: Proceedings of the 15th ACM international conference on Multimedia, pp 78–86

Goodfellow IJ, Shlens J, Szegedy C (2014) Explaining and harnessing adversarial examples. arXiv:1412.6572

Güera D, Wang Y, Bondi L, Bestagini P, Tubaro S, Delp EJ (2017) A counter-forensic method for CNN-based camera model identification. In: Computer vision and pattern recognition workshops (CVPRW). IEEE, pp 1840–1847

Hinton G, Vinyals O, Dean J (2015) Distilling the knowledge in a neural network

Huh M, Liu A, Owens A, Efros AA (2018) Fighting fake news: image splice detection via learned self-consistency. In: Proceedings of the European conference on computer vision (ECCV), pp 101–117

Ian Goodfellow I, Pouget-Abadie J, Mirza M, Xu B, Warde-Farley D, Ozair S, Courville A, Bengio Y (2014) Generative adversarial nets. In: Advances in neural information processing systems, pp 2672–2680

Ilyas A, Santurkar S, Tsipras D, Engstrom L, Tran B, Madry A (2019) Adversarial examples are not bugs, they are features

Isola P, Zhu JY, Zhou T, Efros AA (2018) Image-to-image translation with conditional adversarial networks

Jin Y, Zhang J, Li M, Tian Y, Zhu H, Fang Z (2017) Towards the automatic anime characters creation with generative adversarial networks

Kang X, Liu J, Liu H, Wang ZJ (2016) Forensics and counter anti-forensics of video inter-frame forgery. Multimed Tools Appl 75(21):13833–13853

Karaküçük A, Dirik AE (2015) Adaptive photo-response non-uniformity noise removal against image source attribution. Digit Investig 12:66–76

Karras T, Aila T, Laine S, Lehtinen J (2017) Progressive growing of gans for improved quality, stability, and variation. arXiv:1710.10196

Kim D, Jang HU, Mun SM, Choi S, Lee HK (2017) Median filtered image restoration and anti-forensics using adversarial networks. IEEE Signal Process Lett 25(2):278–282

Kirchner M, Bohme R (2008) Hiding traces of resampling in digital images. IEEE Trans Inf Forensics Secur 3(4):582–592

Kirchner M, Böhme R (2009) Synthesis of color filter array pattern in digital images. In: Media forensics and security, vol 7254. International Society for Optics and Photonics, p 72540K

Kurakin A, Goodfellow I, Bengio S, et al (2016) Adversarial examples in the physical world

Laijie L, Gaobo Y, Ming X (2013) Anti-forensics for unsharp masking sharpening in digital images. Int J Digit Crime Forensics (IJDCF) 5(3):53–65

Ledig C, Theis L, Huszar F, Caballero J, Cunningham A, Acosta A, Aitken A, Tejani A, Totz J, Wang Z, Shi W (2017) Photo-realistic single image super-resolution using a generative adversarial network

Liu Z, Luo P, Wang X, Tang X (2015) Deep learning face attributes in the wild. In: Proceedings of international conference on computer vision (ICCV)

Lukas J, Fridrich J, Goljan M (2005) Determining digital image origin using sensor imperfections. In: Image and video communications and processing 2005, vol 5685. International Society for Optics and Photonics, pp 249–260

Madry A, Makelov A, Schmidt L, Tsipras D, Vladu A (2017) Towards deep learning models resistant to adversarial attacks. arXiv:1706.06083

Mao X, Li Q, Xie H, Lau RY, Wang Z, Paul Smolley S (2017) Least squares generative adversarial networks. In: Proceedings of the IEEE international conference on computer vision, pp 2794–2802

Mayer O, Stamm MC (2015) Anti-forensics of chromatic aberration. In: Media watermarking, security, and forensics 2015, vol 9409. International Society for Optics and Photonics, p 94090M

Mayer O, Stamm MC (2020) Exposing fake images with forensic similarity graphs. IEEE J Sel Top Signal Process 14(5):1049–1064

Mayer O, Stamm MC (2020) Forensic similarity for digital images. IEEE Trans Inf Forensics Secur 15:1331–1346

Milani S, Fontani M, Bestagini P, Barni M, Piva A, Tagliasacchi M, Tubaro S (2012) An overview on video forensics. APSIPA Trans Signal Inf Process 1

Miroslav Goljan, Jessica Fridrich, and Mo Chen. Sensor noise camera identification: Countering counter-forensics. In *Media Forensics and Security II*, volume 7541, page 75410S. International Society for Optics and Photonics, 2010

Mirza M, Osindero S (2014) Conditional generative adversarial nets

Papernot N, McDaniel P, Jha S, Fredrikson M, Celik ZB, Swami A (2016) The limitations of deep learning in adversarial settings. In 2016 IEEE European symposium on security and privacy (EuroS&P). IEEE, pp 372–387

Park T, Liu MY, Wang TC, Zhu JY (2019) Gaugan: semantic image synthesis with spatially adaptive normalization. In: ACM SIGGRAPH 2019 Real-Time Live!, p 1

Pasquini C, Boato G (2013) JPEG compression anti-forensics based on first significant digit distribution. In: 2013 IEEE 15th international workshop on multimedia signal processing (MMSP). IEEE, pp 500–505

Pathak D, Krahenbuhl P, Donahue J, Darrell T, Efros AA (2016) Context encoders: feature learning by inpainting

Piva A (2013) An overview on image forensics. ISRN Signal Process 2013

Poursaeed O, Katsman I, Gao B, Belongie S (2018) Generative adversarial perturbations. In: Proceedings of the IEEE conference on computer vision and pattern recognition, pp 4422–4431

Radford A, Metz L, Chintala S (2015) Unsupervised representation learning with deep convolutional generative adversarial networks. arXiv:1511.06434

Stamm MC, Lin WS, Liu KR (2012a) Temporal forensics and anti-forensics for motion compensated video. IEEE Trans Inf Forensics Secur 7(4):1315–1329

Stamm MC, Lin WS, Liu KR (2012b) Forensics versus anti-forensics: a decision and game theoretic framework. In: 2012 IEEE international conference on acoustics, speech and signal processing (ICASSP). IEEE, pp 1749–1752

Stamm MC, Liu KR (2011) Anti-forensics of digital image compression. IEEE Trans Inf Forensics Secur 6(3):1050–1065

Stamm MC, Tjoa SK, Lin WS, Liu KR (2010) Anti-forensics of jpeg compression. In: 2010 IEEE international conference on acoustics, speech and signal processing. IEEE, pp 1694–1697

Stamm MC, Wu M, Liu KR (2013) Information forensics: an overview of the first decade. IEEE Access 1:167–200

Tero Karras T, Laine S, Aila T (2019) A style-based generator architecture for generative adversarial networks. In: Proceedings of the IEEE/CVF conference on computer vision and pattern recognition, pp 4401–4410

Tero K, Samuli L, Miika A, Janne H, Jaakko L, Timo A (2020) Analyzing and improving the image quality of style GAN. In: Proceedings of the CVPR

Vondrick C, Pirsiavash H, Torralba A (2016) Generating videos with scene dynamics

Wang SY, Wang O, Zhang R, Owens A, Efros AA (2020) CNN-generated images are surprisingly easy to spot... for now. In: Proceedings of the IEEE conference on computer vision and pattern recognition, vol 7

Wu ZH, Stamm MC, Liu KR (2013) Anti-forensics of median filtering. In: 2013 IEEE international conference on acoustics, speech and signal processing. IEEE, pp 3043–3047

Wu J, Kang X, Yang J, Sun W (2021) A framework of generative adversarial networks with novel loss for jpeg restoration and anti-forensics. Multimed Syst, pp 1–15

Wu J, Zhang C, Xue T, Freeman WT, Tenenbaum JB (2016) Learning a probabilistic latent space of object shapes via 3d generative-adversarial modeling. arXiv:1610.07584

Wu H, Zheng S, Zhang J, Huang K (2019) Towards realistic high-resolution image blending, Gp-gan

Xiao C, Li B, Zhu JY, He W, Liu M, Song D (2018) Generating adversarial examples with adversarial networks. arXiv:1801.02610

Yu F, Seff A, Zhang Y, Song S, Funkhouser T, Xiao J (2016) Lsun: construction of a large-scale image dataset using deep learning with humans in the loop

Zhang H, Goodfellow I, Metaxas D, Odena A (2019) Self-attention generative adversarial networks. In: International conference on machine learning. PMLR, pp 7354–7363

Zhang X, Karaman S, Chang SF (2019) Detecting and simulating artifacts in GAN fake images. In: 2019 IEEE international workshop on information forensics and security (WIFS), pp 1–6

Zhang H, Xu T, Li H, Zhang S, Wang X, Huang X, Metaxas DN (2017) Stackgan: text to photo-realistic image synthesis with stacked generative adversarial networks. In: Proceedings of the IEEE international conference on computer vision, pp 5907–5915

Zhao X, Chen C, Stamm MC (2021) A transferable anti-forensic attack on forensic CNNs using a generative adversarial network

Zhao X, Stamm MC (2020) The effect of class definitions on the transferability of adversarial attacks against forensic CNNs. Electron Imaging 2020(4):119–1

Zhao X, Stamm MC (2021) Making GAN-generated images difficult to spot: a new attack against synthetic image detectors

Zhu JY, Park T, Isola P, Efros AA (2017) Unpaired image-to-image translation using cycle-consistent adversarial networks. In: Proceedings of the IEEE international conference on computer vision, pp 2223–2232

Printed in the United States
by Baker & Taylor Publisher Services